AF771033

Advances in Multivariate Statistical Methods

Statistical Science and Interdisciplinary Research

Series Editor: Sankar K. Pal *(Indian Statistical Institute)*

Description:
In conjunction with the Platinum Jubilee celebrations of the Indian Statistical Institute, a series of books will be produced to cover various topics, such as Statistics and Mathematics, Computer Science, Machine Intelligence, Econometrics, other Physical Sciences, and Social and Natural Sciences. This series of edited volumes in the mentioned disciplines culminate mostly out of significant events — conferences, workshops and lectures — held at the ten branches and centers of ISI to commemorate the long history of the institute.

Vol. 1 Mathematical Programming and Game Theory for Decision Making
edited by S. K. Neogy, R. B. Bapat, A. K. Das & T. Parthasarathy
(Indian Statistical Institute, India)

Vol. 2 Advances in Intelligent Information Processing:
Tools and Applications
edited by B. Chandra & C. A. Murthy
(Indian Statistical Institute, India)

Vol. 3 Algorithms, Architectures and Information Systems Security
edited by Bhargab B. Bhattacharya, Susmita Sur-Kolay,
Subhas C. Nandy & Aditya Bagchi
(Indian Statistical Institute, India)

Vol. 4 Advances in Multivariate Statistical Methods
edited by A. SenGupta (Indian Statistical Institute, India)

Vol. 5 New and Enduring Themes in Development Economics
edited by B. Dutta, T. Ray & E. Somanathan
(Indian Statistical Institute, India)

Vol. 6 Modeling, Computation and Optimization
edited by S. K. Neogy, A. K. Das and R. B. Bapat
(Indian Statistical Institute, India)

Platinum Jubilee Series

Statistical Science and
Interdisciplinary Research — Vol. 4

Advances in Multivariate Statistical Methods

Editor

Ashis SenGupta

Indian Statistical Institute, India

Series Editor: ***Sankar K. Pal***

World Scientific

NEW JERSEY • LONDON • SINGAPORE • BEIJING • SHANGHAI • HONG KONG • TAIPEI • CHENNAI

Published by

World Scientific Publishing Co. Pte. Ltd.
5 Toh Tuck Link, Singapore 596224
USA office: 27 Warren Street, Suite 401-402, Hackensack, NJ 07601
UK office: 57 Shelton Street, Covent Garden, London WC2H 9HE

British Library Cataloguing-in-Publication Data
A catalogue record for this book is available from the British Library.

Statistical Science and Interdisciplinary Research — Vol. 4
ADVANCES IN MULTIVARIATE STATISTICAL METHODS

Copyright © 2009 by World Scientific Publishing Co. Pte. Ltd.
All rights reserved. This book, or parts thereof, may not be reproduced in any form or by any means, electronic or mechanical, including photocopying, recording or any information storage and retrieval system now known or to be invented, without written permission from the Publisher.

For photocopying of material in this volume, please pay a copying fee through the Copyright Clearance Center, Inc., 222 Rosewood Drive, Danvers, MA 01923, USA. In this case permission to photocopy is not required from the publisher.

ISBN-13 978-981-283-823-0
ISBN-10 981-283-823-6

Printed in Singapore by World Scientific Printers

Foreword

The Indian Statistical Institute (ISI) was established on 17th December, 1931 by a great visionary Prof. Prasanta Chandra Mahalanobis to promote research in the theory and applications of statistics as a new scientific discipline in India. In 1959, Pandit Jawaharlal Nehru, the then Prime Minister of India introduced the ISI Act in the parliament and designated it as an Institution of National Importance because of its remarkable achievements in statistical work as well as its contribution to economic planning.

Today, the Indian Statistical Institute occupies a prestigious position in the academic firmament. It has been a haven for bright and talented academics working in a number of disciplines. Its research faculty has done India proud in the arenas of Statistics, Mathematics, Economics, Computer Science, among others. Over seventy five years, it has grown into a massive banyan tree, like the institute emblem. The Institute now serves the nation as a unified and monolithic organization from different places, namely Kolkata, the Head Quarter, Delhi and Bangalore, two centers, a network of six SQC-OR Units located at Mumbai, Pune, Baroda, Hyderabad, Chennai and Coimbatore, and a branch (field station) at Giridih.

The platinum jubilee celebrations of ISI have been launched by Honorable Prime Minister Prof. Manmohan Singh on December 24, 2006, and the Government of India has declared 29th June as the "Statistics Day" to commemorate the birthday of Prof. Mahalanobis nationally.

Prof. Mahalanobis was a great believer in interdisciplinary research, because he thought that this will promote the development of not only Statistics, but also the other natural and social sciences. To promote interdisciplinary research, major strides were made in the areas of computer science, statistical quality control, economics, biological and social sciences, physical and earth sciences.

The Institute's motto of "unity in diversity" has been the guiding principle of all its activities since its inception. It highlights the unifying role of statistics in relation to various scientific activities.

In tune with this hallowed tradition, a comprehensive academic programme, involving Nobel Laureates, Fellows of the Royal Society, Abel Prize winner and other dignitaries, has been implemented throughout the Platinum Jubilee year, highlighting the emerging areas of ongoing frontline research in its various scientific divisions, centers, and outlying units. It includes international and national-level seminars, symposia, conferences and workshops, as well as series of special lectures. As an outcome of these events, the Institute is bringing out a series of comprehensive volumes in different subjects under the title Statistical Science and Interdisciplinary Research, published by World Scientific, Singapore.

The present volume titled "Advances in Multivariate Statistical Methods" is the fourth one in the series, and is also a special tribute to mark the birth centenary of Prof. S.N. Roy, who has greatly contributed to the aforesaid theme area of the book and enriched the training programs of ISI as a member of the research team of Prof. Mahalanobis. The volume consists of twenty seven chapters, written by eminent scientists and experienced researchers from different parts of the world. The chapters deal with both theory and applications covering the topics like large dimensional data analysis, directional data analysis, clustering, estimation, distribution theory, Bayesian inference, reliability analysis, sample surveys, and bioinformatics. Excellent reviews mentioning the challenging problems in some thrust areas e.g., large dimension-small sample size which has tremendous applications in data mining and bioinformatics, are also addressed. I believe the state-of-the art studies presented in this book will be very useful to readers.

Thanks to the contributors for their excellent research contributions, and to the volume editor Prof. Ashis SenGupta and the invited editor Prof. Barry Arnold for their sincere effort in bringing out the volume nicely in time. Initial design of the cover by Mr. Indranil Dutta is acknowledged. Sincere efforts by Prof. Dilip Saha and Dr. Barun Mukhopadhyay for editorial assistance are appreciated. Thanks are also due to World Scientific for their initiative in publishing the series and being a part of the Platinum Jubilee endeavor of the Institute.

April 2008
Kolkata

Sankar K. Pal
Series Editor and
Director, ISI

Preface

The year 2006 marked the Platinum Jubilee of Indian Statistical Institute (ISI). ISI has played a seminal role in promoting statistics in general worldwide. A leading and significant contribution of ISI has been in the area of Multivariate Statistical Methods as presented by Prof. P.C. Mahalanobis and his team of researchers from the very genesis of ISI. Of them, Prof. Samarendra Nath Roy produced real gems, through his research on difficult and interesting problems, such as p-statistics and joint distribution of roots of certain Wishart matrices, Union-intersection testing principle, Growth curve analysis, Categorical data analysis. The year 2006 also marked the birth centenary of Professor S.N. Roy. It was thus felt that a celebration of these two simultaneous events be made in the form of a Workshop and a Conference on Multivariate Statistical Methods, focusing on its recently emerging areas of theory and applications. The International Conference was held during December 28-29, 2006, at Kolkata. The conference was organized jointly by Profs. Barry C. Arnold, University of California, Riverside, USA, and Ashis SenGupta, Indian Statistical Institute, Kolkata, India, and was attended by about 150 researchers. The topics, in tune with the theme of the conference, covered a wide spectrum - a snapshot includes Distribution Theory, Stochastic Eigen Analysis, Multiple Testing Procedures, Longitudinal Data Analysis, Large p - Small n, Robust Inference, Directional Data Analysis, Bayesian Inference, Bioinformatics, Categorical Data and Clinical Trials, Ranking and Selection, Spatio-Temporal Data Analysis, Econometrics, Statistical Process Control.

This volume is number 4 in the series formed from the events celebrating the Platinum Jubilee of ISI which is being published by World Scientific. It presents a collection of 27 invited papers by international luminaries in the said spirit, which emanated from the Conference, and were peer-reviewed and edited. The chapters reflect advances in both theory and applications of multivariate statistical methods. The first five chapters deal with the analysis of large dimensional data. The first chapter appeals to the Union-Intersection principle developed by Prof. S.N. Roy, and is thus a fitting tribute to him also. The challenging area of Large p - Small n, which is currently receiving great attention, is dealt with in Chapters 1 and 2. These are followed by the chapter on Clustering methods and two chapters on Bioinformatics. Chapters 6 and 7 deal with models and inference for Directional Data. Chapter 8 presents some elegant results on Characterization using regression. Chapters 9-11, present reviews and recent results related to multivariate Distribution Theory. Chapters 12-14, consider Estimation, while Chapters 15-17 deal with Testing of Hypotheses problems, illustrated by various examples from applied sciences. Bayesian Methods are having a great impact on analysis of multivariate data and this fact is enhanced through Chapters 18-21, which reflect on both theory and applications. The last five chapters deal mainly with multivariate statistical methods in several real-life applications. Chapters 22-24 consider the area of Reliability Analysis, Chapters 25 and 26 deal with applications of multivariate techniques for analysis of some special types of time series data which are of recently emerging nature and interest. Finally, Chapter 27 enhances multivariate methods in Sample Surveys.

We are thankful to Indian Statistical Institute for partial financial support and also to Institute of Mathematical Statistics, USA, for co-sponsoring it. We acknowledge with thanks the support received from Calcutta Mathematical Society, Forum for Interdisciplinary Mathematics, Indian Bayesian Society, Indian Association of Productivity, Quality and Reliability, and International Indian Statistical Association. It is a pleasure to thank, Prof. Sankar K. Pal, Director of ISI, for his support and encouragements. We also thank the Members of the International Advisory Committee, Professors J.K. Ghosh, K.V. Mardia, S.P. Mukherjee, P.K. Sen and K. Shimizu, for their advice with the editorial work of this volume. We are indeed grateful to the referees for their cooperation and scholarly reports on the papers they have reviewed. The list of such referees includes: Professors Arun Adhikary, Sourabh Bhattacharya*, Subir Bhandari*, Arijit Chakraborty, Asis Chattopadhyay, Ajit Chaturvedi, Samarjit Das, B.K. Kale, Shogo Kato, Arnab Laha*, H.K. Tony Ng, Chandranath Pal, Arthur Pewsey*, R.N. Rattihalli*, T. Samanta, P.G. Sankaran, M.S. Sreehari and T.S. Vasulu (*indicates Referee for multiple papers). We also thank Mr. Prabhat Roy and Mr. Indranil Dutta for their help with the processing of this volume. Last but not the least, we are indeed thankful to all authors for comtributing scholarly papers which have greatly enriched this volume.

We present this volume, in the series celebrating the Platinum Jubilee of ISI, as a special tribute to mark the birth centenary of Prof. S.N. Roy, who has greatly enriched the research and teaching programs of ISI. It is hoped that this volume will serve as a useful and timely resource monograph to the researchers and practitioners for advances in the area of Multivariate Statistical Methods.

Barry C. Arnold
Invited Editor
Distinguished Professor
Department of Statistics
University of California
Riverside, USA
April 2008
Riverside

Ashis SenGupta
Editor
Professor and Head
Applied Statistics Unit
Indian Statistical Institute
Kolkata, India
April 2008
Kolkata

Contents

Chapter 1

High-Dimensional Discrete Statistical Models: UIP, MCP and CSI in Perspectives

Pranab Kumar Sen

Departments of Biostatistics, and Statistics & Operations Research, University of North Carolina at Chapel Hill, NC 27599-7420, USA

E-mail: pksen@bios.unc.edu

The ongoing informatics evolution has posed some challenging statistical problems. Most statistical models arising in applications in bioinformatics, data mining and a variety of other computer-intensive interdisciplinary research fields are complex in their designs, sampling plans, and associated probability laws. The curse of dimensionality is so overwhelming that conventional likelihood ratio based statistical inference may not be much useful. Further, such models are typically constrained by inequality, order, functional, shape or other restraints, often on functional parameters, and as a result, optimal statistical inference procedures are hard to find or may not even exist. Although. the use of restricted-, profile- ,pseudo-, partial-, or quasi-likelihood has been advocated in such constrained environments, their optimality properties are generally not known, and at least, are hard to establish. S. N. Roy's ingenious union-intersection principle along with innovations in discrete multivariate analysis provide an alternative avenue, often having some computational advantages, increased scope of adaptability, and flexibility beyond conventional likelihood paradigms. This scenario is appraised here with some illustrative examples of high-dimensional discrete multivariate analysis of covariance models arising in genomic studies in parametrics as well as beyond parametrics setups.

Contents

1.1. Introduction

During the last twelve years before his premature demise, Professor S. N. Roy made most significant and innovative contributions in three important areas in multivariate statistical analysis, namely, *multiple comparisons procedures* (MCP), *union-intersection principle* (UIP) and discrete multivariate analysis. During 1930 - 1960's, classical multivariate analysis stole the limelight of most innovative and sophisticated statistical analysis, albeit being mostly confined either to samples from finie-dimensional multivariate normal populations or simple and low multidimensional categorical data models. The present study reflects some of Roy's innovative perspectives in finite dimensional data models with an eye on their fruitful incorporation in high dimensional perspectives.

Statistical models usually advocated for complex problems arising in some interdisciplinary research and many real life applications are rarely simple so as to make room for routine incorporation of conventional or standrad (i.e., likelihood based) statistical inference tools. The ongoing evolution of genomics (and bioinformatics, in general) has indeed posed some enormously large dimensional statistical models where the sample size may often be relatively much smaller. Such high-dimensional low sample size (HDLSS) models generally involve complex designs, sampling plans, and the underlying stochastics relate to probability laws which, typically, not only involve a multitude of parameters but also with the parameters subjected to various nonlinear restraints. Inequality, order, functional and shape constraints are commonly encountered, in probability as well as sample spaces, where the HDLSS perspectives may complicate the models as well as their statistical resolutions considerably. For beyond parametrics (i.e., nonparametrics and semiparametrics) setups, often, there could be more complex restraints involving functional constraints. Stochastic ordering (dominance) of functional pasrameters in categorical data models, arising in genomic studies, particularly, in single nucleotide proliferation (SNP) models being a notable example of this kind (Sen et al. 2007). In conventional statistical inference, likelihood, sufficiency and invariance principles play a key role in finite sample methodology, and some of the finie-sample optimality properties usually transpire in large sample cases even without sufficiency, invariance or some other regularity conditions. Nevertheless, even in such asymptotic cases, lacking support of suitable regularity assumptions, particularly in constrained environments, optimal statistical inference may encounter roadblocks of diverse types.

Generally, complex statistical models create impasses for computation of *maximum likelihood estimators* (MLE) and *likelihood ratio tests* (LRT) in closed, explicit or manageable forms; often, this may become a formidable task. Even so, various algorithms have been developed for such computational convenience, though the finite sample optimality properties of MLE and LRT ranging over the exponential family of densities may not automatically transpire in more complex models where the underlying probability laws are rarely bonafide members of such regular families; even if they are, underlying constraints or implicit structural restraints may take away such optimality properties. (Restricted) RMLE and (restricted) RLRT along with various modifications of the likelihood function have therefore been advocated for such complex models, albeit they may not have any universal optimality property parallel to that in simple models. S. N. Roy's (1953) ingenious UIP, having its genesis in the *likelihood principle* (LP), has emerged as a viable alternative, often having some computational advantages, increased scope of applicability (beyond the likelihood paradigm), greater adaptability to nonstandard situations (beyond the parametrics), and good robustness perspectives.

Roy, Gnanadesikan and Srivastava (1971) contains a thorough treatise of UIP in multivariate models with due emphasis on simultaneous confidence sets and multiple hypothesis testing problems relating to the domain of MCP. Though their treatise is mostly confined to continuous multinormal distributional models, the past three decades have witnessed a steady flow of research on UIP in a variety of beyond parametric models. For a general treatise of *constrained statistical inference* (CSI) we refer to Silvapulle and Sen (2004) which recaptured the prior developments in in the classical monograph of Barlow et al. (1972) and its follow-up by Robertson et al. (1988). The major emphasis in Barlow et al. (1972) has been the finite sample methodology with due consideration of the basic role of the likelihood function in such formulations. More in-depth computational aspects are additionally reported in Robertson et al. (1988). The Silvapulle-Sen (2004) treatise goes beyond that into more general setups with adequate asymptotics to simplify the methodology; in line with the Wald-type tests, in CSI such procedures are elaborated along with the basic role of UIP in some important problems. The present study is devoted to a display of the basic role of UIP in *high-dimensional discrete statistical models* (HDDSM) with due emphasis on CSI as well as MCP; this development being very useful in the evolving field of genomics or bioinformatics. A key factor in this respect is the most innovative treatise of some multidimensional categorical data models in Roy (1957) and its subsequent ramification during the past 50 years. Functional parameters arising in HDDSM, as is commonly perceived in genomics studies (Tsai and Sen 2005, Sen at el. 2007), is a basic perspective deserving thorough appraisals. In Section 2, we outline the preliminary notion. In Section 3 an overview of classical categorical multivariate analysis of covariance (CATMANOCOVA) in low to moderate dimensional setups (where the sample size n is still $>> K$, the dimension) is made. In Section 4, the basic formulation of UIP is considered along with some illustrations in some simple simple models. Section 5 is devoted to HDDSM with due emphasis on some SNP models. Section 6 deals with fruitful incorporation of the Roy (1953) UIP and MCP in such HDDSM's with due emphasis on some applications in genomic models (Sen et al. 2007).

1.2. Preliminary Notion

The scenario of statistical modeling and inference changes drastically from parametric to beyond parametric perspectives, and even in the parametric setup, from the single parameter to multiparameter models. The more complex a model is, the more likely is the feature that optimal statistical inference may not exist, and even it exists, it may be harder to implement. The evolving field of genomics is a pertinent illustration of the enormous difficulties which conventional statistical inference tools are encountering in this high-dimensional low-sample size (HDLSS) setups. For (continuous, discrete, as well as, categorical) distributions belonging to the so called exponential family, generally optimal statistical inference procedures, based on the sufficiency principle, work out well. However, even for the exponential family, if there are too many parameters or if the parameters are constrained in some way, such optimality properties may not transpire. Often invariance structures (with respect to some group of transformations mapping the sample space onto itself, inducung conjugate transformations on the parameter space)allow us to formulate invariant statistical procedures where within the class of such invariant procedures, an optimal one could be found out in a rational way. Sans such exponential families of densities, usually, finite-sample optimal statistical inference procedure may not exist, although, in the conventional asymptotic setup :$K << n$ (generally, K being fixed but n is large), asymptotically optimal statistical inference procedures have been prescribed. Whereas in the univariate case, in many situations, a moderately large sample size provides adequate asymptotic theory based methodology for inference tools, as the dimension becomes large, a reasonably good asymptotic approximation may generally require a much larger sample size; the rate of increase of the required sample size being usually much faster than the dimension increase. This basic limitation stands in the way of valid and efficient statistical inference for HDLSS data models.

As an illustration, consider the classical multivariate analysis of variance

(MANOVA) problem in the setup of (multi-)normally distributed errors. Let

$$\mathbf{Y} = \mathbf{X}\beta + \mathbf{E}, \tag{1.1}$$

where the observable stochastic matrix $\mathbf{Y} = (\mathbf{Y}_1, \ldots, \mathbf{Y}_n)'$, with each $\mathbf{Y}_i$ being a stochastic p-vector, is related to a known (nonstochastic) matrix $\mathbf{X}$ (of order $n \times m$) of regression constants, through an unknown (regression) parametric matrix β (of order $m \times p$), and where $\mathbf{E}' = (\mathbf{e}_1, \ldots, \mathbf{e}_n)'$ with each (p-vector) $\mathbf{e}_i$ having a multivariate normal distribution with null mean vector and a positive definite (p.d.) but unknown dispersion matrix Σ. Symbolically, we write

$$\mathbf{E} \sim \text{MN}(\mathbf{0}, \mathbf{I}_n \otimes \Sigma), \tag{1.2}$$

where $\otimes$ stands for the Kronecker product of the two matrices. In this setup, consider the most simple null hypothesis

$$H_0 : \beta = \mathbf{0}, \tag{1.3}$$

against (global) alternatives that

$$H_1 : \beta \neq \mathbf{0}. \tag{1.4}$$

In this normal theory setup, the MLE of β is given by

$$\hat{\beta}_n = (\mathbf{X}'\mathbf{X})^{-1}(\mathbf{X}'\mathbf{Y}). \tag{1.5}$$

The residuals are then defined by

$$\begin{aligned} \hat{\mathbf{Y}}_n &= \mathbf{Y} - \mathbf{X}\hat{\beta}_n \\ &= (\mathbf{I}_n - \mathbf{X}(\mathbf{X}'\mathbf{X})^{-1}\mathbf{X}')\mathbf{E}. \end{aligned} \tag{1.6}$$

The residual sum of product matrix (of order $p \times p$) is defined by

$$\mathbf{S}_E = (\hat{\mathbf{Y}}_n)'(\hat{\mathbf{Y}}_n). \tag{1.7}$$

Side by side, the sum of product matrix (of order $p \times p$) due to regression is defined as

$$\mathbf{S}_H = (\hat{\beta}_n)'(\mathbf{X}'\mathbf{X})(\hat{\beta}_n) \tag{1.8}$$

The rank of $\mathbf{S}_E$ (and $\mathbf{S}_H$) is equal to $\min(p, n-m)$ (and $\min(m, p)$). Thus, in order that $\mathbf{S}_E$ is of full rank, it is tacitly assumed that $n - m \geq p$, while the other matrix is nonsingular when $m \geq p$.

First, even for this simple model (belonging to the exponential family), if p and m are greater than one, there may not be a uniformly most powerful (UMP) test for H_0 vs. H_1. As a matter of fact, since Σ is nuisance, we need to confine ourselves to UMP similar regions. To resolve this problem, one considers the class

of tests which are invariant under nonsingular transformations on the observation vectors: $\mathbf{Y} \to \mathbf{Z} = \mathbf{YB}$, for some nonsingular $p \times p$ matrix $\mathbf{B}$. Within this class of invariant tests, the usual likelihood ratio test is best (i.e., uniformly most powerful invariant (UMPI)) test when the rank of $\mathbf{S}_H$ is equal to 1; for this particular case, all the three test statistics are equivalent and enjoy the same optimality property. However, even in this case, we need the condition that $n - m \geq p$. For the general case where both m, p are greater than 1, there are several test statistics including the classical likelihood ratio test based on $|\mathbf{S}_E + \mathbf{S}_H|/|\mathbf{S}_\mathbf{E}|$, Hotelling-Lawley trace criterion based on Trace($\mathbf{S}_H\mathbf{S}_E^{-1}$), Roy's largest root criterion based on $ch_{max}(\mathbf{S}_H(\mathbf{S}_E)^{-1}$, and other ramifications. None of these would be UMPI. Since all these test criteria depend on the maximal invariant, the characteristic roots of $\mathbf{S}_H(\mathbf{S}_E)^{-1}$, it is imperative that $n \geq m + p$.
Instead of the simple null hypothesis H_0 considered above, one may consider a more general null hypothesis where for suitable prespecified matrices $\mathbf{A}$ and $\mathbf{C}$ of order $r \times m$ and $p \times q$ respectively with $r \leq m$ and $q \leq p$, we set $H_0 : \mathbf{A}\beta\mathbf{C} = \mathbf{0}$ and the alternatives relate to nonnull matrices on the right hand side. For simultaneous (or multiple) hypotheses testing, we allow $\mathbf{A}$ to be arbitrary within a class, say $\mathcal{A}$, and also $\mathbf{C}$ arbitrary within another class, say $\mathcal{C}$, and desire to control the type I error over these classes as a whole. Similarly, we may want to set a simultaneous confidence set for all $\mathbf{A}\beta\mathbf{C}$, allowing $\mathbf{A} \in \mathcal{A}$ and $\mathbf{C} \in \mathcal{C}$, with the property that the coverage probability is some specified $1 - \alpha$, for some $\alpha \in (0, 1)$. The genesis of UIP lies in this complex, and this will be elaborated in Section 1.4.
Like the univariate one sided alternatives, we could have in the multivariate case some one sided alternatives, or more generally, alternatives specified by some inequality, order or other constraints. CSI typically pertains to such complex statistical environments. The affine invariance structure mentioned earlier may not pertain to such restricted alternatives, and hence, the classical (unrestricted) likelihood based tests sketched above may not be ideal in such CSI problems. Much of the development in CSI rests on RMLE and RLRT, which takes into account the underlying restraint(s), although not much optimality property may be retained in this setup. There could be computational complexities too. Therefore, it may be natural to appraise the interactive role of UIP and CSI. This will be considered in Section 1.4.
The above discussion pertains specifically to multinormal distributions, In many fields of application, such an assumption may be very untenable. In some cases, the random vectors are continuous, and hence, nonparametric models are more reasonable to adopt. In some other cases, we may be confronted with count variables while in many other cases we have categorical (and possibly qualitative) data models. The likelihood principal (LP) may generally encounter roadblocks

in such general setups. In the next section, we consider simple multidimentional categorical data models, albeit in the conventional asymptotic setups, to illustrate some of these basic difficulties with the LP. In Section 1.5, we shall introduce the HDLSS discrete multivariate setups, and appraise these models in the light of MCP and CSI perspectives.

1.3. CATANOCOVA

Categorical ANOVA models typically involve product-multinomial distributions which can be presented in their simplest form as follows. Consider G independent populations, each involving a set of C categorical responses not necessarily quantitative or even ordered in some way, and let $\pi_g(c),\ c = 1,\dots,C$ be the cell probabilities for the gth population, for $g = 1,\dots,G$. Let there be n_g observations from the gth population and let n_{gc} be the cell frequency of the cth cell, for $c = 1,\dots,C$, and $g = 1,\dots,G$, all these samples being drawn independently. The joint probability function of these n_{gc}, known as the product multinomial law, is given by

$$\prod_{g=1}^{G} \frac{n_g!}{n_{g1}!\cdots n_{gC}!} \prod_{c=1}^{C} \{\pi_g(c)\}^{n_{gc}}, \tag{1.9}$$

where n_{gc} are nonnegative integers such that $\sum_{c=1}^{C} n_{gc} = n_g$, and the $\pi_g = (\pi_g(1),\dots,\pi_g(C))'$, $g = 1,\dots,G$ all belong to the simplex $\mathcal{S}_{C-1} = \{\mathbf{x} \in [0,1]^C : \mathbf{x}'\mathbf{1} = 1\}$. In this nonparametric formulation, an ANOVA model relates to the homogeneity of the π_g. In many cases, we may be interested in various MHT testing problems, possibly in CSI setups, for this CATANOVA model. As an illustration, first, consider the (opinion) response model for

Reduction of US Military Involvement in Iraq

where we have the following (ordered) response categories: HS = highly supportive, S = generally supportive, N = No opinion, O = opposed and TO = totally opposed. Note that despite an inherent ordering, there is no linear or precise mathematical scale for the ordering. The probability space is the simplex S_4, generated by the vector of the 5 cell probabilities. Let us now consider the following classification of respondents: Democratic and Republican. The probability vector for the two groups are denoted by π_D and π_R respectively, each belonging to the common simplex S_4. We frame the null hypothesis of homogeneity as

$$H_0 : \pi_D = \pi_R = \pi \text{ (unknown)}.$$

We denote the corresponding cumulative probability vectors by Π_D and Π_R respectively (where the last element in each vector is equal to 1). To reflect a one-sided preference of the Democrats to the Republicans, we consider the following restricted alternative

$$H_1^< : \Pi_D \geq \Pi_R$$

with at least one (of the four coordinates) bearing a strict inequality sign. The conventional 2×5 contingency table based χ^2-test or even the Fisher exact randomisation test may not be ideal, as they address global alternatives of lack of homogeneity and thereby are more likely to be less powerful for such restricted alternatives. This one-sided hypothesis testing in a multiparameter setup could be even more complicated if there are some covariates or explanatory variables (such as male / female and young, middle-age and senior people, type of employment, educatiobnal and racial diversity etc). Some of these problems are discussed in Silvapulle and Sen (2004, Ch.6) in detail.

Suppose that instead of one such query, we have a questionnaire involving K basic questions, for each of which, we have a set of C ordered categories of responses. Thus there will be a totality of C^K possible response category-combinations. The K questions could be related (as in the classical *item analysis* schemes) when typically they refer to different aspects of a composite health or psychiatric problem. As such, the K response vectors are expected to be stochastically dependent. Although some restricted alternative hypothesis testing problems may relate to the K marginal probability laws, without taking into account their interdependence, such hypothesis testing problems can not be treated efficiently. In the context we are more interested the categories may not have any partial ordering and hence, hypotheses are to be formulated in a somewhat different way.

In a parametric formulation, the $\pi_g(c)$ are expressed in terms of some unknown parameters (vectors) θ_g of dimension less than C (and possibly involving some explanatory or concomitant variates), and as in the classical ANOVA problem, we may like to test for the homogeneity of these parametric vectors, as well as, associated MHT along with CSI formulations. In passing, we may remark that if the response is quantal (i.e., all or nothing) where $C = 2$, regression on the explanatory variables may not be in conventional linear models, and a logit or probit transformation is advocated for using generalized linear models for drawing statistical conclusions. On the other hand, if $C \geq 3$ and the response categories are qualitative, such transformations are usable, and more general regression models are to be sought. Due to such model complexities, the finite sample treatment of the nonparametric CATANOVA model may not generally be tenable in such parametric formulations. Usually, BAN (best asymptotically normal) estimators are incorporated in an asymptotic setup wherein the n_g are all taken to be large. Wald-type tests are more commonly used in this large sample size (relative to the dimension) context, including CSI setups; we again refer to Silvapulle and Sen (2004, Ch. 6). The UIP based approach will be outlined in later sections. Our main contention is to appraise HDDSM problems arising in this context, and this will be done in a later section.

1.4. UIP and CSI

S.N. Roy (1953) motivated the UIP through multivariate models with due emphasis on multiple comparisons and simultaneous statistical inference. We illustrate UIP with a general composite hypothesis testing problem that lends itself naturally to CSI as well as HDDSM, typically involving multiparameter models. Consider a general hypothesis testing problem, not necessarily the multinormal or multinomial models treated in earlier sections, or even a parametric model. Let H_0 be the null hypothesis of interest and let H_1 be the alternative one; both of them are composite so that the likelihood function is not completely specified under either of them. As it is the case with composite hypotheses testing problems, there may not be in general an optimal test for testing H_0 vs. H_1, and in many case, even finding out a similar region may restrict attention to a subclass of tests like invariant tests, conditional tests, etc.. This situation is likely to be worse in CSI where the conceived restraints may preempt the relevance of invariant or conditional tests. However, for a general class of testing problems, including in CSI, it might be possible to express

$$H_0 = \cap_{j \in \mathcal{J}} H_{0j}, \; H_1 = \cup_{j \in \mathcal{J}} H_{1j} \tag{1.10}$$

where $\mathcal{J}$ is a suitable index set, and for each $j \in \mathcal{J}$, there exists a suitable (and often optimal in a certain sense) test for testing H_{0j} vs H_{1j}. In a parametric framework, such a test could be the UMP test whenever the latter exists, could be LMP (locally most-powerful) test in some other case, and in beyond parametrics setups, such a test can be decided on the basis of robustness, validity and efficiency considerations. Further, the index set $\mathcal{J}$ can be a finite (discrete) set, or it may even be a set in continuium. In this way, there is flexibility in the decomposition of the hupotheses and choice of appropriate test statistics. Bearing in mind the genesis of UIP in LP (Roy, 1953), we consider first the following illustrative example pertaining to multinormal populations where the connection of UIP and LP can be identified easily.

Let $\mathbf{X}_1, \ldots, \mathbf{X}_n$ be n independent and idenetically distributed random vectors having a p-variate (multi-)normal distribution with unknown mean vecor μ and dispersion matrix Σ, unknown but positive definite (p.d.). Consider first the null hypothesis $H_0 : \mu = \mathbf{0}$ versus $H_1 : \mu \neq \mathbf{0}$, treating Σ as a nuisance parameter (matrix). There is no UMP test for this hypotheses testing problem. The likelihood ratio test statistic for this problem is a monotone function of the Hotelling T^2-statistic

$$T^2 = n(\bar{\mathbf{X}}_n)'\mathbf{S}_n^{-1}(\bar{\mathbf{X}}_n), \tag{1.11}$$

where

$$\bar{\mathbf{X}}_n = \frac{1}{n}\sum_{i=1}^{n}\mathbf{X}_i, \; \mathbf{S}_n = \frac{1}{n-1}\sum_{i=1}^{n}(\mathbf{X}_i - \bar{\mathbf{X}}_n)(\mathbf{X}_i - \bar{\mathbf{X}}_n)'. \tag{1.12}$$

The test is affine-invariant and within the class of affine-invariant tests it is UMP. The affine invariance is defined in terms of the invariance of the test under the class of transformations $\mathbf{X} \to \mathbf{Y} = \mathbf{B}\mathbf{X}$, $\mathbf{B}$ being nonsingular. Let us look into this picture from a different angle.

Let $\mathbf{a} \in \mathbf{R}^p$ be an arbitrary p-vector, and let $H_{0\mathbf{a}} : \mathbf{a}'\mu = 0$ and $H_{1\mathbf{a}} : \mathbf{a}'\mu > 0$. Then note that

$$H_0 = \cap_{\mathbf{a}\in\mathbf{R}^p} H_{0\mathbf{a}}, \; H_1 = \cup_{\mathbf{a}\in\mathbf{R}^p} H_{1\mathbf{a}}. \tag{1.13}$$

Further note that for testing the null hypothesis $H_{0\mathbf{a}}$ against $H_{1\mathbf{a}}$, a UMP test is based on the Student t-statistic

$$t(\mathbf{a}) = \sqrt{n}(\mathbf{a}'\bar{\mathbf{X}}_n)/(\mathbf{a}'\mathbf{S}_n\mathbf{a})^{1/2} \tag{1.14}$$

using the right hand side critical region marked by the Student t-distribution with $n-1$ degrees of freedom (DF). The overall hypothesis H_0 is only accepted when all the component hypotheses are accepted and H_1 is accepted when at least one component alternative hypothesis is deemed acceptable, thus implementing the UIP. Therefore, the UIT is based on the test statistic

$$t^* = \sup\{t(\mathbf{a}) : \mathbf{a} \in \mathbf{R}^p\} \tag{1.15}$$

and some routine computations yield that

$$(t^*)^2 = T_n^2. \tag{1.16}$$

Thus, the UIT and LRT are isomorphic for this hypotheses testing problem. If, however, we go to the K-sample problem of testing the homogeneity of the mean vectors for $K \geq 3$ or the general MANOVA problem, the UIT and LRT statistics may not be generally isomorphic, albeit both are affine-invariant (and none being UMP invariant). The UIT relates to Roy's largest root criterion.

Let us considere the same one-sample model under a CSI setup. Namely, we consider the same null hypothesis that $\mu = \mathbf{0}$ against one-sided alternative $H_1^{>}$: $\mu \geq \mathbf{0}$, i.e., the mean vector lies in the positive orthant $\mathbf{R}^{p+} = \{\mathbf{x} \in \mathbf{R}^p : \mathbf{x} \geq \mathbf{0}\}$. The RLRT for this problem, treated by Perlman (1969) and others, rests on the test statistic

$$\lambda_n = \log L_n(\hat{\theta}_n^{(1)}) - \log L_n(\hat{\theta}_n^{(0)}), \tag{1.17}$$

where $\hat{\theta}_n^{(0)}$ is the RMLE of $\theta = (\mu, \Sigma)$ under the null hypothesis (which is simple), and $\hat{\theta}_n^{(1)}$ under the alternative $H_1^{>}$, and this ic complex. By the use of the KTL-point formula, one can obtain an expression of the estimates, though not verey simple. Further, the estimator of Σ (under the alternative) is not orthogonal to the RMLE of μ and it does not have the anticipated Wishart distribution. To eliminate this problem, Perlman (1969) suggested a conservative test by allowing a further maximisation over all Σ in the class of p.d. matrices.

The UIT can be constructed as follows: The alternative hypothesis can be equivaletly expressed as $H_1 = \cup\{\mathbf{a}'\mu > 0, \forall \mathbf{a} \geq \mathbf{0}\} = \cup_{\mathbf{a} \in \mathbf{R}^{p+}\}} H_{1(\mathbf{a})}$. Then, for testing $H_{0\mathbf{a}}$ versus $H_{1\mathbf{a}}$, we have an optimal one-sided test based on $t_n(\mathbf{a})$, as defined above. Therefore, the UIT test statistic is given by

$$t_n^+ = \sup\{t_n(\mathbf{a}) : \mathbf{a} \in \mathbf{R}^{p+}\}, \tag{1.18}$$

an explicit expression for this UIT test statistic, also based on the KTL-point formula theorem, comparable to the RLRT test statistic is available in the literature (Silvapule and Sen 2004, Ch. 5). We also refer to Sen and Tsai (1999) for a detailed treatise of the LRT and UIT for one-sided alternatives for multivariate normal mean testing problem when the dispersion matrix is nuisance (i.e., positive definite but arbitrary). It is shown that the UIT statistic can not be smaller than the RLRT statistic, and both of them have a certain amount of conservativeness due to the nuisance dispersion matrix. They suggested a two-stage LRT and UIT to overcome this problem. If we look into the corresponding simultaneous confidence sets, the procedures based on the LP and UIP have possibly different forms, and it is known (Wijsmann, 1979) that the UIP has certain advantages over the LP in this setup.

The UIP has also received due attention in other parametric CSI problems along with new interpretations and motivations. While a comprehensive account of some of these developments is available in the literature (viz., Silvapulle and Sen, 2004), it should be noted that in specific CSI problems, often these UIP motivated procedures have much simplicity and rationality to offer; we refer to Mudholkar and McDermott (1989), McDermott and Mudholkar (1993), Mudholkar et al. (1993,1995), and Srivastava and Mudholkar (2001) and Mudholkar et al. (2001) for some interesting developments including some robust tests for the multivariate orthant restricted alternative testing problem. Tsai (1995) considered the estimation of covariance matrices under Lōwner order restriction, and Tsai (2004) considered the covariance matrices estimation problem under simple tree ordering. Das and Sen (1994) considered the restricted canonical correlation inference problem. In the context of genomics, Sen et al. (2007) have developed restricted alternative tests based on Hamming distance, and have discussed their plausibility for the high-dimension low sample size contexts. Also, nonparametric tests for ordered diversity in a genomic sequence have been considered by Sen (2005). Although, these developments relate to some specific distributional setups or pseudo-likelihood problems, in principle, that goes over to the more general case of densities not belonging to the so called exponential family, as well as, to beyond parametrics. However, computational and distributional complexities may mar the simple appeal of the UIP to a certain extent.

In a parametric setup with quantitative multifactor multiresponse experiments, a detailed account of UIP is available in Roy et al. (1971). This interesting monograph, a culmination of the fundamental ideas of Roy (1953, 1957), has clearly illustrated the basic appeal of UIP in various statistical inference problems, albeit in traditional setups without so much emphasis on CSI. The evolution of CSI during the past four decades has added more impetus to look into UIP in CSI for parametric as well as beyond parametric setups. Towards this end, we consider first the same one-sample multivariate problem as in above but without assuming that the underlying distribution (F) is multinormal or of some specified form. Simply assume that F is diagonally symmetric about its location μ (in the sense that both $\mathbf{X}-\mu$ and $\mu-\mathbf{X}$ have the same distribution). Consider the hypotheses testing problem: $H_0 : \mu = \mathbf{0}$ vs. $H_1^+ : \mu \geq \mathbf{0}$. If we write $H_{0j} : \mu_j = 0, j = 1, \ldots, p$ and $H_{1j} : \mu_j \geq 0, j = 1, \ldots, p$, then H_0 and H_1^+ can be expressed as the (finite) intersection and union of the H_{0j} and H_{1j} respectively. This is therefore a finite union-intersection formulation (see J. Roy 1958) for a step-down procedure) that makes sense since we are not seeking for affine or similar invariance in our resolution. For the jth marginal, we have a univariate testing problem for which (locally or globally) optimal or at least desirable signed rank tests are known to exist. The crux of the problem is however to find the distribution theory for the maximum of these p possibly correlated statistics. Unfortunately, this distribution depends on the unknown F, even under the null hypothesis. An easy way to eliminate this impasse is to take recourse to the permutation distribution theory generated by the $n!2^n$ conditionally equally likely sign-inversions and column permutations, a detailed treatise of this being available in Sen and Puri (1967). Silvapulle and Sen (2004, Sec.5.5) have considered some parallel UIT statistics based on derived

R-estimators of location, and have shown that in an asymptotic setup the multinormal setup is retained in this formulation as well.

In the context of multiparameter/multivariate hypothesis testing problems, restricted alternatives, often, there is no UMP test, even along specific directions. This feature complicates the construction of usual LRT and RLRT. In a majority of cases, a locally most powerful (LMP) test can be constructed for specific directions, and thus, a UIT based on such LMP statistics can be constructed. The concept extends easily for nonparametric tests where LMPR tests are advocated. Some of these procedures have been discussed in detail in Silvapulle and Sen (2004, Ch. 5). As such, we omit most of these discussions here.

The extension of statistical reasoning from simple parametrics to more complex beyond parametrics setups has been fortified with less emphasis on the likelihood and more on asymptotics to accomodate workable resolutions. This is no exception in CSI too, and the where UIP plays a special role. In most of the complex statistical inference problems, the usual likelihood formulation stumbles into methodological as well as computational difficulties, even in asymptotic setups. For example, in semiparametric inference, the celebrated Cox (1972) proportional hazards model, a partial likelihood approach was innovated to accomodate censoring in a meaningful way, and suitable counting processes along with martingale methodology provided the needed methodological support. Yet, the very basic assumption of proportional hazards may often appear to be rather untenable. More general semiparametrics may require further modification of the LP, and along this line, profile-, partial-, penalized-, pseudo-, and quasi-likelihoods have been advocated in the literature. The usual estimating equations appearing in likelihood formulations have been extended to "generalized estimating equation" (GEE), hence linking the generalised linear models in this broader setup. Empirical likelihoods have also been advocated, along the lines of conventional resampling methods. In all these developments, the validity of exact statistical inference is questioned, loss of information is assessed, and robustness aspects have been focused.

As we look into CSI in such a complex setup, we observe that statistical inference perspectives become even more unclear. For example, in many such problems, suitable scores statistics based on appropriate modification of the likelihood formulation are used in the formulation of test statistics or estimating equations. In constrained evvironments, these score statistics often require considerable modifications to satisfy the set constraints, and thus resulting in a different distributional problem. Silvapulle (1995) and Silvapulle and Silvapulle (1995) formulated a somewhat different approach, termed the Wald-type tests. Recall that in a conventional regular model, Wald (1943) considered a modification of the likelihood ratio test by considering the unrestricted MLE of the associated parameters and exploiting their asymptotic normality in a formulation of a quadratic form which is asymptotically equivalent to the LRT for the same hypothesis testing problem. This approach has been systematically explored in Silvapulle and Sen (2004) covering some parametrics as well as beyond parametrics CSI problems. This approach has also been extended to beyond parametrics situations. The only point in this approach is the need for the computation of the restricted estimators as well as the unrestricted ones, even in an asymptotic setup and that in general requires extensive computational tools. It has been observed (Silvapulle and Sen 2004) that in some simple CSI problems it might be more convenient to incorporate the UIP to derive parallel inference tools which may be computationally less cumbersome and yet asymptotically equivalent. In the rest of this study, we shall elaborate this feature of UIP with some specific CSI problems.

1.5. Statistical Reasoning for HDDSM

Low dimensional discrete statistical models in general CSI setups have been treated in Silvapulle and Sen (2004), Section 6.5 (pp.306-313) in a general case of $r \times c$ contingency tables, for $r, c \geq 2$. In the LP based formulation (viz., Dardanoni and Forcina, 1998), it is necessary to find the MLE of π_D and π_R under the null as well as alternative hypotheses. The computation of the MLE under H_0 is simple, namely, the pooled group marginal proportions in the r or c categories. However, analytical computation of the MLE under general restricted alternatives may be usually quite cumbersome, requiring extensive computational algorithms. Further, once these are done, one has to appeal to the large sample distribution theory of RLRT as customarily given by the conventional chi-square bar distribution. There is a further complication due to the nature of the dispersion matrix (unknown and not of full rank), and hence, as in Perlman's (1969) multinormal mean testing problem against positive orthant alternatives (with unknown arbitrary p.d. dispersion matrix), one has to deal with the least favorable configuration to obtain a conservative p-value. Side by side, let us consider the UIP approach. We need to find out the (unrestricted) UMLE of the two probability vectors, and the task is comparatively simpler, specially in the same asymptotic setup. We may motivate the approach through the asymptotic joint normality of these UMLE, again a well known result. Once this is done, we have a set of inequality constraints for which the classical Kūhn- Tūcker- Lagrange (KTL) point formula (viz., Silvapulle and Sen, 2004, pp. 166 - 168), under the Shapiro (2000) regularity coditions, could be used to formulate the appropriate test statistic. This may generally require less intensive computational schemes. Its asymptotic null hypothesis distribution is given by a chi square bar distribution, a convex mixture of chi-square distributions with degrees of freedom ranging from 0 to p (here $p = 4$), similar to the case treated in Silvapulle and Sen (2004, p.157). Finally, we note that under the null hypothesis, we have the homogeneity of the two probability vectors, and hence, the exact conditional probability law (given the marginal totals) can be effectively used to find a conditional test. For sample sizes not too large, this provides a better control of the type I error than a conservative testing procedure based on the asymptotics solely. For some details, we may refer to Tsai and Sen (2005) where a more general case of restricted alternative hypothesis testing problem has been treated under the Shapiro (2000) regularity conditions, exploiting the UIP there to a greater extent. Whereas the LP based approach exploits the Wald formulation, the UIP based approach does so through the Rao score statistics, under constrained environments.

The second motivating example relates to a statistical comparison of 4 epidemiologic groups with respect to their SARSCoV genomes, treated in Sen et al. (2007). Following the origin of SARS (severe acute respiratory syndrome) in Southern China, the global epidemic resulted in 8,422 infected people with 916 deaths. The SARS causative agent was identified as a novel coronavirus (SARSCoV), as a signle-stranded and positive sense RNA virus with large genome sise (around 30kb). Compared to other RNA viruses, the mutation rate in the SARSCoV is moderate but still several orders of magnitude higher than in DNA virus. An appraisal of variations on the viral RNA was therefore sought for molecular, clinical and therapeutic studies. After preliminary data handling, 25 SARS genome with single nucleotide variation were identified including 6 from Beijing, 3 from Hong Kong, 4 from Singapore and 12 from Taiwan. There were 192 screened genes, so that we have 4 groups of 6, 3, 4 and 12 sequences, each with 192 positions and at each position 4 possible response: Nucleotides A, C, G and T. This perfectly fits with our contemplated HDDSM.

Motivated by the above, we consider G groups of sequences, where in the gth group, there are n_g sequences, for $g = 1,\ldots,G$. In the kth position, let n_{gkc} denote the number of sequences in the gth group with the response categoy c, where c ranges over $1,\ldots,C$ and k over $1,\ldots,K$. In the above example, $K = 192$ and $C = 4$. Thus, we have G stochastic matrices $((n_{gkc}))_{K\times C}$, for $g = 1,\ldots,G$. Although, it could be assumed that the sequences are independent, it is unreasonable to assume that the responses at the K positions are stochastically independent. To capture the HDDSM, we let $\mathbf{X}_{gi} = (X_{gi,1},\ldots,X_{gi.K})'$, $i = 1,\ldots,n_g$, where $X_{gi,k}$ can take on lebels $1,\ldots,C$ for each $k = 1,\ldots,K$. Therefore, there is a set $\mathcal{C}$ of C^K possible joint labels $\mathbf{c} = (c_1,\ldots,c_K)$, where each c_k can take on the labels $1,\ldots,C$, so that the cardinality of $\mathcal{C}$ is C^K. The corresponding cell probability is denoted by $\pi_g(\mathbf{c})$, for $\mathbf{c} \in \mathcal{C}$. Note that $\sum_{\mathbf{c}\in\mathcal{C}} n_{gkc} = n_g$ and $\sum_{\mathbf{c}\in\mathcal{C}} \pi_g(\mathbf{c}) = 1$, for every $g(=1,\ldots,G)$. The full multisample, multidimensional, multinomial law is given by

$$\prod_{g=1}^{G} \frac{n_g!}{\prod_{\mathbf{c}\in C} n_g(\mathbf{c})!} \prod_{\mathbf{c}\in C} [\pi_g(\mathbf{c})]^{n_g(\mathbf{c})}, \tag{1.19}$$

defined over the product simplex $S_{G\times(C^K-1)}$.
Since here the categories $1,\ldots,C$ relate to purely qualitative characteristics (without any implicit ordering), conventional measures of variability are not usable. Rather variation is viewed in the light of mutation rates or other diversity measures, which are functions of the cell probabilities. If we consider the full multinomial model, formulated above, when K is large, even if C may not be, for each group there being C^K-1 cell probabilities, we need the individual $n_g(\mathbf{c})$ to be at least moderately large, so that n_g should be $>> C^K$, a condition rarely tenable in HDDSM, specially in genomics context where experiments are excessively costly. For this reason, a full likelihood based statistical inference procedure is impractical in use in HDDSM, and alternative approaches are to be advocated.
We consider here a pseudo-marginal approrach wherein for each of the K marginal probability laws, a convenient measure of diversity is used in a composite way to formulate an overall measure of diversity, Among plausible measures of diversity, we may consider two important ones, namely, the Gini-Simpson index and the entropy measure. For a simple multinomial law with cell probabilities $\pi_1,\ldots,\pi_C$, the Gini-Simpson Index (GSI) (Gini 1912, Simpson 1949) is defined as

$$I_{GS}(\pi) = 1 - \pi'\pi, \tag{1.20}$$

which attains a maximum value $(C-1)/C$ when all the π_j are equal (to C^{-1}), and a minimum value 0 when only one of the π_j is equal to 1 and the rest 0. Thus, if we consider a simplex S_{C-1} then π being defined on this simplex, minimum diversity occurs at the C vertexes of the simplex while a maximum occurs at the centroid of this simplex. Sen et al. (2007) have exhibited diversity contours based on the GSI. The basic idea is the following: If some gene (position) is not associated with a specific disease/disorder then its variation, as measured by some diversity indenx, will be stable. On the other hand, for disease-genes, the variation would stochastically different with more concentration in some specific transitions. Therefore, the average of the marginal diversity measures has an interpretable role to play in this study. Actually, CSI comes in good handy form in such a marginal approach. Let π_{gk} be the vector of cell probabilities of the gth group at the kth position, for $k=1,\ldots,K;\ g=1,\ldots,G$. Let us then define

$$\theta_g = K^{-1} \sum_{k=1}^{K} I_{GS}(\pi_{gk}),\ g=1,\ldots,G. \tag{1.21}$$

It is also possible to express θ_g as $K^{-1}\sum_{k=1}^{K} P\{X_{gi,k} \neq X_{gj,k}\}$, and this is known as the Hamming distance for the probability law π_g. The sample counterparts are easily shown to be (suitable U-statistics)

$$U_{n_g} = \binom{n_g}{2}^{-1} \sum_{1 \leq i < j \leq n_g} \phi(\mathbf{X}_{gi}, \mathbf{X}_{gj}), \; g = 1, \ldots, G, \tag{1.22}$$

where $\phi(\mathbf{a}, \mathbf{b}) = K^{-1}\sum_{k=1}^{K} I(a_k \neq b_k)$ is the Hamming distance between $\mathbf{a}$ and $\mathbf{b}$ both being K vectors. Pinheiro et al. (2005) have incorporated these Hamming distances for the G groups to test for the homogeneity of the θ_g. A subgroup decomposability property (Sen 1999) underlies their formulation.
We combine the G groups into a single one with n sequences. Define

$$U_n = \binom{n}{2}^{-1} \sum^{*} \phi(\mathbf{X}_{gi}, \mathbf{X}_{g'i'}), \tag{1.23}$$

where the summation $\sum^{*}$ extends over all possible pairs of vectors from the pooled sample. Further, let

$$U_{n_g, n_{g'}} = (n_g n_{g'})^{-1} \sum_{i=1}^{n_g} \sum_{j=1}^{n_{g'}} \phi(\mathbf{X}_{gi}, \mathbf{X}_{g'i'}), \; g \neq g' = 1, \ldots, G; \tag{1.24}$$

these are the generalised U-statistics for pairs of samples. Then the subgroup decomposability relates to the following:

$$U_n = \sum_{g=1}^{G} (n_g/n) U_{n_g} + \sum_{1 \leq g < g' \leq G} \frac{n_g n_{g'}}{n(n-1)} \{2U_{n_g, n_{g'}} - U_{n_g} - U_{n_{g'}}\}, \tag{1.25}$$

where the second term, termed the adjusted between group Hamming distance, has nonnegative expectation when the null hypothesis is not true and 0 expectation under the null hypothesis. Thus, we may consider a regular ANOVA-type test based on this decomposition, rejecting the null hypothesis for large positive values. The crux of the problem is to determine the distribution theory of this test statistic under the null hypothesis. We refer to Pinheiro et al. (2005) and Sen et al. (2007) for some details.

1.6. UIP and MCP in HDDSM

Sen et al. (2007) have considered CSI problems relating to the Hamming distances for the G groups presented in the preceding section. Basically, they considered suitable ordered alternatives against the null hypothesis of homogeneity. Thus, the null hypothesis H_0 relates to the homogeneity of the π_g, $g = 1, \ldots, G$

against alternatives relating to suitable ordering of the Hamming distances θ_g having interpretable biological implications. For example, we might have $H_1^{>}: \theta_1 \leq \theta_2 \leq \cdots \leq \theta_G$ with at least one strict inequality sign being true. Note that there is no linear ordering involved, and moreover, the θ_g are all defined on the interval $[0,(C-1)/C]$ so that translation or scale euivariance property may not hold.

For the normal theory models, for such ordered alternative hypothesis testing problems, as treated in detail in Silvapulle and Sen (2004), even if translation or scale equivariance may hold, we may not have an optimal test, specially for partial ordering when there are nuisance scale parameters or dispersion matrices. Roy's UIP has been incorporated to formulate suitable union-intersection test statistics. Further, for small sample sizes and relatively large K, the exact distribution theory of such test statistics may be difficult to obtain, and the problem becomes even more unmanageable for our contemplated HDDSM. In order to bypass some of these technical difficulties, the UIP is incorporated to formulate suitable test statistics (without necessarily claiming that these are optimal), and a permutation approach is advocated for good approximation for critical levels. If the n_g were large, Tsai and Sen (2005) prescription for asymptotic CIS would apply here well. However, if the n_g are relatively small, copared to K, such approximations are not adequate, and the proposed permutation-sampling provides a better resolution. Recall that under H_0, all the n sequences conform to a common multidimensional multinomial probability law, so that all possible partitioning into G subsets, namely,

$$M_n = \frac{n!}{\prod_{g=1}^{G} n_g!} \tag{1.26}$$

are equally likely, each having the common probability M_n^{-1}. This provides the basis for the permutation distribution.

Even if we do so, M_n could be prohibitively large. For example, in the SARSCoV dataset, we have $n = 25$, $n_1 = 6, n_2 = 3, n_3 = 4, n_5 = 12$ so that M_n is unmanageably large. As such, what was done to draw a random sample of 5,000 pertitionings from this set, and for each drawn permutation, the UIT statistic was computed. Thus, we arrive at a set of 5,000 values of the test statistic against which the actual sample realisation was compared. This procedure gives a fairly good apprpximation of the critical level based on the permutation distribution. Of course, this would not give us the exact critical level, but the procedure remains valid for the HDDSM whereas conventional chi-squared bar distributional approximations are much less reliable.

As has been noted earlier, one of the basic problems in high-dimensional data models is the abundance of hypotheses or comparisons, often outnumbering the sample size. We therefore consider some pertinent remarks on the role of UIP in MCP studies role of UPI in meta analysis, or pooling of information from composite sites, as is currently adapted in in various fields of application, specially in genomic studies. In passing, we should note, howver, in such high-dimensional perspectives, the test statistics for different subhypotheses may not be stochastically independent. For multi-center clinical trials, generally conducted under not so homogeneous environment (e.g., different geographical or demographic strata, age / cultural differences), inter-center heterogeneity may account for some extra variation, although all the centers may have a common objective of drawing statistical conclusions that pertain to a broader population. The picture is similar in genomics studies where inter-species heterogeneity may mar the simplicity of usual UIP or MCP approaches. Typically with a huge number of genes and with relatively smaller number of replications, there is a high level of degeneracy of statistical models so that conventional MCP formulations may generally encounter serious roadblocks. At the present there is considerable emphasis on the use of individual gene based statistical analysis and then combining these marginal statistics into some rational statistical inference scheme. Although this is typically in line with conventional meta analysis, possible stochastic dependence among the genes may vitiate standard meta analysis tools. We discuss these problems briefly here.

For motivation, we briefly take a detour to multi-center clinical trials where the clinics can be taken as independent. Consider in this vein, $C(\geq 2)$ centers, each one conducting a clinical trial with the common goal of comparing a new treatment with an existing one or a control or placebo. Since such centers pertain to patients with possibly different clutural, racial, demographic profiles, diet and physical exercise habits etc. and they may have somewhat different clinical norms too, the intra-center test statistics $\mathcal{L}_c$, $c = 1,\ldots,C$, used for CSI/RST, though could be statistically independent, might not be homogeneous enough to pull directly. This feature may thus create some impasses in combining these statistics values directly into a pooled one to enhance the statistical information. Meta analysis, in the context of MCP, based on *observed significance levels* (OSL) or p-values, is commonly advocated in this context. Recall that under the null hupothesis (which again can be interpreted as the intersection of all the center null hypotheses), the p-values have the common uniform $(0,1)$ distribution, providing more flexibility to adopt UIP in meta analysis. Under restricted alternatives, these OSL values are left-tilted (when appropriate UIT are used) in the sense that the probability density is postively skewed over $(0,1)$ with high density at the lower tail and low at the upper. Let us denote the p-values by

$$P_c = P\{\mathcal{L}_c \geq \text{the observed value } |H_0\}, c = 1,\dots,C. \tag{1.27}$$

The well-known Fisher's test is based on the statistic

$$F_n = \sum_{c=1}^{C}\{-2\log P_c\}, \tag{1.28}$$

which, under the null hypothesis, has the central chi-square distribution with $2C$ degrees of freedom. This test has some desirable asymptotic (in n) properties, albeit in HDLSSM such properties may not be tenable. There are many other tests based on the OSL values. The well known step-down procedure (Roy 1958) has also been adapted in this vein (cf. Mudholkar and Subbaiah 1980, Sen 1983), and they have been amended for CSI and RST as well (cf. Sen 1988). One technical drawback observed in this context is the insensitivity (to small to moderate departures from the null hypothesis) of such tests (including the Fisher's) when C is large, resulting in nonrobust and, to a certain extent, inefficient procedure. In multi-center clinical trials, typically, C may not be too large, and hence, the extent of nonrobustness and inefficacy of the Fisher method as well as other conventional ones might not be that significant. However, as C becomes large, these deficiencies can be more apparent. Thus, alternative approaches based on the OSL values have been explored more recently in the literature.

In the evolving field of bioinformatics and genomics, generally, we encounter an excessively high dimensional data set with inadequately small sample size creating impasses for the applicability of standard CSI or even conventional statistical inference tools. On top of that, the OSL values to be combined (corresponding to different genes) may not be independent, and in many cases, due to actual distributional assumptions (e.g., nonparametric ones) may not have strictly uniform distribution under the null hypothesis, creating another layer of difficulty with conventional meta analysis. This led to the development of multiple hypotheses testing in large dependent data models based on OSL values. This field is going through an evolution, and much remains to be accomplished. In this spectrum, the Simes (1986) theorem occupies a focal point. Let there be K null hypotheses (not necessarily independent) $H_{0k},\ k = 1,\dots,K$ with respective alternatives (which possibly could be restricted or constrained as in clinical trials or microarray studies) $H_{1k},\ k = 1,\dots,K$. We thus come across the same UIP scheme by letting H_0 as the intersection of all the component null hypotheses, and H_1 as the union of the component alternatves.Let $P_k, k = 1,\dots,K$ be the OSL values associated with the hypotheses testing H_{0k} vs. H_{1k}, for $k = 1,\dots K$. We denote the ordered values of these OSL values by

$$P_{K:1}, \cdots, P_{K:K}. \tag{1.29}$$

The basic idea is to exploit the information contained in these ordered p-values in a more creative way. If the individual tests have continuous null distributions then the ties among the P_k (and hence, among their ordered values) can be negelected, in probability. Assuming independence of the P_k, and uniform distribution for the unordered ones, Simes theorem states that

$$P\{P_{K:k} > k\alpha/K, \forall k = 1, \ldots, K|H_0\} = 1 - \alpha. \tag{1.30}$$

Interestingly enough, the Simes theorem is a restatement of the classical Ballot theorem, developed some twenty years earlier (cf. Karlin, 1969): Let $U_1, \ldots, U_K$ be i.i.d. r.v.'s having the Unif.(0, 1) distribution and let $G_K(u) = K^{-1} \sum_{k=1}^{K} I(U_k \leq u)$, $u \in (0,1)$ be the associated empirical d.f. Then, for evey $\gamma \geq 1$, and every $K \geq 1$,

$$P\{G_K(u) \leq \gamma u, \forall u \in (0,1)\} = 1 - \gamma^{-1}. \tag{1.31}$$

In any case, granted the unawareness, it is a nice illustration how the UIP is linked to the extraction of extra statistical information through ordered OSL values.
It did not take long time for applied mathematical statisticians to make good uses of the Simes-Ballot theorem in CSI and multiple hypothesis testing problems. The above results pertains to tests for an overall null hypothesis in the UIP setup. Among others, Hochberg (1988) incorporated a variant of the above result:

$$P\{P_{K:j} \geq \alpha/(K-j+1), \forall j = 1, \ldots, K|H_0\} = 1 - \alpha, \tag{1.32}$$

in a multiple testing framework. Benjamini and Hochberg (1995) introduced the concept of false discovery rate (FDR) in the context of multiple hypothesis testing, and illustrated the role of the Simes-Ballot theorem in that context. Just to point out how difficult may be such procedures, let us consider the following illustrative example (Sen et al. 2007). Suppose that $K = 192$ and there are 4 groups with sample sizes 4,6,3 and 12 respectively. Basically then one has 192 tests in a multiple hypotheses testing setup. Even if we assume multinormality, the sample sizes are so small that the recording of the actual p-values from appropriate tables could be very sensitive for values close to zero (or 1); the tabular values could be quite different from the actual ones. A permutation approach, as has been prescribed in Sen et al. (2007), may be quite conservative and thereby subject to the same limitation. On top of that for any chosen value of α, the numbers α/K or $\alpha/(K-j+1)$, for small values of $j(\geq 1)$ will be so small that these MTP will have too little power. For example, for $\alpha = 0.05$ and $K = 192$,we have $\alpha/K = 0.00025$

so that we need to have a fairly accurate recording of the actual p-values, especially near the lower end-point 0. From robustness point of view, this is often a challenging task.

The past ten years have witnessed a phenomenal growth of research literature in this subfield with applications to genomics and bioinformatics. The basic restraint in this respect is the assumption of independence of the $P_j, j = 1, \ldots, K$, and in bioinformatics, this is hardly the case. Sarkar (1998) and Sarkar and Chang (1997) incorporated the MTP_2 (multivariate total positivity of order 2)property to relax the assumption of independence to a certain extent. Sarkar (2000, 2002, 2004) has added much more to this development with special emphasis on controlling of FDR in some dependent cases. The literature is too large to cite adequately, but our primary emphasis here is to stress how UIP underlies some of these developments and to focus on further potential work.

Combining OSL values, in whatsoever manner, may generally involve some loss of information when the individual tests are sufficiently structured to have coherence that should be preserved in the meta analysis. We have seen earlier how guided by the UIP, progressive censoring in clinical trials provided more efficient and interpretable testing procedures. The classical Cochran-Mantel-Haenszel (CMH) procedure is a very notable example of this line of attack. In a comparatively more general multiparameter CSI setting, Sen (1999b) has emphasized the use of the CMH procedure in conjunction with the OSL values to induce greater flexibility. The field is far from being saturated with applicable research methodology. The basic assumption of independence or specific type of dependence is just a part of the limitations. A more burning question is the curse of dimensionality in CSI problems. Typically, there K is large and the sample size n is small, i.e., $K >> n$. In the context of clinical trials in genomics setups, Sen (2006) has appraised this problem with due emphasis on the UIP. Conventional test statistics (such as the classical LRT) have awkward distributional problems so that usual OSL values are hard to compute and implement in the contemplated CSI problems. Based on the Roy (1953) UIP but on some nonconventional statistics, it is shown that albeit there is some loss of statistical information due to the curse of dimensionality, there are suitable tests which can be implemented relatively easily in high-dimension low sample size environments. In CSI for clinical trials in the presence of genomics undercurrents, there is a tremendous scope for further developments along this line.

References

1. Barlow, R.E., Bartholomew, D.J., Bremner, J.M. and Brunk, H.D. (1972). *Statistical Inference under Order Restrictions*, John Wiley, New York.
2. Benjamini, Y. and Hochberg, Y. (1995). *Controlling the false discovery rate: A practical and powerful approach to multiple testing.* Jour. Roy. Statist. Soc. B57,, 289-300.
3. Cox, D. R. (1972). *Regression models and life tables (with discussion).* Jour. Roy. Statist. Soc.B 34, 187-220.
4. DeMets, D.L. and Lan, K.K.G. (1983). *Discrete sequential boundaries for clinical trials.*Biometrika 70, 659-663.
5. Dardanoni, V. and Forcena, A, (1998). *A unified approach to likelihood inference on stochastic ordering in a nonparametric context.* Jour. Amer. Statist. Assoc.93, 1112 - 1123.
6. Hochberg, Y. (1988). *A sharfer Bonferroni procedure for multiple tests of significance.* Biometrika 75, 800 - 802.
7. Karlin, S. (1969). *A first Course in Stochastic Processes*, Academic Press, New York.
8. McDormott, M. P. and Mudholkar, G. S. (1993). *A simple approach to testing homogeneity of order-constrained means.* Jour. Amer. Statist. Assoc. 88, 1371 - 1379.
9. Mudholkar, G. S., Kost, J. and Subbaiah, P. (2001). *Robust tests for orthant - restricted mean vector.* Commun. Statist. Theor. Meth. 30, 1789 - 1810.
10. Mudholkar, G. S. and McDermott, M. P. (1989). *A class of tests for equality of ordered means.* Biometrika 76, 161 - 168.
11. Mudholkar, G. S. and Subbaiah, P. (1980). *Testing significance of a mean vector - a possible alternative to Hotelling* T^2. Ann. Institut. Statist. Math. 32, 43 - 52.
12. Perlman, M. D. (1969). *One-sided problems in multivariate analysis.* Ann. Math. Statist. 40, 549-567.
13. Pinheiro A.S., Pinheiro, H.P. and Sen, P.K. (2005) *Comparison of genomic sequences by Hamming distance.* Jour. Statist. Plan. Infer. 130, 325-339.
14. Robertson, T., Wright, F.T. and Dykstra, R. (1988). *Order Restricted Statistical Inference*, John Wiley, New York.
15. Roy, J. (1958). *Step-down procedures in multivariate analysis.* Ann. Math. Statist. 29, 1177 - 1188.
16. Roy, S. N. (1953). *On a heuristic method of test construction and its use in multivariate analysis.* Ann. Math. Statist.24, 220-238.
17. Roy, S.N. (1957). *Some Aspects of Multivariate Analysis*, John Wiley, New York, and Asia Publ. House, Bombay.
18. Roy, S.N., Gnanadesikan, R. and Srivastava, J.N. (1971). *Analysis and Design of Certain Quantitative Multiresponse Experiments*, Pergamon Press, New York.
19. Sarkar, S.K. (1998). *Some probability inequalities for ordered* MTP_2 *random variables: a proof of the Simes conjecture.* Ann. Statist.26, 494-504.
20. Sarkar, S.K. (2000). *A note on the monotonicity of the critical values of a step-up test.* Jour. Statist. Plann. Infer. 87, 241-249.
21. Sarkar, S.K. (2002). *Some results on false discovery rate in multiple testing procedures.*Ann. Statist. 30, 239-257.
22. Sarkar, S.K. (2004). *FDR-controlling stepwise procedures and their false negativesc rates.*Jour. Statist. Plann. Infer. 125, 119-137.

23. Sarkar, S.K. and Chang, C.-K. (1997). *The Simes method for multiple hypothesis testing with positively dependent test statistics.* Jour. Amer. Statist. Assoc. 92, 1601-1608.
24. Sen, P.K. (1981). *Sequential Nonparametrics: Invariance Principles and Statistical Inference*, John Wiley, New York.
25. Sen, P.K. (1983). *A Fisherian detour of the step-down procedure. In Contributions to Statistics: Essays in honour of Norman L. Johnson*, North Holland, Amsterdam, pp. 367-377.
26. Sen, P.K. (1988). *Combination of statistical tests for multivariate hypotheses against restricted alternatives. In Advances in Multivariate Statistical Analysis* (eds. S. Dasgupta and J.K. Ghosh), Ind. Statist. Inst. pp. 377-402.
27. Sen, P. K. (1999 a). *Multiple comparisons in interim analysis.* Jour. Statist. Plann. Infer. 82, 5-23.
28. Sen, P.K. (1999 b). *Some remarks on the Stein-type multiple tests of significance.* Jour. Statist. Plann. Infer. 82, 139-145.
29. Sen, P. K. (2001). *Survival analysis: Parametrics to semiparametrics to pharmacogenomics.* Brazilian Jour. Probab. Statist. 15, 201 - 220.
30. Sen, P. K. (2005). *Nonparametric tests for ordered diversity in a genomic sequence.* Jour. Statist. Res. 39, No. 2, 7 - 21.
31. Sen, P. K. (2006). *Robust Statististical inference for high-dimension low sample size problems with applications to genomics.* Austrian Jour. Statist. 35 197-214.
32. Sen, P. K. (2007). *Union-intersection principle and constrained statistical inference.* Jour.Stat. Plan. Infer. 1bc, in press.
33. Sen, P.K. and Puri, M. L. (1967). *On the theory of rank order tests in the multivariate one-sample problem.* Ann. Math. Statist. 136, 3741-3752.
34. Sen, P. K. and Tsai, M.-T. (1999). *Two-stage likelihood ratio and union-intersection tests for one-sided alternatives multivariate mean with nuisance dispersion matrix.* Jour. Multivar. Anal. 68, 264-282.
35. Sen, P. K., Tsai, M.-T. and Jou, Y.-S. (2007). *High dimension low sample size perspectives in constrained statistical inference: The SARSCoV genome in illustration.* Jour. American Statist. Assoc., 102, in press.
36. Shapiro. A. (2000). *On the asymptotics of constrained local M-estimators.* Ann. Statist. 28, 685-694.
37. Silvapulle, M.J. (1995). *A Hotelling T^2-type statistic for testing against one-sided hypotheses.* Jour. Multivar. Anal. 55, 312-319.
38. Silvapulle, M.J. and Sen, P. K. (2004). *Constrained Statistical Inference: Inequality, Order and Shape Restrictions*, John Wiley, New York.
39. Silvapulle, M. J. and Silvapulle, P. (1995). *A score test against one-sided alternatives.* Jour. Amer. Statist. Assoc. 90, 342-349.
40. Simes, R.J. (1986). *An improved Bonferroni procedure for multiple tests of significance.* Biometrika 73, 751-754.
41. Tsai, M.-T. and Sen, P.K. (2005). *Asymptotically optimal tests for parametric functions against ordered functional alternatives.* Jour. Multivar. Anal.95, 37-49.
42. Wald, A. (1943). *Tests of statistical hypotheses concerning several parameters when the number of observations is large.* Trans. Amer. Math. Soc. 54, 426-482.
43. Wijsmann, R.A. (1979). *Constructing all smallest simultaneous confidence sets in a general class with applications to MANOVA.* Ann. Statist. 7, 1003-1018.

Chapter 2

A Review of Multivariate Theory For High Dimensional Data With Fewer Observations

Muni S. Srivastava

University of Toronto

E-mail: srivasta@utstat.toronto.edu

In this article, we review available methods for analyzing multivariate data with fewer observations than the dimension. It includes verifying the assumptions made on the covariance matrix before making inference on the mean vector/vectors in one-sample, two-sample and MANOVA. The problem of classifying an observation vector into one of several groups is also considered.

Contents

2.1. Introduction

In DNA microarray data, gene expressions are available on thousands of genes of an individual but there are only few individuals in the dataset. Although, these genes are correlated, most of the statistical analyses carried out in the literature ignore this correlation without any justification. For example if in one-sample problem it is found that the data supports the spherecity hypothesis about the covariance matrix Σ, that is $\Sigma = \sigma^2 I_p$,for some unknown σ^2 and $p \times p$ identify matrix I_p, then any inference on the mean vector may ignore the correlations between the genes, and use the univariate methods. However, it has not been done in the literature. Similarly,if the covariance matrix is a diagonal matrix, then univariate methods with unequal variances for the components may be used for any inference. For example, Dudoit et al. (2002) assumed that the covariance matrix is a diagonal matrix in their classification procedure without verifying that the data support this assumption. In fact, it is shown in Srivastava (2006b) that the data do not support this assumption. This also implies that if the global hypothesis $H = \cap_{i=1}^{p} H_i$, is rejected, the False Discovery rate (FDR) of Benjamini and Hochberg (1995) cannot be used to determine the components that may have caused the rejection of the hypothesis H unless the test statistics used in testing the hypothesis H_i are positively dependent, such as under the assumption of normality, the covariance matrix Σ is of the intraclass correlation form

$$\Sigma = \sigma^2 \left[(1-\rho) I_p + \rho \mathbf{1}_p \mathbf{1}_p' \right]$$

where $\rho \geq 0$ and $\mathbf{1}_p'$ is a row vector whose all entries are one $\mathbf{1}_p' = (1, \ldots, 1)$. A test for this hypothesis is also given.

Throughout this article, we shall assume that the sample size N is less than the dimension p. This in turn implies that no tests invariant under the nonsingular linear transformations exist for testing the hypothesis on the mean vector in one-sample, and mean vectors in two-sample and many samples, the so-called MANOVA problem, see Lehmann (1959, p.318). Similarly, the likelihood ratio tests for the hypotheses on the covariance matrix or the covariance matrices in two or more than two populations do not exist. Thus various test criteria have been recently proposed to verify the assumptions made on the covariance matrix or matrices in two or more than two populations. Similarly, several test criteria have been proposed for the inference on the mean vectors. In this article, we review these procedures. The tests on the mean vector in one-sample, and mean vectors in two-sample, and MANOVA are considered in Section 2.2, 2.3 and 2.4 respectively.

In Section 2.5, we discuss the classification and discriminant analysis. The tests on the covariance matrix or matrices are described in Section 2.6. The paper concludes in Section 2.7. It may be mentioned that Fujikoshi (2004) recently reviewed inference procedures for high-dimensional data but mostly for the case when $n > p$ and $(p/n) \to c < 1, n = N-1$.

2.2. Inference on the Mean Vector in One-Sample

Let $\mathbf{x}_1,\ldots,\mathbf{x}_N$ be independently and identically distributed (hereafter, iid) as multivariate normal with mean vector μ and unknown positive definite (hereafter, $>$ 0) covariance matrix Σ, denoted $N_p(\mu,\Sigma)$, where p denotes the dimension of the random vector $\mathbf{x}$ on which N observations have been obtained and $N \leq p$. The sample mean vector and the sample covariance matrix are defined by

$$\overline{\mathbf{x}} = N^{-1}\sum_{i=1}^{N}\mathbf{x}_i\ , \quad n = N-1, \tag{2.1}$$

and

$$S = n^{-1}V = n^{-1}\sum_{i=1}^{N}(\mathbf{x}_i-\overline{\mathbf{x}})(\mathbf{x}_i-\overline{\mathbf{x}})'\ , \tag{2.2}$$

respectively. Since $N \leq p$, $n < p$, and hence S is a singular matrix with probability one, the distributions connected with singular S have been given by Srivastava (2003).

Thus for testing the hypothesis

$$H : \mu = \mathbf{0},\ \Sigma > 0, \text{ vs } A : \mu \neq \mathbf{0},\ \Sigma > 0, \tag{2.3}$$

we need to consider tests other than the Hotelling's T^2 since the inverse of S does not exist. In fact as mentioned in the introduction since $N \leq p$, no tests invariant under the transformation by an element of the group Glp of $p\times p$ nonsingular matrices exist. Thus various tests have been proposed in the literature which are invariant under the transformation of some smaller groups. We will describe them in the next few subsections.

2.2.1. *Tests invariant under orthogonal and a non-zero scalar transformations*

Dempster (1958,1960) proposed the test statistic

$$T_D = (N\overline{x}'\overline{\mathbf{x}})/(\text{tr } S) \tag{2.4}$$

for testing the hypothesis $H : \mu = \mathbf{0}$ against the alternative $A : \mu \neq \mathbf{0}$. This test is invariant under the transformation by an element $c\Gamma$, $c \neq 0$, where Γ belongs to the group of $p\times p$ orthogonal matrices O_p and c belongs to the group of scalars $R_{(0)} = R - \{0\}$, where $R_{(0)}$ is the real line without the element 0. Thus, if $\Sigma = \sigma^2 I$, T_D is the best uniformly most powerful invariant test under the transformation $c\Gamma, c \neq 0$.

The power depends on $N\mu'\mu/\sigma^2$, and from Samaika (1943), it follows that it is the best test among all tests whose power depends on $N\mu'\mu/\sigma^2$. Under the hypothesis and when $\Sigma = \sigma^2 I$, T_D has an F-distribution with p and np degrees of freedom, denoted $F_{p,np}$. However, when $\Sigma \neq \sigma^2 I$, the above property of uniformly most powerful invariant test does not exist. Nevertheless, Dempster (1958) proposed the test T_D for testing the hypothesis $H : \mu = 0$ against the alternative that $\mu \neq 0$, for all Σ. From the normal theory, it follows that

$$T_D = \frac{nQ_1^2}{Q_2^2 + , \ldots , Q_N^2} ,$$

where Q_i are iid but not necessarily distributed as chi-square unless $\Sigma = \sigma^2 I$, in which case it will be χ_p^2, a chi-square random variable with p degrees of freedom. To obtain the distribution of T_D when $\Sigma \neq \sigma^2 I$, Dempster (1958) assumed that Q_i are distributed as $m\chi_r^2$ for some unknown m and r. Since the statistic T_D does not depend on m, Dempster gave two iterative equations to solve to get the value of r. Thus $F_D \sim F_{[\hat{r}],[n\hat{r}]}$, where $\hat{r}$ is the solution of any of the two iterative equations given by Dempster (1958) and [a] denotes the largest integer contained in [a]. Alternatively, under the assumption that Q_i are iid $m\chi_r^2$, it follows that by equating the first two moments of $m\chi_r^2$ to the corresponding two moments of Q_i^2, we find that under the hypothesis

$$E(m\chi_r^2) = mr = E(Q_i) = \text{tr}\Sigma$$

and

$$\text{Var}(m\chi_r^2) = 2m^2 r = 2\text{tr}\Sigma^2$$

Thus, r is given by

$$\begin{aligned} r &= \frac{(\text{tr}\Sigma)^2}{(\text{tr}\Sigma^2)} = p\frac{(\text{tr}\Sigma/p)^2}{(\text{tr}\Sigma^2/p)} = p\left(\frac{a_1^2}{a_2}\right) \\ &\equiv pb , \end{aligned} \tag{2.5}$$

where

$$a_i = (\text{tr}\Sigma^i/p) , i = 1,2 \tag{2.6}$$

and

$$b = (a_1^2/a_2) . \tag{2.7}$$

A consistent estimator of b and a ratio consistent estimator of r as $n \to \infty$ or if $n = O(p^\delta)$, $0 < \delta \leq 1$ and $(n,p) \to \infty$, have been given by Srivastava (2005) as

$$\hat{b} = (\hat{a}_1^2/\hat{a}_2), \text{ and } \hat{r} = p\hat{b} , \tag{2.8}$$

where

$$\hat{a}_1 = (\mathrm{tr}S/p) \tag{2.9}$$

and

$$\hat{a}_2 = \frac{n^2}{(n-1)(n+2)}\frac{1}{p}\left[\mathrm{tr}S^2 - \frac{1}{n}(\mathrm{tr}S)^2\right] \tag{2.10}$$

provided

$$0 < \lim_{p\to\infty} a_i = a_{i0} < \infty \ , i = 1,\ldots,4 \tag{2.11}$$

Thus we get the following theorem.
Theorem 2.1 Under the assumption (2.2.11), the T_D statistic defined in (2.4) has approximately an $F_{[\hat{r}],[n\hat{r}]}$ distribution under the hypothesis that $\mu = \mathbf{0}$,where $\hat{r}$ is given in (2.2.8).
The asymptotic non-null distribution of F_D has been given by Bai and Sarandosa (1996). It is given by

$$\lim_{(n,p)\to\infty}\left[P\{(\frac{r}{2})^{\frac{1}{2}}(F_D - 1) > z_{1-\alpha}|\mu = (nN)^{-\frac{1}{2}}\delta\} - \Phi\left(-z_{1-\alpha} + \frac{n\mu'\mu}{\sqrt{2pa_2}}\right)\right] = 0 \ , \tag{2.12}$$

where Φ denotes the cdf of a standard $N(0,1)$ random variable, and

$$\Phi(z_{1-\alpha}) = 1 - \alpha \ . \tag{2.13}$$

Bai and Saranadasa (1996) also proposed another test statistic

$$T_{BS} = \frac{N\overline{\mathbf{x}}'\overline{\mathbf{x}} - \mathrm{tr}S}{[(n+1)/n]^{\frac{1}{2}}[2p\hat{a}_2]^{\frac{1}{2}}} \tag{2.14}$$

for testing the hypothesis $H_1 : \mu = \mathbf{0}$ vs $A_1 : \mu \neq \mathbf{0}$. The asymptotic distribution of the test statistic T_{BS} is given in the following theorem, for proof, see Bai and Saranadasa (1996) or Srivastava (2007).
Theorem 2.2 Let $\lambda_i = O(p^\gamma)$, $0 \le \gamma < \frac{1}{2}$, where λ_i are the eigenvalues of Σ,then under the hypothesis H and condition (2.2.11),

$$\lim_{(n,p)\to\infty} P_0[T_{BS} \le z] = \Phi(z) \ ,$$

An asymptotic distribution of the test statistic T_{BS} under local alternatives is given in Bai and Saranadasa (1996) or Srivastava (2007), and it is the same as that of Dempster's test T_D. That is

$$\lim_{(n,p)\to\infty}\left[P\left\{T_{BS} > z|\mu = (nN)^{-\frac{1}{2}}\delta\right\} - \Phi\left(-z + \frac{n\mu'\mu}{\sqrt{2pa_2}}\right)\right] = 0 \tag{2.15}$$

Thus, asymptotically the two tests T_D and T_{BS} have the same power. In fact from most simulated power comparisons of the two tests carried out by Bai and Sarandasa (1996), and Srivastava and Du (2006), it appears that the two tests T_D and T_{BS} are almost equivalent. In any case, it would appear reasonable to consider only one of them, say T_D for comparing its power with other tests.
Another test that is also invariant under the transformation by an element $c\Gamma$, $c \neq 0, \Gamma \in O_p$ and $c \in R_{(0)}$ has recently been proposed by Srivastava (2007). This test is based on the statistic

$$T^{+2} = N\overline{\mathbf{x}}' S^{+} \overline{\mathbf{x}} , \tag{2.16}$$

where S^{+} is the Moore-Penrose inverse (see Srivastava and Khatri, 1979, pp12) of S given by

$$S^{+} = H'L^{-1}H , \tag{2.17}$$

where

$$S = H'LH , \ HH' = I_n , \tag{2.18}$$

and

$$L = \text{diag}(l_1, \ldots, l_n) , \tag{2.19}$$

an $n \times n$ diagonal matrix of n non-zero eigenvalues of the $p \times p$ singular matrix S of rank n. Let

$$F^{+} = \frac{p-n+1}{n^2} T^{+2} . \tag{2.20}$$

Then the distribution of the F^{+} statistics under the hypothesis is given in the following theorem
Theorem 2.3 Let $n = O(p^{\delta})$, $0 \leq \delta < 1$, and the condition (2.2.11) be satisfied. Then, with $\hat{b}$ defined in (2.2.8),

$$\lim_{(n,p)\to\infty} P_0 \left[c_{n,p} \left(\frac{n}{2} \right)^{\frac{1}{2}} (\hat{b}F^{+} - 1) \leq z \right] = \Phi(z) ,$$

where $c_{n,p}$ is chosen to speed up the convergence to normal distribution for moderate n and p. We have chosen

$$c_{n,p} = \left[1 - \frac{n}{p+1} \right]^{\frac{1}{2}} . \tag{2.21}$$

The asymptotic power of the F^{+} test is given by

$$\lim_{(n,p)\to\infty} \left[P_1 \left\{ c_{n,p} \left(\frac{n}{2} \right)^{\frac{1}{2}} (\hat{b}F^{+} - 1) > z_{1-\alpha} \right\} - \Phi \left[-z_{1-\alpha} + (n/p)(n/2)^{\frac{1}{2}} \frac{\mu' \wedge \mu}{a_2} \right] \right] = 0 .$$

Thus, the asymptotic power of F^+ test is given by

$$\beta(F^+) \simeq \Phi\left[-z_{1-\alpha} + (n/p)(n/2)^{\frac{1}{2}}\frac{\mu' \wedge \mu}{a_2}\right],$$

where $\wedge = \text{diag}(\lambda_1, \ldots, \lambda_p)$ is the covariance matrix of $\mathbf{x}$.

2.2.2. *A Test invariant under scalar transformation of each component*

In this subsection, we consider a test which is invariant under the transformation $\mathbf{x} = C\mathbf{x}$, where $C = \text{diag}(c_1, \ldots, c_p)$ and $c_i \neq 0$, $i = 1, \ldots, p$. A test having this property has been proposed by Srivastava and Du (2006), it is given by

$$T_{SD} = \frac{N\bar{\mathbf{x}}' D_s^{-1} \bar{\mathbf{x}} - (n/(n-2))p}{q_{n,p}\sqrt{2(\text{tr}R^2 - \frac{1}{n}p^2)}}, \tag{2.22}$$

where

$$D_s = \text{diag}(s_{11}, \ldots, s_{pp}),\ S = (s_{ij}), \tag{2.23}$$

$$R = D_s^{-\frac{1}{2}} S D_s^{-\frac{1}{2}}, \tag{2.24}$$

and $q_{n,p} \xrightarrow{p} 1$, and chosen to speed up the convergence of the statistic T_{SD}. Srivasatava and Du (2008) choose

$$q_{n,p} = \left[1 + p^{-\frac{3}{2}} \text{tr}R^2\right]^{\frac{1}{2}} \tag{2.25}$$

which goes to one in probability for $n = O(p^\delta)$, $\frac{1}{2} < \delta \leq 1$. The asymptotic distribution of the test statistic T_{SD} is given in the following theorem.

Theorem 2.4 Let $q_{n,p}$ be a sequence of random variables that goes to one in probability as $(n,p) \to \infty$, Then, under the condition (2.2.11)

$$\lim_{(n,p)\to\infty} [P_0(T_{SD} < z_{1-\alpha})] = \Phi(z_{1-\alpha}),$$

where T_{SD} is defined in (2.2.22). The asymptotic power of the test statistic T_{SD} is given by

$$\lim_{(n,p)\to\infty}\left[P_1(T_{SD} > z_{1-\alpha}) - \Phi\left(-z_{1-\alpha} + \frac{\delta' D_\sigma^{-1}\delta}{n\sqrt{2\text{tr}\mathcal{R}^2}}\right)\right] = 0, \tag{2.26}$$

when

$$\mu = \left(\frac{1}{nN}\right)^{\frac{1}{2}} \delta, \tag{2.27}$$

$$D_\sigma = \text{diag}(\sigma_{11}, \ldots, \sigma_{pp}),\ \Sigma = (\sigma_{ij}), \tag{2.28}$$

and

$$\mathcal{R} = D_\sigma^{-\frac{1}{2}} \Sigma D_\sigma^{-\frac{1}{2}} \tag{2.29}$$

2.2.3. *Comparison of Power*

The asymptotic power of all the four tests are given in the previous subsections. From this, it is clear that asymptotically T_D and T_{BS} have the same power. Also, it is known that if $\Sigma = \sigma^2 I_p$, then the Dempster's test T_D is the best invariant test under the transformation $\mathbf{x} \to c\Gamma\mathbf{x}$, $c \neq 0$, $\Gamma\Gamma' = I_p$, among all tests whose power depends on $\mu'\mu/\sigma^2$ irrespective of the sizes of n and p and so better than the Hotelling's T^2-test (when $n > p$). However, when $\Sigma \neq \sigma^2 I$, then no ordering between four tests exist. Thus, we shall compare the power of the three tests T_D, T_{SD} and F^+ by simulation. The simulation is based on 1000 replications. We consider the case when the covariance matrix is a diagonal matrix, the p elements of which are obtained by simulation, taking p iid observations from their distributions. But once these values are obtained, it remains fixed for the rest of the simulation. In the same way, two values of the mean vector are obtained where the components of the mean vector are iid from two distributions.

We first obtain the cut off points for the distribution of the statistic under the hypothesis. For example, for the statistic F^+, we obtain F_α^+ such that

$$\frac{(\#\text{ of } F^+ \geq F_\alpha^+)}{1000} = \alpha$$

where F^+ is calculated from the $(n+1)$ samples from $N_p(\mathbf{0}, \Sigma)$ for each 1000 replications. The power is then calculated from the $(n+1)$ samples from $N_p(\mu, \Sigma)$ replicated again 1000 times. In this way, all the three tests have the same significance level α which we have chosen to be 0.05. The power of the three tests are shown in the following Tables, Tables 2.1 - 2.4. The mean vectors for the alternative are shown as

$$\begin{aligned} \mu_1 &= \{x_1, \ldots, x_p\},\ i = 1, \ldots, p,\ x_i \sim U(-0.5, 0.5). \\ \mu_2 &= \{x_1, \ldots, x_p\},\ i = 1, \ldots, p,\ \text{for } i = 2k,\ x_i \sim U(-0.5, 0.5); \\ &\quad \text{for } i = 2k+1,\ x_i = 0,\ k = 0, 1, \ldots, p-1. \end{aligned}$$

2.3. Two-sample Tests

Consider a sample of N_1 iid observation vectors $\mathbf{x}_{11}, \ldots, \mathbf{x}_{1N_1}$ from $N_p(\mu_1, \Sigma)$ and a sample of N_2 iid observation vectors $\mathbf{x}_{21}, \ldots, \mathbf{x}_{2N_1}$ from $N_p(\mu_2, \Sigma)$ where both

Table 2.1. Power, $\Sigma = I_p$

		μ_1			μ_2		
p	n	T_{SD}	T_D	F^+	T_{SD}	T_D	F^+
60	30	1.0000	1.0000	0.9780	0.9987	0.9997	0.8460
100	40	1.0000	1.0000	1.0000	1.0000	1.0000	0.9530
	60	1.0000	1.0000	1.0000	1.0000	1.0000	1.0000
	80	1.0000	1.0000	1.0000	1.0000	1.0000	0.9970
150	40	1.0000	1.0000	1.0000	1.0000	1.0000	0.9830
	60	1.0000	1.0000	1.0000	1.0000	1.0000	1.0000
	80	1.0000	1.0000	1.0000	1.0000	1.0000	1.0000
200	40	1.0000	1.0000	1.0000	1.0000	1.0000	0.9870
	60	1.0000	1.0000	1.0000	1.0000	1.0000	1.0000
	80	1.0000	1.0000	1.0000	1.0000	1.0000	1.0000
400	40	1.0000	1.0000	1.0000	1.0000	1.0000	0.9960
	60	1.0000	1.0000	1.0000	1.0000	1.0000	1.0000
	80	1.0000	1.0000	1.0000	1.0000	1.0000	1.0000

Table 2.2. Power, $\Sigma = D$, $D = \text{diag}\{d_1, \ldots, d_p\}$, and $d_i \sim U(2,3)$, $i = 1, \ldots, p$

		μ_1			μ_2		
p	n	T_{SD}	T_D	F^+	T_{SD}	T_D	F^+
60	30	0.6477	0.6424	1.0000	0.2868	0.2909	1.0000
100	40	0.9504	0.9417	1.0000	0.6221	0.5934	1.0000
	60	0.9983	0.9979	1.0000	0.8800	0.8608	1.0000
	80	1.0000	1.0000	1.0000	0.9730	0.9622	1.0000
150	40	0.9703	0.9606	1.0000	0.6982	0.6605	1.0000
	60	0.9994	0.9991	1.0000	0.9308	0.9127	1.0000
	80	1.0000	1.0000	1.0000	0.9911	0.9829	1.0000
200	40	0.9878	0.9845	1.0000	0.8311	0.7892	1.0000
	60	1.0000	0.9998	1.0000	0.9787	0.9668	1.0000
	80	1.0000	1.0000	1.0000	0.9987	0.9981	1.0000
400	40	0.9998	1.0000	1.0000	0.9255	0.9032	1.0000
	60	1.0000	1.0000	1.0000	0.9979	0.9951	1.0000
	80	1.0000	1.0000	1.0000	1.0000	0.9998	1.0000

samples are independently distributed. Let

$$\overline{\mathbf{x}}_1 = N_1^{-1} \sum_{i=1}^{N_1} \mathbf{x}_{1i} \,, \; \overline{\mathbf{x}}_2 = N_2^{-1} \sum_{i=1}^{N_2} \mathbf{x}_{2i} \,, \tag{2.30}$$

$$nS = V = \sum_{i=1}^{N_1} (\mathbf{x}_{1i} - \overline{\mathbf{x}}_1)(\mathbf{x}_{1i} - \overline{\mathbf{x}}_1)' + \sum_{i=1}^{N_2} (\mathbf{x}_{2i} - \overline{\mathbf{x}}_2)(\mathbf{x}_{2i} - \overline{\mathbf{x}}_2)' \,, \tag{2.31}$$

Table 2.3. Power, $\Sigma = D$, $D = \text{diag}\{d_1, \ldots, d_p\}$, and $d_i \sim \chi_3^2$, $i = 1, \ldots, p$

		μ_1			μ_2		
p	n	T_{SD}	T_D	F^+	T_{SD}	T_D	F^+
60	30	1.0000	0.9541	1.0000	0.9757	0.5420	1.0000
100	40	1.0000	0.9998	1.0000	1.0000	0.9394	1.0000
	60	1.0000	1.0000	1.0000	1.0000	0.9983	1.0000
	80	1.0000	1.0000	1.0000	1.0000	1.0000	1.0000
150	40	1.0000	0.9999	1.0000	1.0000	0.9622	1.0000
	60	1.0000	1.0000	1.0000	1.0000	0.9999	1.0000
	80	1.0000	1.0000	1.0000	1.0000	1.0000	1.0000
200	40	1.0000	1.0000	1.0000	1.0000	0.9925	1.0000
	60	1.0000	1.0000	1.0000	1.0000	1.0000	1.0000
	80	1.0000	1.0000	1.0000	1.0000	1.0000	1.0000
400	40	1.0000	1.0000	1.0000	1.0000	0.9996	1.0000
	60	1.0000	1.0000	1.0000	1.0000	1.0000	1.0000
	80	1.0000	1.0000	1.0000	1.0000	1.0000	1.0000

and

$$n = N_1 + N_2 - 2 \,. \tag{2.32}$$

For the two-sample problem, we wish to test the hypothesis

$$H : \mu_1 = \mu_2 \text{ vs } A : \mu_1 \neq \mu_2 \,. \tag{2.33}$$

The four tests for the two-sample problem are given by

$$\begin{aligned} T_D &= \frac{N_1 N_2}{N_1 + N_2}(\bar{\mathbf{x}}_1 - \bar{\mathbf{x}}_2)'(\bar{\mathbf{x}}_1 - \bar{\mathbf{x}}_2)/\text{tr}S \,, \\ T_{BS} &= \frac{\frac{N_1 N_2}{N_1 + N_2}(\bar{\mathbf{x}}_1 - \bar{\mathbf{x}}_2)'(\bar{\mathbf{x}}_1 - \bar{\mathbf{x}}_2) - \text{tr}S}{[(n+1)/n]^{\frac{1}{2}}[2p\hat{a}_2]^{\frac{1}{2}}} \,, \\ F^+ &= \frac{p-n+1}{n^2} T^{+2} \,. \end{aligned}$$

where

$$T^{+2} = \left(\frac{N_1 N_2}{N_1 + N_2}\right)(\bar{\mathbf{x}}_1 - \bar{\mathbf{x}}_2)' S^+ (\bar{\mathbf{x}}_1 - \bar{\mathbf{x}}_2) \,,$$

and

$$T_{SD} = \frac{\left(\frac{N_1 N_2}{N_1 + N_2}\right)(\bar{\mathbf{x}}_1 - \bar{\mathbf{x}}_2)' D_s^{-1} (\bar{\mathbf{x}}_1 - \bar{\mathbf{x}}_2)}{\sqrt{2[\text{tr}R^2 - (p^2/n)]}} \,.$$

The asymptotic distributions under the hypothesis $H : \mu_1 = \mu_2$ remain the same as in the case of one-sample discussed in Section 2.2. Also, the asymptotic non-null distribution can be obtained from the results in Section 2.2.

2.4. Multivariate Analysis of Variance (MANOVA)

We consider the multivariate linear regression model in which the $N \times p$ observation matrix Y is related by

$$Y = X\Xi + E \,, \tag{2.34}$$

where X is the $N \times k$ design matrix of rank $k < N$, assumed known, and Ξ is the $k \times p$ matrix of unknown parameters. We shall assume that the N row vectors of E are iid multivariate normal with mean vector zero and covariance matrix Σ, denoted as $\mathbf{e}_i \sim N_p(\mathbf{0}, \Sigma)$, where $E' = (\mathbf{e}_1, \ldots, \mathbf{e}_N)$. Similarly, we write $Y' = (\mathbf{y}_1, \ldots, \mathbf{y}_N)$, where $\mathbf{y}_1, \ldots, \mathbf{y}_N$ are independently distributed as multivariante normal with common covariance matrix Σ. We shall assume that

$$N \leq p \,. \tag{2.35}$$

The maximum likelihood or the least equares estimate of Ξ

$$\hat{\Xi} = (X'X)^{-1}X'Y : k \times p \,. \tag{2.36}$$

The $p \times p$ covariance matrix Σ can be unbiasedly estimated by

$$\hat{\Sigma} = n^{-1}W, \; n = N - k \,,$$

where

$$W = (Y - X\hat{\Xi})'(Y - X\hat{\Xi}) \,. \tag{2.37}$$

and often called as the matrix of the sum of squares and products due to error or simply 'within' matrix. The $p \times p$ matrix W is, however, singular matrix of rank n which is less than p. We consider the problem of testing the linear hypothesis

$$H : C\Xi = 0 \text{ vs } A : C\Xi \neq 0 \,, \tag{2.38}$$

where C is a $q \times k$ matrix of rank $q \leq k$ of known constants. The matrix of the sum of squares and products due to the hypothesis, or, simply 'between' matrix is given by

$$B = N(C\Xi)'(CGC')^{-1}(C\hat{\Xi}) \,, \tag{2.39}$$

where

$$G = (N^{-1}X'X)^{-1} \,. \tag{2.40}$$

When normality is not assumed, it is often required that G converges to a $k \times k$ positive definite matrix for asymptotic normality to hold, and although a weaker condition than (2.2.7) has been given by Srivastava (1968, 1972) for the asymptotic normality to hold, we will assume that G is positive definite. Under the assumption of normality,

$$W \sim W_p(\Sigma, n) , \tag{2.41}$$

and

$$B \sim W_p(\Sigma, q, N\eta\eta') \tag{2.42}$$

are independently distributed as Wishart and non-central Wishart respectively, where

$$\eta = (\eta_1, \ldots, \eta_q) = (C\Xi)'(CGC')^{-\frac{1}{2}} . \tag{2.43}$$

Thus, we may write

$$B = ZZ' \tag{2.44}$$

where $Z = (\mathbf{z}_1, \ldots, \mathbf{z}_q)$ and $\mathbf{z}_i$ are independently distributed as $N_p(N^{\frac{1}{2}}\eta_i, \Sigma)$. The tests corresponding to the T_D, T_{BS}, F^+ and T_{SD} are respectively given by

$$T_D = \frac{n\mathrm{tr}B}{q\mathrm{tr}W} \tag{2.45}$$

$$T_{BS} = \left[\frac{p}{2q\hat{a}_2(1+n^{-1}q)}\right]^{\frac{1}{2}} \left[p^{-1}\mathrm{tr}B - q\hat{a}_1\right] , \tag{2.46}$$

$$T_S^+ = -p\hat{b}\log|I + BW^+|^{-1} , \tag{2.47}$$

$$T_{SD} = \frac{n\mathrm{tr}BD_W^{-1} - npq(n-2)^{-1}}{q_{n,p}\sqrt{2q(\mathrm{tr}R^2 - p^2/n)}} , \tag{2.48}$$

where $q_{n,p}$ is defined in (2.25).
The tests T_S^+ and T_{SD} have been proposed by Srivastava (2007) and Srivastava and Du (2008) respectively. The tests T_D and T_{BS} have been proposed by Srivastava and Fujikoshi (2006). A theoretical power comparison of the first three tests is also given in Srivastava and Fujikoshi (2006).
The approximate null distribution of the test statistic F_D is an F-distribution with $[\hat{r}q]$ and $[n\hat{r}]$ degrees of freedom, where $\hat{r}$ is given in (2.2.8). The asymptotic null distributions of the test statistics T_{BS} and T_{SD} are both $N(0,1)$. And for the asymptotic distribution of T_S^+ under the hypothesis, we have

$$\lim_{n\to\infty}\lim_{p\to\infty} P_0\left[\frac{T_S^+ - nq}{\sqrt{2nq}} < z\right] = \Phi(z) . \tag{2.49}$$

Next, we give the non-null distribution of the statistics T_D, T_{BS} and T_S^+ under local alternatives. Since these three statistics are invariant under an orthogonal transformation, we shall assume without any loss of generality that the covariance matrix Σ is a diagonal matrix.

$$\wedge = \text{diag}(\lambda_1, \ldots, \lambda_p) . \tag{2.50}$$

For local alternatives, we assume that

$$\eta = (nN)^{\frac{1}{2}}\delta, \ \delta = O(1), \text{ and } n = O(\sqrt{p}) . \tag{2.51}$$

The asymptotic non-null distributions of the statistics T_D and T_{BS} are the same, and thus we give the non-null distribution of T_{BS}. It is given by

$$\lim_{n\to\infty}\lim_{p\to\infty}\left[P_1(T_{BS} > z_{1-\alpha}) - \Phi\left(-z_{1-\alpha} + \frac{\text{tr}\delta\delta'}{n\sqrt{2pqa_2}}\right)\right] = 0 . \tag{2.52}$$

The asymptotic non-null distribution of the statistic T_S^+ is given by

$$\lim_{n\to\infty}\lim_{p\to\infty}\left[P_1\left(\frac{T_S^+ - nq}{\sqrt{2nq}} > z_{1-\alpha}\right) - \Phi\left(-z_{1-\alpha} + \frac{\text{tr}\wedge\delta\delta'}{pa_2\sqrt{2nq}}\right)\right] = 0 . \tag{2.53}$$

The asymptotic non-null distribution of T_{SD} is not available.

2.5. Discriminant Analysis

In discriminant analysis, we consider the problem of classifying or assigning a $p \times 1$ observation vector $\mathbf{x}_0$ as coming from one of the several groups Π_i , $i = 1, \ldots, k$, where it is assumed that Π_i is distributed as $N_p(\mu_i, \Sigma_i)$. We shall consider the case when N_i observation vectors $\mathbf{x}_{i1}, \ldots, \mathbf{x}_{iN_i}$, are obtained from the group Π_i. The mean vector μ_i and the covariance matrices Σ_i are estimated by the sample mean vectors $\overline{\mathbf{x}}_i$ and the sample covariance matrices S_i given by

$$\overline{\mathbf{x}}_i = (1/N_i)\sum_{j=1}^{N_i} \mathbf{x}_{ij} , \tag{2.54}$$

and

$$S_i = (1/n_i)\sum_{j=1}^{N_i}(\mathbf{x}_{ij} - \overline{\mathbf{x}}_i)(\mathbf{x}_{ij} - \overline{\mathbf{x}}_i)' , n_i = N_i - 1, \ i = 1, \ldots, k . \tag{2.55}$$

If all the parameters μ_i and Σ_i are known, then the Mahakanobis squared distance between the observations vector $\mathbf{x}_0$ and the group Π_i is defined by

$$D^2 = (\mathbf{x}_0 - \mu_i)'\Sigma_i^{-1}(\mathbf{x}_0 - \mu_i) , \ i = 1, \ldots, k . \tag{2.56}$$

However, when μ_i and Σ_i are not known, the sample Mahalanobis squared distance is obtained by substituting their estimates. While the sample mean vector $\overline{x}_i$ can be substituted in place of μ_i, S_i is a singular matrix and cannot be substituted as such. Thus, we need an estimator for the precision matrices Σ_i^{-1}. Srivastava (2006a) substituted the Moore-Penrose inverse of S_i in place of Σ_i and defined the sample distance by

$$D^{+2}_{ii} = \left(1+\frac{1}{N_i}\right)^{-1} (\mathbf{x}_0 - \bar{\mathbf{x}}_i)' S_i^{+} (\mathbf{x}_0 - \bar{\mathbf{x}}_i) \ , \ i = 1,\ldots,k \ . \tag{2.57}$$

When $\Sigma_i = \Sigma$, then the sample distance is defined by

$$D^{+2}_{i} = \left(1+\frac{1}{N_i}\right)^{-1} (\mathbf{x}_0 - \bar{\mathbf{x}}_i)' S^{+} (\mathbf{x}_0 - \bar{\mathbf{x}}_i) \ , \ i = 1,\ldots,k \ , \tag{2.58}$$

where

$$S = \left(\sum_{i=1}^{k} n_i S_i\right) \Big/ \left(\sum_{i=1}^{k} n_i\right) \ . \tag{2.59}$$

Srivastava and Kubokawa (2007) and Kubokawa and Srivastava (2008) also considered an empirical Bayes estimator of the precision matrix Σ_i^{-1}. This empirical Bayes estimator of the precision matrix Σ_i^{-1} is given by

$$\hat{\Sigma}^{-1}_{\mathrm{EB}i} = c\left[S_i + \hat{\lambda}_i I_p\right]^{-1} \ , \tag{2.60}$$

where $\hat{\lambda}_i$ is either an arithmetic mean or the harmonic mean of the eigenvaluers of S_i. The harmonic mean, however, has some theoretical advantage in the sense that it has smaller risk with respect to the loss function

$$L(\delta_1, \Sigma^{-1}) = \mathrm{tr}(\delta - \Sigma^{-1})^2 S^2 \tag{2.61}$$

as compared to the estimator aS^{+} for any constant a, where δ denotes an estimator of Σ^{-1}. It has been shown in Srivastava and Kubokawa (2007) and Kubokawa and Srivastava (2008) through simulation that the minimum distance classification method with empirical Bayes estimator $\hat{\Sigma}^{-1}_{\mathrm{EB}}$ in place of Σ^{-1} provides lower error of misclassification than using aS^{+} in place of Σ^{-1}. Next, we consider a special case of two groups with common covariance matrix Σ. The estimator of the common covariance matrix Σ is given by the pooled estimate

$$S = n^{-1}\left[n_1 S_1 + n_2 S_2\right] \ , \ n = n_1 + n_2 = N_1 + N_2 - 2 \tag{2.62}$$

Since $n < p$, the pooled sample covariance matrix is singular. The sample squared distance between the observation vector $\mathbf{x}_0$ to be classified and the group Π_i using the Moore-Penrose of S, as defined by Srivastava (2006a) is given by

$$D_i^{+2} = \left(1+\frac{1}{N_i}\right)^{-1} (\mathbf{x}_0 - \bar{\mathbf{x}}_i)' S^{+} (\mathbf{x}_0 - \bar{\mathbf{x}}_i) \ , \ i = 1,2 \tag{2.63}$$

Srivastava (2006a) proposed to classify $\mathbf{x}_0$ into Π_1 if $D_1^{+2} < D_2^{+2}$; otherwise $\mathbf{x}_0$ is classified into Π_2. The probability of misclassifying $\mathbf{x}_0$ into Π_2 while actually it belongs to Π_1 is given by

$$e_1 = P\left\{a(\mathbf{x}_0 - \bar{x}_1)'S^+(\mathbf{x}_0 - \bar{\mathbf{x}}_1) > (\mathbf{x}_0 - \bar{\mathbf{x}}_2)'S^+(\mathbf{x}_0 - \bar{\mathbf{x}}_2)|\mathbf{x}_0 \in \Pi_1\right\} , \qquad (2.64)$$

where

$$a = \left(1 + N_1^{-1}\right)^{-1}\left(1 + N_2^{-1}\right) . \qquad (2.65)$$

Similarly an expression for e_2, the error of misclassifying $\mathbf{x}_0$ into Π_1 when actually it comes from Π_2, can be obtained. An asymptotic expression for e_1, when

$$\mu_1 - \mu_2 = n^{\frac{1}{2}}\delta ,$$

where δ is a non-null vector of constants, has been obtained by Srivastava (2006a) and is given by

$$\lim_{(n,p)\to\infty}(e_1) = \Phi(-l) - \frac{l^2-1}{6}\phi(l)\frac{12k_2^{-1}k_1^{-1}\theta^2}{\left[n + 2(k_1^{-2} + k_2^{-2})\theta^2\right]^{\frac{3}{2}}} + o(n^{-\frac{1}{2}}) \qquad (2.66)$$

where

$$\begin{aligned} l &= \frac{4k_2^{-1}k_1^{-1}\theta^2}{2\left[n + 2(k_1^{-2} + k_2^{-2})\theta^2\right]^{\frac{1}{2}}} , \\ \theta^2 &= \left(\delta' \wedge \delta / pa_2\right) , \\ k_1^2 &= 2\left[\left(1 + N_2^{-1}\right) - a^{\frac{1}{2}}\right] , \\ k_2^2 &= 2\left[\left(1 + N_2^{-1}\right) + a^{\frac{1}{2}}\right] , \end{aligned} \qquad (2.67)$$

and Φ and ϕ denotes the standard $N(0,1)$ cdf and pdf respectively. So far no asymptotic expression like in (2.66) for the error of misclassification is available for the distance function defined by (2.63) with S^+ replaced by $\hat{\Sigma}_{EB}^{-1}$. However, when S^+ is replaced by I_p, an asymptotic expression like (2.66) has been given by Saranadasa (1993).

2.6. Tests of Hypotheses on Covariance Matrices

The performance of many procedures discussed in Section 2.2-2.5 depend heavily on the assumption made on the covariance matrix or matrices. For example if $\Sigma = \sigma^2 I$, the T_D test is the best test among all tests (including Hotelling's T^2-test when $n > p$) whose power depends on $\mu'\mu/\sigma^2$, where μ is the mean vector of the observation vector in one-sample problem. Similarly in the two-sample case, it is assumed that both populations from which samples have been drawn have the same common covariance matrix. In this section, we review available tests to verify or tests these assumptions. We begin with one-sample case.

2.6.1. *One-sample case*

We follow the notation and assumptions described in Section 2.2. For the positive definite covariance matrix Σ, we consider the following four testing of hypotheses problems

1. $H_1 : \Sigma = \sigma^2 I,\ \sigma^2 > 0,$ vs $A_1 \neq H_1$,
2. $H_2 : \Sigma = I$, vs $A_2 \neq H_2$,
3. $H_3 : \Sigma = \text{diag}(\lambda_1, \ldots, \lambda_p) = \wedge$ vs $A_3 \neq H_3$.
4. $H_4 : \Sigma = \sigma^2 \left[(1-\rho)I_p + \rho \mathbf{1}_p \mathbf{1}_p'\right]$.

The hypothesis H_1 is called the 'sphericity' hypothesis. There are three tests available for this hypothesis which we describe next. All the three tests perfrom equally well in the limited simulation study carried out by Srivastava (2006b).

2.6.1.1. *Tests for the Sphericity Hypothesis*

Let $\tilde{l}_1, \ldots, \tilde{l}_n$ be the non-zero eigenvalues of the matrix $n\hat{S} = V$. That is,$(\tilde{l}_i/n)$ are the non-zero eigenvalues of S. Let

$$L_1 = (\Pi_{i=1}^n \tilde{l}_i) \Big/ \left(\sum_{i=1}^n \tilde{l}_i/n\right)^n . \tag{2.68}$$

$$Q_1 = -m_1 \log L_1 , \tag{2.69}$$

$$m_1 = p - \frac{2n^2+n+2}{6n} ,\ g_1 = \frac{1}{2}n(n+1) - 1 , \tag{2.70}$$

$$c_1 = (n+1)(n-1)(n+2)(2n^3+6n^2+3n+2)/288n^2 . \tag{2.71}$$

Then, we have the following theorem

Theorem 6.1 Let $n = O(p^\delta)$, $0 < \delta < 1$. Then under the hypothesis $H_1 : \Sigma = \sigma^2 I$, asymptotically as $(n,p) \to \infty$,

$$P\{Q_1 \geq z\} = P\{\chi^2_{g_1} \geq z\} + c_1 m_1^{-2}\left[P\left(\chi^2_{g_1+4} \geq z\right) - P\left(\chi^2_{g_1} \geq z\right)\right] + O(m_1^{-3}) ,$$

The Q_1-test for testing the spherecity has been proposed by Srivastava (2006b). This test requires that $(n/p) \to 0$ as $p \to \infty$.

Another test that requires $n = O(p)$, that is, $(n/p) \to c$ has been proposed by Ledoit and Wolf (2002). It is given by

$$U_1 = \left(\frac{n}{2}\right)\left[\frac{(1/p)\mathrm{tr}S^2}{\hat{a}_1^2} - 1\right] - \frac{p+1}{2} \tag{2.72}$$

The asymptotic distribution of U_1 under the hypothesis H_1 is given in the following theorem.

Theorem 6.2 Let $n = O(p)$. Then under the hypothesis that $\Sigma = \sigma^2 I$, σ^2 unknown, U_1 is asymptotically distributed as $N(0,1)$.

The asymptotic non-null distribution of the test statistics U_1 and Q_1 are not available.

A test that holds for all values of n and p has been proposed by Srivastava (2005). It is given by

$$T_1 = \left(\frac{n}{2}\right)\left[\frac{\hat{a}_2}{\hat{a}_1^2} - 1\right] \equiv \left(\frac{n}{2}\right)(\hat{\gamma}_1 - 1) \tag{2.73}$$

The distribution of the test statistic W_1 is given in the following theorem.

Theorem 6.3 Let $n = O(p^\delta)$, $\delta > 0$. Then under the hypothesis that $\Sigma = \sigma^2 I$, asymptotically as $(n,p) \to \infty$, $T_1 \sim N(0,1)$. The non-null distribution of the statistic W_1 is given by

$$\lim_{(n,p)\to\infty}\left[P(T_1 > z_{1-\alpha}) - \Phi\left\{\left(\frac{n}{2}(\gamma_1 - 1) - z_{1-\alpha}\right)\Big/\tau\right\}\right] = 0\,,$$

where

$$\gamma_1 = \left(a_2/a_1^2\right), \text{ and } \tau^2 = \frac{2n(a_4a_1^2 - 2a_1a_2a_3 + a_2^3)}{pa_1^6} + \frac{a_2^2}{a_1^4} \tag{2.74}$$

That is, the power of the T_1-test is given by

$$\beta(T_1) \simeq \Phi\left[\left\{\frac{n}{2}(\gamma_1 - 1) - z_{1-\alpha}\right\}/\tau\right] \tag{2.75}$$

2.6.1.2. *Testing that the covariance matrix is an identity matrix*

We now consider the second problem of testing the hypothesis $H_2 : \Sigma = I$ vs $A_2 : \Sigma \neq I$. For this case also, we have three tests which we describe next. Let

$$L_2 = \left(\frac{e}{p}\right)^{\frac{1}{2}pn} \left(\prod_{i=1}^{n} \tilde{l}_i\right)^{\frac{1}{2}p} e^{-\frac{1}{2}\Sigma_{i=1}^{n}\tilde{l}_i} \tag{2.76}$$

$$Q_2 = -\left(\frac{2m_2}{p}\right)\log L_2 \tag{2.77}$$

$$m_2 = p - \frac{2n^2+3n+1}{6(n+1)}, \; g_2 = \frac{1}{2}n(n+1) \tag{2.78}$$

$$c_2 = \frac{n}{288(n+1)}(2n^4+6n^3+n^2-12n-13) \tag{2.79}$$

which $\tilde{l}_i$ are non-zero eigenvalues of nS. Then we have the following theorem.

Theorem 6.4 Let $n = O(p^{\delta}), 0 < \delta < 1$. Then under the hypothesis $H_2 : \Sigma = I$, asymptotically as $(n, p) \to \infty$,

$$P\{Q_2 \geq z\} = P\{\chi^2_{g_2} \geq z\} + c_2 m_2^{-2}\left[P(\chi^2_{g_2+4} \geq z) - P(\chi^2_{g_2} \geq z)\right] + O(m_2^{-3}) .$$

The test Q_2 has been proposed by Srivastava (2006b). The second test given by Ledoit and Wolf (2002) requires that $n = O(p)$. This test is given by

$$U_2 = \left(\frac{n}{2}\right)\left[\frac{1}{p}\mathrm{tr}S^2 - \frac{p}{n}\hat{a}_1^2 - 2\hat{a}_1 + \frac{p+n}{n}\right] - \frac{p+1}{2} \tag{2.80}$$

The asymptotic null distribution of the test statistic U_2 is given in the following theorem.

Theorem 6.5 Let $n = O(p)$. Then under the hypothesis that $\Sigma = I$, asymptotically as $(n, p) \to \infty$, $U_2 \sim N(0, 1)$.

The asymptotic non-null distributions of the statistics Q_2 and U_2 are not available. The third test statistic which has no restriction on n and p has been proposed by Srivastava (2005). The test statistic is given by

$$T_2 = \left(\frac{n}{2}\right)(\hat{\gamma}_2 + 1) , \tag{2.81}$$

where

$$\hat{\gamma}_2 = \hat{a}_2 - 2\hat{a}_1 \tag{2.82}$$

which is a consistent estimator of

$$\gamma_2 = a_2 - 2a_1 , \tag{2.83}$$

as $(n, p) \to \infty$. The following theorem gives the asymptotic distribution of T_2 under the hypothesis $\Sigma = I$.

Theorem 6.6 Let $n = O(p^\delta)$, $\delta > 0$. Then under the hypothesis that $\Sigma = I$, asymptotically as $(n,p) \to \infty$, $T_2 \sim N(0,1)$. The asymptotic non-null distribution of the statistic T_2 is given by

$$\lim_{(n,p)\to\infty} P\{T_2 > z_{1-\alpha} | \gamma_1 + 1 > 0\} = \lim_{(n,p)\to\infty} \Phi\left[\left\{\frac{n}{2}(\gamma_2+1) - z_{1-\alpha}\right\} \Big/ \tau_2\right] \tag{2.84}$$

where

$$\tau_2^2 = (2n/p)(a_2 - 2a_3^3 + a_4) + a_2^2 \tag{2.85}$$

Thus, the asymptotic power of the T_2-test is given by

$$\beta(T_2) \simeq \Phi\left[\left\{\frac{n}{2}(\gamma_2+1) - z_{1-\alpha}\right\} \Big/ \tau_2\right] \tag{2.86}$$

2.6.1.3. *Testing that the Covariance Matrix is a Diagonal Matrix*

In this section, we consider the problem of testing the hypothesis H_3 vs A_3. Two tests have been proposed by Srivastava (2005, 2006b). We describe them next. let

$$z_{ij} = \frac{1}{2}\log\left(\frac{1+r_{ij}}{1-r_{ij}}\right),\ i \neq j$$

and

$$Q_3 = \frac{(n-2)\sum_{i<j}^{p} z_{ij}^2 - \frac{1}{2}p(p-1)}{\sqrt{p(p-1)}}$$

Then, the test Q_3 is based on Fisher's z-transform. The following theorem gives the distribution under the hypothesis.

Theorem 6.7 Under the hypothesis that Σ is a diagonal matrix, Q_3 is asymptotically distributed as $N(0,1)$ as n and $p \to \infty$.

The test Q_3 is based on the pair-wise correlations. A test that is based on the pair-wise covariances is given by

$$T_3 = \left(\frac{n}{2}\right) \frac{(\hat{\gamma}_3 - 1)}{\left[1 - \frac{1}{p}\left(\frac{\hat{a}_{40}}{\hat{a}_{20}^2}\right)\right]^{\frac{1}{2}}},$$

where

$$\hat{\gamma}_3 = (\hat{a}_2/\hat{a}_{20}) ,$$
$$\hat{a}_{20} = \frac{n}{p(n+2)} \sum_{i=1}^{p} s_{ij}^2$$
$$\hat{a}_{40} = \frac{1}{p} \sum_{i=1}^{p} s_{ij}^4 ,$$

and $\hat{a}_2$ has been defined in (2.10). The distribution under the hypothesis is given in the following theorem.

Theorem 6.8 Let $n = O(p^{\delta})$, $\delta > 0$. Then, under the hypothesis that the covariance matrix is a diagonal matrix, asymptotically as $(n,p) \to \infty$, T_3 is distributed as $N(0,1)$. The asymptotic non-null distribution of the statistic T_3 is given by

$$\lim_{(n,p)\to\infty} \left[P\{T_3 > z_{1-\alpha}\} - \Phi\left[-\frac{z_{1-\alpha} + \delta_3}{\tau_3} \right] \right] = 0 ,$$

where

$$\tau_3^2 = (a_2^2 - p^{-1}a_4)/(a_{20}^2 - p^{-1}a_{40})$$
$$\delta_3 = \frac{n}{2}(\gamma_3 - 1) \Big/ \left[1 - \frac{1}{p}\frac{a_{40}}{a_{20}^2} \right]^{\frac{1}{2}}$$
$$\gamma_3 = (a_2/a_{20})$$
$$a_{20} = \left(\sum_{i=1}^{p} \sigma_{ii}^2/p \right)$$
$$a_{40} = \left(\sum_{i=1}^{p} \sigma_{ii}^4/p \right)$$

That is, the asymptotic power of the test based on the statistic T_3 is given by

$$\beta(T_3) \simeq \Phi[(-z_{1-\alpha} + \delta_3)/\tau_3] .$$

The asymptotic power of the test Q_3 is not available.

2.6.1.4. *Testing that the covariance matrix is of intraclass correlation structure*

In this section, we consider the problem of testing the hypothesis

$$H_4 : \Sigma = \sigma^2 \left[(1-\rho)I_p + \rho \mathbf{1}_p \mathbf{1}_p' \right]$$

against the alternative $A_4 \neq H_4$. When $n > p$, the likelihood ratio test is available, see for example, Srivastava and Carter (1983) and Srivastava (2002). However,

when $n < p$, no such test exists. In order to obtain a test for testing the hypothesis H_4 vs A_4 when $n < p$, we proceed as follows. Let G be a $p \times p$ known orthogonal matrix, $GG' = I_p$ such that the first row of G is given by $\mathbf{1}'/\sqrt{p}$.
Then

$$\mathbf{u}_i = G\mathbf{x}_i$$

are iid $N_p(G\mu, G\Sigma G')$, where we can write

$$\begin{aligned} G\Sigma G' &= \begin{pmatrix} \mathbf{1}'/\sqrt{p} \\ G_2 \end{pmatrix} \Sigma(\mathbf{1}/\sqrt{p}, G_2') \\ &= \begin{pmatrix} \mathbf{1}'\Sigma\mathbf{1}/\sqrt{p} & \mathbf{1}'\Sigma G_2'/\sqrt{p} \\ G_2\Sigma\mathbf{1}/\sqrt{p} & G_2\Sigma G_2') \end{pmatrix} \\ &= \begin{pmatrix} \Lambda_{11} & \Lambda_{12}' \\ \Lambda_{12} & \Lambda_{22} \end{pmatrix} \\ &= \Lambda \,. \end{aligned}$$

where $G_2 : (p-1) \times p$, $G_2G_2' = I_{p-1}$, and $G_2\mathbf{1} = 0$. Thus $GVG' \sim W_p(\Lambda, n)$, and $G_2VG_2' \sim W_{p-1}(\Lambda_{22}, n)$.
Under the hypothesis H,

$$\Lambda_{22} = \gamma^2 I_{p-1}, \; \gamma^2 = \sigma^2(1-\rho) \,.$$

Thus, we test for the sphericity of the covariance matrix of the (p-1) dimensional random vectors $G_2\mathbf{x}_i, i = 1, \ldots, p$. Clearly, the acceptance of this hypothesis does not imply the acceptance of the hypothesis H_4. However, the rejection of this hypothesis implies the rejection of the hypothesis H_4.
Assuming that $n < (p-1)$, let $\tilde{l}_1, \ldots, \tilde{l}_n$ be the n non-zero eigenvalues of G_2VG_2'. Then a sphericity test is based on the statistic (2.68) with $\tilde{l}_i$ replaced by $\tilde{l}_i^*$. Similarly, other two tests corresponding to (2.72) and (2.73) can be obtained by using G_2SG_2' in place of S, $S = n^{-1}V$.
It may be noted that this is a test for sphericity and thus no further discussion of this test is pursued. It may also be noted that the ith row of the $(p-1) \times p$ matrix G_2 may be chosen as in Helmet's matrix, that is,

$$\begin{matrix} 1^{st} & 2^{nd} & & (j+1)^{th} & (j+2)^{nd} & & p^{th} \\ \Bigg(\dfrac{i}{\sqrt{i(i+1)}}, & -\dfrac{1}{\sqrt{i(i+1)}}, & \ldots, & -\dfrac{1}{\sqrt{i(i+1)}}, & 0 & ,\ldots, & 0 \Bigg), \end{matrix}$$

i = 1,...,(p-1) .

2.6.1.5. *Simulation results of power comparison and attained significance level*

In this subsection, we carry out simulation for tests T_i and Q_i, $i = 1,2,3$ when $n = O(p^\delta)$, $O \leq \delta < 1$, and thus excluding the tests U_i, $i = 1,2$. We first check how good the normal approximations are. For this, we draw an independent sample of size $N = n+1$ from the hypothesis distributions. Then it is replicated 1000 times. We calculate

$$ASL(T_i) = \frac{(\#\ T_i > z_{1-\alpha})}{1000} \text{ and } ASL(Q_i) = \frac{(\#\ Q_i > z_{1-\alpha})}{1000},$$

denoting the attained significance level of T_i and Q_i respectively, where $z_{1-\alpha}$ is the upper $100\alpha\%$ point of the standard normal dsitribution. We have chosen $\alpha = 0.05$. Tables 6.1-6.6 give the ASL for six tests. It shows that the tests Q_1 and Q_2 attain their significance levels for n reasonally smaller than p. For comparison of powers, we first carry out simulations from the hypothesis distributions to obtain the significance points $T_{i\alpha}$ and $Q_{i\alpha}$ where α is chosen to be 0.05. A second simulation is carried out by taking samples from alternative distributions by calculating

$$\beta(T_i) = \frac{(\#T_i > T_{i\alpha}|A_i)}{1000},$$

and

$$\beta(Q_i) = \frac{(\#Q_i > Q_{i\alpha}|A_i)}{1000},$$

These powers are given in Tables 2.10-2.15.

Table 2.4 ASL^* of T_1 Test Under H_1
Sample from N(0,1), n=N-1

	n=20	n=30	n=60	n=100
p=60	0.053	0.050	0.052	0.048
p=100	0.050	0.045	0.049	0.041
p=150	0.050	0.058	0.053	0.048
p=200	0.046	0.058	0.053	0.048
p=250	0.070	0.051	0.046	0.048
p=300	0.043	0.058	0.055	0.059
p=400	0.048	0.055	0.049	0.047

*ASL-Attained Significance Level

Table 2.5 ASL^* of Q_1 Test Under H_1
Sample from N(0,1), n=N-1

	n=20	n=30	n=60	n=100
p=60	0.060	0.062	**NA**	0.147
p=100	0.053	0.045	0.056	**NA**
p=150	0.057	0.045	0.056	0.571
p=200	0.051	0.052	0.052	0.157
p=250	0.045	0.059	0.058	0.064
p=300	0.032	0.045	0.052	0.066
p=400	0.049	0.046	0.054	0.056

*ASl-Attained Significance Level

Table 2.6 ASL^* of T_2 Test Under H_2
Sample from N(0,1), n=N-1

	n=20	n=30	n=60	n=100
p=60	0.057	0.051	0.042	0.064
p=100	0.067	0.046	0.047	0.056
p=150	0.043	0.060	0.056	0.049
p=200	0.049	0.055	0.050	0.046
p=250	0.048	0.054	0.045	0.047
p=300	0.063	0.061	0.049	0.065
p=400	0.058	0.055	0.053	0.047

$^*ASL$-Attained Significance Level

Table 2.7 ASL^* of Q_2 Test Under H_2
Sample from N(0,1), n=N-1

	n=20	n=30	n=60	n=100
p=60	0.057	0.065	**NA**	0.178
p=100	0.050	0.060	0.163	**NA**
p=150	0.052	0.056	0.066	0.536
p=200	0.049	0.040	0.056	0.169
p=250	0.066	0.044	0.060	0.088
p=300	0.048	0.049	0.053	0.066
p=400	0.054	0.051	0.055	0.057

$^*ASL$-Attained Significance Level

Table 2.8 ASL^* of T_3 Test Under H_3
Sample from N(0,1), n=N-1

	n=20	n=30	n=60	n=100
p=60	0.054	0.050	0.044	0.037
p=100	0.050	0.051	0.049	0.049
p=150	0.061	0.037	0.050	0.055
p=200	0.048	0.056	0.059	0.054
p=250	0.055	0.055	0.060	0.049
p=300	0.057	0.049	0.045	0.048
p=400	0.044	0.054	0.051	0.051

$^*ASL$-Attained Significance Level

Table 2.9 ASL^* of Q_3 Test Under H_3
Sample from N(0,1), n=N-1

	n=20	n=30	n=60	n=100
p=60	0.061	0.067	0.055	0.052
p=100	0.051	0.056	0.044	0.061
p=150	0.055	0.053	0.057	0.044
p=200	0.043	0.059	0.055	0.052
p=250	0.060	0.054	0.044	0.050
p=300	0.038	0.046	0.052	0.060
p=400	0.042	0.049	0.067	0.045

$^*ASL$-Attained Significance Level

Table 2.10 Power of T_1 Test Under A_1
(1) 1000 sample from N(0,1) to simulate T_{1a}, a=0.05, for each pair of (p,n).
(2) 1000 sample from N(0,D) to obtain $T_1, P(T_1 > T_{1a})$=power.
(3) D=diag($d_1,\ldots,d_p$), where $d_i \sim U(0.5,1.5)$.

	n=20	n=30	n=60	n=100
p=60	0.284	0.350	0.495	0.979
p=100	0.207	0.436	0.887	0.940
p=150	0.165	0.400	0.725	0.986
p=200	0.194	0.362	0.730	0.993
p=250	0.177	0.334	0.839	0.992
p=300	0.221	0.352	0.771	0.984
p=400	0.182	0.329	0.721	0.989

Table 2.11 Power of Q_1 Test Under A_1
(1) 1000 sample from N(0,1) to simulate Q_{1a}, a=0.05, for each pair of (p,n).
(2) 1000 sample from N(0,D) to obtain $Q_1, P(Q_1 > Q_{1a})$=power.
(3) D=diag($d_1,\ldots,d_p$), where $d_i \sim U(0.5,1.5)$.

	n=20	n=30	n=60	n=100
p=60	0.162	0.352	0.169	0.786
p=100	0.208	0.258	0.605	0.285
p=150	0.194	0.344	0.693	0.923
p=200	0.234	0.290	0.755	0.953
p=250	0.188	0.336	0.818	0.963
p=300	0.218	0.337	0.735	0.985
p=400	0.199	0.340	0.752	0.988

Table 2.12 Power of T_2 Test Under A_2
(1) 1000 sample from N(0,1) to simulate T_{2a}, a=0.05, for each pair of (p,n).
(2) 1000 sample from N(0,D) to obtain $T_2, P(T_2 > T_{2a})$=power.
(3) D=diag$(d_1,\ldots,d_p)$,where $d_i \sim U(0.5,1.5)$.

	n=20	n=30	n=60	n=100
p=60	0.159	0.389	0.748	0.909
p=100	0.275	0.360	0.682	0.992
p=150	0.261	0.300	0.727	0.985
p=200	0.244	0.339	0.675	0.984
p=250	0.165	0.287	0.714	0.994
p=300	0.193	0.361	0.724	0.992
p=400	0.196	0.376	0.735	0.991

Table 2.13 Power of Q_2 Test Under A_2
(1) 1000 sample from N(0,1) to simulate Q_{2a}, a=0.05, for each pair of (p,n).
(2) 1000 sample from N(0,D) to obtain $Q_2, P(Q_2 > Q_{2a})$=power.
(3) D=diag$(d_1,\ldots,d_p)$,where $d_i \sim U(0.5,1.5)$.

	n=20	n=30	n=60	n=100
p=60	0.189	0.237	0.148	0.896
p=100	0.168	0.300	0.618	0.215
p=150	0.209	0.283	0.689	0.899
p=200	0.186	0.354	0.626	0.981
p=250	0.256	0.348	0.671	0.946
p=300	0.276	0.386	0.780	0.977
p=400	0.179	0.307	0.760	0.987

Table 2.14 Power of T_3 Test Under A_3
(1) 1000 sample from N(0,1) to simulate T_{3a}, a=0.05, for each pair of (p,n).
(2) 1000 sample from N(0,D) to obtain $T_3, P(T_3 > T_{3a})$=power.
(3) $\Sigma = D_1 * R_1 * D_1$ which is the variance matrix for sample generating.
(4) D=diag$(d_1,\ldots,d_p)$,where $d_i \sim U(0.5,1.5)$. R_1 is correlation matrix with $e_{ij} = (k/6**(|i-j|/2)$, where k=1.

	n=20	n=30	n=60	n=100
p=60	0.861	0.987	1.000	1.000
p=100	0.807	0.988	1.000	1.000
p=150	0.849	0.992	1.000	1.000
p=200	0.849	0.989	1.000	1.000
p=250	0.842	0.989	1.000	1.000
p=300	0.833	0.993	1.000	1.000
p=400	0.867	0.988	1.000	1.000

Table 2.15 Power of Q3 Test Under A3
(1) 1000 sample from N(0,1) to simulate Q_{3a}, a=0.05, for each pair of (p,n).
(2) 1000 sample from N(0,D) to obtain $Q_3, P(Q_3 > Q_{3a})$=power.
(3) $\Sigma = D_1 * R_1 * D_1$ which is the variance matrix for sample generating.
(4) D=diag$(d_1,\ldots,d_p)$,where $d_i \sim U(0.5,1.5)$. R_1 is correlation matrix with $e_{ij} = (k/6**(|i-j|/2)$, where k=1.

	n=20	n=30	n=60	n=100
p=60	0.933	1.000	1.000	1.000
p=100	0.960	1.000	1.000	1.000
p=150	0.963	1.000	1.000	1.000
p=200	0.966	1.000	1.000	1.000
p=250	0.952	1.000	1.000	1.000
p=300	0.962	1.000	1.000	1.000
p=400	0.953	1.000	1.000	1.000

2.6.2. *Two-Sample case*

In the two-sample case, we have independent observations of sizes N_1 and N_2 from the two groups which are assumed to be normally distributed as $N_p(\mu_i,\Sigma_i)$, $i = 1,2$. The tests developed in Section 2.3, assume that the two groups have common covariance matrix Σ, that is $\Sigma_1 = \Sigma_2 = \Sigma$, say, where Σ is unknown. It would therefore be necessary to ascertain this assumption from the data. We now describe the test statistic used to check this asumption. In the notation of Section 3, let

$$V_i = n_i S_i\ ,\ \ n_i = N_i - 1,\ i = 1,2\ ,$$

and

$$T_4^{(i)} = \left[p\hat{b}\ \mathrm{tr}V_i^+ V_{(3-i)} - n_1 n_2\right] / (2n_1 n_2)^{\frac{1}{2}}\ , i = 1, 2\ ,$$

Then, it is shown in Srivastava (2004) that $T_4^{(i)}$ are normally distributed. Let

$$R_4 = \max\left(\left|T_4^{(1)}\right|, \left|T_4^{(2)}\right|\right)\ .$$

Then the hypothesis of the equality of two covariances is rejected if

$$R_4 > z_{1-\alpha/4}$$

We may also choose to compare $\left|T_4^{(1)}\right|$ or $\left|T_4^{(2)}\right|$ with $z_{1-\alpha/2}$.

2.6.3. *Testing the equality of covariances in MANOVA*

In a special case of MANOVA, we compare the means of k populations under the assumption that all the k populations are normally distributed with common covariance matrix Σ. That is, if $\mathbf{x}_{ij}$ are independently distributed as $N_p(\mu_i, \Sigma_i)$, $j = 1, \ldots, N_i$, $i = 1, \ldots, k$, we compare the means under the assumption that $\Sigma_1 = \ldots = \Sigma_k$. A test proposed by Srivastava (2007) for testing the equality of k covariances is described below. Let S_i be the sample covariances and

$$V_i = n_i S_i,\ n_i = N_i - 1,\ i = 1, \ldots, k\ .$$

Define

$$V_{(i)} = V_1 + \ldots + V_{i-1} + V_{i+1} + \ldots + V_k,\ i = 1, \ldots, k\ ,$$

and

$$T_5^{(i)} = \frac{p\hat{b}\ \mathrm{tr}V_i V_{(i)}^+ - n_i n_{(i)}}{\sqrt{2n_i n_{(i)}}}\ , i = 1, \ldots, k\ ,$$

where

$$n_{(i)} = n_1 + \ldots + n_{i-1} + n_{i+1} + \ldots + n_k\ .$$

Then, it has been shown by Srivastava (2004) that $T_5^{(i)}$ are asymptotically normally distributed. Let

$$R_5 = \max\left\{\left|T_5^{(1)}\right|, \ldots, \left|T_5^{(k)}\right|\right\}\ .$$

Then the hypothesis of the equality of k covariances is rejected if

$$R_5 > z_{1-\alpha/2k}\ ,$$

where $z_{1-\alpha}$ is the upper $100\alpha\%$ point of the standard $N(0,1)$ random variables.

2.7. Conclusion

In this article, we have reviewed the literature on various tests of hypotheses on mean vectors and covariance matrices when there are fewer observations than the dimension of the observation vector. From power comparison of tests on the mean vector in the one-sample problem, it is clear that the two recently proposed tests, namely F^+ test and F_{SD} tests proposed by Srivastava (2007) and Srivastava and Du (2006) respectively perform better than the F_D test proposed by Dempster (1958, 1960) or equivalently the F_{BS} test proposed by Bai and Sarandasa (1996). Even when the spherecity hypothesis for the covariance matrix holds, which can be verified for the data by tests given in Section 6, the performance of the Dempster's test F_D does not appear superior to F^+ or T_{SD} tests. Tests for other hypotheses on the covariance matrix or covariance matrices in two or more than two populations have been reviewed. The problem of classifying an observation vector into two or more than two populations has also been reviewed.

References

1. Bai, Z. and Saranadasa, H. (1996). *Effect of high dimension: by an example of a two sample problem.*Statistica Sinica.,6, 311-329.
2. Benjamini, Y. and Hochberg, Y. (1995). *Controlling the false discovery rate: a practical and powerful approach to multiple testing.* J. Roy. Statist. Soc., B 57, 289-300.
3. Benjamini, Y. and Yekuteli, D. (2001). *The control of the false discovery rate in multiple testing under dependency.* Ann. Statist., 29, 1165-1181.
4. Dempster, A.P. (1958). *A high dimensional two sample significance test.*Ann. Math. Stat., 29, 995-1010.
5. Dempster, A.P. (1960). *A significance test for the separation of two highly multivariate small samples.* Biometrics, 16, 41-50.
6. Fujikoshi, Y. (2004) *Multivariate analysis for the case when the dimension is large compared to the sample size.* J. Korean Statist. Soc., 33, 1-24.
7. Fujikoshi, Y., Himeno, T. and Wakaki, H. (2004). *Asymptotic results of a high dimensional MANOVA test and power comparison when the dimension is large compared to the sample size.* J. Japan Statist. Soc., 34, 19-26.
8. Kubokawa, T. and Srivastava, M.S. (2008). *Estimation of the precision matrix of a singular Wishart distribution and its application in high dimensional data.* J. Multivariate Analy, 99, 1906-1928.
9. Ledoit, O., and Wolf, M. (2002).*Some hypothesis tests for the covariance matrix when the dimension is large compared to the sample size.* Ann. Statist., 30, 1081-1102.
10. Lehmann, E. L. (1959). *Testing Statistical Hypotheses.* Wiley, New York.
11. Rao, C. R. (1973). *Linear Statistical Inference and Its Applications.* Wiley, New York.
12. Saranadosa, H. (1993). *Asymptotic expansion of the misclassification probabilities of D- and A- criteria for discrimination from two high-dimensional populations using the theory of large-dimensional random matrices* J. Multivariate Analy., 46, 154-174.

13. Simaika, J. B. (1941). *On an optimum property of two important statistical tests.* Biometrika, 32, 70-80.
14. Srivastava, M.S. (2007). *Multivariate theory for analyzing high dimensional data.* J. Japan Stat. Soc., 37, 53-86.
15. Srivastava, M.S. (2006a). *Minimum distance classification rules for high dimensional data.* J. Multivariate Analy., 97, 2057-2070
16. Srivastava, M.S. (2006b). *Some tests criteria for the covariance matrix with fewer observations than the dimension.* Acta Et Commentationes Universitatatis Tartuensis De Mathematica., 10, 77-93.
17. Srivastava, M.S. (2005). *Some tests concerning the covariance matrix in high-dimensional data.* J. Japan Stat. Soc., 35, 251-272.
18. Srivastava, M.S. (2003). *Singular Wishart and Multivariate beta distributions.* Ann. Statist., 31, 1537-1560.
19. Srivastava, M.S. (1972). *Asymptotically most powerful rank tests for regression parameters in MANOVA.* Ann. Inst. Math. Statist., 24, 285-297.
20. Srivastava, M.S. (1968). *On a class of nonparametric tests for regression parameters.* Ann. Math. Statist.(Abstract), 39, 697.
21. Srivastava, M.S. and Carter, E.M. (1983). *An introduction to Applied Multivariate Statistics.* North-Holland, New York.
22. Srivastava, M.S. and Du, M (2008). *A test for the mean vector with fewer observations than the dimension.* Jour. Multivariate Analys., 99, 386-402.
23. Srivastava, M.S. and Fujikoshi, Y. (2006). *Multivariate analysis of variance with fewer observations than the dimension.* J. Multivariate Analy., 97, 1927-1940.
24. Srivastava, M.S. and Kubokawa, T. (2007). *Comparison of discriminant methods for high dimensional data.* J. Japan Stat. Soc., 37, 123-134.
25. Srivastava, M.S. and Khatri, C. G. (1979). *An Introduction to Multivariate Statistics.* North-Holland, New York.

Chapter 3

Model Based Penalized Clustering for Multivariate Data

Samiran Ghosh[1] and Dipak K. Dey[2]

[1]*Department of Mathematical Sciences, Indiana University-Purdue University, Indianapolis*

Email: samiran@math.iupui.edu

[2]*Department of Statistics, University of Connecticut, Storrs*

E-mail: dipak.dey@uconn.edu

Over the last decade a variety of clustering algorithms have evolved. However one of the simplest (and possibly overused) partition based clustering algorithm is K-means. It can be shown that the computational complexity of K-means does not suffer from exponential growth with dimensionality rather it is linearly proportional with the number of observations and number of clusters. The crucial requirements are the knowledge of cluster number and the computation of some suitably chosen similarity measure. For this simplicity and scalability, among large data sets K-means remains an attractive alternative when compared to other competing clustering philosophies especially for high dimensional domain. However being a deterministic algorithm, traditional K-means have several drawbacks. It only offers hard decision rule, with no probabilistic interpretation. In this paper we have developed a decision theoretic framework by which traditional K-means can be given a probabilistic footstep. This will not only enable us to do a soft clustering, rather the whole optimization problem could be recasted into Bayesian modeling framework, in which the knowledge of cluster number could be treated as an unknown parameter of interest, thus removing a severe constrain of K-means algorithm. Our basic idea is to keep the simplicity and scalability of K-means, while achieving some of the desired properties of the other model based or soft clustering approaches.

Contents

3.1. Introduction

Clustering [Kaufman and Rousseeuw,1989] is a general data analytic problem, in which our main aim is to partition group of objects into meaningful sub-classes or clusters. Clustering methodologies and related algorithms cover a wide range of data compression and structurer detection in machine learning and artificial intelligence. Recently, clustering methods have gained tremendous popularity in high dimensional data found in Bioinformatics, E-commerce and Financial domain. Unlike classifcation, clustering does not assume any class label or group membership to begin with. Consequently it is a rather unsupervised learning problem. In clustering, our basic objective is to find groups so that the intra group variation is low while inter group variation is high. There exists a variety of clustering algorithms for example, partitioning, hierarchical, grid and model based approaches. From a statistical perspective we can divide these techniques either being deterministic or probabilistic. The model based approach developed on the basis of mixture modeling has some desirable properties. It offers soft clustering where each member is assigned a probability of being included in a cluster, which often seems an attractive property rather than a hard classification rule where each member either belongs to a cluster or not. Traditional K-means [MacQueen,1967] is a distance based deterministic clustering approach, which has clear advantages in high dimension as it only depends upon some dissimilarity measure like pair wise distance or distance from cluster center. This is often very inexpensive to compute irrespective of high dimension. In this paper we have recasted the original optimization problem related to K-means algorithm in a probabilistic framework. Basic idea is to exploit desirable properties of both the realm. Each cluster point is being considered as a particular realization of a stochastic framework. Since each clusters point is random the distance between any two cluster points is also a random quantity. We have considered generalized distance measure to incorporate robustness and constructed likelihood function to get a new loss function based approaches for traditional K-means clustering algorithm. A model based K-means clustering algorithm is embedded in Hofman and Buhmann [Hofmann and Buhmann,1997] where the authors used stochastic optimization technique by maximum entropy inference. For unknown number of clusters a regularized (or penalized) "loss + penalty" version of the problem is formulated with the introduction of a novel penalty term. This "loss + penalty" based approach is then integrated within the Bayesian framework. It should be noted that regularization is a critical step for successful statistical modeling of high-dimensional data. Saharon [Saharon,2003] defined regularization as, "an integral part of model building which takes into account finiteness and imperfection of the data and the limited information in it". Many of the traditional statistical and machine learning tools for examples support vector machine (SVM), kernel logistic regression (KLR) and Boosting etc. can be shown to follow "regularized path" of solutions in an abstract sense. In this paper a similar and quite general regularized version of the K-means clustering problem is formulated.

Notably, for the unknown cluster number, a non-heuristic Bayesian solution is to perform reversible jump Markov Chain Monte Carlo (RJMCMC). This may be appropriate and desirable for small dimensional problem. However even for the fairly moderate dimensional domain, implementing RJMCMC approach successfully is a computationally challenging task and for the data sets where dimension regularly exceeds thousands in number, it is simply infeasible. However Rafetry and Dean [Raftery and Dean,2006] developed a model-based clustering by recasting the problem of comparing two nested subsets of feature selection as a model comparison problem using approximate Bayes factor. For clustering with unknown cluster number a practical (in terms of efficiency and scalability) and non-heuristic algorithm is needed. We propose a simple technique to achieve this, parallel to BClass algorithm [Medrano-Soto et al., 2005] developed elsewhere for our regularized probabilistic K-means algorithm. This eliminates additional computational complexity that comes with RJMCMC mechanism, while keeping cluster numbers as an unknown quantity to begin with. We also propose a novel algorithm based on the analysis of cluster assignment correlations. This correlation based algorithm is fairly general enough to be applicable for discovering hidden group structure even outside clustering domain. The main focus of this paper is three fold. First, we formulate a regularized version of the traditional K-means clustering with introduction of a penalty term. We term it as regularized K-means. Next we show the exact duality between regularized K-means and a stochastic model framework and finally, to develop a scalable and efficient clustering algorithm which does not suffer from the curse of dimensionality for unknown number of clusters.

The rest of the paper is organized as follows. We have developed a decision theoretic methodology in section 3.2. The optimization problem is formulated into likelihood based framework in section 3.3. This section also introduces an alternative to reversible jump as well as Bayes factors based approach and its corresponding optimization problem from clustering perspective. Choice of prior and development of posterior sampling scheme have been discussed in section 3.4. Section 3.5 deals with the application domain. We consider two data sets to show the efficacy and capture the versatility of our methodology. In this section we have analyzed two benchmark data sets, namely Old Faithful data [Hardle,1991] and Fisher's Iris data [Fisher,1936]. An algorithm to analyze cluster correlation is also introduced. We conclude the article with a brief discussion in section 3.6.

3.2. Optimization Framework for Regularized K-means Clustering

Rather than starting model based clustering from usual mixture model framework, we begin our development from K-means clustering and its central choice of distance function. The duality between loss and negative log-likelihood (NLL) then exploited to derive underlying stochastic model framework. This helps introduction of the "loss+penalty" based regularization scheme for clustering in an unified framework.

Note that, K-means algorithm essentially tries to partition N, p dimensional samples $(x_i; i = 1, ..., N)$, into K classes depending upon some distance based measure in an iterative fashion. Most commonly employed distance measure is the Euclidean distance. The algorithm starts with K randomly selected cluster centers and then allocate each observation to a cluster center from which it has minimum distance. This can be described by means of an indicator function between data vector x_i and the cluster center m_k,

$$I(x_i \in j) = \begin{cases} 1 \; if j = argmin_{1 \leq k \leq K} D(x_i, m_k) \\ 0 \; otherwise, \end{cases}$$

where D(.,.) denotes a metric distortion error. Generally EM-type algorithm is being employed [McLachan and Basford,1988] to update $m_k, k = 1, ..., K$ (cluster centers) until some convergence criterion is satisfied. Clearly, corresponding optimization problem would be to minimize the distortion measure of loss,

$$L(x, m) = \sum_{i=1}^{N} \sum_{k=1}^{K} I(x_i \in k) D(x_i, m_k).$$

A common question is about the knowledge of K i.e., the cluster number itself. The algorithm will give a wrong result in case if K is misspecified. Also above optimization problem will have a trivial solution in case we treat $K \in [1, N]$ as a variable quantity. As for $K = N$, $\underset{K \in [1,N]}{argmin} \; \underset{m_k \in \mathbb{R}\Xi^p, \, 1 \leq k \leq K}{argmin}$ L$(x, m) = 0$, in other words each point will become a separate cluster. This fact reveals that there is a need for introduction of a penalty term. The idea of introducing penalty term is to guard against choosing too many cluster centers. For our purpose we choose $\gamma(m) = \sum_{1\leq i,j\leq K} D(m_i, m_j)$ as the penalty, which also denotes distortion measure of the cluster center. Note that $\gamma(m)$ will have trivial minimum when K = 1, or in other words all observations will fall in a single cluster. So our enlarged and regularized optimization problem is

$$\underset{K\in[1,N]}{argmin} \quad \underset{m_k\in\mathbb{R}^p,\ 1\leq k\leq K}{argmin} \quad \underset{\alpha_k\in\mathbb{R}^+,\ 1\leq k\leq K}{argmin}$$

$$\left[\sum_{i=1}^{N}\sum_{k=1}^{K}\alpha_k D(x_j, m_k) + \lambda \sum_{1\leq i,j\leq K} D(m_i, m_j)\right] \tag{3.1}$$

where λ is the regularization parameter and α_k is a function of k-th cluster for $k = 1, ..., K$. We call the above penalized version of K-means as regularized K-means. Note for the choice of $\alpha_k = I(x \in k)$, (1) reduces to the vanilla flavored K-means clustering [MacQueen,1967]. If we restrict $\alpha_k \in [0,1]$, and $\sum_{k=1}^{K}\alpha_k = 1$, then we could interpret $L(x,m) = \sum_{i=1}^{N}\sum_{k=1}^{K}\alpha_k D(x_i, m_k)$ as the weighted loss. This additional constrain on α_k could be removed if we rewrite the loss function as $L(x,m) = \sum_{i=1}^{N}\sum_{k=1}^{K}\frac{\alpha_k}{\sum_{j=1}^{K}\alpha_j}D(x_i, m_k)$, however for notational convenience let us denote $w_k = \frac{\alpha_k}{\sum_{j=1}^{K}\alpha_j}$. Note that when $w_k \in [0,1]$ with $\sum_{k=1}^{K} w_k = 1$ then the form of loss function $L(x,m)$ follows form Hofmann and Buhmann [Hofmann and Buhmann,1997]. In our version we consider $w_k \in [0,1]$ with $\sum_{k=1}^{K} w_k = 1$ for $k = 1, ..., K$, as an extension of a discrete probability on a simplex. So we can interpret $P(x \in \text{k-th cluster}) = w_k$.

Selection of Norm: An appropriate distortion measure depends on the application domain. In K-means algorithm, generally Euclidean norm is chosen as a distortion measure from the cluster center. A more general and robust choice will be Mahalanobis generalized distance given by

$$D(x, m_k) = [(x - m_k)' \, (\sum_k)^{-1}(x - m_k)]^s, \text{ for } s > 0,$$

where $\sum_k$ is a positive definite matrix and s is possibly an unknown power parameter. Note that for $\sum_k = I_k$ and s = 1 this reduces to Euclidean norm. We will see later that the choice of this generalized distance measure expands our modeling effort outside traditional multivariate normal setup. For penalty term involving cluster center again Mahalanobis generalized distance can be defined as a distortion measure,

$$D(m_i, m_j) = [(m_i - m_j)'(2\sum_m)^{-1}(m_i - m_j)]^s, \text{ for } s > 0,$$

where $\sum_m$ is a positive definite matrix. Again note that s = 1 reduces to the usual Mahalanobis distance and introduction of $s > 0$ brings additional robustness.

3.3. Likelihood Formulation

Notably our optimization problem (1) is of the form "loss + penalty". Regarding "loss" as the negative of the log likelihood (NLL), which also follows from maximum entropy principle [Rose et al.,1990]. We like to integrate above problem in a Bayesian framework. The duality between viewing "loss" as the negative of the "log-likelihood" is referred as the "logarithmic scoring rule" in Bayesian literature (Bernardo, page 688) [Bernardo,1979]. In a Bayesian framework we may assume $p(x|\Theta) \propto exp(-L(x,m))$ as the likelihood and $p(m) \propto exp(-\lambda\gamma(m))$ as the prior distribution. Note $|\Theta = \{m, \sum_k, w, s\}$ is the complete parameter space. Thus data likelihood using above choice of distortion measure, will be reduced to,

$$p(x|\Theta, K) \propto \prod_{i=1}^{N}\prod_{k=1}^{K} exp(-w_k[(x_i - m_k)'(\Sigma_k)^{-1}(x_i - m_k)]^s) \qquad (3.2)$$

Note that above density represents multivariate exponential power (MVEP) family, which is com- monly used in robustness criterion. Hence the choice of $\sum_k$ is very important in determining the geometry of the clusters. We like to refer to Fraley and Raftery [Fraley and Raftery, 2002] for details about the different choices of $\sum_k$. Correspondingly we may express prior distribution as

$$p(m \mid K, s) \propto \prod_{1 \le i,j \le K} exp(-\lambda[(m_i - m_j)'(2\sum_m)^{-1}(m_i - m_j)]^s), \qquad (3.3)$$

where each of the cluster centers are assumed to follow $N(\mu_m, \sum_m)$. Note that the prior distribution also arises from MVEP family. However the penalty function is defined in such a way that all the terms corresponding to the prior are not necessarily independent. Notably, above prior is location invariant or rather independent of the choice of the hyperparameter μ_m.

Similar to Richardson and Green [Richardson and Green,1997], we like to introduce a group level indicator variable z_i, for $i = 1, 2, ..., N$, where each z_i is independently distributed and drawn with probability distribution,

$$P(z_i = j) = P(x_i \in j^{th}\text{cluster}) = w_j; \ \ \text{for} j = 1, ..., K. \qquad (3.4)$$

Note that

$$1(z_i = j) = \begin{cases} 1 & \text{with probability} \quad w_j, \\ 0 & \text{with probability} \quad 1 - w_j, \end{cases}$$

where z_i's are "unobserved" cluster allocation variable. It can be assumed that $(1_{(z_i=k)}; k = 1, \ldots, K)$ follows a Multinomial distribution with parameter $(w_k, k = 1, ..., K)$ which reduces to the new likelihood function as

$$p(x|\Theta, z, K) \propto \prod_{i=1}^{N} \prod_{k=1}^{K} [f_{MVEP}(x_i|\Theta)]^{1(z_i=k)}, \tag{3.5}$$

where $f_{MVEP}(x_i|\Theta) = \frac{C_k}{\sqrt{|\Sigma_k|}} exp\{-w_k[(x_i - m_k)' \Sigma_k^{-1}(x_i - m_k)]^s\}$ is the multivariate exponential power family and C_k is the normalizing constant for k-th cluster involving w_k. Again following Richardson and Green [Richardson and Green,1997], the formulation given by equations 3.4 and 3.5 are convenient for calculation and interpretation and will be used in the rest of the paper. Integration of z from (3.4) and (3.5) brings back to (3.2). A similar idea of introducing latent allocation variable (z) is also explained in Fraley and Raftery [Fraley and Raftery, 2002].

3.4. Choices of Prior

The above formulation represents a very general setup. However from a practical point of view, implication of keeping K as an unknown parameter poses a variable dimensional optimization problem and usually requires implementation of RJMCMC. Though RJMCMC is becoming quite popular in practice for different transdimensional computation problem, almost all of the present example of successful implementation of it considers about implementations for few dimensions (say less than six or seven at most). While novelty of the RJMCMC is unquestionable at the same time this trend points out that RJMCMC is not tractable for very high dimensional clustering problem. We like to note that it is very common in bioinformatics and other new emerging application domains where high dimensional data is ubiquitous. For these application domains RJMCMC is not a practical approach to solve clustering problem with unknown cluster number. We have integrated a simple to use and highly scalable indicator variable based technique motivated by the example of B-Class [Medrano-Soto et al.,2005] algorithm developed earlier. In the case of absence of any prior information regarding cluster number, we like to choose K = N. This is an extreme case since with N observations it is not possible to obtain more than N clusters. However in case there is some prior information available regarding the upper bound (say K') of cluster number, we like to use that information by choosing $K = K'(< N)$. This is intuitive as computational complexity of K-means is linearly proportional with the number of observations and number of clusters. A "dimensionality variable" is introduced as ϕ_j for the j-th cluster, which is defined as $\phi_j = 1$ if j-th cluster is alive, (0 if dormant), for $j = 1, 2, ..., K$. Number of alive components tell us the actual number of clusters. At this point we like to emphasize again regarding the novelty of the penalty term, which when combined with above indicator function makes RJMCMC redundant. The revised regularized optimization problem incorporating dimensionality variable can be written as

$$\underset{m_k \in \mathbb{R}^P,\ 1 \le k \le K}{argmin} \quad \underset{\alpha_k \in \mathbb{R}^+,\ 1 \le k \le K}{argmin}$$

$$\left[\sum_{i=1}^{N} \sum_{k=1}^{K} \frac{\alpha_k \phi_k}{\sum_{j=1}^{K} \alpha_j} D(x_i, m_k) + \lambda \sum_{1 \le i,j \le K} \phi_i \phi_j D(m_i, m_j) \right] \tag{3.6}$$

Again for notational convenience we can redefine $w_k \propto \alpha_k \phi_k$. Note that the contribution to the loss or penalty term is only coming from live clusters, which seems intuitive. It can be easily checked that the likelihood constructed before will again go through with this revised "loss + penalty" framework. We may assign priors to the unknown parameters $m, \Sigma, \alpha, z, \phi, s$ as follows:

$$\begin{array}{ccc} m_k|K & \overset{ind}{\sim} & N_p(\mu_m, \Sigma_m) \\ \Sigma_k & \sim & IW(\psi, Q) \\ \alpha|k & \overset{ind}{\sim} & Dirch(\delta \mathbb{1}^K), \\ z|\alpha, K & \propto & \prod_{i=1}^{N} \prod_{k=1}^{K} (\alpha_k \phi_k)^{1_{(z_i=k)}}, \\ s & \sim & Gamma, (c, d) \end{array}$$

Initialization Mechanism	Update Mechanism
Randomly allocate N points into	Variation of grouped Gibbs Sampling,
K-clusters as,	1. Update α as weight variable
1. Set $\alpha_k = \frac{1}{K}$ for $\forall k = 1, 2, ..., K$	2. Update z as allocation variable
2. Set $\phi_k = 1$ for $\forall k = 1, 2, ..., K$	3. Update ϕ as dimension variable
(clusters are active)	4. Update cluster centers
3. Randomly allocate each point x_i in different cluster	5. Update $\sum_k$ or cluster geometry
a. draw u $\sim$ Multinomial$(1; \alpha_1, \alpha_2, ...\alpha_K)$	6. Update $\sum_m$ or center covariance matrix
b. if u[k] = 1, put x_i into k-th cluster and set $z_i = k$	7. Update s or power variable
4. For k-th Cluster center $m_k = (m_{k1}, m_{k2}, ..., m_{kp})$	8. Update λ or regularization parameter
and estimate $m_{kj}^{\wedge} = \frac{\sum_{i=1}^{N} x_{ij} 1_{(z_i=k)}}{\sum_{i=1}^{N} 1_{(z_i=k)}}$, for $k = 1, 2, ..., K$	
5. For each cluster estimate $\sum_k$	
a. if no points inside a cluster set $\sum_k = 0$	
b. if single point inside a cluster set $\sum_k = 1$	
6. Estimate $\sum_m$ from cluster centers	
7. Set $s = 1$	

Schema 1

and $P[\phi_j = 1] = 0.5$ (non-informative choice) for $j = 1, 2, ..., K$. As noted earlier, K is a suitably chosen large constant (a conservative approach, to guard against inappropriately choosing too few clusters). This completes the Bayesian hierarchical modeling framework. For the sake of space, details of the sampling scheme and acceptance probability of each parameter is given in the appendix. Initialization and sampling scheme of the different parameters are described in Schema 1. We would like to note that for cluster center covarainace matrix ($\sum_m$) we have not specified any prior distribution. In each MCMC iteration we have computed empirical estimate of $\sum_m$ from updated cluster centers (m). While complete prior specification of $\sum_m$ is another possibility, we choose empirical estimate to reduce computational burden. These two variance covariance matrices ($\sum_k$ and $\sum_m$) are algebraically dependent on each other and we put more emphasis on $\sum_k$, as for determining cluster geometry $\sum_k$ is more important than $\sum_m$. Also constructing a joint prior on $(\sum_k, \sum_m)$ is very prohibitive in terms of prior elicitation.

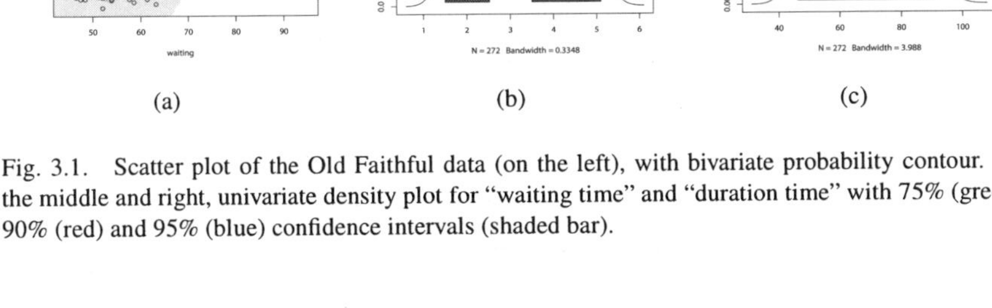

(a) (b) (c)

Fig. 3.1. Scatter plot of the Old Faithful data (on the left), with bivariate probability contour. On the middle and right, univariate density plot for "waiting time" and "duration time" with 75% (green), 90% (red) and 95% (blue) confidence intervals (shaded bar).

3.5. Clustering Data Examples

Clustering is an ubiquitous data analytic technique which finds its usefulness in almost all of the applications domains. For the testing purpose, we consider two benchmark data sets. The first benchmark data set considered here is the Old Faithful data, which was considered earlier by Hardle [Hardle,1991] and Stephens [Stephens,1997]. Next we have explored famous Fisher's Iris [Fisher,1936] data set, considered earlier by McLachlan et al. [see for 14, 15] for clustering purpose. Details about each of them will be described latter in respective subsections.

3.5.1. *Old faithful data: clustering analysis*

The present data set consists of data on 272 eruptions of the Old Faithful geyser in the Yellowstone National Park. Each observation consists of two dimensions, namely the duration (in minutes) of the eruption, and the waiting time (in minutes) before the next eruption. When dimension is less than or equal to three it is always helpful to draw scatter plots, which give us some rough idea about the cluster number. However this advantage may not be present for the higher dimension. Simple scatter plot and corresponding highest density contours are presented in Figure 3.1. The data is clearly bimodal in nature, even when considered in each dimension separately as depicted in Figure 3.1.
For the present problem we choose K = 5. A higher value of K is also possible. We want to choose K large enough so that the actual (unknown, but probably 2) cluster number is smaller than K. In the MCMC sampling scheme first 5,000 iterations are thrown out as burn-in period, then an additional 70,000 iterations are obtained out of which we accepted every 50-th iteration, creating 1400 samples from the posterior distribution for each model parameter. For hyper parameters

Algorithm 1 Get Cluster Vector From Cluster Assignment Correlation Matrix

Input: $\Omega = \{x_1, \ldots, x_N\}, U = \{x_j, 0\}_{j=1}^{N} = \{U_1, U_2\}, C_m =$ Cut of value (user provided)
Initialization: Set $i = 1; A_i = \phi; \Omega_i = \Omega$
while $(\Omega_i \neq \phi)$ do
Choose any $x \in \Omega_i$
repeat
$A_i \Leftarrow A_i \cup x; S \Leftarrow \phi$,
Define $C \Leftarrow \{x_j, Cor(x_i, x\}_{j=1}^{N} = \{C_1, C_2\}$
Update $U \Leftarrow \{U_1 \cap C_1, C_2\} = \{U_1, U_2\}$
Sort U by U2 Keep only those elements in U such that $C_2 \geq C_m$
Define $S = \{y | y \in U_1$ and $y \notin A_i\}$
Choose $x \in S$ such that $U_2(x)$ is maximum
until $(x = \phi)$
$i \Leftarrow i + 1$
$\Omega_i \Leftarrow \Omega_i - 1 - A_{i-1}$
$U = \{x_j, 0\}(x_j \in \Omega_i)$
$A_i \Leftarrow \phi$
end while
Output: i = K and Cluster Vector $A_1, A_2, \ldots, A_K$.

we choose $\delta = 1, c = 4$ and $d = 20$. We have also tried with several combinations of hyperparameter values and results are fairly similar. Underlying guideline for choosing hyperparameter is that choice of prior should not drive our inference. In fact the above choice of hyperparameter values reasonably worked well for both the data examples considered in this paper. Since $p \ll N$, an unconstrained specification of $\sum_k$ is a possibility. However for computational simplicity, we assume $\sum_k = \sigma^2 Diag(\varsigma)$ where $\sigma \sim IG(2,10)$ and each $\varsigma_i \sim IG(2,10)$ for $i = 1,2,...,p$, while $\sum_m$ is being estimated empirically as explained earlier. For convenience λ is chosen to be fixed at 0.5, however it is possible to put an uniform prior on λ. We use two chains with separate set of initial values for each MCMC computation. Note that for the exchangeable prior case, mixture components become non-identifiable under the permutations in labeling of the parameters. This is termed as label switching problem in the literature [see for 24, 4]. For this reason, convergence of the MCMC chains in the mixture context is a complicated issue. Following suggestions in Jasra et al. [Jasra et al, 2005], we put minimum requirement of convergence is such that, all the possible labeling of the parameters are explored by the MCMC sampler. To ensure above, we have introduced tempering mechanism within each MCMC sampler as suggested by Celeux [Celeux et al. 2000]. Regarding details of effcient construction of the MCMC sampling scheme we would like to refer Celeux [Celeux et al. 2000] and other references cited in [Jasra et al., 2005]. For Old Faithful data is presented, at left. A dark red spot is an

(a) (b)

Fig. 3.2. Heat map of the correlations obtained from posterior cluster allocation of MCMC output

indicative of high correlation and those observations should be clustered together. A specific region inside the heat map at left is zoomed in to give a clear view of the correlation matrix, at right.

3.5.2. *Clustering result through correlation decomposition*

Observations Notably Likelihood developed in equation (3.2) is not an usual mixture model. For that reason standard parameter constraint based on label switching methodology to obtain clustering is not an option in the present setup. In fact in an honest exploration under the constraint $\alpha_1 < \alpha_2 < \ldots < \alpha_k$ we are able to classify those points which are not strongly associated with any of the clusters, as represented in Figure 3.3(a). However the other points inside any of the clusters are not identified properly. We have developed our result through the analysis of correlation matrix of cluster assignment. Philosophically speaking, similar observation should be assigned to the same cluster in different iterations of MCMC. Though clusters may change their labels from iteration to iteration but similar observations should stick together, while dissimilar ones would not. For that reason cluster assignment of similar observation in different MCMC itearion should be highly correlated. We have presented the correlation matrix as a heat map in Figure 3.2 (a). A detailed observation level view is presented in Figure 3.2 (b). When the number of observations are not very high, this kind of visual or rather exploratory analysis holds great potential. We have developed an algorithm through which this correlation matrix can be used to figure out which observations have highly correlated cluster assignment. The algorithm is user driven in the sense that we can specify a "cutoff" (C_m) value. It will consider all observations having inter-correlation more than the cutoff value in the same cluster and continues to do so until it exhausts all observations. Let Ω denotes set of all observations. The algorithm starts with a randomly chosen observation from Ω as a seed and selects all observations having correlation more than C_m with the seeded observation and makes a set S. Next it chooses the observation from the set S having highest correlation with the seeded observations. This newly chosen observation becomes current seed and the previously seeded observation is pushed into a linked list called A_1. This continues and as a result the list A_1 grows until the set S becomes empty. The set A_1 is a cluster whose members have inter-correlation more than C_m. The algorithm then update Ω as $\Omega = \Omega - A_1$, and selects a new seed from this set and continues in an iterative fashion. Details of the algorithm is presented in pseudo code form as Algorithm 1. We would like to note that this algorithm can be easily modified to accommodate expert specified seed rather than the randomly chosen one. The randomly chosen seed may sometimes produce poor cluster assignment if the seed point itself is isolated. The algorithm also produces few linked lists (A_i's) with only one or two (very few) members. These points are isolated points with no specific cluster membership.

The output of this correlation matrix is presented in Figure 3.3(b). Different contours are represented for different values of C_m as concentric regions. We have chosen $C_m = 0.4, 0.5$ and 0.6. The output of Algorithm 1 clearly indicates the existence of two clusters in the Old Faithful data. It is obvious as C_m value decreases each of the clusters grow and thus include more and more points in its neighborhood. Similarly $C_m = 1$, will result in too many clusters. An optimum value of C_m should be expert driven. We have tried with different values of C_m, but results for only few values are reported here. We would also like to point out that the clustering result obtained through our algorithm is agreeing with Stephens [Stephens, 1997] result reported elsewhere for Old Faithful data. However number of observations correctly allocated to proper cluster is dependent upon expert chosen value of C_m. For well chosen value of C_m our algorithm will beat Stephen's output at the same time for a poorly chosen one it will fail to do so. From the computational complexity point of view our algorithm is fast enough and comparable to other mixture model based MCMC approach [see for 23, 20]. However for

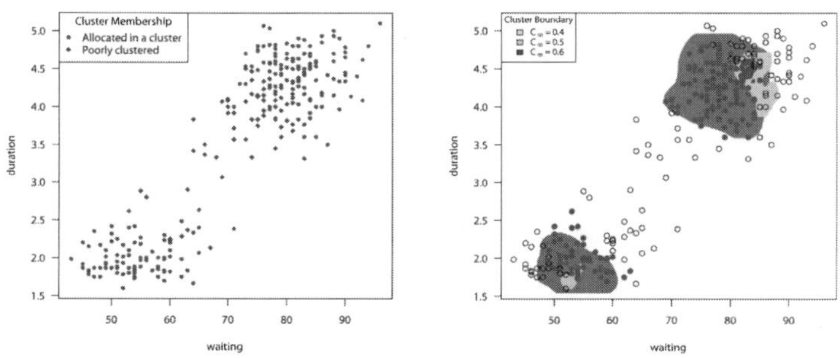

(a) (b)

Fig. 3.3. Figure (a) represents clustering after relabeling constraint $\alpha_1 < \alpha_2 < \ldots < \alpha_5$. This constraint is able to gure out those points that are poorly allocated. Figure (b) represents results from Algorithm 1 for different values of C_m (0.4, 0.5 and 0.6). Different concentric areas represent the size of the clusters, which increases as Cm value decreases. In figure (b), "o" denotes unallocated points, while "•" denotes allocated points to a cluster.

small dimensional problem like Old Faithful data this computational advantage is nonsignificant. We will see in next few examples that the real computational scalability of our proposed algorithm becomes clearly evident as dimension explodes rapidly. Also for high (or even low) dimension, our algorithm does not require any expertise of the user (as well as computational burden) regarding reversible jump scheme, which is a major advantage as pointed out earlier.

3.5.3. *Fisher's Iris Set: Clustering Analysis*

We consider another famous benchmark data set namely Iris data, originally collected by Anderson [Anderson, 1935] and first analyzed by Fisher [Fisher, 1936].

The original data set consists of length and width measurements of both the sepals and the petals (i.e., dimension equals to 4) of 50 plants for each of three types of Iris species namely setosa, versicolor and virginica. We would like to mention that a specific subset of this data set is analyzed from frequentist mixture model perspective by McLachlan and Peel [MaLachan and Peel, 1996]. They derived an algorithm namely MIXFIT [MaLachan and Peel, 1998] which uses mixture model based bootstrap approach to test user specified cluster number (say

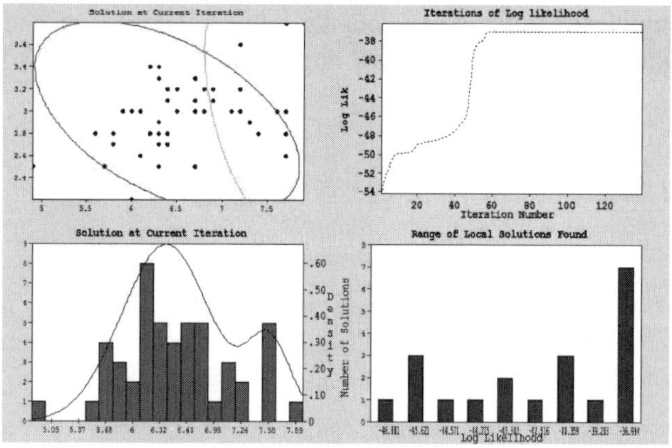

Fig. 3.4. Result through EMMIX algorithm. Two clusters are obtained (top left). Histogram of the most significant dimension (sepal length) is preened (at bottom left). Nine log likelihoods corresponding to the roots of the likelihood equation are also presented (bottom right).

k) verses some alternative value (k + 1) in hypothesis testing setup. MIXFIT algorithm can automatically select a starting cluster number in case no cluster number being specified. The algorithm then uses EM approach to find unknown parameters. Bootstrapping is carried out to calculate likelihood ratio statistics for testing the user provided hypothesis regarding the cluster number. In spirit MIXFIT and its successor EMMIX [Malachan and Peel, 1999] offers greater flexibility than regular K-means as user may progress sequentially to get the optimum cluster number. Our method though a fully Bayesian approach is targeted towards high dimensional data, and is compared with MIXFIT algorithm through Iris data example in this subsection. Following McLachlan et al. [1998] we have selected 50 observations of the Iris virginica set. We have analyzed the data set by the EMMIX [Mclachan and Peel,1999] algorithm obtained through journal website and the result is represented in Figure 3.4. Current version of the EMMIX [McLachan and Peel,1999] algorithm available through world wide web is still in developmental phase. It is bit restrictive as it allows analysis of few data sets only and fortunately Iris data set is one of them. Results obtained through EMMIX algorithm for Iris data set in our test bed is very much comparable with that of its predecessor MIXFIT algorithm reported elsewhere [McLachan et al, 1997]. In their original paper McLachlan et al. [1997] suggested different criteria apart from the P-value of the test to suggest existence of two clusters corresponding to the log likelihood -36:994 and the observations corresponding to the smaller cluster are $6, 8, 18, 19, 23, 26, 30, 31, 32$. For details of the clustering result through MIXFIT algorithm we refer to [McLachan et al., 1997]. At this point we mention that the criterion suggested by McLachlan et al. [1996] though intuitive however is subjective in nature. In fact for any unsupervised learning problem it is difficult to produce results which is completely objective in nature. Next we apply our Regularized K-means algorithm for this specific data set. For the present

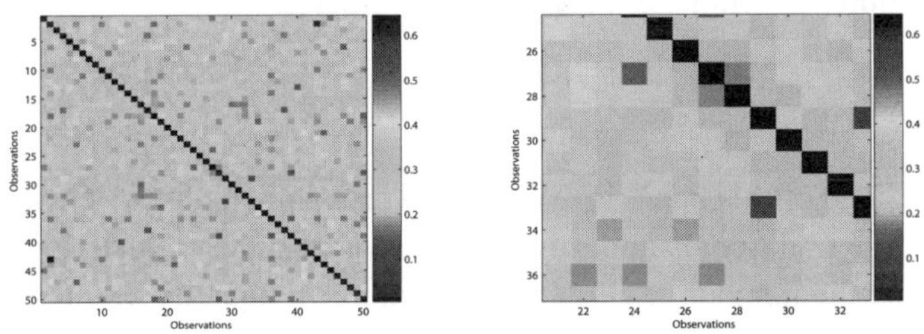

(a) (b)

Fig. 3.5. Heat map of the correlations obtained from posterior cluster allocation of MCMC output for Iris data is presented, at left. A dark red spot is an indicative of high correlation and those observations should be clustered together. A specific region inside the heat map at left is zoomed in to give a clear view of the correlation matrix, at right.

clustering problem we have chosen K = 5. Notably a higher value of K is also possible, however for the present study, reported cluster number seems well less than 5 (only 2 large clusters). In the MCMC sampling scheme first 5,000 iterations are thrown out as burn-in period, then an additional 70,000 iterations are obtained out of which we accepted every 50-th iteration, creating 1400 samples from the posterior distribution for each model parameter. For hyper parameters, we choose $\delta = 1, c = 4$ and $d = 20$. For computational simplicity, we have assumed $\Sigma_k = \sigma^2 Diag(\varsigma)$ where $\sigma \sim IG(2, 10)$ and each $\varsigma_i \sim IG(2, 10)$ for $i = 1, 2, ..., p$. We have presented our results in Figure 3.5, which represents the heat map of the correlation matrix of cluster assignment of the 50 observations. Figure 3.5 (b) shows a zoomed down version of the same heat map. Note that from Figure 3.5 (b) observation number 26 has relatively high correlation with observation 30 and 31.

Similar point could be made between observation numbered 33 and 29. Finally we have analyzed the cluster allocation correlation matrix through Algorithm 1. For $C_m = 0.25$ the algorithm gives four large clusters and for $C_m = 0.2$ it gives only three clusters. For $C_m = 0.2$ we found following observations in two clusters namely Group1 = $\{18, 19, 32\}$, Group2=$\{6, 22, 23, 26, 27, 28, 30, 31\}$. All other observations are placed in a large cluster. Notably these two small clusters encompass all members in the smallest cluster observed in McLachlan et al. [1996] and results are more or less on agreement. On the test bed our MCMC based algorithm took only 142 seconds while EMMIX algorithm in default setup took almost double the time. However this time comparison may not be significant as it heavily depends upon the programming language in which both algorithms are implemented and other software related issues. Nevertheless our algorithm is fast enough and results are comparable even for relatively small dimensional data sets with that of other benchmarked clustering techniques.

3.6. Conclusion

The original idea of this paper is to develop a probabilistic "soft" clustering algorithm out of the traditional similarity measure based on hard clustering algorithm, while keeping the number of clusters as an unknown quantity of interest to begin with. K-means and its different variants are by far the most heavily used similarity measure based clustering algorithm. We have exploited both K-means and the mixture model based approach to come up with a new algorithm, namely "Regularized K-means". A regularization framework for clustering is also provided with the introduction of a novel penalty term. This regularized framework when integrated with "dimensionality variable", avoids using RJMCMC approach. This will be highly advantageous for high dimensional clustering, which we are going explore somewhere else. This paper explicitly shows the duality between "Regularized K-means" and model based approach. We have shown efficacy of our method for two test data sets.

Two open questions arise from this paper. First, how to construct an efficient penalty function when data distribution is non-normal, i.e., when Euclidean norm is not a proper similarity measure. Recently Banerjee et al. [Banerjee et al., 2005] have shown that, for known cluster number a large class of similarity measure based clustering algorithm can be thought of as a special case of Bregman divergence based clustering. However for unknown cluster number only solution is to perform RJMCMC, which is prohibitive, as the dimension of the problem grows rapidly. We have provided a scalable alternative approach for such a scenario in this paper, by constructing innovative penalty function. Secondly, original "loss

+ penalty" based optimization problem for clustering automatically gives rise to model based likelihood. We have presented an algorithm to analyze the cluster assignment of different observations and hence relabeling of the clusters. This approach though informative, is somewhat subjective in the sense that user need to specify the cluster cutoff value. It should be noted though, all clustering algorithm is somewhat subjective in determining what is cluster and what is not.

Appendix

A Sampling Scheme for Weight parameter

Let us define $n_j = \{i : z_i = j\}$, for $j = 1,2,...,K$ and we use $| \ldots$ throughout to denote conditioning of all other variables. It can be shown that for k-th ($k = 1,2,...,K$) cluster,

$$\alpha_k|\ldots \propto \phi_k Gamma(B_k; \tfrac{p}{2s} + k_k + \delta + 1);$$
$$\text{where } B_k = \frac{\phi_k}{\sum_{j=1}^{K}\alpha_j\phi_j}\sum_{i|z_i=k}[(x_i - m_k)'\Sigma_k^{-1}(x_i - m_k)]^s;$$

with $\phi_k = 1$ if the cluster is alive, 0 if dormant. In any stage of the iteration, the denominator part in B_k involving α_k, namely $\sum_{j=1}^{K}\alpha_j\phi_j$ is constant for all α's, so it may be ignored. Once we have drawn samples from this posterior we need to scale it up so that it sums up to 1. If a dormant group comes to life in some stage of the iteration, we need to assign a small probability to give it a fair chance of survival. For our case we have given a fixed probability $\frac{1}{2K}$.

B Sampling Scheme for Allocation parameter

It can be shown that for $k = 1,2,...,K$ and i-th observation,

$$p_k = p(z_i = k|\ldots) \propto \phi_k(\alpha_k)^{\frac{p}{2s+1}}\exp\{-\alpha_k D_k\};$$
$$\text{where} D_k = \frac{\phi_k[(x_i - m_k)'\Sigma_k^{-1}(x_i - m_k)]^s}{\sum_{j=1}^{K}\alpha_j v\phi_j}.$$

Each p_k should be normalized to define $p_k^* = \frac{p_k}{\sum_{j=1}^{K}p_j}$. Note that if a cluster is dormant then $\phi_k = 0$, implying $p_k = 0$. To allocate i-th observation, draw $\mu \sim$ Multinomial $(1, p_1^*, p_2^*, ..., p_{K'}^*)$. Note that μ will be K' dimensional vector, with only one of the elements (say k-th) equals to 1 and all others are zero. If u[k] = 1, put x_i into k-th cluster and set $z_i = k$. K' denotes number of live clusters.

C Sampling Scheme for Dimension parameter

Much of the work is inspired from B-Class algorithm developed earlier by Medrano-Soto et al. [18]. We like to refer them for detailed calculations. Basic principle is that if a cluster has at least one allocated observation then it survives into the next iteration. A cluster which is empty (without any observation) should either die or stay alive with a survival probability $P[\phi_i = 1] = \in$. It can be shown that for the present setup if we choose $P[\phi_i = 1] = 0.5$, the acceptance probability in the Metropolis step is always 1, or in other words for an empty cluster we always accept a proposal ϕ_k^* over ϕ_k (old value).

D Sampling Scheme for Cluster Center and Cluster Dispersion

Specific calculations of the acceptance probability for these parameters are straight forward, though full conditional distribution of none of these parameters can be expressed in a standard form. We are only describing important strategic steps. Each of the K cluster parameters (m_k and $\sum_k$) are updated in turn. If a cluster is dormant in an iteration its center and covariance parameter are not updated. If a cluster comes into life at any stage of an iteration a random point is chosen as the cluster center with global variance covariance matrix of the centers ($\sum_m$) as cluster geometry parameter. If a cluster dies all its information are lost.

E Sampling Scheme for Power Variable

The full conditional distribution for the power parameter in the Mahalanobis generalized distance is,

$$s|\ldots \propto \prod_{k=1}^{K} \frac{s(2w_k)^{\frac{p}{2s}}}{\Gamma(p/2s)} exp(-\alpha_k B_k) \times \prod_{1\le i,j\le K} \frac{s(2w_k)^{\frac{p}{2s}}}{\Gamma(p/2s)} exp(-\lambda\phi_i\phi_j[(m_i - m_j)'2 \\ (2\sum_m)^{-1}(m_i - m_j)]^s) exp(-ds)s^{c-1}.$$

Since full conditionals are not in conventional form we approximate them by sampling with a Metropolis step within the Gibbs sampler [Gelfand and Smith, 1990]. We use gamma proposal density with mean around 1 for our purpose.

References

1. Andesron, E., The Irises of the Gaspe Peninsula. Bulletin of the American Iris Society, 593-5,1935.
2. Banerjee, A. S. Merugu, I. Dhillon, and J. Ghosh. Clustering with Bregman divergences. Journal of Machine Learning Research, 6:1705-1749, 2005.
3. Bernardo, J.M, Expected information as expected utility. The Annals of Statistics, 7:686-690, 1979.
4. Celeux, G. M. Hurn, and C. P. Robert. Computational and inferential diculties with mixture posterior distributions. Journal of the American Statistical Association, 95:957-970, 2000.
5. Fisher, R.A. The use of multiple measurements in taxonomic problems. Annals of Eugenics, 7:179-188, 1936.
6. Fraley,C. and A. E. Raftery. Model-Based clustering, discriminant analysis and density estimation. Journal of the American Statistical Association, 97:611-631, 2002.
7. Gelfand,A.E. and A. F. M. Smith. Sampling-based approaches to calculating marginal densities. Journal of the American Statistical Association, 85:398-409, 1990.
8. Hardle.,W. Smoothing techniques with implementation in S. Springer-Verlag, 1991.

9. Hofmann,T. and J. M. Buhmann. Pairwise data clustering by deterministic annealing. IEEE Transactions on Pattern Analysis and Machine Intelligence, 19, 1997.
10. Jasra,A., C. C. Holmes, and D. A. Stephens. Markov Chain Monte Carlo methods and the label switching problem in Bayesian mixture modeling. Statistical Science, 20:50-67, 2005.
11. Kaufman,L., and P. J. Rousseeuw. Finding groups in data: An introduction to cluster analysis. John Wiley and Sons, Inc., 1989.
12. MacQueen.J. Some methods for classication and analysis of multivariate observations. In Proc. of the 5th Berkeley Symposium on Mathematical Statistics and Probability, pages 281-297, 1967.
13. McLachan,G.J. and K. E. Basford. Mixture Models: Infernce and Applications to Clustering. Marcel Decker, New York, USA, 1988.
14. McLachlan,G. and D. Peel. An algorithm for unsupervised learning via normal mixture models. In D. L. Dowe, K. B. Korb, and J. J. Oliver, editors, ISIS: Information, Statistics and Induction in Science, pages 354-363. Singapore: World Scientific, 1996.
15. McLachlan,G. and D. Peel. MIXFIT: An algorithm for the automatic fitting and testing of normal mixture models. In Journal of Statistical Software, volume 4, pages 553-557, 1998.
16. McLachlan,G. and D. Peel. The EMMIX algorithm for the fitting of normal and t-components. In Proc. of the 4th International Conference on Pattern Recognition, volume 1, pages 1506-1524, 1999.
17. McLachlan,G. D. Peel, and P. Prado. Clustering via normal mixture models. In Proceedingsof the American Statistical Association (Bayesian Statistical Science Section), pages 98-103. American Statistical Association, 1997.
18. Medrano-Soto,A. J. A. Christen, and J. Collado-Vides. BClass: A Bayesian approach based on mixture models for clustering and classification of heterogeneous biological data. Journal of Statistical Graphics, 13:398-409, 2005.
19. Raftery,E.A. and N. Dean. Variable selection for model-based clustering. Journal of the American Statistical Association,, 101:168-178, 2006.
20. Richardson,S. and P. J. Green. On Bayesian analysis of mixtures with an unknown number of components. Journal of the Royal Statistical Society, B, 59:731-792, 1997.
21. Rose,K., E. Gurewitz, and G. Fox. Statistical mechanics and phase transitions in clustering. Physical Review Letters, 65:945-948, 1990.
22. Saharon.,R., Topics in regularization and boosting. Ph.D. dissertation, Stanford University, 2003.
23. Stephens,M. Bayesian methods for mixtures of normal distributions. Ph.D. dissertation, University of Oxford, 1997.
24. Stephens.M., Dealing with label-switching in mixture models. Journal of the Royal Statistical Society, B, 62:795-809, 2000.

Chapter 4

Jacobians Under Constraints and Statistical Bioinformatics

K. V. Mardia
University of Leeds
Email:sta6kvm@maths.leeds.ac.uk

Roy's theorem on Jacobians under constraints is applied to obtain an explicit form of the Haar measure. This is a key result for the matrix Fisher distribution on rotations. The distributions have recently appeared for some fundamental problems in Bioinformatics.

Contents

4.1. Introduction

Suppose in p dimensions, the orientation of the object is specified by n directions $x_1, \ldots, x_n$ with

$$XX^T = I_n, \tag{4.1}$$

where $x_1^T, \ldots, x_n^T$ are n rows of X and I_n is the $n \times n$ identity matrix, $n \leq p$ (for simplicity). The Riemann space whose elements are X is called the Stiefel manifold, and we shall denote it by $O(x,p)$. For $n = p$, the Stiefel manifold becomes the orthogonal space, $O(p)$. A Haar measure of unit mass on $O(n,p)$ will be written as

$$[dX], X \,\varepsilon\, O\,(n,p)$$

One of the most common distributions on X is the matrix Fisher distribution (Downs, 1972; Khatri and Mardia, 1977), defined as

$$a(F)\exp(\mathrm{tr}FX^T)[dX], X \varepsilon O\,(n,p),$$

where F is an $n \times p$ parameter matrix.
This distribution is becoming increasingly important with its new applications in Bioinformatics (see Green and Mardia, 2006). The focus has shifted on how to obtain a suitable parametrization of X. An Euler-angle representation of X turns to be effective for calculating the Haar measure $[dX]$ as well as simulating the matrix Fisher distribution. We will here concentrate on deriving $[dX]$. This follows basically from a powerful theorem of Roy (1957, Thm A.5.5., p.165) for Jacobians under constraints.

4.2. Jacobians Under Constraints and the Haar Measure

Let H be an $p \times p$ orthogonal matrix and $0(p)$ be an orthogonal space. This space is of dimension $p(p-1)/2$ and it can be generated from functionally independent elements H_I of H. The unit invariant Haar measure $[dH]$ over $0(p)$ will be called a uniform distribution over an orthogonal space $0(p)$. In terms of random elements H_I which spans the whole space $0(p)$ this uniform distribution will be denoted by

$$[dH] = \{\Gamma_p(p/2)\}\ dH_I/g(H_I)\pi^{\frac{1}{2}p^2} \text{ for } HH' = I, \tag{4.2}$$

where $\Gamma_p(x) = \pi^{\frac{1}{4}p(p-1)} \Pi_{i=1}^p \Gamma(x - \frac{i-1}{2})$ and $g(H_I)$ is the Jacobian of the transformation by Roy (1957, Theorem A-5.5, p.165) is

$$g(H_I) = J(HH^T; H_D) \tag{4.3}$$

H_D being the set of functionally dependent elements of H on H_I or if dH denotes a matrix in the differential elements of H, then $H(dH)' = S$ is a skew symmetric matrix and

$$J(S;(dH)_I) = \{g(H_I)\}^{-1} \tag{4.4}$$

where $(dH)_I$ denotes the set of differential elements of H_I and $J(S;(dH)_I)$ denotes the Jacobian of the transformations from S to $(dH)_I$.

Now, let us consider a sub-matrix H_n of H having n rows and p columns. Then $H_nH_n^T = I_n$ and an invariant Haar measure corresponding to H_n will be denoted by $[dH_n]$ and which will be equivalent in terms of random elements $H_{n(I)}$ of H_n as

$$[dH_n] = \pi^{-\frac{1}{2}pn}\{\Gamma_n(p/2)\}\ dH_{n(I)}/g(H_{n(I)}) \text{ for } H_nH_n' = I, n \leq p \tag{4.5}$$

where $g(H_{n(I)}) = J(H_nH_n^T \ ; \ H_{n(D)})$ or if $H_n(dH_n)^T = S_{11}$ and $H_{(p-n)}(dH_n)^T = S_{12}$, $H^T = (H_n^T, H_{(p-n)}^T)$ being an orthogonal matrix, then

$$J(S_{11}, S_{12} \ ; \ H_{n(I)}) = \{g(H_{n(I)})\}^{-1}. \tag{4.6}$$

For example, suppose that in H_n, we have

$$H_{n(I)} = \{h_{12}, \ldots, h_{1p}, h_{23}, \ldots, h_{2p}, \ldots, h_{n,n+1}, \ldots, h_{n,p}\}$$

and taking H_{ii} as the sub-matrix obtained from H_n by taking the first i rows and columns and assuming $|H_{ii}| > 0$, then

$$g(H_{n(I)}) = \Pi_{i=1}^n |H_{ii}| = 2^{-n} J(HH^T; H_D)$$

because $|H_{ii}|$ takes only one positive sign. Hence (4.2) becomes

$$[dH_n] = \pi^{-\frac{1}{2}pn}\{\Gamma_n(p/2)\} \, dH_{n(I)}/(\Pi_{i=1}^n |H_{ii}|) \tag{4.7}$$

where $H_nH_n' = I_n$ and $|H_{ii}| > 0$ for $i = 1, 2, \ldots, n$. An alternative representation of this Jacobian is given by Srivastava and Khatri (1979, Theorem, A.5.5, p.165).

4.3. The Euler-Angle Representation

Now suppose that the orthogonal matrix is characterized by the generalised Euler's angles as follows; see for example, Khatri and Mardia (1977).

Let $H_{ij}(\theta_{ij})$ be an orthogonal matrix such that in the diagonal places, there are unities except at (i,i)th and (j,j)th places in which there is $\cos\theta_{ij}$, and in the off-diagonal places, there are zeros except at (i,j)th and (j,i)th places in which there are $\sin\theta_{ij}$ and -$\sin\theta_{ij}$ respectively $(j > i)$. Define

$$H^{(j)} = H_{j-1,j}\ (\theta_{j-1,j}) \ldots H_{1j}(\theta_{1j}) \tag{4.8}$$

and

$$H = H^{(p)}H^{(p-1)} \ldots H^{(2)}. \tag{4.9}$$

We can have different types of orthogonal matrices which can be obtained by permuting the orders of multiplications mentioned in (4.7) and (4.8). These are such $\left(\frac{p(p-1)}{2}\right)!$ different orthogonal matrices which will give different (or same) jacobian of transformations. The domain of θ_{ij}'s mentioned is

$$\{-\pi \le \theta_{i,i+1} \le \pi \ , \ -\frac{1}{2}\pi \le \theta_{ij} \le \frac{1}{2}\pi \text{ for } i = 1, 2, \ldots, p \text{ } and \text{ } j = i+2, \ldots, p\}. \tag{4.10}$$

For simplicity, we shall consider $p = 4$ and write

$$H^{(4)} = \begin{pmatrix} \cos\theta_{14} & 0 & 0 & \sin\theta_{14} \\ -\sin\theta_{14}\sin\theta_{14} & \cos\theta_{24} & 0 & \cos\theta_{14}\sin\theta_{14} \\ -\sin\theta_{14}\cos\theta_{24}\sin\theta_{34} & -\sin\theta_{24}\sin\theta_{34} & \cos\theta_{34} & \cos\theta_{14}\cos\theta_{24}\sin\theta_{34} \\ -\sin\theta_{14}\cos\theta_{24}\cos\theta_{34} & -\cos\theta_{34}\sin\theta_{24} & -\sin\theta_{34} & \cos\theta_{14}\cos\theta_{24}\cos\theta_{34} \end{pmatrix}$$

$$H^{(3)} = \begin{pmatrix} \cos\theta_{13} & 0 & \sin\theta_{13} & 0 \\ -\sin\theta_{13}\sin\theta_{23} & \cos\theta_{23} & \cos\theta_{13}\sin\theta_{23} & 0 \\ -\sin\theta_{13}\cos\theta_{23} & -\sin\theta_{23} & \cos\theta_{13}\cos\theta_{23} & 0 \\ 0 & 0 & 0 & 1 \end{pmatrix}$$

and

$$H^{(2)} = \begin{pmatrix} \cos\theta_{12} & \sin\theta_{12} & 0 & 0 \\ -\sin\theta_{12} & \cos\theta_{12} & 0 & 0 \\ 0 & 0 & 1 & 0 \\ 0 & 0 & 0 & 1 \end{pmatrix}.$$

Then

$$H = H^{(4)}H^{(3)}H^{(2)}.$$

4.4. The Jacobian

We first obtain the Jacobian of the transformation using (4.6) for $p = 4$. The general results become then fairly obvious (For $p = 2,3$, the Jacobian can be written down easily but does not give insight into its extension). We have for $p = 4$,

$$\Pi_{i=1}^{4}|H_{ii}| = \cos^3\theta_{14}\cos^2\theta_{13}\cos^2\theta_{24}\cos\theta_{23}\cos\theta_{34}\cos\theta_{12},$$

and

$$h_{12} = \cos\theta_{13}\cos\theta_{14}\sin\theta_{12}\ ,\ h_{13} = \cos\theta_{14}\sin\theta_{13}\ ,\ h_{14} = \sin\theta_{14},$$

$$h_{23} = -\sin\theta_{13}\sin\theta_{14}\sin\theta_{24} + \cos\theta_{24}\cos\theta_{13}\sin\theta_{23}\ ,\ h_{24} = \cos\theta_{14}\sin\theta_{24},$$

and

$$h_{34} = \cos\theta_{14}\cos\theta_{24}\sin\theta_{34}.$$

$$J(h_{ij}(j > i = 1,2,3,4);\theta_{ij}(j > i = 1,2,3,4)) = \Pi_{j>i=1}^{4} J(h_{ij};\theta_{ij})$$

$$= \cos^5\theta_{14}\cos^3\theta_{13}\cos^3\theta_{24}\cos\theta_{23}\cos\theta_{34}\cos\theta_{12}.$$

Thus the required Jacobian as

$$J(S;\theta_{ij}(j>i=1,2,3,4)) = \Pi_{i=1}^{2}\Pi_{j=i+2}^{4}(\cos^{j-i-1}\theta_{ij}), \tag{4.11}$$

if $-\pi/2 \le \theta_{j-1,j} \le \pi/2$ for $j=2,3,4$, otherwise we have to multiply it by 2^3.

In general, we shall have the following Jacobian of the transformation

$$J(S;\theta_{ij}(j>i=1,2,\ldots,n)) = \Pi_{i=1}^{p-2}\Pi_{j=i+2}^{p}(\cos^{j-i-1}\theta_{ij}). \tag{4.12}$$

If we have H_n matrix from H, then the Jacobian is

$$J(S_{11},S_{12};\theta_{ij}(j>i=1,2,\ldots,n \;\&\; j=2,\ldots,p)) = \Pi_{i=1}^{n}\Pi_{j=i+2}^{p}(\cos^{j-i-1}\theta_{ij}). \tag{4.13}$$

This gives the uniform distribution (4) in terms of angles as

$$\pi^{-\frac{1}{2}pn}\{\Gamma_n(p/2)\}\Pi_{i=1}^{n}\Pi_{j=i+1}^{p}\left[(\cos^{j-i-1}\theta_{ij})d\theta_{ij}\right] \tag{4.14}$$

for $-\pi/2 \le \theta_{ij} \le \pi/2$, $j>i=1,2,\ldots,n$ and $j=2,3,\ldots,p$. Notice that we have changed the range of $\theta_{j-1,j}$ from $(-\pi,\pi)$ to $(-\pi/2,\pi/2)$, on account of the assumption $|H_{ii}|>0$ for all $i=1,2,\ldots,n$.

Further, we note that if we define the orthogonal matrix by

$$H_0 = \Pi_{i=1}^{p-1}\Pi_{j=i+1}^{p}H_{ij}(\theta_{ij}), \tag{4.15}$$

then, we shall get the same Jacobian as mentioned in (4.11) and (4.12) even though the matrices H_0 and H are different.

4.5. Discussion

Green and Mardia (2006) have given some applications of the Fisher Matrix distribution in Bioinformatics. Bioinformatics applications have revived the subject of Directional Statistics; the previous innovations came mainly from Earth Science applications. Directional Statistics has also appeared in the new field of Shape Analysis (see Dryden and Mardia, 1998).

4.6. Acknowledgements

This paper is dedicated to Chunni Khatri with whom the above approach was developed and was used in Khatri and Mardia (1977) but was left out since it was thought to be too technical!

References

1. Downs, T. D. (1972). Orientation statistics. *Biometrika*, 59, 665–676.
2. Dryden, I.L., and Mardia, K.V. (1998). *Statistical Shape Analysis*, Wiley, New York.
3. Green, P.J. and Mardia, K.V. (2006). Bayesian alignment using hierarchical models, with applications in protein bioinformatics. *Biometrika*, 93, 235-254.
4. Khatri, C.G. and Mardia, K.V. (1977). The von Mises-Fisher Matrix distribution in orientation statistics. *J. Roy. Statist. Soc., Ser. B*, 39, 95-106.
5. Mardia, K.V. and Jupp, P.E. (2000). *Directional Statistics*. Wiley, Chichester.
6. Roy, S.N. (1957). *Some Aspects of Multivariate Analysis*. Wiley, New York.
7. Srivastava, M.S. and Khatri, C.G. (1979). *An Introduction to Multivariate Statistics*. North-Holland, New York.

Chapter 5

Cluster Validation for Microarray Data: An Appraisal

V. Pihur, G. N. Brock and S. Datta

Bioinformatics and Biostatistics, University of Louisville USA
E-mail: v0pihu01@louisville.edu
guy.brock@louisville.edu
susmita.datta@louisville.edu
somnath.datta@louisville.edu

Clustering is routinely used in the post genomic era, notably, in microarray studies. It is realized that some form of validation of the clustering results is essential. Numerous performance or validation measures have been proposed in the literature which bring out various aspects of the clustering procedures including their statistical properties, cluster quality, biological relevance of the produced clusters and so on.

This paper gives an up to date account of the cluster validation research as applied to microarray data. In particular, we review a number of such validation measures some of which are fairly classical and some are very new. We illustrate these measures with a numerical example involving a number of clustering procedures. We also provide an account of currently available software for cluster validation including our own R package called *clValid* [Brock et al, 2008]. The paper also has a brief discussion on ways to summarize and reconcile the findings using various validation procedures.

Contents

5.1. Introduction

Clustering could refer to any method that places objects into groups. It is often very exploratory in nature since the group membership information is not assumed to be known for any of the objects. Because of that, clustering is also known as *unsupervised learning* in the machine learning literature. Generally speaking, a clustering algorithm groups together objects that are "close" to one another in a multidimensional feature space, usually for the purpose of uncovering some inherent structure which the data possesses. In particular, clustering is commonly used in the analysis of high-throughput genomic data arising from the microarray or SAGE platforms, with the aim of grouping together genes or proteins which have similar expression patterns and possibly share common biological pathways [see refereces numbered 2,5,8,7]. A number of clustering algorithms shown some promise in the analysis of genomic data [see references numbered 7,15,21,31]. Ideally, the resulting clusters should not only have good statistical properties (compact, well-separated, connected, and stable), but also give results that make good biological sense.

A variety of measures aimed at validating the results of a clustering analysis and determining which clustering algorithm performs the best for a particular experiment been proposed [see references numbered 4,27,39]. This validation can be based solely on the internal properties of the data or on some external reference, that is, on the expression data alone or in conjunction with relevant biological information [see references numbered 3,6,16,17]. The article by [Handl et al., 2005], in particular, gives an excellent overview of cluster validation with post-genomic data and provides a synopsis of many of the available validation measures.

In this work we provide an up to date appraisal of the cluster validation research as applied to microarray data analysis. While we review a number of such validation measures (section 5.2), the list is not exhaustive. We illustrate these measures with a numerical example involving a number of clustering procedures (section 5.3). We also provide an account of available software for cluster validation that are currently available including our own R package called *clValid* (section 5.4). The article ends with a discussion section (section 5.5) which reflects on the difficulty in validating clustering results and presents a new research direction in this problem.

5.2. Validation Measures

The *clValid* package [Brock et al, 2008] offers three types of cluster validation, "internal", "stability", and "biological". Internal validation measures take only the dataset and the clustering partition as input and use intrinsic information in the data to assess the quality of the clustering. The stability measures are a special version of internal measures. They evaluate the consistency of a clustering result by comparing it with the clusters obtained after each column is removed, one at a time. Biological validation evaluates the ability of a clustering algorithm to produce biologically meaningful clusters. We have measures to investigate both the biological homogeneity and stability of the clustering results.

5.2.1. *Internal measures*

For internal validation, we selected measures that reflect the compactness, connectedness, and separation of the clustering partitions. Connectedness relates to what extent observations are placed in the same cluster as their nearest neighbors in the data space, and is here measured by the connectivity [Handl et al., 2005]. Compactness assesses cluster homogeneity, usually by looking at the intra-cluster variance, while separation quantifies the degree of separation between clusters (usually by measuring the distance between cluster centriods). Since compactness and separation demonstrate opposing trends, popular methods combine the two measures into a single score. The Dunn Index and Silhouette Width [Rousseeuw,1987] are both examples of non-linear combinations of the compactness and separation. The IGP (In-Group Proportion) incorporates the idea of predictive power. The details of each measure are given below, and for a good overview of internal measures in general see [Handl et al., 2005].

5.2.1.1. *Connectivity*

Let N denote the total number of observations to be clustered. Define $nn_{i(j)}$ as the jth nearest neighbor of observation i, and let $x_{i,nn_{i(j)}}$ be zero if i and j are in the same cluster and $1/j$ otherwise. Then, for a particular clustering partition $C = \{C_1, \ldots, C_K\}$ of the N observations into K disjoint clusters, the connectivity is defined as

$$Conn(C) = \sum_{i=1}^{N} \sum_{j=1}^{\ell} x_{i,nn_{i(j)}} ,$$

where ℓ is user-selectable. The connectivity has a value between zero and ∞ and should be minimized.

5.2.1.2. *Silhouette Width*

The Silhouette Width is the average of each observation's Silhouette value. The Silhouette value measures the degree of confidence in the clustering assignment of a particular observation, with well-clustered observations having values near 1 and poorly clustered observations having values near -1. For observation i, it is defined as

$$S(i) = \frac{b_i - a_i}{max(b_i, a_i)},$$

where a_i is the average distance between i and all other observations in the same cluster

$$a_i = \frac{1}{n(C(i)) - 1} \sum_{j \neq i \in C(i)} d(i, j),$$

where $C(i)$ is the cluster containing observation i, $d(i, j)$ is the distance (e.g. Euclidean, Manhattan) between observations i and j, $n(C)$ is the cardinality of cluster C, and b_i is the average distance between i and the observations in the "nearest neighboring cluster", i.e.

$$b_i = \min_{C_k \in C \setminus C(i)} \frac{1}{n(C_k)} \sum_{j \in C_k} d(i, j).$$

The Silhouette Width thus lies in the interval $[-1, 1]$, and should be maximized. For more information, see the help page for the *silhouette()* function in package *cluster*[Rouseeuw et al., 2006].

5.2.1.3. *Dunn Index*

The Dunn Index is the ratio of the smallest distance between observations not in the same cluster to the largest intra-cluster distance. It is computed as

$$D(C) = \frac{\min\limits_{C_k, C_l \in C, C_k \neq C_l} \left(\min\limits_{i \in C_k, j \in C_l} dist(i, j) \right)}{\max\limits_{C_m \in C} diam(C_m)},$$

where $diam(C_m)$ is the maximum distance between observations in cluster C_m.The Dunn Index has a value between zero and ∞, and should be maximized.

5.2.1.4. *In-Group Proportion (IGP)*

The In-Group Proportion (IGP) is the proportion of observations in a cluster whose nearest neighbors are in the same cluster. IGP captures the idea of prediction accuracy and quantifies the degree to which points close to each other are predicted to belong to the same cluster. It is computed as

$$IGP(C_k) = \frac{n(i : i \in C_k \text{ and } i^* \in C_k)}{n(C_k)},$$

where $n(A)$ is the cardinality of a set A and i^* is the nearest neighbor of i defined as $i^* = argmin_{x \neq i}\, d(i,x)$, in which d is a distance function. IGP scores, obviously, take values between 0 and 1 and larger scores correspond to a better predictive ability. The IGP was introduced recently by [Kapp et al., 2007]. We compute the overall IGP score for the partitioning as

$$IGP(C) = \sum_k IGP(C_k).$$

5.2.2. *Stability measures*

The stability measures compare the results from clustering based on the full data to clustering based on removing each column, one at a time. These measures work especially well if the data are highly correlated, which is often the case in high-throughput genomic data. The included measures are the average proportion of non-overlap (APN), the average distance (AD), the average distance between means (ADM), and the figure of merit (FOM) [see references numbered 4,39]. In all cases the average is taken over all the deleted columns, and all measures should be minimized.
Let N denote the total number of observations (rows) in a dataset and M denote the total number of columns, which are assumed to be numeric (e.g., a collection of samples, time points, etc.).

5.2.2.1. *Average Proportion of Non-overlap (APN)*

The APN measures the average proportion of observations not placed in the same cluster by clustering based on the full data and clustering based on the data with a single column removed. Let $C^{i,0}$ represent the cluster containing observation i using the original clustering (based on all available data), and $C^{i,\ell}$ represent the cluster containing observation i where the clustering is based on the dataset with column ℓ removed. Then, with the total number of clusters set to K, the APN measure is defined as

$$APN(C) = \frac{1}{MN}\sum_{i=1}^{N}\sum_{\ell=1}^{M}\left(1 - \frac{n(C^{i,\ell} \cap C^{i,0})}{n(C^{i,0})}\right).$$

The APN is in the interval $[0, 1]$, with values close to zero corresponding with highly consistent clustering results.

5.2.2.2. *Average Distance (AD)*

The AD measure computes the average distance between observations placed in the same cluster by clustering based on the full data and clustering based on the data with a single column removed. It is defined as

$$AD(C) = \frac{1}{MN}\sum_{i=1}^{N}\sum_{\ell=1}^{M}\frac{1}{n(C^{i,0})n(C^{i,\ell})}\left[\sum_{i \in C^{i,0}, j \in C^{i,\ell}} dist(i, j)\right].$$

The AD has a value between zero and ∞, and smaller values are preferred.

5.2.2.3. *Average Distance between Means (ADM)*

The ADM measure computes the average distance between cluster centers for observations placed in the same cluster by clustering based on the full data and clustering based on the data with a single column removed. It is defined as

$$ADM(C) = \frac{1}{MN}\sum_{i=1}^{N}\sum_{\ell=1}^{M} dist(\bar{x}_{C^{i,\ell}}, \bar{x}_{C^{i,0}}),$$

where $\bar{x}_{C^{i,0}}$ is the mean of the observations in the cluster which contain observation i, when clustering is based on the full data, and $\bar{x}_{C^{i,\ell}}$ is similarly defined. Currently, ADM only uses the Euclidean distance. It also has a value between zero and ∞, and again smaller values are preffered.

5.2.2.4. *Figure of Merit (FOM)*

The FOM measures the average intra-cluster variance of the observations in the deleted column, where the clustering is based on the remaining (undeleted) samples. This estimates the mean error using predictions based on the cluster averages. For a particular left-out column ℓ, the FOM is

$$FOM(\ell, C) = \sqrt{\frac{1}{N} \sum_{k=1}^{K} \sum_{i \in C_k(\ell)} dist(x_{i,\ell}, \bar{x}_{C_k(\ell)})},$$

where $x_{i,\ell}$ is the value for observation (gene) i in the ℓ-th column in cluster $C_k(\ell)$, and $\bar{x}_{C_k(\ell)}$ is the average of cluster $C_k(\ell)$. These are summed over the columns ℓ to obtain the overall FOM. Currently, the only distance available for FOM is Euclidean. The FOM is multiplied by an adjustment factor $\sqrt{\frac{N}{N-K}}$, to alleviate the tendency to decrease as the number of clusters increases. The final score is averaged over all the removed columns, and has a value between zero and ∞, with smaller values equalling better performance.

5.2.3. *Biological validation measures*

Biological validation evaluates the ability of a clustering algorithm to produce biologically meaningful clusters. A typical application of biological validation is in microarray data, where observations correspond to genes (where "genes" could be open reading frames (ORFs), express sequence tags (ESTs), serial analysis of gene expression (SAGE) tags, etc.). There are two measures available, the biological homogeneity index (BHI) and biological stability index (BSI), both originally presented in [Datta et al., 2006].

5.2.3.1. *Biological Homogeneity Index (BHI)*

As its name implies, the BHI measures how homogeneous the clusters are biologically. Let $\mathcal{B} = \{B_1, \ldots, B_F\}$ be a set of F functional classes, not necessarily disjoint, and let $B(i)$ be the functional class containing gene i (with possibly more than one functional class containing i). Similarly, we define $B(j)$ as the function class containing gene j, and assign the indicator function $I(B(i) = B(j))$ the value 1 if $B(i)$ and $B(j)$ match (any one match is sufficient in the case of membership to multiple functional classes), and 0 otherwise. Intuitively, we hope that genes placed in the same statistical cluster also belong to the same functional classes. Then, for a given statistical clustering partition $C = \{C_1, \ldots, C_K\}$ and set of biological classes $\mathcal{B}$, the BHI is defined as

$$BHI(C, \mathcal{B}) = \frac{1}{K} \sum_{k=1}^{K} \frac{1}{n_k(n_k - 1)} \sum_{i \neq j \in C_k} I(B(i) = B(j)).$$

Here $n_k = n(C_k \cap B)$ is the number of annotated genes in statistical cluster C_k. The BHI is in the range $[0, 1]$, with larger values corresponding to more biologically homogeneous clusters.

5.2.3.2. *Biological Stability Index (BSI)*

The BSI is similar to the other stability measures, and inspects the consistency of clustering for genes with similar biological functionality. Each sample is removed one at a time, and the cluster membership for genes with similar functional annotation is compared with the cluster membership using all available samples. The BSI is defined as

$$BSI(C,\mathcal{B}) = \frac{1}{F}\sum_{k=1}^{F}\frac{1}{n(B_k)(n(B_k)-1)M}\sum_{\ell=1}^{M}\sum_{i\neq j\in B_k}\frac{n(C^{i,0}\cap C^{j,\ell})}{n(C^{i,0})},$$

where F is the total number of functional classes, $C^{i,0}$ is the statistical cluster containing observation i based on all the data, and $C^{j,\ell}$ is the statistical cluster containing observation j when column ℓ is removed. The BSI is in the range $[0,1]$, with larger values corresponding to more stable clusters of the functionally annotated genes.

5.3. A Numerical Illustration

We are going to illustrate some of the validation measures using a well-known yeast sporulation data collected and analyzed by [Chu et al., 1998]. It records mRNA levels at seven different time points during the sporulation process in budding yeast. Only a subset consisting of 513 genes that were positively expressed is analyzed in this example. Functional class annotations are necessary to calculate the two biological validation measures, BHI and BSI. They were mined through the FunCat webtool available at `http://fatigo.bioinfo.cipf.es/`. 503 of the 513 genes were annotated into the following sixteen functional classes: metabolism (138), energy (27), cell cycle and DNA processing (152), transcription (50), protein synthesis (10), protein fate (72), protein with binding function or cofactor requirement (81), protein activity regulation (16), transport (63), cell communication (12), defense (36), interaction with environment (33), cell fate (17), development (13), biogenesis (77), and cell differentiation (82).

The R statistical computing project [R: Language,2006] has a wide variety of clustering algorithms available in the base distribution and various add-on packages. We make use of six algorithms from the base distribution and add-on packages *cluster* [see references numbered 25,35], *kohonen* [R Package,2006], and *mclust02* [see references numbered 15,16], and in addition include a function for implementing SOTA in the *clValid* package. A brief description of each clustering method and its availability is given below.

5.3.1. *UPGMA*

Unweighted Pair Group Method with Arithmetic Mean is probably the most frequently used clustering algorithm [Kaufman et al., 1990]. It is an agglomerative, hierarchical clustering algorithm that yields a dendrogram which can be cut at a chosen height to produce the desired number of clusters. Each observation is initially placed in its own cluster, and the clusters are successively joined together in order of their "closeness". The closeness of any two clusters is determined by a dissimilarity matrix, and can be based on a variety of agglomeration methods. UPGMA is included with the base distribution of R in function *hclust()*, and is also implemented in the *agnes()* function in package *cluster*.

5.3.2. *K-means*

K-means is an iterative method which minimizes the within-class sum of squares for a given number of clusters [Hartigan et al., 1979]. The algorithm starts with an initial guess for the cluster centers, and each observation is placed in the cluster to which it is closest. The cluster centers are then updated, and the entire process is repeated until the cluster centers no longer move. Often another clustering algorithm (e.g., UPGMA) is run initially to determine starting points for the cluster centers. K-means is implemented in the function *kmeans()*, included with the base distribution of R.

5.3.3. *Diana*

Diana is a divisive hierarchical algorithm that initially starts with all observations in a single cluster, and successively divides the clusters until each cluster contains a single observation. Along with SOTA, Diana is one of a few representatives of the divisive hierarchical approach to clustering. Diana is available in function *diana()* in package *cluster*.

5.3.4. *PAM*

Partitioning around medoids (PAM) is similar to K-means, but is considered more robust because it admits the use of other dissimilarities besides Euclidean distance. Like K-means, the number of clusters is fixed in advance, and an initial set of cluster centers is required to start the algorithm. PAM is available in the *cluster* package as function *pam()*.

5.3.5. *SOM*

Self-organizing maps [Kohonen, 1997] is an unsupervised learning technique that is popular among computational biologists and machine learning researchers. SOM is based on neural networks, and is highly regarded for its ability to map and visualize high-dimensional data in two dimensions. SOM is available as the *som()* function in package *kohonen*.

5.3.6. *Model based clustering*

Under this approach, a statistical model consisting of a finite mixture of Gaussian distributions in which each mixture component represents a cluster is fit to the data [Fraley et al., 2001]

$$P(x|\theta_K) = \sum_{k=1}^{K} \pi_k \Phi(x|\mu_k, \Sigma_k),$$

where π_k denotes the proportion of a mixture component k, $\Phi(\mu_k, \Sigma_k)$ denotes the normal distribution function with mean μ_k and variance-covariance matrix Σ_k, and θ_K denotes the set of unknown parameters $\{\pi_i, \mu_i, \Sigma_i : i = 1, \ldots, K\}$.
The mixture components and group memberships are estimated using maximum likelihood (EM algorithm). The Bayesian Information Criterion (BIC) is used to compare different models, not necessarily nested, with different number of mixture components and/or underlying component densities, thus, solving the problem of selecting the number of clusters in the data. The function *Mclust()* in package *mclust02* implements model based clustering.

5.3.7. *Bayesian clustering*

Bayesian techniques have been widely applied to cluster analysis and numerous clustering algorithms based on them exist in the literature. In this section, we will briefly describe two of the Bayesian clustering methods, namely Bayesian Hierarchical Clustering (BHC) and Bayesian Clustering by Dynamics (BCD) .
The BHC approach developed by [Heller et al., 2005] implements a similar strategy to the UPGMA method discussed above in a sense that it is a 1-pass bottom-up technique which constructs a binary tree structure (a dendrogram). Important differences between the methods, however, exist regarding the ways in which methods handle merging of the observations (or sets of observations). As opposed to

using an arbitrary distance measure of the UPGMA and merging the closest sets of observations as judged by the chosen distance measure, the BHC method imposes a probabilistic model on the data and uses the established theory of Bayesian hypothesis testing for agglomeration purposes. The merging decision is based on testing the hypothesis that all the observations in the merge under investigation come from a single mixture component.

[Ramoni et al. 2001] proposed the Bayesian Clustering by Dynamics method designed specifically for the time series data. Microarray data with temporal gene profiles would be an example of such data in the bioinformatics context. The BHD makes an assumption that the observed data comes from a set of underlying stochastic processes. It captures these processes by the Markov Chains (MC's) constructed from the time series data and represents them in the form of transition matrices. Thus, each observation has a corresponding transition matrix (and thus MC) associated with it. Then, the problem of clustering the obtained MC's is handled within the framework of Bayesian model selection. The scoring metric used to select the best partitioning is based on the posterior probability of the partitioning given the data. To help guide the search for a partitioning with the maximum posterior probability, a heuristic algorithm is used which merges the closest MC's as judged by the symmetrized Kullback-Liebler distance between the corresponding elements of the transition matrices. The resulting clusters represent the unobserved underlying processes that generated the data.

5.3.8. *SOTA*

Self-organizing tree algorithm (SOTA) is an unsupervised network with a divisive hierarchical binary tree structure. It was originally proposed by [Dopazo et al.,1997] for phylogenetic reconstruction, and later applied to cluster microarray gene expression data in [Herrer0 et al., 2001]. It uses a fast algorithm and hence is suitable for clustering a large number of objects. SOTA is included with the *clValid* package as function *sota()*.

5.3.9. *Cluster validation results*

First, we cluster the yeast sporulation data using the above seven clustering algorithms (excluding the BHC and BCD clustering algorithms), four of which are used with both Euclidean and correlation-based dissimilarities. Standardized validation scores of the five validation measures selected from different categories, ADM, BHI, FOM, Dunn Index, and Silhouette Width, are plotted in Figure 5.1. Since the true number of clusters is unknown, we vary the number of clusters from

6 to 12.

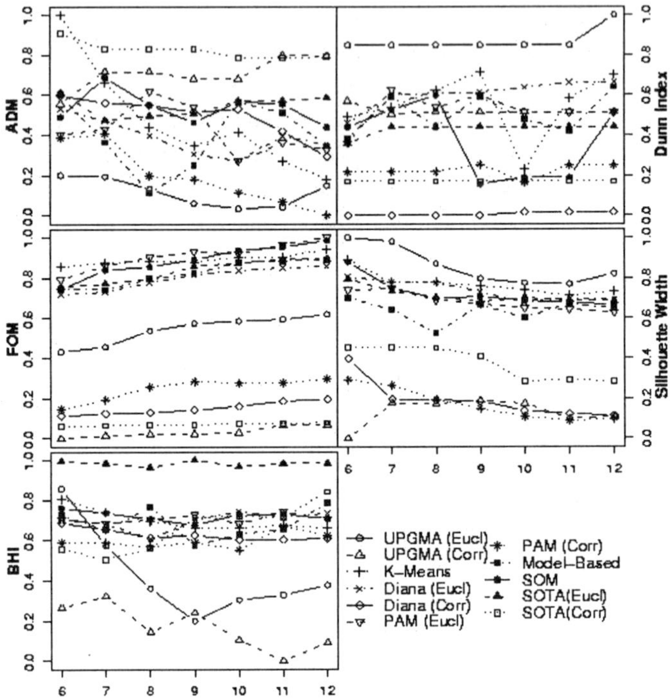

Fig. 5.1. Standardized cluster validation scores for five cluster validation measures across a range of different number of clusters. Scores closer to 1 indicate a better performance under the measure.

Looking at each measure separately, we can see that UPGMA with the Euclidean distance performs best as judged by the two internal measures: the Dunn Index and the Silhouette Width. However, its performance is poor under the BHI and the ADM measures. SOTA is performing well under both the ADM and the BHI, but is not doing great under the internal measures. No clear winner exists for the FOM validation measure. As one can see, performing cluster validation with multiple validation measures does not necessarily leave us with a complete picture. Based on these five measures, it is quite difficult to determine the overall winner for this dataset. Nevertheless, important information is captured by each individual measure and, if utilized properly, this information can lead to an improved clustering analysis.

5.4. Software for cluster validation

5.4.1. *The clValid package*

We have developed an R package, *clValid* [Brock et al, 2008], which contains measures for validating the results from a clustering procedure. We categorize the measures into three distinct types, "internal", "stability", and "biological", and provide plot, summary, and additional methods for viewing and summarizing the validation scores and extracting the clustering results for further analysis. In addition to the object-oriented nature of the language, implementing the validation measures within the R statistical programming framework provides the additional advantage in that it can interface with numerous clustering algorithms in existing R packages, and accommodate further algorithms as they are developed and coded into R libraries. Currently, *clValid()* accepts up to nine different clustering methods. This permits the user to simultaneously vary the number of clusters and the clustering algorithms to decide how best to group the observations in her/his dataset.

5.4.2. *Other packages*

There are several R packages that also perform cluster validation and are available from `http://www.r-project.org` (CRAN) or `http://www.bioconductor.org` (Bioconductor) . Examples include the *clustIndex()* function in package *cclust* [2006], which performs 14 different validation measures in three classes, *cluster.stats()* and *clusterboot()* in package *fpc* [Hennig, 2001], the *clusterRepro* [Kapp et al., 2006] and *clusterSim* [Walesiak et al., 2007] packages, and the *clusterStab* [MacDonald et al., 2006] package from Bioconductor. The *cl_validity()* function in package *clue* [Hornik, 2005] does validation for both paritioning methods ("dissimilarity accounted for") and hierarchical methods ("variance accounted for"), and function *fclustIndex()* in package *e1071* [Dimitriadou et al., 2006] has several fuzzy cluster validation measures. However, to our knowledge none of these packages offers biological validation or the unique stability measures which we present here. [Handl et al., 2005] provides C++ code for the validation measures which they discuss, and the *Caat* tool available in the GEPAS software suite (`http://gepas.bioinfo.cipf.es/`) offers a web-based interface for visualizing and validating (using the Silhouette Width) cluster results. However, neither of these two methods are as flexible for interfacing with the variety of clustering algorithms that are available in the R language. Hence, the *clValid* package is a valuable addition to the growing collection of cluster validation software which is available for researchers.

5.5. Discussion

Validation of clustering results is a challenging task. Over the years, numerous cluster validation measures have been proposed in the literature. We reviewed some of the conventional validation measures along with a few recently developed ones. As seen from the numerical illustration, different validation measures often produce contradicting results making it very difficult to determine the winning algorithm. Very Recently, [Pihur et al. 2007] developed a Monte Carlo rank aggregation algorithm for reconciling the outputs from several validation measures which produces a combined ranking of clustering algorithms based on the individual measure rankings. The weighted rank aggregation procedure greatly simplifies the tedious and subjective visual inspection of cluster validation scores and allows for a far more objective assessment of the cluster quality.

Numerous software packages have been developed which implement the whole array of cluster validation techniques. Our *clValid* package complements the existing software by including recently developed stability and biological validation indexes that may be of interest to researchers analyzing high throughput data.

References

1. Al-Shahrour, F. and Diaz-Uriarte, R. and Dopazo, J.,(2004) *FatiGO: a web tool for finding significant associations of Gene Ontology terms with groups of genes*,Bioinformatics,20,4,578-80.
2. Bhattacherjee, V. and Mukhopadhyay, P. and Singh, S. and Johnson, C. and Philipose, J. T. and Warner, C. P. and Greene, R. M. and Pisano, M. M.,(2007)*Neural crest and mesoderm lineage-dependent gene expression in orofacial development*,Differentiation.
3. Bolshakova, N. and Azuaje, F. and Cunningham, P., (2005)*A knowledge-driven approach to cluster validity assessment*,Bioinformatics,21,10,2546-7.
4. Brock, G. and Pihur, V. and Datta, S. and Datta,S., clValid: Validation of Clustering Results, 2008, R-package version 0.5-7, `http://cran.r-project.org/web/packages/clValid/index.html`.
5. Brock, G., Pihur, V., Datta, S. and Datta, S. clValid , an R package for cluster validation. Journal of Statistical Software, 25, 4, 2008, `http://www.jstatsoft.org/v25/i04`.
6. Chu, S. and DeRisi, J. and Eisen, M. and Mulholland, J. and Botstein, D. and Brown, P. O. and Herskowitz, I.,(1998) *The transcriptional program of sporulation in budding yeast*, Science,282,5389,699-705.
7. Datta, S. and Datta, S.,(2003)*Comparisons and validation of statistical clustering techniques for microarray gene expression data*, Bioinformatics,19,4,459-66.
8. Datta, S. and Datta, S.,(2006)*Methods for evaluating clustering algorithms for gene expression data using a reference set of functional classes*,BMC Bioinformatics,7,397.

9. Dembele, D. and Kastner, P.(2003) *Fuzzy C-means method for clustering microarray data*, Bioinformatics,19,8,973-80.
10. DeRisi, J. L. and Iyer, V. R. and Brown, P. O.,(1997) *Exploring the metabolic and genetic control of gene expression on a genomic scale*,Science,278,5338, 680-6.
11. Dimitriadou, E., Convex Clustering Methods and Clustering Indexes, ,2006,R package version 0.6-13,http://www.R-project.org
12. Dopazo, J. and Carazo, J. M.,(1997) *Phylogenetic reconstruction using a growing neural network that adopts the topology of a phylogenetic tree*,Journal of Molecular Evolution,44,226-233.
13. Eisen, M. B. and Spellman, P. T. and Brown, P. O. and Botstein, D.,(1998)*Cluster analysis and display of genome-wide expression patterns*, Proc Natl Acad Sci U S A,95, 25,14863-8.
14. Fraley, C. and Raftery, A. E.,(2001) *Model-based clustering, discriminant analysis, and density estimation*, Journal of the American Statistical Association, 17,126-136.
15. Fraley, C. and A. E. Raftery, A. E.,(2003)*Enhanced model-based clustering, density estimation, and discriminant analysis software: MCLUST*,Journal of Classification,20,2,263-286.
16. Fraley, C. and Raftery, A. E.,2007,*Model-Based Clustering / Normal Mixture Modeling*, R package version 3.1-1,http://www.stat.washington.edu/mclust
17. Fu, L. and Medico, E., *FLAME, a novel fuzzy clustering method for the analysis of DNA microarray data*, BMC Bioinformatics, 8,3.
18. Gat-Viks, I. and Sharan, R. and Shamir, R.,(2003) *Scoring clustering solutions by their biological relevance*, Bioinformatics, 19,18,2381-9.
19. Gibbons, F. D. and Roth, F. P.,(2002) *Judging the quality of gene expression-based clustering methods using gene annotation*, Genome Res,12,10,1574-81.
20. Handl, J. and Knowles, J. and Kell, D. B.,(2005)*Computational cluster validation in post-genomic data analysis*, Bioinformatics,21,15,3201-12.
21. Hartigan, J. A. and Wong, M. A.,(1979)*A k-means clustering algorithm*, Applied Statistics,28,100-108.
22. Hennig, C., fpc: Fixed point clusters, clusterwise regression and discriminant plots, 2006, R package version 1.2-2, http://www.R-project.org
23. Herrero, J. and Valencia, A. and Dopazo, J.,(2001) *A hierarchical unsupervised growing neural network for clustering gene expression patterns*, Bioinformatics, 17,2,126-36.
24. Hornik, K.,A CLUE for CLUster Ensembles, *Journal of Statistical Software*,14,12,September 2005b,http://www.jstatsoft.org/v14/i12/
25. Kapp, A. and Tibshirani, R.,*Reproducibility of gene expression clusters*, 2006, R package version 0.5-1,http://www.R-project.org.
26. Kapp, A. and Tibshirani, R.,(2007) *Are clusters found in one dataset present in another dataset?*, Biostatistics,8,1,9-31.
27. Kaufman, L. and Rousseeuw, P. J., (1990)*Finding Groups in Data. An Introduction to Cluster Analysis*,Wiley,New York.
28. Katherine A. Heller and Zoubin Ghahramani,(2005)*Bayesian hierarchical clustering*, ICML '05: Proceedings of the 22nd international conference on Machine learning, 297–304.
29. Kerr, M. K. and Churchill, G. A.,(2001) *Bootstrapping cluster analysis: assessing the*

reliability of conclusions from microarray experiments, Proc Natl Acad Sci U S A, 98,16,8961-5.

30. Kohonen, T.,(1997)*Self-Organizing Maps*,Second, Springer-Verlag.
31. MacDonald, J.W. and Ghosh, D. and Smolkin, M.,*Compute cluster stability scores for microarray data*, 2006, R package version 1.6.0,http://www.R-project.org
32. Marco Ramoni and Paola Sebastiani and Paul Cohen,(2002)*Bayesian Clustering by Dynamics*, Mach. Learn.,47,1,0885-6125,91–121.
33. McLachlan, G. J. and Bean, R. W. and Peel, D.,(2002)*A mixture model-based approach to the clustering of microarray expression data*,Bioinformatics,18,3, 413-22.
34. Pihur, V. and Datta, S. and Datta, S., (2007) *Weighted rank aggregation of cluster validation measures: A Monte Carlo cross-entropy approach*, Bioinformatics,23,1607-1615.
35. Robert C Gentleman and Vincent J. Carey and Douglas M. Bates and Ben Bolstad and Marcel Dettling and Sandrine Dudoit and Byron Ellis and Laurent Gautier and Yongchao Ge and Jeff Gentry and Kurt Hornik and Torsten Hothorn and Wolfgang Huber and Stefano Iacus and Rafael Irizarry and Friedrich Leisch Cheng Li and Martin Maechler and Anthony J. Rossini and Gunther Sawitzki and Colin Smith and Gordon Smyth and Luke Tierney and Jean Y. H. Yang and Jianhua Zhang,*Bioconductor: Open software development for computational biology and bioinformatics*, Genome Biology,5,2004,R80,http://genomebiology.com/2004/5/10/R80.
36. Rousseeuw, P. J.,(1987) *Silhouettes: a graphical aid to the interpretation and validation of cluster analysis*,Journal of Computational and Applied Mathematics, 20,53-65.
37. Rousseeuw, P. and Struyf, A. and Hubert, M. and Maechler, M.,*Cluster Analysis Extended Rousseeuw et al.*, 2006,R package version 1.11.4,http://www.R-project.org.
38. Ruepp, A. and Zollner, A. and Maier, D. and Albermann, K. and Hani, J. and Mokrejs, M. and Tetko, I. and Guldener, U. and Mannhaupt, G. and Munsterkotter, M. and Mewes, H. W.,(2004)*The FunCat, a functional annotation scheme for systematic classification of proteins from whole genomes*,Nucleic Acids Res,32,18,5539-45.
39. Walesiak, M. and Dudek, A., *Searching for optimal clustering procedure for a data set*, 2007, R package version 0.30-7,http://www.R-project.org
40. Wien TU, Dimitriadou, E. and Hornik, K. and Leisch, F. and Meyer, D. and Weingessel, A.*Misc Functions of the Department of Statistics (e1071)* ,2006,R package version 1.5-16,http://www.R-project.org
41. Yeung, K. Y. and Haynor, D. R. and Ruzzo, W. L., (2001) *Validating clustering for gene expression data*, Bioinformatics, 17, 4,309-18.
42. R: A Language and Environment for Statistical Computing, R Development Core Team, R Foundation for Statistical Computing, Vienna, Austria, 2006, ISBN 3-900051-07-0,http://www.R-project.org
43. *Supervised and unsupervised self-organising maps*,Wehrens, Ron,2006,R package version 1.1.1,http://www.R-project.org

Chapter 6

Flexible Bivariate Circular Models

Barry C. Arnold[1] and Ashis SenGupta[2]

[1] *University of California, Riverside, USA. E-mail:barry.arnold@ucr.edu*
[2] *Indian Statistical Institute, Kolkata, India. E-mail:ashis@isical.ac.in*

Observations on related random directions are frequently encountered. In this paper, we first survey some of the models that have been proposed for bivariate directional data. Typically, bivariate versions of the circular normal (or von Mises) distribution have been used in such situations. Next, a general bivariate von Mises distribution is considered together with a proposed flexible alternative model based on generalized cardioid distributional components. Multivariate extensions are briefly discussed.

Contents

6.1. Introduction

A classical directional data set is one which deals with observations on the directions of flight of released pigeons. If we let $\Theta_1^{(i)}$ be the random direction taken by the $i'th$ pigeon on day 1 and denote by $\Theta_2^{(i)}$ the direction taken by the same pigeon on day 2, then the set of observations $\underline{\Theta}^{(i)} = (\Theta_1^{(i)}, \Theta_2^{(i)})$ can be considered to be a sample of size n from some bivariate distribution with support $(0, 2\pi)^2$. Our focus is on models suitable for the analysis of such directional data sets. Needless to say, data of this nature arises in many other settings. Mardia (1972) and Jammalamadaka and SenGupta (2001) may be consulted for listings of a great variety of situations in which univariate and multivariate directional data can be encountered and SenGupta (2004) describes some unified approaches to deriving probability

models for such data.
We begin with a brief survey of the most popular univariate directional models. We will include a quite general univariate directional model called the generalized cardioid distribution. It and the classical von Mises circular model (and its extensions) will be utilized in subsequent sections as building blocks for the construction of bivariate (and eventually multivariate) models using the concept of conditional specification.

6.2. One Dimensional Circular Models

To model random directions measured in radians, we require distributions with support $[0,2\pi)$. We will only consider absolutely continuous models and we will impose the quite natural restriction that, if $f_\Theta(\theta)$ is the density of a random direction Θ, it will satisfy

$$\lim_{\theta\to 2\pi} f_\Theta(\theta) = f_\Theta(0) \tag{6.1}$$

This seems appropriate since the direction 0 and the direction 2π are indistinguishable.
The most popular univariate circular model is the von Mises (or circular normal) distribution. Its density is of the form

$$f_\Theta(\theta;\mu,\kappa) \propto \exp[\kappa(\theta-\mu)]I(0 \le \theta < 2\pi) \tag{6.2}$$

where $\mu \in [0,2\pi)$ corresponds to the mode of the density and $\kappa \in (0,\infty)$ is a precision parameter. If a random direction Θ has (6.2) as its density, we will write $\Theta \sim VM(\mu,\kappa)$. There is an alternative parameterization available that has certain computational advantages, though the new parameters lack the attractive interpretations available for (μ,κ). The family of densities (6.2) is equivalent to the following model:

$$f_\Theta(\theta;\alpha,\beta) \propto \exp[\alpha\cos\theta+\beta\sin\theta]I(0 \le \theta < 2\pi) \tag{6.3}$$

where $\alpha,\beta \in (-\infty,\infty)$. The obvious advantage of the representation (6.3) is that it makes it transparent that the von Mises distributions form a two parameter exponential family. Consequenty a sample $\Theta_1,\Theta_2,\ldots,\Theta_n$ of size n from from the density (6.3) (or equivalently, from (6.2)) will have a complete minimal sufficient statistic given by

$$\left(\sum_{i=1}^{n} \cos\Theta_i, \sum_{i=1}^{n} \sin\Theta_i\right). \tag{6.4}$$

The fact that a simple sufficient statistic exists for von Mises samples, goes a long way toward explaining the popularity of this model. A representative von Mises density is displayed in Figure 6.1.

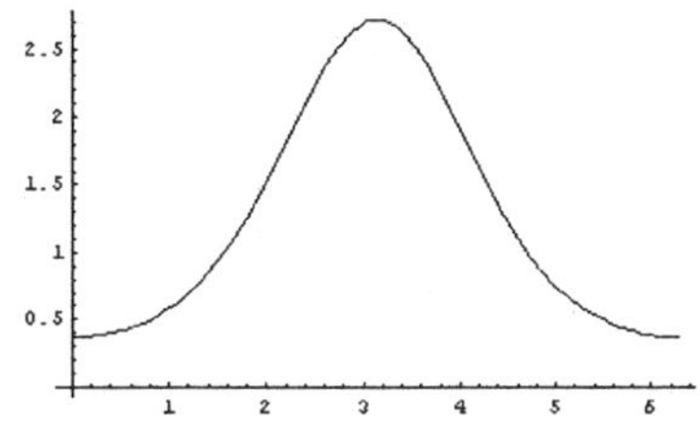

Fig. 6.1. Von Mises Density for $\mu = \pi, \kappa = 1$.

A second source of models for directional data involves "wrapping" a real variable or equivalently evaluating that variable *modulo* 2π to concoct a variable with support $[0, 2\pi)$. To this end, we begin with a random variable X with support $(-\infty, \infty)$ and define the corresponding "wrapped" version of X by

$$\Theta = X \text{ mod } (2\pi). \tag{6.5}$$

It is not clear, in general, why such wrapped models might be appropriate for directional data. However, this is clearly a rich source of models since the distribution of X can be quite arbitrary.

A popular choice for the distribution of the random variable to be "wrapped" is a symmetric stable distribution with characteristic function

$$\varphi_X(t) = \exp[i\mu t - \tau^\alpha |t|^\alpha] \tag{6.6}$$

where $-\infty < \mu < \infty$ and $\tau > 0$. It may be verified that the density of the wrapped version of X with density (6.6) is of the form

$$f_\Theta(\theta; \alpha, \mu, \rho) = \frac{1}{2\pi} + \frac{1}{\pi} \sum_{j=1}^{\infty} \rho^{j^\alpha} \cos[j(\theta - \mu)] \tag{6.7}$$

where $\rho = \exp[-\tau^\alpha]$. The case $\alpha = 2$, corresponding to a wrapped normal variable has received considerable attention. However setting $\alpha = 2$ in (6.7) does not result in any simplification. In contrast, when $\alpha = 1$, corresponding to a wrapped Cauchy variable, the density simplifies serendipitously to the form:

$$f_\Theta(\theta; \alpha, \beta) \propto (1 + \alpha \cos\theta + \beta \sin\theta)^{-1} I(0 \leq \theta < 2\pi) \tag{6.8}$$

where α, β are real valued parameters satisfying $\alpha^2 + \beta^2 < 1$. Note that the choice $\alpha = \beta = 0$ yields the "straw man" model corresponding to a uniform distribution on $[0, 2\pi)$. The von Mises distribution includes the uniform distribution only as a limiting case, as $\kappa \to 0$ in (6.2). A representative wrapped Cauchy density is displayed in Figure 6.2. There are other univariate circular models that have been

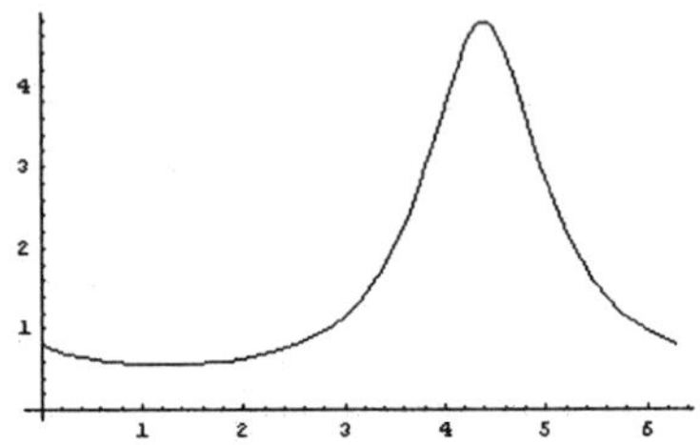

Fig. 6.2. Wrapped Cauchy Density for $\alpha = 0.25, \beta = 0.75$

discussed, but the lion's share of modelling papers seem to have focussed on the von Mises model and the wrapped Cauchy model. Though the wrapped normal model, despite its unattractive density ((6.7) with $\alpha = 2$) has also received a fair amount of attention. The attraction of models involving the classical normal distribution can never be discounted.

There is one other distribution which is sometimes considered for univariate directional data. The cardioid distribution has density of the form

$$f_\Theta(\theta; \alpha, \beta) \propto [1 + \alpha \cos\theta + \beta \sin\theta] I(0 \leq \theta < 2\pi) \tag{6.9}$$

in which α, β are real valued parameters satisfying $\alpha^2 + \beta^2 < 1$. A representative cardiod density is displayed in Figure 6.3. Comparison of (6.8) and (6.9) immediately suggests consideration of what we will call a generalized cardioid density, given by

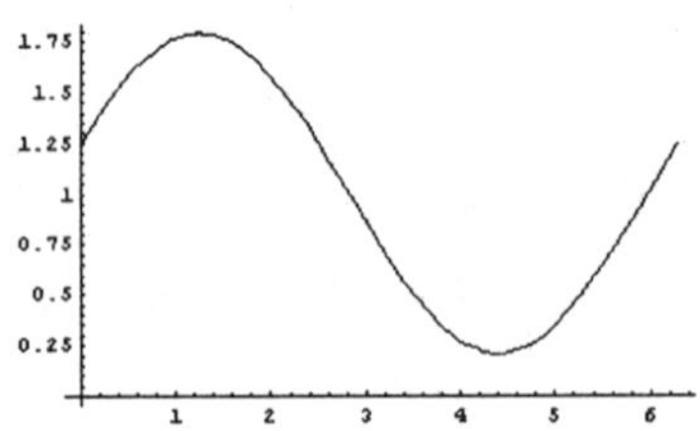

Fig. 6.3. Cardioid Density for $\alpha = .25, \beta = .75$

$$f_\Theta(\theta;\alpha,\beta) \propto [1+\alpha\cos\theta+\beta\sin\theta]^\gamma I(0 \le \theta < 2\pi) \qquad (6.10)$$

in which α, β, γ are real valued parameters satisfying $\alpha^2+\beta^2 < 1$. The introduction of the third parameter γ provides us with a flexible family of models which includes the wrapped Cauchy, the uniform and the cardioid distributions as special cases corresponding to $\gamma = -1, 0$ and 1 respectively. Some representative generalized cardioid densities are displayed in Figures 6.4-6.6. Such densities are unimodal. To a certain extent, γ controls the precision of the density.

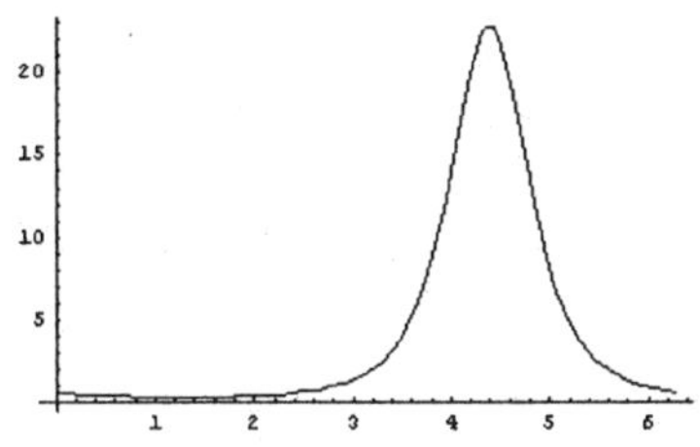

Fig. 6.4. Generalized Cardioid Density for $\alpha = 0.25, \beta = 0.75$ and $\gamma = -2$ respectively

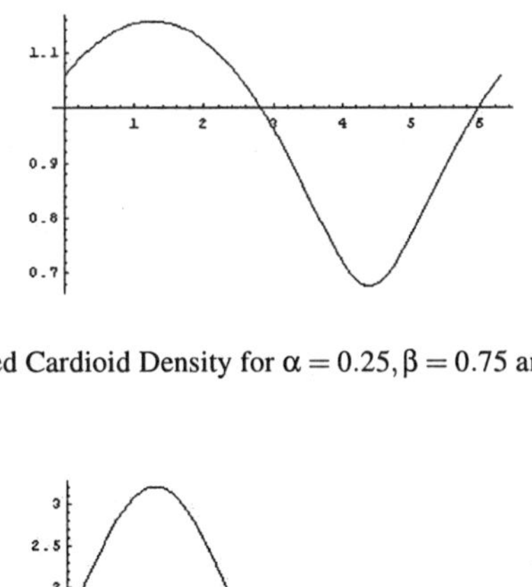

Fig. 6.5. Generalized Cardioid Density for $\alpha = 0.25, \beta = 0.75$ and $\gamma = .25$ respectively

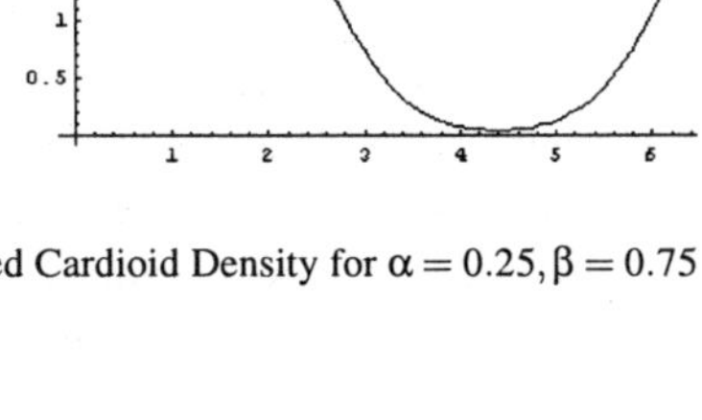

Fig. 6.6. Generalized Cardioid Density for $\alpha = 0.25, \beta = 0.75$ and $\gamma = 2$ respectively

In subsequent Sections , because of the attractive simplicity of their densities, the von Mises and the generalized cardioid distributions will be utilized in the development of bivariate and multivariate directional models.

NOTE: The family of densities (6.10), which we call generalized cardioid densities, coincides with the family of symmetric distributions on the circle introduced by Jones and Pewsey (2005). Jones and Pewsey provide an expression for the required normalizing constant in (6.10) in terms of a Legendre function of the first kind.

6.3. To Higher Dimensions Via Conditional Specification

Consider distributions for (Θ_1, Θ_2) such that all conditional densities of Θ_1 given $\Theta_2 = \theta_2$ belong to a given parametric family, say $g_1(\theta_1; \underline{\eta}(\theta_2))$, and all conditional densities of Θ_2 given $\Theta_1 = \theta_1$ belong to a second (possibly different) given parametric family, say $g_2(\theta_2; \underline{\tau}(\theta_1))$. Such a joint distribution is said to be conditionally specified. Classic examples are those in which $g_1(\theta_1; \underline{\eta}(\theta_2))$ and $g_2(\theta_2; \underline{\tau}(\theta_1))$ are exponential families (Arnold and Strauss (1991)).
In particular, let us consider bivariate densities with von Mises conditionals. We will use the parameterization of equation (6.3) for the von Mises density, i.e.

$$f_\Theta(\theta; \alpha, \beta) \propto \exp[\alpha \cos\theta + \beta \sin\theta] I(0 \le \theta < 2\pi). \tag{6.11}$$

We wish to identify all bivariate densities for (Θ_1, Θ_2) which have all of their conditional densities in the family (6.11). Using the result of Arnold and Strauss (1991), the class of such densities is itself an exponential family and is given by

$$f_{\Theta_1,\Theta_2}(\theta_1, \theta_2) = \exp\{(1, \cos\theta_1, \sin\theta_1) A (1, \cos\theta_2, \sin\theta_2)'\} \tag{6.12}$$

where A is a matrix of real parameters i.e.

$$A = \begin{pmatrix} a_{00} & a_{01} & a_{02} \\ a_{10} & a_{11} & a_{12} \\ a_{20} & a_{21} & a_{22} \end{pmatrix} \tag{6.13}$$

Actually there are only 8 parameters. a_{00} is a function of the other a_{ij}'s chosen to make the density integrate to 1. It is easy to see that the model (6.12) does indeed have von Mises conditionals. The model (6.12) includes most of the popular bivariate von Mises distributions as special cases. Note that it is only in the case of independence that it will have von Mises marginals. The model (6.12) may be found , with different notation in Mardia and Patrangenaru(2005), however it was not derived there using conditional specification and it was not observed to correspond to the class of all distributions with von Mises conditionals. Some representative densities corresponding to the model (6.12) are displayed in Figures 6.7-6.10.

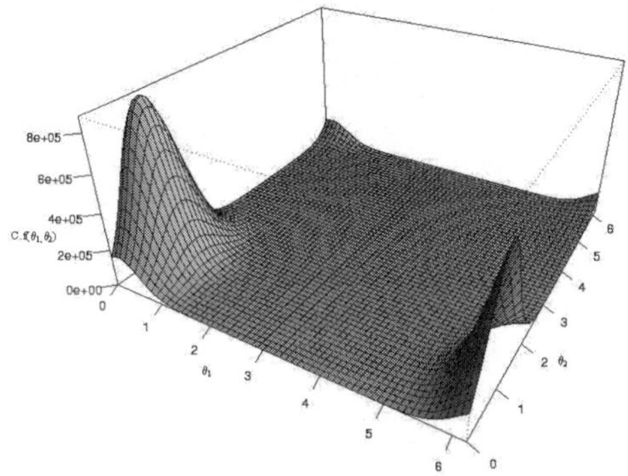

Fig. 6.7. Bivariate Generalized Cardioid conditional density for $a_{00} = 4, a_{01} = 2, a_{02} = 3, a_{10} = 6, a_{11} = 0, a_{12} = 0, a_{20} = 1, a_{21} = 0, a_{22} = 0$

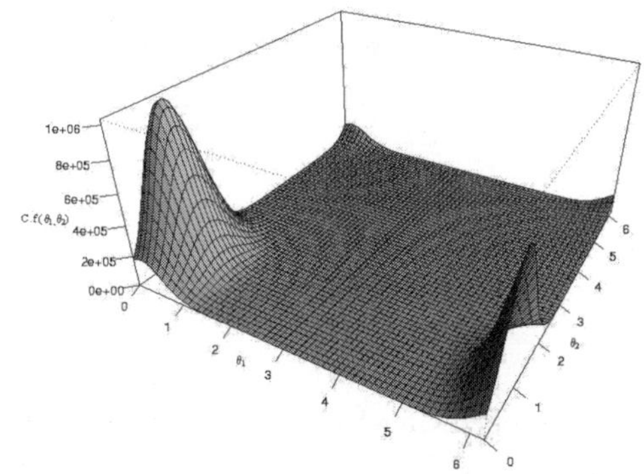

Fig. 6.8. Bivariate Generalized Cardioid conditional density for $a_{00} = 4, a_{01} = 2, a_{02} = 3, a_{10} = 6, a_{11} = 0.1, a_{12} = 0.1, a_{20} = 1, a_{21} = 0.1, a_{22} = 0.1$

In a similar fashion, we might seek all bivariate densities that have wrapped Cauchy or cardioid conditionals. Since we are not dealing with conditionals in exponential families, we cannot make use of the Arnold and Strauss (1991) theorem. Indeed it does not seem to be possible to completely characterize the class of all

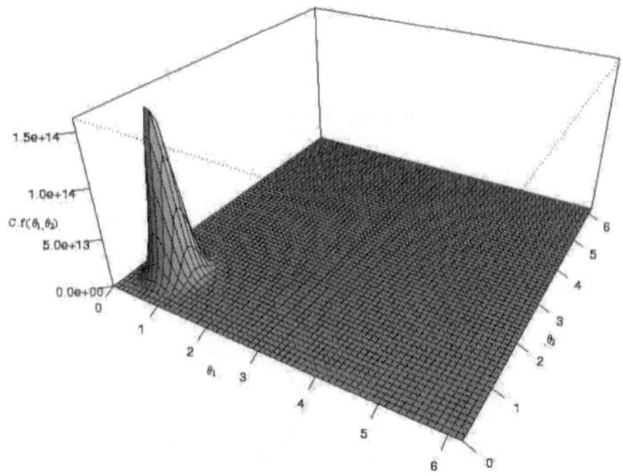

Fig. 6.9. Bivariate Generalized Cardioid conditional density for $a_{00} = 4, a_{01} = 2, a_{02} = 3, a_{10} = 6, a_{11} = 10, a_{12} = 10, a_{20} = 1, a_{21} = 10, a_{22} = 10$

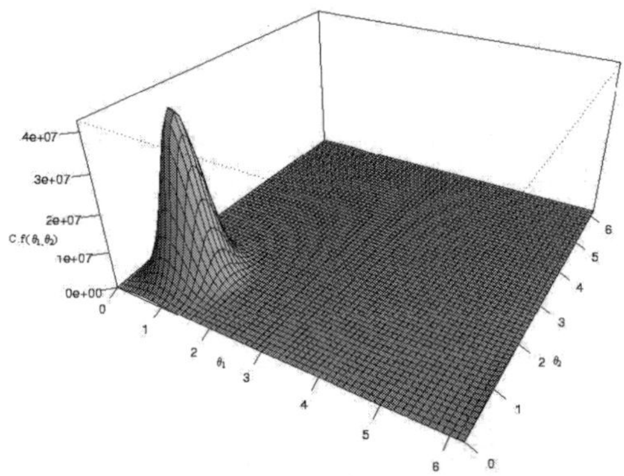

Fig. 6.10. Bivariate Generalized Cardioid conditional density for $a_{00} = 4, a_{01} = 2, a_{02} = 3, a_{10} = 6, a_{11} = 0.1, a_{12} = 0.8, a_{20} = 1, a_{21} = 3.5, a_{22} = 5.3$

bivariate densities with wrapped Cauchy conditionals. What we can and will do is identify a flexible model with the desired conditional characteristics. There is however no reason to not deal with the more general case of models with generalized cardioid conditionals, since such models will of course include the wrapped Cauchy and cardioid cases.
Recall that the generalized cardioid density is of the form

$$f_\Theta(\theta;\alpha,\beta) \propto [1+\alpha\cos\theta+\beta\sin\theta]^\gamma I(0 \leq \theta < 2\pi) \tag{6.14}$$

We wish to construct bivariate densities with all of their conditionals in this family. A rich family of such densities is given by

$$f_{\Theta_1,\Theta_2}(\theta_1,\theta_2) \propto [1+(1,\cos\theta_1,\sin\theta_1)A(1,\cos\theta_2,\sin\theta_2)']^{\gamma} \qquad (6.15)$$

It is not clear whether the family (6.15) includes all densities with conditionals in the family (6.14) for a given fixed value of γ, but it is a very extensive class and it does include cases in which Θ_1 and Θ_2 are independent. The entries in the parameter matrix A must be constrained to ensure that the expression (6.15) is non-negative over the region $(0,2\pi)^2$, but except for that, they may be quite arbitrary. The case $\gamma=-1$ corresponds to wrapped Cauchy conditionals, while $\gamma=1$ yields cardiod conditionals. Some representative densities with generalized cardiod conditionals (i.e. of the form (6.15)) are displayed in Figures 6.11-6.14.

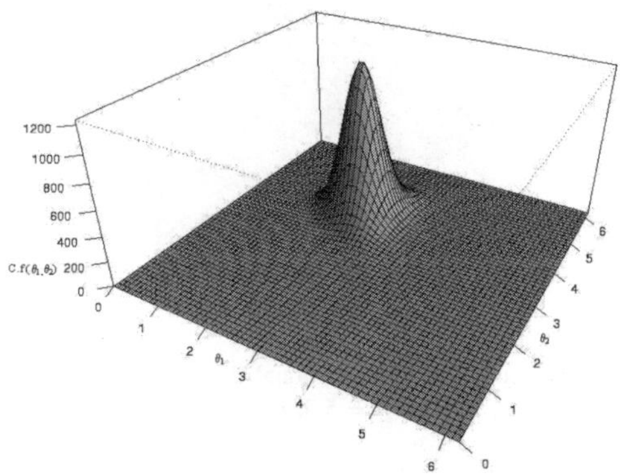

Fig. 6.11. Bivariate Generalized Cardioid conditional density for $a_{00}=4, a_{01}=2, a_{02}=3, a_{10}=6, a_{11}=0.1, a_{12}=0.8, a_{20}=1, a_{21}=3.5, a_{22}=5.3, \gamma=-1$

6.4. Inference

The bivariate von Mises conditionals density (6.12) is a regular 8 parameter exponential family. The minimal sufficient statistics based on a sample of size n, $\{(\Theta_1^{(i)},\Theta_2^{(i)}\}_{i=1}^n$, are:

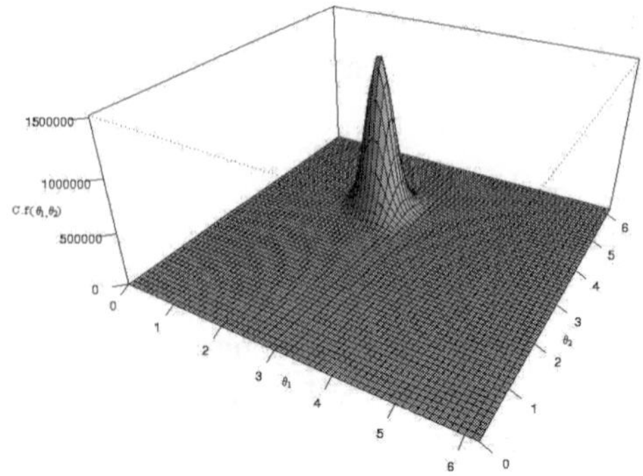

Fig. 6.12. Bivariate Generalized Cardioid conditional density for $a_{00} = 4, a_{01} = 2, a_{02} = 3, a_{10} = 6, a_{11} = 0.1, a_{12} = 0.8, a_{20} = 1, a_{21} = 3.5, a_{22} = 5.3, \gamma = -2$

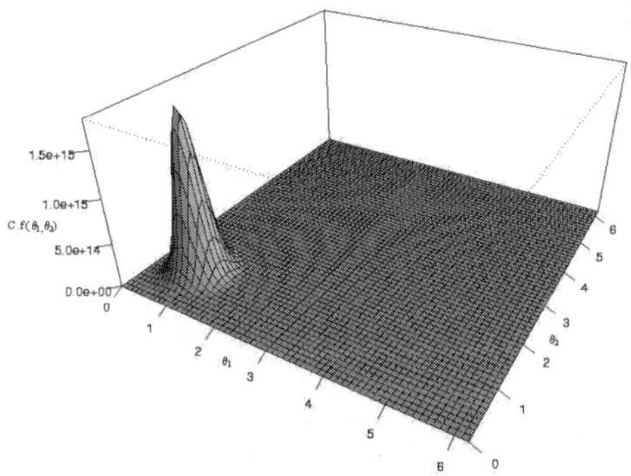

Fig. 6.13. Bivariate Generalized Cardioid conditional density for $a_{00} = 4, a_{01} = 2, a_{02} = 3, a_{10} = 6, a_{11} = 0.1, a_{12} = 0.8, a_{20} = 1, a_{21} = 3.5, a_{22} = 5.3, \gamma = 2$

$$\begin{aligned}
&\textstyle\sum_{i=1}^{n} \cos\Theta_1^{(i)}, \quad \sum_{i=1}^{n} \sin\Theta_1^{(i)}, \quad \sum_{i=1}^{n} \cos\Theta_2^{(i)},\\
&\textstyle\sum_{i=1}^{n} \sin\Theta_2^{(i)}, \quad \sum_{i=1}^{n} \cos\Theta_1^{(i)} \cos\Theta_2^{(i)}, \quad \sum_{i=1}^{n} \cos\Theta_1^{(i)} \sin\Theta_2^{(i)}, \qquad (6.16)\\
&\textstyle\sum_{i=1}^{n} \sin\Theta_1^{(i)} \cos\Theta_2^{(i)}, \quad \sum_{i=1}^{n} \sin\Theta_1^{(i)} \sin\Theta_2^{(i)}.
\end{aligned}$$

Maximum likelihood estimation is possible here, although implementation is complicated by the presence of an awkward normalizing constant which requires repeated numerical evaluation. Alternatively, variations of the method of moments

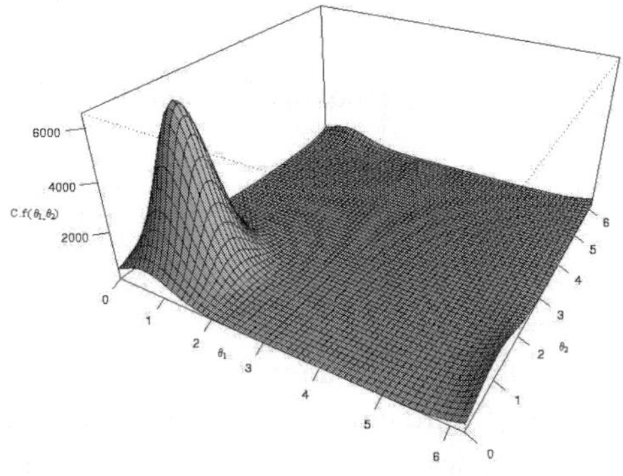

Fig. 6.14. Bivariate Generalized Cardioid conditional density for $a_{00} = 4, a_{01} = 2, a_{02} = 3, a_{10} = 6, a_{11} = 0.1, a_{12} = 0.8, a_{20} = 1, a_{21} = 3.5, a_{22} = 5.3, \gamma = 0.5$

can be successfully employed for estimation purposes (SenGupta and Arnold (2008)). Note that independence will occur if $a_{11} = a_{12} = a_{21} = a_{22} = 0$. Also an asymtotically optimal test can be derived to test this hypothesis (SenGupta and Arnold (2008)). Tests for exchangeability and for other simpler submodels can be of the generalized likelihood ratio type. Obtaining likelihood and restricted likelihood estimates will require creative programming.

Turning to consider the model (6.15) with generalized cardioid conditionals, we find ourselves faced with a much more daunting task. Clearly (6.15) does not constitute an exponential family. Maximum likelihood estimation and likelihood ratio tests may need to utilize effective multivariate search algorithms. This will be the subject of a future report.

6.5. Multivariate Extensions And Open Questions

The k-dimensional version of the von Mises conditionals distribution has a density on $[0, 2\pi)^k$ of the form

$$f_{\underline{\Theta}}(\underline{\theta}) = \exp\{\sum_{\underline{i} \in T_k} m_{\underline{i}} \prod_{t=1}^{k} (\cos\theta_t)^{I(i_t=1)} (\sin\theta_t)^{I(i_t=2)}\} \tag{6.17}$$

where T_k is the set of all vectors of 0's , 1's and 2's of dimension k.

Analogously we have a a k-dimensional version of the generalized cardioid conditionals density on $[0, 2\pi)^k$

$$f_{\underline{\Theta}}(\underline{\theta}) = \{\sum_{\underline{i}\in T_k} m_{\underline{i}} \prod_{t=1}^{k} (\cos\theta_t)^{I(i_t=1)} (\sin\theta_t)^{I(i_t=2)}\}^{\gamma} \tag{6.18}$$

The parameters in (6.17), the $m_{\underline{i}}$'s, for $\underline{i} \neq (0,\ldots,0)$, can take on any real values. This is not true for the parameters in (6.18). They must be constrained to ensure non-negativity of the kernel of the density (6.18). The exact nature of these constraints has not been determined.

Another open problem is the determination of the conditions on the parameters in (6.17) or (6.18) that will ensure unimodality of the resulting density. A related question asks how many modes will the densities (6.17) and (6.18) possess in multimodal cases.

References

1. Arnold, B.C. and Strauss, D. (1991), Bivariate distributions with conditionals in prescribed exponential families. *J. Roy. Stat. Soc., Ser. B.* , **53**, 365375
2. Jammalamadaka, S.R. and SenGupta, A. (2001), *Topics in Circular Statistics.* World Scientific,Singapore.
3. Jones, M.C. and Pewsey, A. (2005), A family of symmetric distributions on the circle. *J. Amer. Stat. Assoc.*, **100**, 14221428
4. Mardia, K.V. (1972), *Statistics of Directional Data.* Academic Press, London.
5. Mardia, K.V. and Patrangenaru V.(2005), Direction s and projective shapes. *Ann. Statist.* **33**,1666-1699.
6. SenGupta, A. (2004), On the construction of probability distributions for Directional Data, *Bulletin of Calcutta Mathematical Society* ,96,138-154.
7. SenGupta, A. and Arnold, B.C. (2008), Distributions with circular normal consitionals and tests for independence on the Torus, To appear.

Chapter 7

Optimal Text Space Representation of Student Essays Using Latent Semantic Analysis

Alexander Villacorta and Sreenivasa Rao Jammalamadaka

Statistics and Applied Probability, University of California,
Santa Barbara, CA 93106, USA
E-mail: alex.villacorta@clearcapital.com
Email: jammalam@pstat.ucsb.edu

This paper provides a framework for optimally representing student written essays in a vector space, based upon Latent Semantic Analysis and instructor evaluated grades. Comparing student essays to an authoritative source, a ranking scheme is optimized that allows for a unique vector space representation on the unit circle. Once such a representation has been found, traditional methods of circular data analysis and inference can be applied, as we demonstrate.

Contents

7.1. Introduction

In many academic settings it is important to understand the textual writing styles of students to measure how a certain lesson makes its way into the students' writings. The focus of this article is to present a new semantic analysis technique that allows for instructors to better quantify written essays in a manner comparable to traditional numerical approaches. A typical question an instructor may be interested in is whether there is a significant amount of variability in the responses to a given question. Another question an instructor may be interested in is how deeply the course content has been retained and to know how far the student responses are, in a semantic sense, from the original source and whether this is influenced by other outside factors. Clearly, some insight into these questions is available from the grade evaluation of the teacher, although at a rough resolution. Typically, grades are given on an interval scale of $1 \ldots 10$ or on the letter grade scale of *A, B, C, D, F*. At this resolution of evaluation, inferences into the structure and variability within grade classes is difficult to assess and quantify. In addition, many educators often normalize their expectations of grades and in doing so internally adjust their grading scale, which makes comparison to other sets of essays problematic. This issue is further exaggerated when there are multiple evaluators of written text as is the case with the essay writing portion of the Standardized Aptitude Test (SAT) in the United States. To dig deeper into these questions we attempt to use current text mining approaches and circular data analysis theory to extract more specific semantic information. In this analysis, we provide a tool which can further highlight the differences in textual responses and which goes much deeper than the grades.

7.2. Textual Decomposition and Circular Analysis

Decomposition of a collection of text documents into a vector space is an active area of research with encouraging results in a wide variety of applications. In particular, a large body of effort has been dedicated to text-mining with applications to educational scoring of student essays and is best represented by the subdisciplines of eLearning and Computational Linguistics.

7.2.1. *Vector space model and latent semantic analysis*

We begin with a brief introduction to the text processing involved in setting up our method. The main vehicle for text analysis used in this work is based on the Vector Space Model (VSM). Under the VSM, each document is simply considered to be a collection of words regardless of syntax, capitalization, or ordering of the words. In essence, the Vector Space Model creates a high dimensional document space, where each word is considered a dimension and each document lies somewhere in this space. Under this setting, document vectors may be compared just as in any other vector space. When put in matrix form with rows corresponding to unique terms and columns associated with unique documents, the matrix X_{TD} is typically called a term-document matrix. The entries (i, j) in X_{TD} can be taken to be any function of the frequency of term j in document i, such as binary indicator function, raw frequency, or the more commonly used term frequency $\times$ inverse document frequency (TFIDF) scheme , a weighting scheme which normalizes the local frequency with its global relative frequency.

The basic VSM models offer a convenient way to represent textual documents in a mathematical framework, but has several limitations in most practical applications. For instance, as more documents with disparate topics are included the size of the dictionary grows, and thus the dimensionality of the original term-document matrix also grows. Even for a modest collection of documents, the dictionary size can be very large. In addition, the distribution of words in most text collections follow a Zipf distribution implying that most words occur only once and a small set accounts for most frequency. Similarly, in most document collections there are many common words that are ambiguous or offer no semantic inference.

To solve the problems associated with the original VSM, Deerwester et al. proposed a secondary step of decomposing the term-document matrix to find a lower rank approximation to the document structure that in theory would isolate the major semantic structure of the document space. The main driving force for this is to approximate the original term document matrix with a lower order approximation based on the eigenvalues of X_{TD}. The basic algorithm of Latent Semantic Analysis (LSA) is to first construct the Term-Document Matrix X_{TD}. In the next step a Singular Value Decomposition is performed on X_{TD} such that $X_{TD} = U \times S \times V'$. The top k singular values of S are then retained and a lower rank approximation $X_{TD}^{(k)}$ is then calculated. For further explanation the reader is referred to the seminal work of Deerwester et. al. The benefits of this method are that it removes the noise associated with the collection and retains only the most prominent themes. More importantly, because of the approximation, generally the entries will not be sparse. As one can show, $X_{TD}^{(k)}$ is the best rank k model with least squares fit to X_{TD}.

In this work, we are mainly interested in measuring how far a student's writing is from the source author. Naturally, to do this we need a formal definition of what constitutes distance in the document space. For this we use the common method of measuring distance between two document vectors by the cosine of the angle which separates them. A linear distance between two points is not used since only those texts with approximately equal number of unique terms *and* frequency would be considered close. Since we do not wish to penalize student essays that do not match the frequency of term usage, we use angular distances that better reflect similarity in concept. In n-dimensional space this takes the form of the dot product of the two vectors, when they are normalized. The cosine of the angle between two vectors, x and y, is defined as $C(x,y) = \left(\frac{x}{||x||}\frac{y}{||y||}\right)'$ and thus the angle between x and y is

$$\theta_{x,y} = \arccos(C(x,y)). \tag{7.1}$$

7.2.2. *Circular analysis*

In much of the analysis of textual collections, distances are frequently calculated among documents, with various objectives such as clustering and measuring distance distributions from a given source. As a consequence of converting a collection of documents into a vector space defined by the dictionary of words and using angles between vectors as the distance metric, the document vectors can be considered to be points on the unit hypersphere. Thus, for making inferences on such measurements, we turn to the theory of circular statistics and borrow relevant tools.

Much work has been done in the field of circular statistics, see for instance for comprehensive coverage of the field. As pointed out in these books, there is considerable difference between the treatment of linear variables and the circular case. For a set of observations on circular/angular data $\alpha_1, \ldots, \alpha_n$, a mean direction may be obtained by first treating each observation in terms of its cosine and sine components and obtaining the ***resultant vector*** as defined by

$$\mathbf{R} = (\sum_{i=1}^{n} \cos\alpha_i, \sum_{i=1}^{n} \sin\alpha_i).$$

The direction that this resultant points to, is the mean direction for the data set and at the same time, the length of this resultant vector provides a useful measure of how concentrated the data are. Define $C = \sum\cos\alpha_i$ and $S = \sum\sin\alpha_i$, then the length of the resultant vector, $|\mathbf{R}| = \sqrt{C^2 + S^2}$, is an indicator how concentrated the angles are near this mean direction. It is straightforward to see that the length of the resultant vector reaches a maximum at n and a minimum at 0. The case when $|\mathbf{R}| = 0$ corresponds to the situation where the angles are evenly distributed on the circumference and in this case a mean direction is said to not exist. Conversely, $|\mathbf{R}| = n$ when all the observations take the same value. A measure of dispersion based of the resultant vector is given by $(n - |\mathbf{R}|)$. The direction of the resultant vector, $\hat{\alpha}_0$, is the mean direction for the circular data and is given by

$$\hat{\alpha}_0 = \arctan\left(\frac{S}{C}\right)$$

for $C > 0, S \geq 0$. For the definition of $\hat{\alpha}_0$ for other combinations of C and S see Jammalamadaka and SenGupta (2001.

Specifically, the von Mises distribution, also known as the Circular Normal, will be employed to model distances between documents in the context of analyzing student texts. The main idea behind this distribution is that the angles have unimodal distributions symmetrically distributed around a single mode, on the circle. This has many analogies to the Normal distribution on the real line. The probability distribution function for angles θ following a von Mises distribution with mean direction μ and concentration parameter κ is given below

$$f(\theta;\mu,\kappa) = \frac{1}{2\pi I_0(\kappa)} e^{\kappa\cos(\theta-\mu)} \tag{7.2}$$

for $0 \leq \theta < 2\pi$, $0 \leq \mu < 2\pi$, and $\kappa \geq 0$ where $I_0(\kappa)$ is a modified Bessel function of the first kind. An important result shown in [Jammalamadaka and SenGupta, 2001] for large values of κ is

$$2\kappa(n - |\mathbf{R}|) \overset{approx}{\sim} \chi^2_{n-1}. \tag{7.3}$$

which follows partly because, for large values of κ, the Circular Normal distribution can be well approximated by a Normal distribution. Although the angles between two unit vectors lie in the range $[0, \pi)$, the von Mises distribution defined on the entire circle $[0, 2\pi)$ provides a reasonable model for our case since a large κ ensures that the angles are tightly distributed in a small arc around the mean direction.

7.3. Optimal Vector Space Representation

One of the most critical steps in conducting a Latent Semantic Analysis on any text collection is determining the appropriate number of dimensions, k, to project the original term document matrix on. Even the authors of the seminal paper introducing LSA admit that choosing the appropriate number of dimensions to be an open research issue. In fact, determining this number is an active area of research in the information retrieval community. The reason why this poses such a challenging problem is because of the subjective nature of text. On the one hand, having too few components may lead to a compressed space which does not accurately capture the distinct semantic concepts and on the other hand retaining too many components leads to high dimensional spaces which are known to be inefficient for measuring distances of any type. This phenomenon is known as *the curse of dimensionality*. Furthermore, in many cases a document collection may not have a strict ordering. A common example of this is seen with the results of a World Wide Web search from any search engine, such as Google. The evaluation of the ordering of the returned list is dependent on the initial objectives of the user and cannot be expected to be the same for any arbitrary user.

In the present case of analyzing student essays we are initially faced with the same challenges. However, when one is given the additional information of the instructor assessed grades for the essays, we may take this as the definitive ordering of the documents. Another way to think of this additional information is as the instructor's personal (and often internal) distance measures for each student's essay from an expected optimal essay. The assumption that an instructor's assessment in the form of grades would be available is a reasonable one since it could not be expected that student's performance on a written essay should be determined solely by a computer program.

In this work we show how to choose an optimal number of dimensions for a text space decomposition based upon a set of instructor given grades. To begin, we first discuss some score functions used to evaluate the performance of a particular text space.

7.3.1. *Loss functions*

For each lower rank approximation, $X_{TD}^{(k)}$, $k \in \{1, \ldots, min(rows(X), cols(X))\}$, a loss (or score) function is required to assess how good a match the document space is to the human judged ordering. The overall idea is that we order the semantic angular distances and partition the ordering according the the grading partitioning desired. Then, if a student is categorized in the same partition for both angular semantic distances and grades, a successful match is made. In the subsequent analysis, a document space will be created for every level of dimensionality from the full model with no dimension reduction to the overly simple case of only 1 dimension. With that in mind, we now present the following score and loss functions. For n students and $i \in \{1, \ldots, n\}$, let $d_i^{(k)}$ = angle between author text and ith student text when using a vector space model with k dimensions, where the angle is calculated as in Equation (7.1). Similarly, let y_i = grade for ith student, $y_i \in \{10, 9, 8, 7, 6, 5, 4, 3, 2, 1, 0\}$.

Next define Q_j, $j = 1, \ldots, J$ to be the grading partitions. For example, if three partitions are desired, a possible configuration of partitions is $Q_1 = \{10, 9\}$, $Q_2 = \{8, 7\}$, $Q_3 = \{6, 5, 4, 3, 2, 1, 0\}$, with $Q_0 \equiv \emptyset$. The introduction of Q_j is necessary to compare the rankings of the grades and the document distances which are on different scales. To represent the students who are graded at a particular level define G_{Q_j} to be the set of all students who are graded in the jth partition, i.e.

$$G_{Q_j} = \{i : y_i \in Q_j\}.$$

Finally, we partition the ordered angular distances according to the same level as for the grades

$$T_{Q_j}^{(k)} = \{i : d_i^{(k)} \in \{d_{(|Q_{j-1}|+1)}^{(k)}, \ldots, d_{(|Q_{j-1}|+|Q_j|)}^{(k)}\}\}$$

where $d_{(m)}^{(k)}$ is the mth ordered distance for the kth document space and $|Q_j|$ denotes the cardinality of Q_j. $T_{Q_j}^{(k)}$ is simply the set of students whose angular semantic distances are partitioned in the same fashion as for their grades. Instead of using the usual form of the loss function which assigns a penalty to misclassification, we opt to instead use the equivalent score function which shows the number of correct matches so that a better intuition is gained about its results. Using this notion, we can now state the Zero-One score function for the dimension structure which produces the distances given in d as

$$S(\mathbf{d}^{(k)}, y) = \sum_{j=1}^{J} \left| G_{\{Q_j\}} \bigcap T_{\{Q_j\}}^{(k)} \right| \tag{7.4}$$

where J is the number of grading partitions. Note that for n total students the equivalent loss function is recovered as $L(d,y) = n - S(d,y)$. Figure 7.1 demonstrates the idea behind this scoring function. This function measures the amount of overlap between the assessment of the essays given by the teacher and those generated by the LSA model. However, it is of interest to understand how students interpret the material. There are various theories as to how students retain and comprehend educational material.

One point of view is that a student who understands the material will write 'close' to an authors source text. With this point of view, strong comprehension of an idea would translate to angles which are near the source document vector. Another point of view is that students comprehend the course content when they internalize the material. In this case comprehension would be represented by angles which were furthest from the source document vector, since it can be expected that a students vocabulary would be quite different from the source authors. In the following analysis we test both ideas by considering the ranking of both sets of lists.

For each set of dimensions we calculate the distances from the author as given by the angle between the vectors in the reduced principal component space. When

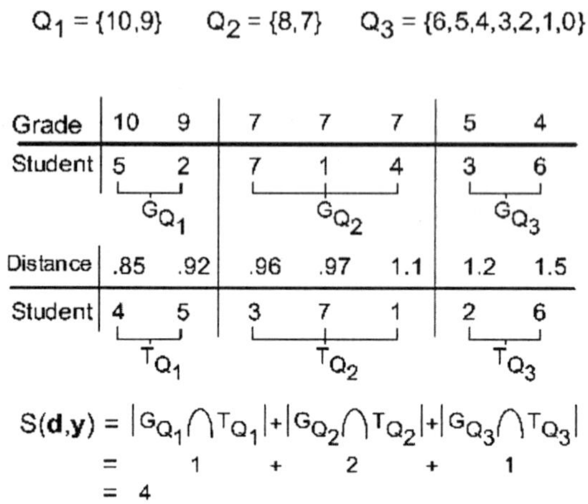

Fig. 7.1. Grading Partition Example

considering the first theory we rank these distances in ascending order from the smallest distance to the largest. In the second theory, which states that students internalize the material, distance ranking is in descending order from largest to smallest. This follows because if a student internalizes the material then they should presumably be the furthest semantically from the source author, because they use their own words.

An alternative loss function used is the more general L_1 distance. Using the notation above, suppose that there are J partitions and that for each student there is an angular distance $d_i^{(k)}$. Recall that $d_i^{(k)}$ is the angle between the source text and a student's essay when represented in a vector space with k dimensions. Then, for $T_{Q_{l_i}}^{(k)}$ under a specified ordering and a instructor assessed grade classified into $G_{Q_{m_i}}$ for the ith student, the the loss associated with these students for the given document structure is given by:

$$L(\mathbf{d}^{(k)}, y) = \sum_i |l_i - m_i|, \qquad \text{for } l_i, m_i \in \{0, 1, 2, \ldots, J\}. \tag{7.5}$$

Equation (7.5) differs from Equation (7.4) in that it measures the severity of the misclassification. These extra penalties help to understand how different a particular angular ordering is from the grade assessments. As with the score function, the outcome from these functions are dependent on the whether ascending or descending orders are used.

For every possible vector space representation of the document collection, the two score functions described above may be applied to assess their validity, and thus identify the optimal representation of the document collection in the vector space. Once an optimal structure is found, further analysis may be based on this document vector space.

7.3.2. *Testing hypotheses using circular analysis of variance*

Through the representation of the document collection by means of the Vector Space Model, a high dimensional vector space is created where the dimensionality is determined by the vocabulary of the text corpus. By normalizing each of these vectors, one may consider that the sample of documents reside on the surface of the unit hypersphere. Further, since we are interested in how course content has disseminated among the students, we wish to investigate the pair wise distances to the source material. In doing this, the vector space is reduced to the circle since we now only consider the angles which separate a student's work from the author's.

In classical linear statistics, comparisons of the treatment effects in a population is most conveniently done using an Analysis of Variance (ANOVA) under some commonly valid assumptions. In this present case, we also wish to test hypothesis concerning the effects of various treatments and accordingly we turn to the circular analogy of the analysis of variance. One of the central assumptions of this approach is that the angular data be approximately distributed as a von Mises distribution (see Equation (7.2)). This is analogous to the linear case in that the data are approximately normally distributed. A further assumption for this approach is that the common concentration parameter κ is reasonably large. These assumptions ensure that the von Mises distribution can be well approximated by a Normal distribution, since a large value of κ means that much of the data are contained in a relatively small band of the circle. As described in Jammalamadaka and SenGupta [2001], an exact test for comparing several mean directions is not available because of the presence of the unknown nuisance parameter κ. Instead, using the assumptions that the data are distributed as a von Mises distribution and that the concentration parameter, κ, be large, an approximate test of hypothesis is used which relies on the resultant vectors.

Formally, suppose we wish to test the hypothesis that for p independent populations, the mean directions are all the same,

$$H_0 : \mu_1 = \cdots = \mu_p$$

where μ_i = mean direction for ith population. The basic intuition is that under the

Table 7.1. Approximate Circular ANOVA Table

Source of Deviation	d.f.	SS	MS	F-Stat
Between	$p-1$	$\sum_{i=1}^{p} R_i - R$	$\frac{SS_B}{p-1}$	$\frac{MS_B}{MS_W}$
Within	$n-p$	$\sum_{i=1}^{p} (n_i - R_i)$	$\frac{SS_W}{n-p}$	
Total	$n-1$	$n-R$		

null hypothesis, the direction of the resultant vector for each of the p populations should be approximately the same. Let $R = |\mathbf{R}|$ and $R_i = |\mathbf{R_i}|$ be the lengths of the result vectors for the set of angles corresponding to the combined total and for the ith population, respectively. Analogous to the linear case, $(n_i - R_i)$ can be thought of as within variation in each population, which in the circular domain translates to the within dispersion. Similarly, the total dispersion (sum of squares total in the linear case) for the data is given by $(n - R)$, where $n = \sum n_i$. It can now be shown that the total dispersion may be decomposed much like in the linear case,

$$\begin{aligned} n - R &= \left(n - \sum R_i\right) + \left(\sum R_i - R\right) \\ &= \left(\sum (n_i - R_i)\right) + \left(\sum R_i - R\right). \end{aligned} \tag{7.6}$$

By multiplying the value 2κ in Equation (7.6), we can recall the result of Equation (7.3), which leads to a similar χ^2 segmentation given by

$$\chi^2_{n-1} = \chi^2_{n-p} + \chi^2_{p-1}.$$

Then by an analogous argument to the linear case, an analysis of variance, or F-Test, may be implemented by comparing the test statistic

$$F = \frac{(\Sigma R_i - R)/(p-1)}{\Sigma(n_i - R_i)/(n-p)}$$

to the upper percentiles of an $F(\nu_1, \nu_2)$ distribution with $\nu_1 = p - 1$ and $\nu_2 = n - p$. The classical ANOVA table can then be represented in the circular case as seen in Table 7.1.

It is important to mention that the arguments for an approximate analysis of variance are based on assumptions that the concentration parameter be large enough, which must be checked prior to the analysis, Jammalamadaka and SenGupta show that reasonable results may be obtained when the sample mean resultant length greater than .45. When concentration assumptions cannot be met alternative techniques may be employed, for example see Jammalamadaka.

Table 7.2. Summary of case study data

	Source Authors		
	Feyerabend	Levi Strauss	Semprini
Word Count	524	634	2146
# of Essays	63	62	57
Avg Word Count	153	167	153
Italian/Non-Italian	45/18	45/17	43/14

For the context of analyzing student essays, we will employ the approximate circular analysis of variance to test for various factors and their effects on the semantic distribution.

7.4. Results

The data used for this analysis was obtained as part of a larger study conducted at the University of Basel, Switzerland by Dr. Terry Inglese and consists of student text essays from a Swiss Political Science course . Each student was required to write three essays with regard to three different philosophers: Paul Feyerabend, Claude Lévi Strauss, and Andrea Semprini. For each author, an authoritative text was available that served as the source of themes for which the students were instructed to comment on. In addition, teacher evaluated scores were also available for a total of 63 students. The data were collected throughout the entire course along with each student's native language. For simplicity we have grouped the non-native speakers into the same classification. The data are summarized in the Table 7.2 below.

To begin, we show how the students perform based on the term document matrix without any post processing but allowing for a stoplist, stemming, and TFIDF weighting scheme. It should also be noted that all of the textual essays were in Italian. Figure 7.2 shows how nearly all of the student responses to each of the three authors are near orthogonal to the original source material which is located at the due East position. The graphs were generated by mapping the high dimensional document vectors to the plane by simply graphing the angle between each essay and the target text.

These figures illustrate the idea that as the number of dimensions increase the document vector space becomes very sparse. It appears that nearly all of the textual responses are perpendicular to the source material, which illustrates the *curse of dimensionality* discussed previously.

Next we implement the LSA approach to remove the noise in the data and to try to extract the top themes in the collection. Recall that choosing the appropriate number of singular values is equivalent to choosing the appropriate number

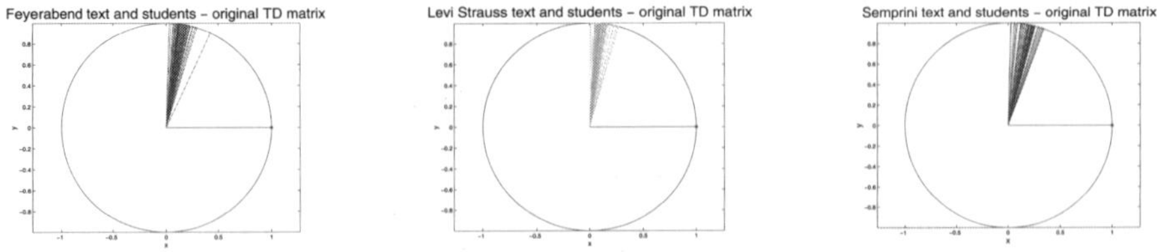

Fig. 7.2. Angular distance using original TD matrix

of principal components. One crude way of choosing an appropriate number of principal components is to investigate a scree plot, which is a graph of the ordered eigenvalues from biggest to smallest. A heuristic approach is to choose the number of components, k, such that there is minimal change in the scree plot for subsequent eigenvalues. This method is related to the percentage of variance explained in the data set by retaining the first k principal components. Figure 7.3 shows the traditional scree plot and percent of variance explained as a function of dimension. Based on the scree plot, one may consider choosing approximately 10 to 12 principal components, since those are the dimensions for which the ordered eigenvalues begin a graduated descent. Similarly, if one wished to retain a set amount of variance in the data, such as 80%, then it would be reasonable to consider using approximately 80 components. Clearly, the choice of criteria significantly affects the number of principal components retained.

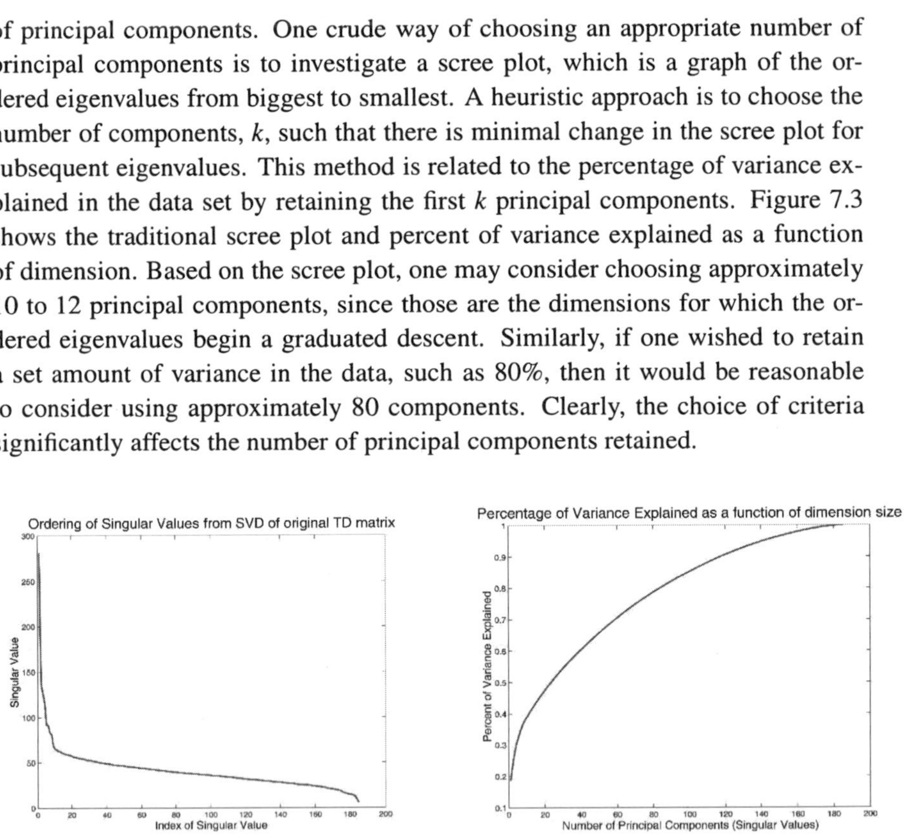

Fig. 7.3. Traditional Scree plot and Percent of Variance explained by first k principal components

First, we show the sensitivity of angular distributions to the number of components used. Figures 7.4, 7.5, and 7.6 shows how the distribution of distances from the three authors changes for two different values of k.

More generally Figure 7.7 shows how the mean distances change as a function

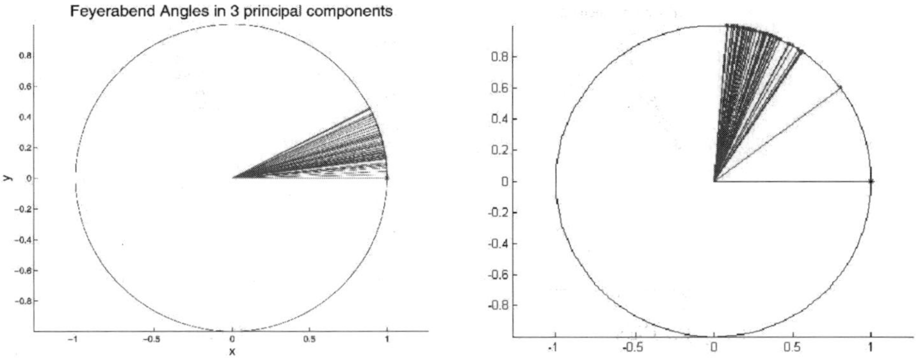

(a) 3 Dims (b) 50 Dims

Fig. 7.4. Distance from Feyerabend to student essays as a function of dimension size

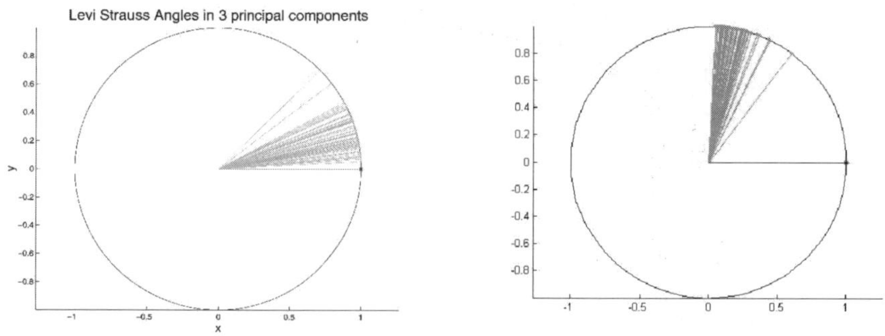

(a) 3 PCs (b) 50 PCs

Fig. 7.5. Distance from Levi Strauss to student essays as a function of dimension size

of the principal components. Clearly, the choice of dimension is critical to the conclusion of this statistical analysis. In the present work we now propose a methodology for choosing the number of principal components, which optimizes the representation of the documents with respect to the human assessed grades. Accordingly, we implement the two score functions discussed in Section 7.3.1 which help to identify when an optimal document structure is reached.

7.4.1. *Strict grade matching*

In certain instances, an instructor may be interested in knowing how varied the textual essays are within every individual grade level. Suppose that grades are

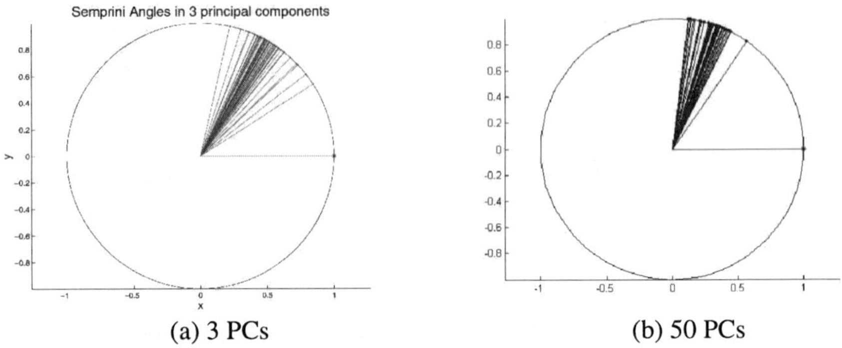

Fig. 7.6. Distance from Semprini to student essays as a function of dimension size

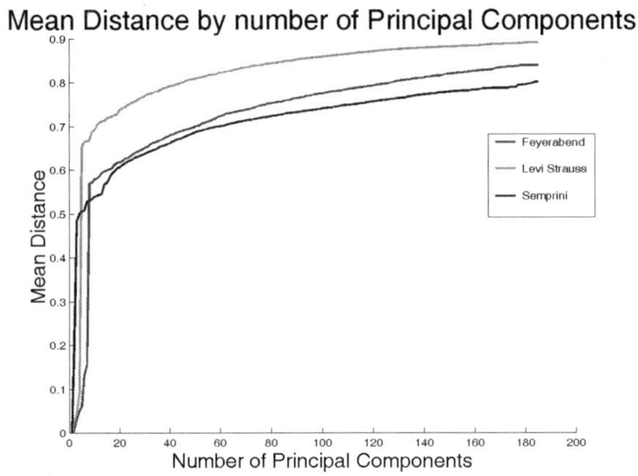

Fig. 7.7. Mean Distance from Author as a function of Principal Component

given by the instructor on an 11 point scale such that a perfect score corresponds to a grade of 10 while the score of 0 is given for incorrect answers with no partial credit. The first partitioning attempted is when the grade level resolution matches the exact assignment of grades as the teacher. That is to say when

$$Q_i = \{(10 - i)\} \quad i = 0, \ldots, 10.$$

Also for each of the partitions considered we implement both ascending and descending ordering. Tables 7.3, 7.4 and 7.5 show the results of the strict partitioning, where the row 'Optimum' refers to the maximum of the Zero-One score function or the minimum of the L_1 loss function. The row 'Dim #' refers to the dimensions where the optimum value is obtained. Figure 7.8 shows a typical

Table 7.3. Summary of Loss functions for Feyerabend on 11 Grade Levels

	Feyerabend			
	$S(d,y)$		$L(d,y)$	
	A	D	A	D
Optimum	14	20	134	102
% correct	22.2%	31.8%	22.2%	20.6%
Dim #	1	167	1	8-9

$Q_j = \{10 - j\} \qquad j = 0, \ldots, 10$

Table 7.4. Summary of Loss functions for Levi Strauss on 11 Grade Levels

	Levi Strauss			
	$S(d,y)$		$L(d,y)$	
	A	D	A	D
Optimum	14	15	108	102
% correct	22.6%	24.2%	19.4%	21%
Dim #	4	5-8	180-185	55,56

$Q_j = \{10 - j\} \qquad j = 0, \ldots, 10$

portrayal of the values of the score and loss function for the author Semprini, as defined in equations (7.4) and (7.5), when considering direct matches with the instructor's evaluation. As can be seen the maximum attainable match in this data set is 31.6%.

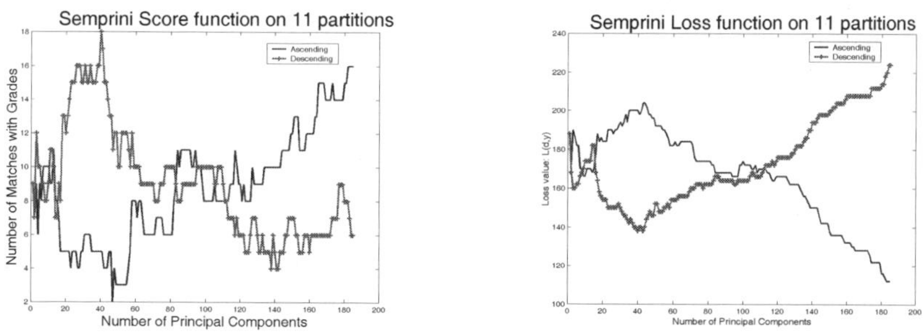

(a) Score Function (b) Loss Function

Fig. 7.8. Values of Score and Loss function for author Semprini with 11 partitions

An immediate pattern in the score and loss functions for all essays written is that the descending order better captures the grade ranking given by the instructor. When considering the 11 strict partitions, the descending order outperforms the ascending order for all three authors. For the essays written about Levi Strauss, there appears to be consistency in both the percentage of correct matches and in

Table 7.5. Summary of Loss functions for Semprini on 11 Grade Levels

	Semprini			
	$S(d,y)$		$L(d,y)$	
	A	D	A	D
Optimum	16	18	112	138
% correct	28%	31.6%	28%	31.6%
Dim #	182-185	40	183-185	40,43

$Q_j = \{10-j\} \quad j=0,\ldots,10$

the dimension. However, for Feyerabend there is a significant difference between the number of matches and the dimensions where they are achieved for ascending and descending orders. The case for the essays which are written about Semprini also shows considerable differences in the optimal dimension. Furthermore, we see that the Zero-One score function consistently outperforms the L_1 loss in terms of finding the dimension with the most percentage of correct matches with the instructor grades, but this is to be expected since the L_1 loss is a generalization of the Zero-One loss. However, the benefit of using L_1 loss is that it shows other dimensions that have matches close to the Zero-One case, which in some cases be at a lower dimension number. An example of this is seen in the essays written about Feyerabend in the decreasing order setup. There we see that at compromise of approximately 11% in overlap with the instructor's grades, we may reduce the number of dimensions by over 150. Alone, this may not seem like a good trade, but when taken in conjugation with all of the other partitions, it may balance in the long run.

7.4.2. *Three level partition*

The next partition investigated is when an instructor wishes to see how the top grades are distributed. In this case, a possible ordering is to set the following partitions to

$$Q_1 = \{10,9\}, Q_2 = \{8,7\}, Q_3 = \{6,5,4,3,2,1,0\}.$$

Tables 7.6, 7.7 and 7.8 show the results of the three level partitioning. Figure 7.9 shows the values of the score and loss function for the author Levi Strauss under three level partitioning.

In this partition setup we observe that an increase in efficiency is achieved, which is to be expected since we are relaxing our restrictions on the distance ordering. However, in this case it is now observable that the descending order is best for all groups under both loss functions. This finding seems to support the case that students who understand the content better are internalizing their interpretation.

Table 7.6. Summary of Loss functions for Feyerabend on 3 Grade Levels

	Feyerabend			
	$S(d,y)$		$L(d,y)$	
	A	D	A	D
Optimum	22	33	52	36
% correct	34.9%	52.4%	34.9%	47.6%
Dim #	1,10-11	161-167	1	8,9

$Q_1 = \{10,9\}, Q_2 = \{8,7\}, Q_3 = \{6,5,4,3,2,1,0\}$

Table 7.7. Summary of Loss functions for Levi Strauss on 3 Grade Levels

	Levi Strauss			
	$S(d,y)$		$L(d,y)$	
	A	D	A	D
Best	28	26	40	40
%	45.1%	41.9%	45.1%	41.9%
Dim #	6-8,180-185	27,49-56,58	180-185	49-56,58

$Q_1 = \{10,9\}, Q_2 = \{8,7\}, Q_3 = \{6,5,4,3,2,1,0\}$

Also, with the exception of the essays written about the theory of Feyerabend, all of the optimal dimensions are well below the full model and seem to be less than 50 principal components. Again we observe that for the case of Feyerabend essays, a compromise of approximately 5% in overlap with the instructor's grades allows one to represent the document space with over 150 reduced dimensions.

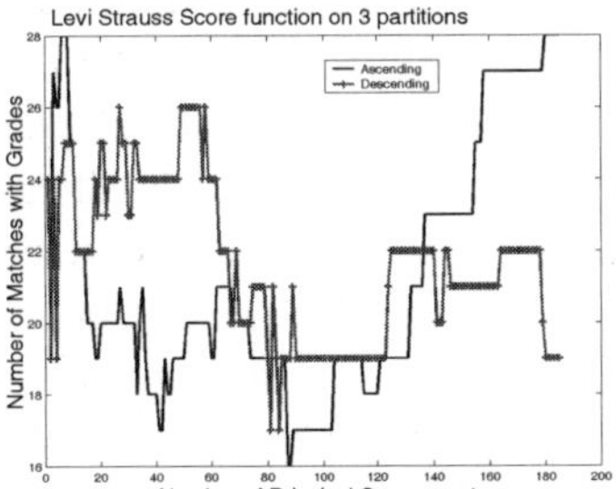

(a) Score Function

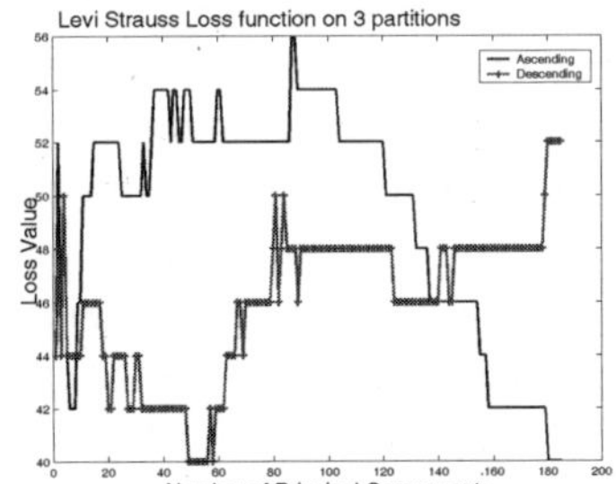

(b) Loss Function

Fig. 7.9. Values of Score and Loss function for author Levi Strauss with 3 partitions

Table 7.8. Summary of Loss functions for Semprini on 3 Grade Levels

	Semprini			
	$S(d,y)$		$L(d,y)$	
	A	D	A	D
Best	31	33	38	38
%	54.4%	57.9%	54.4%	57.9%
Dim #	182-185	40	174-185	34-36,38-40,43-44

$Q_1 = \{10,9\}, Q_2 = \{8,7\}, Q_3 = \{6,5,4,3,2,1,0\}$

Table 7.9. Summary of Loss functions for Feyerabend on 2 Grade Levels

	Feyerabend			
	$S(d,y)$		$L(d,y)$	
	A	D	A	D
Best	33	43	30	20
%	52.4%	68.3%	52.4%	68.3%
Dim #	7,36-50	2	7,36-50	2

$Q_1 = \{10,9,8,7\}, Q_2 = \{6,5,4,3,2,1,0\}$

7.4.3. *Binary partitions*

The final partition considered is when an instructor is simply interested in analyzing the scores for those students who passed the question and those who did not. In this situation, the partition is defined by

$$Q_1 = \{10,9,8,7\}, Q_2 = \{6,5,4,3,2,1,0\}.$$

Tables 9,10 and 11 show the results of the binary partitioning. An immediate observation for this case is that both the Zero-One score function and the L_1 loss function yield identical results for all three authors. Again, this is no coincidence since at a binary partition the largest difference between any two classifications can at most be one. Thus, in this case both the Zero-One and L_1 loss are equivalent. With regards to efficiency of matching the instructor's grading order, the binary partition achieves the best results, with nearly 70% accuracy in most cases. There is again a consistent gain in performance by using the descending order for distances. This further supports the idea that the student who better understands the material is internalizing the content and uses his or her own words. Finally, the dimension number which achieves the best performance is clearly less than the full model and seems to be near 40 and in the case of Feyerabend is at two dimensions.

Table 7.10. Summary of Loss functions for Levi Strauss on 2 Grade Levels

	Levi Strauss			
	$S(d,y)$		$L(d,y)$	
	A	D	A	D
Best	34	42	28	20
%	54.8%	67.7%	54.8%	67.7%
Dim #	4,106, 108-112, 126-170	44,45, 47-49, 51-64	4,106 108-112, 126-170	44,45, 47-49, 51-64

$Q_1 = \{10,9,8,7\}, Q_2 = \{6,5,4,3,2,1,0\}$

Table 7.11. Summary of Loss functions for Semprini on 2 Grade Levels

	Semprini			
	$S(d,y)$		$L(d,y)$	
	A	D	A	D
Optimum	39	39	18	18
% correct	68.4%	68.4%	68.4%	68.4%
Dim #	183-185	37-44	183-185	37-44

$Q_1 = \{10,9,8,7\}, Q_2 = \{6,5,4,3,2,1,0\}$

7.4.4. *Comparison of orderings*

After investigating various outcomes from the different partitioning levels, we now combine the scores of all three partitions to help guide a choice for an optimal dimension regardless of partitions. Figure 7.10 shows the overall distributions of the three partition levels when considering the distance ranking in ascending order. That is, when we consider that small distances relate to closer matching.

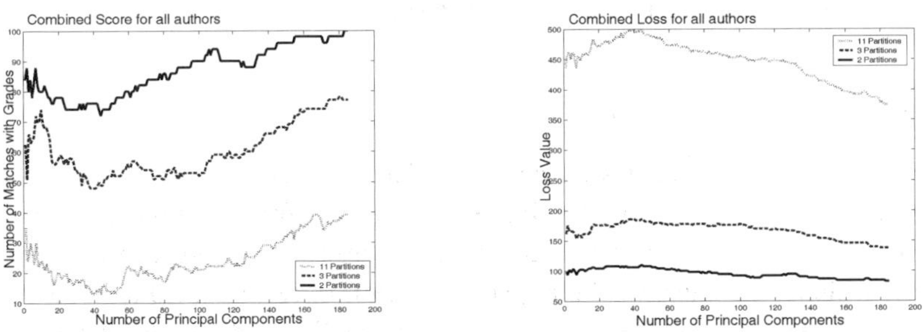

(a) Combined Score (b) Combined Loss

Fig. 7.10. Combined Score and Loss functions at all 3 partitions - Ascending Order

In the combined graph for the score function in ascending order we observe that the optimal dimension is attained at very high levels. Specifically, the optimal dimension for *all* three partitions is at 183 to 185. This is essentially the full model with no dimension reduction. The results shown for the ascending order appear to contradict the notion that the full document space is not appropriate for semantic analysis. Similarly, the results for the L_1 loss function also show an optimal dimension at a high level. The dimension which optimizes all three grade partitions is equal to Zero-One score at levels 183 to 185.

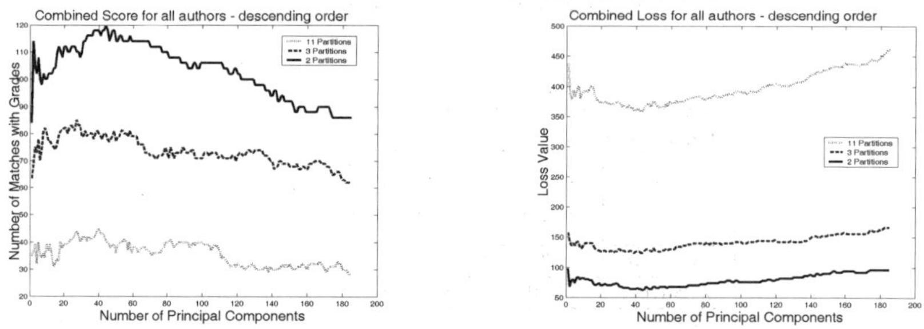

(a) Combined Score (b) Combined Loss

Fig. 7.11. Combined Score and Loss functions at all 3 partitions - Descending Order

Figure 7.11 shows the combined scores for all authors in descending order. In this case, we see a much different story. Clearly, using descending leads to a significantly smaller dimension. The optimal dimension for all three partitions for the score function is 39 to 40. Similarly, for the L_1 loss the optimal dimension is chosen to be 43 to 44. Coupled with the efficiency results seen above, we conclude that the descending ordering appears to bear the most meaningful results in interpreting how distances should be ranked.

7.4.5. *Analysis of variance*

We next turn to an analysis of variance to demonstrate how an instructor could use the document space to make inferences on the effect of a lesson. After being satisfied with the choice of dimension for the text space we proceed to the next phase which is the analysis of the space for effects of certain factors. In the present case of student essays, we determine that descending ordering is optimal with a dimension size of 40 principal components.

Using the results introduced earlier concerning the steps required for a circular ANOVA, we begin by considering the effects of visibility of an author on students' retention of material. The two authors Feyerabend and Levi Strauss are considered

Table 7.12. Approximate Circular ANOVA Table

Source of Deviation	d.f.	SS	MS	F-Stat
Between Groups	2	.4	.2	25
Within Groups	179	1.5	.008	
Total	181	1.9		

the visible authors because of the way their lesson was interactively presented to the class. For each unit, video, images, and audio of the actual author were available for the students to explore. Conversely, the content from the author Semprini was taught in the usual textbook manner with no additional knowledge of the author. Accordingly, there are $p = 3$ populations with $n_1 = 63$, $n_2 = 62$, $n_3 = 57$, and $n = 182$. We wish to test if the mean angular distance from the author is the same for all authors. The hypothesis for this setup is

$$H_0 : \mu_1 = \mu_2 = \mu_3$$

where μ_i is the mean direction for the ith population.
First we check to see if the estimates for the concentration parameter κ are sufficiently large, in this case they are $\hat{\kappa}_1 = 37.8$, $\hat{\kappa}_2 = 75.6$, and $\hat{\kappa}_3 = 89.8$, which are sufficiently large enough to continue with the analysis.
The sample resultant lengths are given by:

$$R_1 = 62.2$$
$$R_2 = 61.6$$
$$R_3 = 56.7$$

and the pooled sample resultant length is

$$R = 180.1.$$

Combining these results with Table 1 we obtain the following table for testing the equality of the mean directions when separated by author.

Clearly, we reject the hypothesis that angular means are all the same and upon inspection of Figure 7.7, we notice that the distances for essays written on the content of author Levi Strauss are significantly different from the other authors. Also, recall that since we accepted the descending ordering for angular distances, the essays for the theory of Levi Strauss appear to be better comprehended than the authors. This suggests that there may be validity for the positive effects of showing students pictures and videos of the authors they are learning. Of course, there are many other affecting factors that may be playing a part here, but the methodology presented above allows for further analysis of factors in a structured approach.

7.5. Conclusions

This aim of this paper is to provide a framework on how to build optimal textual vector space representations of student text essays. Using the principles demonstrated above, an instructor may investigate various hypotheses about a given data set. The key idea is that when the documents are converted to a vector space, which can be thought of as points on the hypersphere, classical linear statistics are not appropriate. In our present case, we performed a further conversion from points on the hypersphere to points on the circle since we are interested in only the distances from a common vector. Depending on the specific applications of the researcher this may not always be the correct decision.

It may be noted that for the current data set, the range of angles were limited and highly concentrated in an arc of small length, characterized by the large estimate of κ. In such a special situation, a linear ANOVA and the usual F-test can perhaps also be justified. But this is not true in general, and one should utilize the circular ANOVA, which is described here.

As a test case we investigated how information in the form of educational content is distributed in the writings of students. Traditional information retrieval approaches offer various suggestions for choosing an appropriate number of dimensions in order to represent the document space. In the work presented here, we use the grading assignment of the teacher as a score function to base the results of any given vector space model decomposition. Our results show, that students who perform well on the assignments, tend to also have writing styles that are more unique in comparison to the source material. This offers further evidence to the theory that students internalize material when they have comfortable grasp on the content.

7.6. Acknowledgements

The authors would like to sincerely thank Dr. Terry Inglese for her helpful input and supply of the test data set. This work was supported by the National Science Foundation's IGERT Program in Interactive Digital Multimedia (Award #DGE-0221713) at the University of California, Santa Barbara.

References

1. Bereiter C. and M. Scardamalia (1987), *The psychology of written composition* Lawrence Erlbaum Associates.
2. Brook Wu Y. fang and X. Chen (2005), elearning assessment through textual analy-

sis of class discussions, in Proceedings of the 5th IEEE International Conference on Advanced Learning Technologies, *ICALT*.

3. Burek G. G., M. Vargas-Vera and E. Moreale (2004), Indexing student essays paragraphs using lsa over an integrated ontological space, in COLING 2004 eLearning for Computa- tional Linguistics and Computational Linguistics for eLearning, *ed. E. H. Lothar Lem- nitzer, Detmar Meurers* (COLING, Geneva, Switzerland, August 28 2004).
4. Deerwester S. C., S. T. Dumais, T. K. Landauer, G. W. Furnas and R. A. Harshman (1990), *Journal of the American Society of Information Science* 41, 391.
5. Haley D., P. Thomas, B. Nuseibeh, J. Taylor and P. Lefrere (2003), *E-assessment using latent semantic analysis*, in LeGE-WG 3.
6. Haley D., P. Thomas, A. D. Roeck and M. Petre (2005), A Research Taxonomy for Latent Semantic Analysis-Based Educational Applications, *Tech. Rep. TR 2005/09*, The Open University.
7. Hand D., H. Mannila and P. Smyth (2000), *Principles of Data Mining* MIT Press.
8. Mardia K.V.and P. E. Jupp(2000), *Directional Statistics* John Wiley.
9. Mayer R. (2001), *Multimedia Learning* Cambridge University Press.
10. Miller T.(2003), *Journal of Educational Computing Research* 29, 495.
11. Inglese T., R. Mayer and F. Rigotti (Feb 2007), *Learning and Instruction* 17, 67.
12. Jammalamadaka S. R.and S. Sengupta (1970), *Journal of Geology* 78, 533.
13. Jammalamadaka S. R.(1967), *Sankhya* 28, 172.
14. Jammalamadaka S. R.and A. SenGupta (2001), *Topics in Circular Statistics* World Scientic.
15. Kontostathis A., W. M. Pottenger and B. D. Davison (2005), *Identication of critical values in latent semantic indexing, in Foundations of Data Mining and Knowledge Discovery*,eds. T. Y. Lin, S. Ohsuga, C. Liau and S. Tsumoto Springer-Verlag.
16. Salton G.(1964), *A document retrieval system for man-machine interaction, in Proceedings of the 1964 19th ACM national conference*, (ACM Press, New York, NY, USA, 1964).
17. Zha H. (1998), *A Subspace-Based Model for Information Retrieval with Applications in Latent Semantic Indexing*, Tech. Rep. TR CSE-98-002, Penn State.

Chapter 8

Linear Regression for Random Measures

M. M. Rao

University of California, Riverside, CA 92521

E-mail:rao@math.ucr.edu

If X and Y are integrable random variables, then Ragnar Frisch raised the characterization problem of linearity for the regression function $g(X) = E(Y|X)$ so that $g(X) = a + bX$. After outlining the basic known work on this problem, an extension is given if the vector (X, Y) is replaced by a random measure $Z : \mathcal{B}_0 \to L^p(P)$ on theδ-ring $\mathcal{B}_0$ of bounded Borel sets of the line R into an$L^p(P), 1 \leq p \leq 2$ where Z takes independent values on disjoint sets, and is$p$-stable valued. The work is extended to measures $Y^f(\cdot)$ defined as indefinite integrals ofboundedly supported functions f relative to Z.

Contents

8.1. Introduction

In 1936 at an Econometrics Conference in London, UK, Ragnar Frisch formulated the following problem and asked for its solution. Let ξ, α, β be independent random variables and

$$X = a\xi + \alpha, \quad Y = b\xi + \beta, \quad a, b \in R. \tag{8.1}$$

When are the regression equations of X on Y (or Y on X) linear in the sense that $E(X|Y) = a_1 Y + b_1$ $[E(Y|X) = a_2 X + b_2]$, $a_i, b_i \in R, i = 1, 2$? The early solutions obtained by

(i) H. V. Allen (1938) and (ii) E. Fix (1949) are worth recalling here and the further extensions due to (iii) M. Kanter (1972) and (iv) C. D. Hardin (1982) explain the nontriviality of the problem. These works (i)–(iv), are as follows.
(i) If β has two moments and α, ξ have all moments then the linearity question has a positive solution iff (=if and only if) α, ξ are normally distributed.
(ii) If X, Y are as in (8.1) and $I \subset R$ is a nonempty compact interval such that either $I \subset R^-$ or $-I \subset R^+$ then $E(X|Y)$ is linear iff either ξ is a constant, $\alpha = 0$ or the characteristic functions $\varphi_\xi, \varphi_\alpha$ of ξ, α are given as follows: ($sgn x = -1, x < 0; = 0, x = 0; = 1, x > 0$)

$$\varphi_\xi(t) = e^{-k^2(u+iv\, sgn t\, sgn a)|t|^\nu}, \tag{8.2}$$

and similarly

$$\varphi_\alpha(t) = e^{-(u+iv\, sgn t)|t|^\nu}, \tag{8.3}$$

where $1 < \nu \leq 2, u > 0, k \neq 0$, or if $I \cap R^- \neq \emptyset \neq I \cap R^+$, then the linearity holde iff

$$\varphi_\xi(t) = e^{-k^2 u}|t|^\nu, \quad \varphi_\alpha(t) = e^{-u|t|^\nu}, \tag{8.4}$$

and ν, u, k satisfy the same conditions as above for (8.2) and (8.3). When α, ξ have finite second moments, then they must be normal.
(iii) The point of the above work is to understand the nontriviality of the general problem. Formulas (8.2) and (8.4) are recognized from the well-known Lévy-Khintchine formula of infinitely divisible characteristic functions, that the random variables ξ, α are of stable type – a subclass of the infinitely divisible family. The result was considered in its general form by M. Kanter (1972), independently of Fix's earlier key special case, and obtained the following. Let $\xi_1, \ldots, \xi_n$ be independent stable and symmetric random variables with the same distribution of exponent $1 < p \leq 2$, and $X = \sum_{i=1}^n a_i \xi_i$, $Y = \sum_{j=1}^n b_j \xi_j$ so that $\varphi_{\xi_j}(t) = e^{-c|t|^p}, c > 0$.
Then $E(Y|X) = \lambda X$ where λ is a constant determined by the a_i and b_j. More generally if X, Y are jointly symmetric stable random variables of index $p \in (1, 2]$, then $E(Y|X) = \lambda X$ where λ is a constant. See Loève (1955) on stable classes.
(iv) A related question, complementing (iii) and is due to C. D. Hardin (1982), will be sketched here as it illuminates the above results. A random vector $X = (X_1, \ldots, X_n)$ is said to be *spherical* if for each $n \times n$ orthogonal matrix A, AX' and X' are identically distributed (here prime denotes transpose) and their common characteristic function $\varphi_X(\cdot)$ is given by (using the dot product notation):

$$\varphi_X(\lambda) = E(e^{i\lambda \cdot X}) = E(e^{i\sum_{j=1}^n \lambda_j X_j}),$$
$$= E(e^{i\lambda \cdot AX'}) = \varphi_X(\|\lambda\|_2), \tag{8.5}$$

where $\|\lambda\|_2^2 = \sum_{j=1}^n \lambda_j^2$ defines the Euclidean norm. A process is termed spherical if each of its finite dimensional distributions is spherical, centered Gaussian processes being examples of spherical processes. Then a process $\{X_t, t \in T\} \subset L^1(P)$ is said to have the linear regression property if for each $t_1, \ldots t_{n+1} \in T$, one has

$$E(X_{t_{n+1}}|X_{t_1}, \cdots, X_{t_n}) = \sum_{j=1}^n a_j X_{t_j}, \quad n \geq 1. \tag{8.6}$$

This may be also be stated as: for each $n, E(X_{t_{n+1}}|X_{t_n}, \cdots, X_{t_1}) \in \mathcal{X}$, the finite linear span of $\{X_t, t \in T\}$.
The main result of Hardin's is:

Theorem 1. If dim $\mathcal{X} \geq 2, \{X_t, t \in T\} \subset L^2(P)$ or if $\{X_t, t \in T\} \subset L^1(P)$ and dim $\mathcal{X} \geq 3$, then the process has the linear regression property iff it is spherical, i.e. $\mathcal{X}$ is spherically generated.
There is a key relation between spherical processes X and symmetric stable processes, namely there is a Gaussian process $Y = \{Y_t, t \in T\}$ whose finite dimensional distributions (or their characteristic functions) are connected as:

$$-\log \varphi_{X_{t_1},\ldots,X_{t_n}} \qquad (\lambda_1, \cdots, \lambda_n)$$
$$= [-\log \varphi_{Y_{t_1},\ldots,Y_{t_n}}(\lambda_1, \cdots, \lambda_n)]^{\frac{p}{2}}, 1 \leq p \leq 2.$$

For an analysis of spherical random variables and their relation to Gaussian processes, the paper by Kelker (1970) is useful.
The preceding account shows the nontriviality of R. Frish's problem, and especially its close relation with the class of p-stable processes in which the Gaussian class is the simplest and most important. But the stable class is an important part of infinitely divisible processes analyzed by P. Lévy and A. Khintchine in the early 1930s.
It shows that $\log \varphi_{X_t}(\cdot)$ is well-defined and its principal branch is taken in the calculations below without comment. There is also an ancillary problem of importance which is related to the conditioning concept. To clarify this special point, which is not really appreciated in the literature although the regression concept itself was initiated by Francis Galton in the late 19th century, it is outlined in the next section to highlight the troublesome points. Then the rest of the paper is devoted to the linearity of regression problem for random measures.

8.2. Conditioning Concepts and Regression

Recall that for an integrable random variable X and $\mathcal{B} \subset \Sigma$, a σ-algebra, the conditional expectation $E^{\mathcal{B}}(X)$ of X given $\mathcal{B}$ is the essentially unique $\mathcal{B}$-measurable random variable $\tilde{X}$ satisfying

$$\int_A X dP = \nu_X(A) = \int_A \tilde{X} dP_{\mathcal{B}} = \int_A \frac{d\nu_X}{dP_{\mathcal{B}}}, A \in \mathcal{B}, \tag{8.7}$$

where ν_X is $P_{\mathcal{B}}$ (restriction)-continuous and $\tilde{X} = \frac{d\nu_X}{dP_{\mathcal{B}}}$ is the Radon-Nikodým derivative. Thus $\tilde{X} = E^{\mathcal{B}}(X)$ and the mapping $X \mapsto E^{\mathcal{B}}(X)$ is linear, positivity preserving, and satisfies $\|E^{\mathcal{B}}(X)\|_1 \leq \|X\|_1$. Moreover, taking $A = \Omega$ in (8.1) one has the *fundamental relation*

$$E(X) = \int_\Omega X dP = \int_\Omega E^{\mathcal{B}}(X) dP_{\mathcal{B}} = E(E^{\mathcal{B}}(X)). \tag{8.8}$$

If $\mathcal{B} = \sigma(Y)$, the σ-algebra generated by Y (i.e.,$\sigma(Y) = Y^{-1}(\mathcal{R}) \subset \Sigma$, where $\mathcal{R}$ is the Borel σ-algebra of $\mathbb{R}$), then $E^{\mathcal{B}}(X) = E^{\sigma(Y)}(X)$ is also written as $E(X|Y)$. The classical Doob-Dynkin lemma (cf., e.g., Rao (1981), p.4) implies that

$$E(X|Y) = g(Y), \tag{8.9}$$

where $g : \mathbb{R} \to \mathbb{R}$ is a Borel function, and $g(Y)$ is termed the *regression of X on Y*. Then by (8..8) $E(g(Y)) = E(X)$ is finite *even though Y itself need not be integrable.* This property is of interest here. If $g(Y) = aY + b, a, b \in \mathbb{R}, Y \in L^1(P)$, then one has the linearity of regression of X on Y. The methodology employed here is based on Kolmogorov's introduction of the conditioning concept by the relation embodied in (8.7). [There is also another procedure assuming only that X is integrable relative to the conditional measure $P(\cdot|Y)$, which is not the same as the above since (8.8), (8.9) need not hold. It is perhaps influenced by Rényi's(1955) approach to conditioning. This will be discussed to explain the difference between these approaches.]

Thus for the alternative procedure, consider the problem starting with conditional distributions in the following sense. Taking $X = \chi_A$ in (8.7) or (8.8) above and setting $P^{\mathcal{B}}(A) = E^{\mathcal{B}}(\chi_A), A \in \Sigma$, it follows from (8.7) that

$$\int_B P^{\mathcal{B}}(A)(\omega) dP_{\mathcal{B}}(\omega) = P(A \cap B), A \in \Sigma, B \in \mathcal{B}, \tag{8.10}$$

and $P^{\mathcal{B}} : \Sigma \to L^p(P_{\mathcal{B}})$ is σ-additive in p-mean $1 \leq p < \infty$, (but this is not true for $p = \infty$). Moreover, if $A_n \in \Sigma, n \geq 1$, are disjoint, then outside a set $N \in \Sigma$, depending on the A_n-sequence, $P(N) = 0$, one has

$$P^{\mathcal{B}}(\cup_{n=1}^\infty A_n)(\omega) = \sum_{n=1}^\infty P^{\mathcal{B}}(A_n)(\omega), \quad \omega \in \Omega - N. \tag{8.11}$$

If there is a fixed null set N_0, as above, and (8.11) holds for all sequences with $N = N_0$, then $P^{\mathcal{B}}(\cdot)$ is termed a *regular conditional measure* and denoted $Q(A,\omega) = P^{\mathcal{B}}(A)(\omega)$. Thus (8.10) can be written as:

$$P(A \cap B) = \int_B Q(A,\omega)dP_{\mathcal{B}}(\omega), \quad A \in \Sigma, B \in \mathcal{B}, \tag{8.12}$$

and $Q(A,\omega) = E(\chi_A|Y)(\omega)$ if $\mathcal{B} = \sigma(Y)$. Under 'standard' conditions of regularity of these measures, it is possible to express the conditional expectation $E(X|Y)$ as the Leb. integral in the form

$$E(X|Y)(\omega') = \int_\Omega X(\omega)Q(\omega,\omega'), \quad a.a.(\omega').$$

In particular, translating this into the image space (R^2 here) one has

$$E(X|Y)(y) = \int_{\mathbb{R}} x dF_{X|Y}(x|y), \tag{8.13}$$

and if the conditional distribution $F_{X|Y}(\cdot|y)$ has a density $f_{X|Y}(\cdot|y)$ relative to the Lebesgue measure, then (8.13) is written in the familiar notation as:

$$E(X|Y = y) = \int_R x f_{X|Y}(x|y)dx,$$

and (8.7) becomes when X is integrable,

$$\begin{aligned} \textstyle\int_R \int_R x f_{X,Y}(x,y)dxdy &= \textstyle\int_R x f_X(x)dx = E(X) \\ &= \textstyle\int_R [\int_R x f_{X|Y}(x|y)dx] f_Y(y)dy (= E(E(X|Y))). \end{aligned} \tag{8.14}$$

It should be emphasized that this equation is derived under the restriction that the conditional measures are regular, so that $f_{X,Y}(x,y) = f_{X|Y}(x|y)f_Y(y)$ holds for almost all (x,y) (Leb.), which is the usual assumption in the elementary treatments of the subject.

However if $f_{X,Y}(x,y)$ is *defined* as the product of $f_{X|Y}(x|y)f_Y(y)$, then the fundamental equation (8.7) [or (8.14)] *need not hold.* But the validity of the equation (8.7) or (8.14) is essential for the definition of a regression function as in (8.8). It is not difficult to construct (conditional) densities $f_{X|Y}(x|y), f_Y(y)$such that a (continuous) function $h(\cdot)$ integrable relative to $f_{X|Y}(\cdot|y)$ for each $y \in R$, but need not satisfy (8.14) if $f_Y(\cdot)$ is chosen suitably. Here is a simple (and known) example illustrating this point. Many others can be constructed for different types of h.

Let $f_{X|Y}$ be given by

$$f_{X|Y}(x|y) = \sqrt{\pi} y^{-\frac{1}{2}} e^{-\frac{x^2}{y}}, y > 0, x \in R$$

so that for any fixed positive y one has

$$E(X^n|Y = y) = 2\sqrt{\frac{\pi}{y}} \int_0^\infty x^n e^{-\frac{x^2}{y}} < \infty, n \geq 1.$$

If $f_Y(y) = (\pi y)^{-\frac{1}{2}} e^{-y}, y > 0$, so that it is a probability density, then it is immediately verified that $f_{X,Y}(x,y) = f_{X|Y}(x|y) f_Y(y)$ is also a probability density and that the marginal $f_X(\cdot)$ is Cauchy for which $E(X^n)$ ($h(x) = x^n$ is continuous!), does not exist and *(8.14) fails for all* $n \geq 1$. From this one can derive using the other marginal the conditional $f_{Y|X}(y|x)$ and they are termed in the literature by different names – the first is a density of X depending on $Y = y$ as a parameter, and the second is termed "posterior", with f_Y "prior". In these applications care is thus needed to interpret results because of the above examples.

Remark. Analyses involving hypotheses on conditional densities that are used to derive properties of measures on product spaces, have to keep in view of the nonuniqueness problems which are often present.

This is not a special situation considered here. In studies of Markov processes, the Chapman-Kolmogorov equation plays an important part. But P. Lévy, W. Feller and others have constructed non-Markov processes (even chains) whose conditional probability functions (or transition probabilities) satisfy the Chapman-Kolmogorov equations! Such problems should be kept in view for applications with regression functions. Thus only the Kolmogorov formulation used in Section 8.1 will be used in the following work.

Now the regression function $g(Y) = E(X|Y)$ is well-defined and $g(Y)$ is $\sigma(Y)$-measurable for any (even infinite) random vector Y, as a consequence of the Doob-Dynkin lemma. This property is crucial for the following work (as it was for all the results discussed in Section 8.1).

In this paper, the case that X, Y are values of a random measure on reasonable sets will be analyzed, motivated by the researches discussed in the first section.

8.3. Regression for Random Measures

Let $\mathcal{B}_0(\mathbb{R}^n)$ be the ring of all bounded Borel sets of $\mathbb{R}^n$. [Recall that a set in $\mathbb{R}^n$ is bounded if its closure is compact.] It is also closed under countable intersections so that it is a δ-ring. A mapping $Z : \mathcal{B}_0(\mathbb{R}^n) \to L^p(P), p \geq 1$, is additive if $Z(A \cup B) = Z(A) + Z(B)$ for disjoint $A, B \in \mathcal{B}_0(\mathbb{R}^n)$, and is independently valued if all such $Z(A), Z(B)$ are independent random variables. This $Z(\cdot)$ is a *random measure* if for all $A_n \in \mathcal{B}_0(\mathbb{R}^n)$, disjoint, and $\cup_n A_n \in \mathcal{B}_0(\mathbb{R}^n)$ one must have

$$Z(\cup_{n=1}^{\infty} A_n) = \sum_{n=1}^{\infty} Z((A_n), \tag{8.15}$$

the series converges in probability, which in the present case of independent values, is equivalent to convergence with probability 1 (by a classical theorem due to P. Lévy). It also converges in p-mean in that $\|Z(\cup_{k=1}^{\infty} A_k) - \sum_{j=1}^{m} Z(A_j)\|_p \to 0$ as

$m \to \infty, 1 \leq p < \infty$.
The random measure is of symmetric p-stable class if $\{Z(A), A \in \mathcal{B}_0(\mathbb{R}^n)\}$ is a symmetric p-stable process so that for each $A \in \mathcal{B}_0(\mathbb{R}^n)$ its characteristic function is given by

$$\varphi_{Z(A)}(t) = \exp\{-c(A)|t|^p\}, \quad c(A) \geq 0. \tag{8.16}$$

It is seen that $c : \mathcal{B}_0(\mathbb{R}^n) \to \mathbb{R}^+$ is σ-additive. In fact let $A_j \in \mathcal{B}_0(\mathbb{R}^n)$ be disjoint, $j = 1, 2$. Then

$$\varphi_{Z(A_1 \cup A_2)}(t) = \varphi_{Z(A_1)}(t)\varphi_{Z(A_2)}(t), \tag{8.17}$$

by independence of $Z(A_j), j = 1, 2..$ Since $Z(A)$ is stable, hence infinitely divisible, the logarithm of its characteristic function is well-defined, and from (8.7) and (8.8), one has on taking (natural) logs

$$-c(A_1 \cup A_2)|t|^p = -(c(A_1) + c(A_2))|t|^p, \quad t \in \mathbb{R}, (c(A_j) \geq 0). \tag{8.18}$$

It follows that $c(\cdot)$ is additive, and if $A_n \downarrow \emptyset, A_n \in \mathcal{B}_0(\mathbb{R}^n)$, then $Z(A_n) \to Z(\emptyset) = 0$ in probability, and hence $\varphi_{Z(A_n)}(t) \to \varphi_0(t) = 1, t \in \mathbb{R}$, which implies by (8.16) that $c(A_n) \downarrow 0$. It results from these two facts that $c(\cdot)$ is σ-additive which is generally a σ-finite measure. This $c(\cdot)$ is usually called the Lévy measure governing $Z(\cdot)$. See Loève (1955, pp.327 ff) on this representation and related matters. A solution of the regression problem for random measures is given by:

Theorem 1. Let $Z : \mathcal{B}_0(\mathbb{R}^n) \to L^p(P), 1 \leq p \leq 2$ be a symmetric p-stable random measure. Then $E(Z(A)|Z(B)) = aZ(B)$ where $a \in \mathbb{R}$ depends only on A, B, so that the regression function is necessarily linear.

Proof. Since $Z(A)$ is integrable and symmetric, $E(Z(A)) = 0$, and its characteristic function is differentiable. First consider the case that $p > 1$. If $X = Z(A), Y = Z(B)$, then equation (8.8) of Section 8.2 holds. To avoid a triviality let $P(Z(B) \neq 0) > 0$, since otherwise $Z(B)$ will be independent of all $Z(A)$ so that $E(Z(A)|Z(B)) = 0, A \in \mathcal{B}_0(\mathbb{R}^n)$.
Consider an initial simplification:

$$\begin{aligned} E(Z(A)|Z(B)) &= E[(Z(A - A \cap B) + Z(A \cap B)|Z(B)] \\ &= E[Z(A) - Z(A \cap B)] + E(Z(A \cap B)|Z(B)), \end{aligned}$$

since $Z(\cdot)$ is a random measure,

$$\begin{aligned} &= [E(Z(A)) - E(Z(A \cap B))] + E(Z(A \cap B)|Z(B)), \\ &= h(A, B), \text{ (say)}. \end{aligned}$$

It will now be shown that the random variable $h(A,B)$ is a constant multiple of $Z(B)$, the constant determined only by A,B, to complete the proof of the theorem. For this one uses the Doob-Dynkin lemma, and the hypothesis that $Z(\cdot)$ is symmetric p-stable so that (8.16) is available with $c(\cdot)$ as a σ-additive function. Letting $i=\sqrt{-1}$ and $D=A\cap B\in\mathcal{B}_0(\mathbb{R}^n)$, consider

$$\begin{aligned} E(Z(D)e^{itZ(B)}) &= E(Z(D)e^{it[Z(D)+Z(B-D)]}) \\ &= E(Z(D)e^{itZ(D)})E(e^{itZ(B-D)}), \end{aligned}$$

since $Z(D)$ and $Z(B-D)$ are independent,

$$\begin{aligned} &= \tfrac{1}{i}E(\tfrac{d}{dt}(e^{itZ(D)}))E(e^{itZ(B-D)}) \\ &= \tfrac{1}{i}\tfrac{d}{dt}(E(e^{itZ(D)}))E(e^{itZ(B-D)}) \end{aligned}$$

since $\frac{d}{dt}$ and E commute,

$$= \tfrac{1}{i}\tfrac{d}{dt}(e^{-c(D)|t|^p})e^{-c(B-D)|t|^p}$$

by the symmetric p-stable property of $Z(\cdot)$,

$$\begin{aligned} &= \tfrac{1}{i}[-c(D)e^{-c(D)|t|^p}p|t|^{p-1}]e^{-c(B-D)|t|^p} \\ &= -\tfrac{c(D)}{ic(B)}[c(B)e^{-(c(D)+c(B-D))|t|^p}p|t|^{p-1}] \\ &= \tfrac{c(D)}{ic(B)}\left[\tfrac{d}{dt}(e^{-c(B)|t|^p}\right], \text{ since } D\subset B, \end{aligned}$$

and the additivity of $c(\cdot)$,

$$= \tfrac{c(D)}{c(B)}E(Z(B)e^{itZ(B)}), \text{ as before.} \tag{8.19}$$

On the other hand, to simplify the left side above, using (8.16) with $Y=Z(B)$ and $X=e^{itZ(B)}Z(D)$, one has

$$\begin{aligned} E(Z(D)e^{itZ(B)}) &= E(E(Z(D)e^{itZ(B)}|Z(B))) \\ &= E(e^{itZ(B)}E(Z(D)|Z(B))) \\ &= E(e^{itZ(B)}g_D(B)), \text{ (say)}, \end{aligned} \tag{8.20}$$

where $g_D(B)$ is the random variable coming from the Doob-Dynkin lemma which is thus a function of $Z(B)$ alone. Hence from (8.19) and (8.20), it follows that

$$E(e^{itZ(B)}[g_D(B) - \frac{c(D)}{c(B)}Z(B)]) = 0, t \in \mathbb{R}, \tag{8.21}$$

and thus the expression in [ ] is a function of $Z(B)$. Consequently writing V for $Z(B)$, since $g_D(B) \in L^1(P)$ and $g_D(B) - \frac{c(D)}{c(B)}Z(B) = k(V) \in L^1(P)$, (18.21) gives

$$\begin{aligned} 0 = & \quad E(e^{itV}k(V)) \\ & = \int_\Omega e^{itV}\, dQ_V, \quad t \in \mathbb{R}, \end{aligned}$$

where Q_V is a measure on the σ-algebra determined by the random element $k(V))$ which is just a Borel function of V (Doob-Dynkin lemma). Hence the measure Q_V must vanish identically, since its Fourier transform vanishes for all $t \in \mathbb{R}$.
It follows therefore that on setting $a = \frac{c(D)}{c(B)} \in \mathbb{R}$, the regression of $Z(A)$ on $Z(B)$ is linear for any $A, B \in \mathcal{B}_0(\mathbb{R}^n)$, the constant depending only on A, B, as asserted. The case of $p = 1$ is obtained by a simple modification as noted in the following remark, and thus the theorem is established.

Remarks. 1. This proposition includes some earlier work related to the topic by Lukacs and Laha (1964) and Rosinski (1984).

2. In the above result, replacing $Z(A)$ by X and Y as any random variable, write $E(X|Y) = g(Y)$ so that $E(g(Y)) = E(X)$, it follows that $g(Y)$ cannot be a polynomial regression of order $k > 1$ where

$$g(Y) = \sum_{j=1}^{k} b_j Y^j,$$

unless $E(Y^k)$ exists to begin with.

Also in the proof the omitted case $p = 1$ follows in just the same way with the understanding that $\frac{d}{dt}(e^{-|t|})$ is taken as $e^{-|t|}sgn(t)$, as usually done in analysis for differentiating an absolute value. Note that $E(Z(A))$ exists by assumption which rules out the Cauchy measures. Thus both the p-stable and the existence of the expectations of all $Z(A)$ effectively concentrates on $p > 1$ in the above work.

In the next section, an extension of the preceding analysis for stochastic integrals of deterministic functions with respect to the random measure Z will be considered after first sketching the relevant concept of integration, since Z need not have finite variation on any nonempty open interval and this work should include Brownian motion.

8.4. Regression for Random Integrals

Since $Z: \mathcal{B}_0(\mathbb{R}^n) \to L^p(P)$ typically does not have finite (Vitali) variation on non-degenerate intervals, the symbol $\int_{\mathbb{R}^n} f(t)Z(dt)$ cannot be defined in the Lebesgue-Stieltjes sense. But now the range of $Z(\cdot)$ is a Banach space, and so one must adapt an extended version of the integral given in Dunford and Schwartz (1968). For immediate use a key aspect of the D-S work generalized to δ-rings such as $\mathcal{B}_0(\mathbb{R}^n)$ from the σ-algebra case will be outlined so that it can be applied to cases including Brownian motion. Observe that $L^p(P), p \geq 1$, is a Banach space and $Z: \mathcal{B}_0(\mathbb{R}^n) \to L^p(P)$ is also σ-additive in p-mean. In case that $\mathcal{B}_0(\mathbb{R}^n)$ is replaced by the Borel σ-algebra $\mathcal{B}(\mathbb{R}^n)$, it was first shown by Bartle, Dunford and Schwartz (1955) that there exists a controlling measure $\mu: \mathcal{B}(\mathbb{R}) \to \mathbb{R}^+$ for Z so that Z is μ-continuous.

Since $\mathcal{B}_0(K)$ is also a σ-algebra for any compact $K \subset \mathbb{R}^n$ and since $\mathbb{R}^n$ is countably compact, the above result of the three authors admits a nontrivial extension to $\mathcal{B}_0(\mathbb{R}^n)$, as shown by Dinculeanu (2000, pp.54-55), and that will be needed here. The procedure adapted to the present situation is as follows. For each simple function $f_m : \mathbb{R}^n \to \mathbb{R}$, representable as $f_m = \sum_{j=1}^{k_m} a_{jm}\chi_{A_{jm}}, A_{jm} \in \mathcal{B}_0(\mathbb{R}^n)$, disjoint, consider

$$\int_A f_m dZ = \sum_{j=1}^{k_m} a_{jm} Z(A \cap A_{jm}). \tag{8.22}$$

It is easily verified that the definition, does not depend on the representation of f_m, and is unique. The result now can be extended nontrivially for general Borel functions $f: \mathbb{R}^n \to \mathbb{R}$ with the following procedure.

By the classical structure theorem there exist (Borel) simple functions $f_m : \mathbb{R}^n \to \mathbb{R}$ such that $f_m(x) \to f(x)$ for all $x \in \mathbb{R}^n$. If $\{\int_A f_m dZ, m \geq 1\}$ is Cauchy in $L^p(P)$, then let $Y^f(A)$ be this limit for each $A \in \mathcal{B}_0(\mathbb{R}^n)$, and set it as:

$$Y^f(A) = \int_A f dZ = \lim_{m \to \infty} \int_A f_m dZ. \tag{8.23}$$

The fact that $Y^f(A)$ does not depend on the sequence $\{f_m, m \geq 1\}$ approximating f is not at all obvious. It is here that the existence of a controlling measure μ of Z is required (cf., Dunford-Schwartz (1958), Theorem IV.10. 8). The result applies to all Banach spaces, not just the $L^p(P)$ class. In the present case, a specialized computation can be made to obtain an equivalent but more convenient controlling measure which can be utilized for the regression problem under consideration.

Since $Y^{f_m}(A) = \int_A f_m dZ$, its characteristic function can be obtained explicitly using (8.22) as:

$$\begin{aligned}\varphi_{Y^{f_m}(A)}(t) &= E(e^{it\sum_{j=1}^{k_m} a_{jm}Z(A\cap A_{jm})}) \\ &= \exp\{-\textstyle\sum_{j=1}^{k_m} |a_{jm}t|^p (c(A\cap A_{jm})\}\end{aligned}$$

by (8.16) of Section 8.3,

$$\begin{aligned}&= \exp\{-|t|^p \textstyle\int_A |\sum_{j=1}^{k_m} a_{jm}|^p dc)\} \\ &\to \exp\{-|t|^p \textstyle\int_A |f|^p(v)dc(v)\},\end{aligned}$$

by the Lévy continuity theorem.
Hence one has

$$\varphi_{Y^f(A)}(t) = E(e^{itY^f(A)}) = \exp\{-|t|^p \int_A |f|^p(v)dc(v)\}, \tag{8.24}$$

so that $Y^f(\cdot)$ is also a p-stable random measure. Since this is seen to hold for all $f \in L^p(c)$, one can conclude (using a result of Schilder (1970)) that

$$\|Y^f(A)\|_p = [\int_A |f|^p dc]^{\frac{1}{p}}. \tag{8.25}$$

It follows that $Y^f(\cdot)$ as well as $Z(\cdot)$ are absolutely continuous relative to $c(\cdot)$ which thus acts as a controlling measure for both random set functions and $\forall f \in L^p(c)$.
The preceding work establishes the following result:
Theorem 1. Let $Z : \mathcal{B}_0(\mathbb{R}^n) \to L^p(P), 1 < p \leq 2$, be a symmetric p-stable random measure with Lévy measure $c : \mathcal{B}_0(\mathbb{R}^n) \to \bar{\mathbb{R}}^+$. Then for each $f \in L^p(c)$, the integrals $\{Y^f(A) = \int_A f dZ, A \in \mathcal{B}_0(\mathbb{R}^n)\}$ are well-defined, and each $Y^f(\cdot)$ is again a symmetric p-stable random measure having a controlling measure $\tilde{c} : A \to \int_A |f|^p dc, A \in \mathcal{B}_0(\mathbb{R}^n)$. Moreover, the closure of $\{Y^f(A), A \in \mathcal{B}_0(\mathbb{R}^n), f \in L^p(c)\} \subset L^p(P)$ and the space $L^p(c)$ are isometrically isomorphic, as given by (8.25).
As a consequence of this theorem and the one in the preceding section, the following result is obtained on the linearity of regression of the class $\{Y^f(A), A \in \mathcal{B}_0(\mathbb{R}^n)\}$.
Proposition 2. Let $Z : \mathcal{B}_0(\mathbb{R}^n) \to L^p(P), 1 < p \leq 2$, be a symmetric p-stable random measure with the controlling measure $c : \mathcal{B}_0(\mathbb{R}^n) \to \bar{\mathbb{R}}^+$. Then for each $f \in L^p(c)$ one has

$$E(Y^f(A)|Y^f(B)) = \alpha Y^f(B), \quad A, B \in \mathcal{B}_0(\mathbb{R}^N), \tag{8.26}$$

so that the regression of this class is also linear, and the constant α is determined just by (A, B, f).

It is desirable to extend this result if the random measures $Y^f(\cdot)$ and $Y^g(\cdot)$ are considered for different $f, g \in L^p(c)$ which applies to 'processes of measures'. However, the preceding work is not adequate for this purpose since Y^f and Y^g may be dependent, although jointly symmetric p-stable. Hence a further extension and an idea, generalized for this situation from Lukacs and Laha (1964, p.118), is needed. The resulting solution can be stated as follows, omitting the details for a later publication.

Proposition 3. Let $Z : \mathcal{B}_0(\mathbb{R}^n) \to L^p(P), 1 < p \leq 2$, be a symmetric p-stable random measure with a controlling Lévy measure $c : \mathcal{B}_0(\mathbb{R}^n) \to \bar{\mathbb{R}}^+$. Then for any pair of elements $f, g \in L^p(c)$ the random measures Y^f, Y^g are jointly symmetric p-stable functions admitting a linear regression

$$E(Y^f(A)|Y^g(B)) = \beta Y^g(B), a.e., A, B \in \mathcal{B}_0(\mathbb{R}^n), \tag{8.27}$$

where β is a constant determined by A, B, f, g only.

A result of this type is needed in considering a jointly symmetric p-stable family of random measures $\{Z_t(\cdot), t \in T\}$ to treat multiple linear regressions generalizing the preceding work.

8.5. Final Remarks

The regression problem, whether linear or not, is inherently related to conditional expectations, and as such, many questions concerning existence, uniqueness and computability are nontrivial. In the above analysis, and in most of the literature on the subject, one follows the Kolmogorov approach. There is an alternative, sometimes considered as a generalization, proposed by Rényi (1955), and it is not immune to these difficulties.

In fact for the most part, the sets of measure zero have been excluded in the latter approach, and special prescriptions were indicated to include some null sets, but the solutions lack uniqueness, depend on the methods used,and other difficulties. More details and the differences of these questions were detailed in the author's book (cf.,Rao (2005)).

The linearity of regression has several other ramifications, including work on weak martingales and stochastic games (cf., e.g., Rao (2007)for some of these developments). On the other hand, Kanter's (1972) results show that this regression problem is closely tied to a study of (symmetric) p-stable processes. A combination of these two lines of attack leads to studying symmetric p-stable processes

$\{X_t, t \in S\}$ representable as integrals

$$X_t = \int_S K(s,t)dZ(s), \quad t \in S,$$

where $Z(\cdot)$ is a random measure studied above, and $K(\cdot,\cdot)$ is called its kernel. Several path properties of such collections are of interest in applications. An investigation of these questions have been undertaken by Rosinski (1986). An extension of this work as well as the analysis of the present paper for not necessarily symmetric classes is of immediate interest. This and related problems are natural candidates for a follow up study. These considerations will be omitted for the present.

References

1. Allen, H.V.(1938), *A theorem concerning the linearity of regression,* Statist. Res. Mem., 2, 60–68.
2. Bartle, R.G., N. Dunford, and J. Schwartz (1955), *Weak compactness and vector measures,* Canad. J. Math., 7, 289–305.
3. Dinculeanu, N. (2000), *Vector Integration and Stochastic Integration in Banach Spaces,* Wiley-Interscience, New York.
4. Dunford, N. and J. T. Schwartz (1958), *Linear Operators, Part I: General Theory,* Wiley-Interscience, New York.
5. Fix E. (1949), *Distributions which lead to linear regression,* Proc.,Berkeley Symp. Math. Statist. and Prob., 79–91.
6. Hardin C.D.(1982), *On the linearity of regression,* Z.Wahrs., 61, 293–302.
7. Kanter, M. (1972), *Linear sample spaces and stable processes,* J. Funct. Anal., 9, 441–459.
8. Kelker D. (1970), *Distribution theory of spherical distributions and a location-scale parameter generalization,* Sankhya, Ser. A, 32, 419–430.
9. Loève,M. (1955), *Probability Theory (3rd Edition)*, D. Van Nostrand, Princeton, N J.
10. Lukacs, E. and R. G. Laha (1964), *Applications of Characteristic Functions*, Hafner Publishing Co., New York.
11. Rao, M.M. (1981), *Foundations of Stochastic Analysis*, Academic Press, New York.
12. Rao M.M.(2005), *Conditional Measures and Applications, (2nd Edition)*, Chapman-Hall/CRC, Boca Raton, FL.
13. Rao M.M.(2007), *Exploring ramifications of the equation $E(Y|X) = X$,* J.Statist. Theory and Practice, 1, 73–88.
14. Rényi A.(1955), *On a new axiomatic theory of probability,* Acta Math. Sci. Hung., 6, 285–335.
15. Rosinski J.(1984), *Random integrals of Banach space valued functions,* Studia Math., 78, 15–38.
16. Rosinski J. (1986), *On a stochastic integral representation of stable processes with sample paths in Banach spaces,* J. Multivar. Anal., 20, 277–302.

17. Schilder M.(1970), *Some structure theorems for the symmetric stable laws,* Ann. Math. Statist., 41, 412–421.

Chapter 9

Mixed Multivariate Models for Random Sums and Maxima

Tomasz J. Kozubowski, Anna K. Panorska and Franco Biondi

University of Nevada at Reno, Reno, NV 89557, USA

E-mail:tkozubow@unr.edu

Motivated by problems arising in hydro-climatology, we consider multivariate distributions connected with X, Y, and N, where N has a discrete distribution on positive integers while X and Y are the sum and the maximum of N independent copies of a random variable W, independent of N. Our focus are bivariate distributions of (X,N) and (Y,N) in case where N and W are geometric and exponential variables, respectively. We present basic properties of these models, including marginal and conditional distributions, joint integral transforms, infinite divisibility, stability with respect to geometric summation, and estimation, and argue why such models are useful in describing the joint behavior of magnitudes, peak values, and durations of hydro-climatic episodes.

Contents

9.1. Introduction

In hydrology and climate research, the problem of estimating total stream flow, precipitation, Palmer Drought Severity Index, El Niño Southern Oscillation (ENSO) or Pacific Decadal Oscillation indexes exceeding a threshold (e.g. long term mean or high percentile) is of primary importance to water resource managers and safe engineering design needing reasonable flood, drought, or water storage estimates [see references numbered 11,12,16,46,61,64]. Long series of observations on environmental or hydrological processes are typically described in terms of positive and negative episodes. An episode contains consecutive observations either above or below a reference (threshold) level. The episodes are quantified in terms of three variables: duration N, which is the number of time intervals in the episode, magnitude X, which is the sum of all process values for a given duration, and the peak value Y, which is the absolute maximum reached by the process within a given episode. In these applications, the joint distributions involving the duration, magnitude, and maximum of an episode are of main interest.

Various univariate [see references numbered 55,56] and multivariate [see references numbered 17,22,42,54,57,62,63] stochastic models of duration, magnitude and maximum of episodes have appeared in the literature in recent years. In the context of flood analysis and ENSO, several bivariate models have been proposed for duration and magnitude [see references numbered 23,31,58,62,63], duration and maximum value [see references numbered 17,57,62,63] and magnitude and maximum value [see references numbered 22,42,54]. However, majority of existing models of episode characteristics have largely been selected according to the empirical fit to the data rather than theoretically derived from the mathematical properties of the process under consideration.

A natural mathematical interpretation of all three episode characteristics is connected with *random* sums and maxima of random variables. Following the work of [Biondi et al. 2005,2008] and [Kozubowski and Panorska 2005,2008] we shall describe general models for the duration, magnitude and maxima of episodes, which follow from the stochastic representation

$$(X,Y,N) \stackrel{d}{=} \left(\sum_{i=1}^{N} E_i, \bigvee_{i=1}^{N} E_i, N \right). \tag{9.1}$$

Here, the $\{E_i\}$ are the excesses of the process values within a positive episode of duration N, and $\vee$ denotes the maximum. We focus on the bivariate distributions of (X,N) and (Y,N) in case where N has a geometric distribution with the PDF

$$P(N=n)=p(1-p)^{n-1},\ n=1,2,\ldots, \tag{9.2}$$

and the $\{E_i\}$ are independent and identically distributed (IID) exponential variables with the PDF

$$f(x)=\beta e^{-\beta x},\ x>0, \tag{9.3}$$

independent of N. In the context of hydroclimatic events, the geometric distribution is often fit to univariate duration data [see references numbered 11,20,46,55,56], which is justified by the theory of runs (Feller, 1957]. The exponential distribution is also a popular model for the magnitude of hydroclimatic episodes, mostly due to its simplicity and good empirical fit to the data [see references numbered 17,42,46,61,64], although in [Biondi et al., 2002] this model was justified within a stochastic framework as a limit of a random number of IID random variables. Another reason why exponential distribution might provide a reasonable description of the values of a process above (or below) a threshold is the POT theory [see references numbered 10,18,49], which shows that suitably normalized excesses (of the process values above the threshold) converge to one of the POT distributions as the threshold increases. The fact that the exponential distribution is one of the three possible limits justifies its use in this context.
Another area where episodes with their duration, magnitude and maximum are of interest is finance and economics. Here, one is interested in *periods of growth* (or decline), where consecutive (say, daily) values of a quantity such as a stock index, interest rate, or a currency exchange rate are increasing (or decreasing). In practice, one typically considers *log returns*, which are the logarithms of two consecutive values, and the growth (decline) periods and episodes correspond to positive (negative) log returns. In these applications, geometric distribution for the duration and exponential for the magnitude of the episodes were proposed in [Kozubuski and Panorska, 2005], following an empirical observation of the conditional stability property of the returns noted in [Kozubuski and Panorska, 2003]. The latter stipulates that the (random) sum of the daily log returns over a growth period has the same distribution (up to the scale) as that of the daily log returns within the period. This stability under geometric compounding is a fundamental property of the exponential distribution [see references numbered 3,21,32]. Let us note that models for growth rates are of interest in many diverse areas, including modeling annual gross domestic product [Lee et al., 1998], stock prices [Madan et al., 1998], interest or foreign currency exchange rates [see references numbered 34,37,38,47], company sizes [see references numbered 1,14,59,60], and other processes [Reed 2001].

Thus, we start our analysis of (9.1) assuming geometric distribution for N and exponential distribution for the $\{E_i\}$. The two main bivariate models derived from (9.1) are the BEG model [see references numbered 12,34] describing the joint distribution of duration and magnitude, and the BTLG model [see references numbered 13,35] for duration and maximum of episodes. These are reviewed in Sections 9.2 and 9.3, respectively, where we present their basic properties, including estimation and their stability related to geometric summation and maxima. The latter are perhaps the most fundamental properties of the two models, extending those of their marginal geometric, exponential, and truncated logistic distributions. Recall that geometric and exponential distributions are both stable with respect to geometric compounding: a geometric sum of IID geometric random variables is geometric and geometric sum of IID exponential random variables is exponential. As shown in [Kozubuski, 2005], this property carries over to the bivariate case: a geometric sum of IID BEG random vectors has BEG distribution. Similarly, a geometric maximum of exponential random variables has a truncated logistic (TL) distribution, and a geometric maximum of IDD TL random variables is TL. One new result presented in Section 9.3 is an extension of this property to the bivariate case. We show that the mixed geometric maximum-sum $\left(\bigvee_{i=1}^{N} Y_i, \sum_{i=1}^{N} N_i\right)$ of IID BTLG random vectors (Y_i, N_i) has again a BTLG distribution whenever N has a geometric distribution independent of the (Y_i, N_i). In Section 9.4 we consider extentions of the two models, connected with deterministic sums and maxima of IID BEG and BTLG random vectors. These generalizations are related to the tri-variate distribution of

$$(X,Y,Z) \stackrel{d}{=} \left(\sum_{i=1}^{n} X_i, \bigvee_{i=1}^{n} Y_i, \sum_{i=1}^{n} N_i \right), \tag{9.4}$$

where the (X_i, Y_i, N_i) are IID random vectors admitting the stochastic representation (9.1). It turns out that (9.4) can be embedded into continuous-time stochastic processes. Our presentation will focus on the bivariate marginal distributions of this process, and include numerous new results related to the distribution of (Y, Z).

9.2. The BEG Model

Let $E_1, E_2, \ldots$ be IID exponential variables with parameter $\beta > 0$ and the probability density function (9.3) and let N be a geometric random variable with the pdf (9.2), independent of the $\{E_i\}$. We denote these distributions by $\mathcal{EXP}(\beta)$ and $\mathcal{GEO}(p)$, respectively. Following [Kozubowski,2005], we say that a random vector (X, N) has a BEG distribution with parameters $\beta > 0$ and $p \in (0,1)$, denoted by $\mathcal{BEG}(\beta, p)$, if

$$(X,N) \stackrel{d}{=} \left(\sum_{i=1}^{N} E_i, N \right). \tag{9.5}$$

As shown in [Kozubowski, 2005], the joint PDF of $(X,N) \sim \mathcal{BEG}(\beta, p)$ is

$$f(x,n) = \frac{p\beta^n}{(n-1)!}[x(1-p)]^{n-1}e^{-\beta x}, \; x > 0, \; n = 1,2,\ldots, \tag{9.6}$$

while the corresponding distribution function $F(x,n) = \mathbb{P}(X \leq x, N \leq n)$ and survival function $S(x,n) = \mathbb{P}(X > x, N > n)$ are

$$F(x,n) = 1 - e^{-p\beta x}\sum_{k=0}^{n-1}\left(e^{-(1-p)\beta x}\frac{[(1-p)\beta x]^k}{k!}\right) - (1-p)^n\left(1 - \sum_{k=0}^{n-1} e^{-\beta x}\frac{[\beta x]^k}{k!}\right) \tag{9.7}$$

and

$$S(x,n) = e^{-p\beta x}\left(1 - \sum_{k=0}^{n-1} e^{-(1-p)\beta x}\frac{[(1-p)\beta x]^k}{k!}\right) + (1-p)^n\sum_{k=0}^{n-1} e^{-\beta x}\frac{[\beta x]^k}{k!}, \tag{9.8}$$

respectively, for any real $x > 0$ and integer $n > 0$. Moreover, the moment generating function (MGF) of $(X,N) \sim \mathcal{BEG}(\beta, p)$ is

$$\mathbb{E}e^{tX+sN} = \frac{p\beta e^s}{\beta - t - \beta(1-p)e^s}, \; t < \beta[1-(1-p)e^s]. \tag{9.9}$$

9.2.1. *Marginal distributions*

By definition, the marginal distribution of N is geometric with pdf (9.2). The marginal distribution of X is exponential with parameter $p\beta$. This follows from the stability of the exponential distribution with respect to geometric compounding [Arnold, 1973]. We see that the name BEG stands for **b**ivariate distributions with **e**xponential and **g**eometric marginals.

9.2.2. *Conditional distributions*

Note that given $N = n$, the variable X is the sum of n IID exponential variables with parameter $\beta > 0$, so that it has a gamma distribution $\mathcal{G}(n, \beta)$ with the pdf

$$f_{X|N=n}(x) = \frac{\beta^n}{(n-1)!}x^{n-1}e^{-\beta x}, \; x > 0. \tag{9.10}$$

Since the marginal distribution of X is $\mathcal{EXP}(\beta p)$, the conditional pdf of N given $X = x > 0$ is

$$f_{N|X=x}(n) = \frac{[x\beta(1-p)]^{n-1}e^{-x\beta(1-p)}}{(n-1)!}, \quad n = 1,2,\ldots. \tag{9.11}$$

This is a Poisson distribution with parameter $\lambda = x\beta(1-p)$ shifted up by one.

9.2.2.1. *The conditional distribution of* (X,N) *given* $N > n$

As shown in [Kozubowski, 2005], for any real $x > 0$ and positive integers m,n, we have

$$\mathbb{P}(X > x, N > m|N > n) = \frac{e^{-p\beta x}}{(1-p)^n}\left(1 - \sum_{k=0}^{m-1}\frac{[(1-p)\beta x]^k}{e^{(1-p)\beta x}k!}\right) + (1-p)^{m-n}\sum_{k=0}^{m-1}\frac{[\beta x]^k}{e^{\beta x}k!}$$

when $m \geq n > 0$ and

$$\mathbb{P}(X > x, N > m|N > n) = \frac{e^{-p\beta x}}{(1-p)^n}\left(1 - \sum_{k=0}^{n-1}\frac{[(1-p)\beta x]^k}{e^{(1-p)\beta x}k!}\right) + \sum_{k=0}^{n-1}\frac{[\beta x]^k}{e^{\beta x}k!}$$

when $0 < m \leq n$. Note that the last expression represents the conditional survival function of X given $N > n$.

9.2.2.2. *The conditional distribution of* (X,N) *given* $X > u$

As shown in [Kozubowski, 2005], for any positive integer $n > 0$ and positive real x,u, we have

$$\mathbb{P}(X > x, N > n|X > u) = e^{-p\beta(x-u)}\left(1 - \sum_{k=0}^{n-1}\frac{[(1-p)\beta x]^k}{e^{(1-p)\beta x}k!}\right) + (1-p)^n e^{p\beta u}\sum_{k=0}^{n-1}\frac{[\beta x]^k}{e^{\beta x}k!}$$

when $x \geq u > 0$ and

$$\mathbb{P}(X > x, N > n|X > u) = 1 - \sum_{k=0}^{n-1}\frac{[(1-p)\beta u]^k}{e^{(1-p)\beta u}k!} + (1-p)^n e^{p\beta u}\sum_{k=0}^{n-1}\frac{[\beta u]^k}{e^{\beta u}k!}$$

when $u \geq x > 0$. The last expression is also the conditional survival function of N given $X > u$.

9.2.3. *Moments*

Routine calculations show that if $(X,N) \sim \mathcal{BEG}(\beta, p)$ and $\eta, \gamma \geq 0$ then

$$\mathbb{E}X^\eta N^\gamma = \left(\frac{1}{\beta}\right)^\eta \sum_{n=1}^{\infty} n^\gamma \frac{\Gamma(n+\eta)}{\Gamma(n)} p(1-p)^{n-1}.$$

In particular, we have $\mathbb{E}X = (\beta p)^{-1}$, $\mathbb{E}N = p^{-1}$, and $\mathbb{E}XN = (2-p)/(\beta p^2)$. The covariance matrix of (X,N) is

$$\Sigma = \begin{bmatrix} \frac{1}{\beta^2 p^2} & \frac{1-p}{\beta p^2} \\ \frac{1-p}{\beta p^2} & \frac{1-p}{p^2} \end{bmatrix}, \tag{9.12}$$

while the correlation coefficient of X and N is $\rho = \sqrt{1-p}$. It is worth noting that as $p \to 0$ (so that the geometric variable N converges to infinity), the correlation coefficient approaches one, which is a reflection of the fact that the distribution of (pX, pN) converges to a degenerate distribution with exponential marginals.

9.2.4. *Representations*

As shown in [Kozubowski, 2005], BEG random vectors admit certain representations involving random and deterministic sums. These generalize similar properties of exponential and geometric distributions, and justify BEG stochastic models in situations where the variable(s) of interest arise from aggregating a deterministic or random number of independent innovations.
The first representation of $(X,N) \sim \mathcal{BEG}(\beta, p)$ is of the form

$$(X,N) \stackrel{d}{=} \sum_{j=1}^{n} (R_j, \nu_j), \ n = 1,2,\ldots, \tag{9.13}$$

where the random vectors (R_j, ν_j) are IID. This shows that the BEG distribution is infinitely divisible. Here, $\nu_j = 1/n + N_j$, where the $\{N_j\}$ are IID negative binomial random variables given by the PDF (9.51) with $t = 1/n$ and

$$R_j = \sum_{i=1}^{N_j} E_{ij} + G_j, \tag{9.14}$$

where all the variables on the right-hand-side of (9.14) are mutually independent. The variables $\{E_{ij}\}$ are IID with the exponential $\mathcal{EXP}(\beta)$ distribution, and the $\{G_j\}$ are IID gamma $\mathcal{G}(1/n, \beta)$ variables (with shape parameter $1/n$ and scale β). The second representation involves a random sum with a Poissonian number of terms,

$$(X,N) \stackrel{d}{=} (E,1) + \sum_{j=1}^{Q_\lambda} (R_j, Z_j). \tag{9.15}$$

Here, E has the $\mathcal{EXP}(\beta)$ distribution, Q_λ has a Poisson distribution with mean $\lambda = -\log p$, the $\{Z_j\}$ are IID variables with logarithmic distribution given by the PDF

$$\mathbb{P}(Z_j = k) = \frac{1}{\lambda} \frac{(1-p)^k}{k}, \ k = 1,2,\ldots, \tag{9.16}$$

and

$$R_j = \sum_{i=1}^{Z_j} E_{ij}. \tag{9.17}$$

The $\{E_{ij}\}$ in (9.17) are IID copies of E, and all the variables on the right-hand-side of (9.15) are mutually independent.

9.2.5. *Stability properties*

Recall that both geometric and exponential distributions are stable with respect to geometric compounding: if N_q is a geometric variable $\mathcal{GEO}(q)$ with $q \in (0,1)$, then the distribution of the random sum $\sum_{i=1}^{N_q} W_i$ is exponential (geometric) if the $\{W_i\}$ are IID exponential (geometric), whenever the $\{W_i\}$ and N_q are independent. As shown in [Kozubowski, 2005], this stability property is shared by the BEG distribution.

Proposition 1 Let $\beta > 0$ and $p,q \in (0,1)$. If N_q is a geometric $\mathcal{GEO}(q)$ variable and (X_i,N_i) are IID $\mathcal{BEG}(\beta,p)$ variables, independent of N_q, then the distribution of the geometric sum $\sum_{i=1}^{N_q}(X_i,N_i)$ is $\mathcal{BEG}(\beta,pq)$.

Note that by Proposition 1, if $(X,N) \sim \mathcal{BEG}(\beta,p)$ then for $0 < p < q < 1$ we have

$$(X,N) \stackrel{d}{=} \sum_{i=1}^{N_q}(\tilde{X}_i,\tilde{N}_i), \tag{9.18}$$

where the $(\tilde{X}_i,\tilde{N}_i)$ are IID $\mathcal{BEG}(\beta,p/q)$ random vectors. However, as shown in [Kozubowski, 2005], when $0 < q < p < 1$ then the relation (9.18) does not hold with any IID variables under the summation, so that the BEG distribution is not *geometrically infinitely divisible* (cf. [Klebanov et al., 1984]).

9.2.6. *Estimation*

Here we summarize method of moments and maximum likelihood estimation of the BEG parameters. Let $(X_1,N_1),\ldots,(X_n,N_n)$ be a random sample from a $\mathcal{BEG}(\beta,p)$ distribution. The log-likelihood function,

$$L(\beta,p) = n\overline{N}_n \log\beta + \sum_{i=1}^{n} \log\frac{X_i^{N_i-1}}{(N_i-1)!} - n\overline{X}_n\beta + n(\overline{N}_n-1)\log(1-p) + n\log p, \tag{9.19}$$

is maximized by a unique pair

$$\hat{\beta}_n = \frac{\overline{N}_n}{\overline{X}_n},\ \hat{p}_n = \frac{1}{\overline{N}_n}, \tag{9.20}$$

where $\overline{X}_n$ and $\overline{N}_n$ are the sample means of the $\{X_i\}$ and the $\{N_i\}$, respectively. It is not hard to see that these are also moment estimators (MMEs), resulting from equating the theoretical moments of X and N with the sample moments. Moreover, the maximum likelihood estimators (MLEs) given above are consistent, asymptotically normal, and efficient, as shown in the following result taken from [Kozubowski, 2005].

Proposition 2 Let $(X_1,N_1),\ldots,(X_n,N_n)$ be IID variables from a $\mathcal{BEG}(\beta,p)$ distribution, such that the sample averages satisfy the relations $\overline{X}_n > 0$, $\overline{N}_n > 1$. Then there exist unique MLEs of β and p, given by (9.20). The vector MLE $(\hat{\beta}_n,\hat{p}_n)'$ is
(i) consistent;
(ii) asymptotically normal, that is $\sqrt{n}[(\hat{\beta}_n,\hat{p}_n)' - (\beta,p)']$ converges in distribution to a bivariate normal distribution with the (vector) mean zero and the covariance matrix

$$\Sigma_{MLE} = \begin{bmatrix} \beta^2 p & 0 \\ 0 & p^2(1-p) \end{bmatrix}; \tag{9.21}$$

(iii) asymptotically efficient - the asymptotic covariance matrix (9.21) coincides with the inverse of the Fisher information matrix.

9.3. The BTLG Model

A random vector (Y,N) with the stochastic representation

$$(Y,N) \stackrel{d}{=} \left(\bigvee_{i=1}^{N} E_i, N \right), \tag{9.22}$$

where again the $\{E_i\}$ are IID exponential variables (9.3) and N is a geometric variable (9.2), independent of the $\{E_i\}$, is said to have a BTLG distribution with parameters $\beta > 0$ and $p \in (0,1)$, denoted by $\mathcal{BTLG}(\beta,p)$. The name BTLG stands for **b**ivariate distribution with **t**runcated logistic and **g**eometric marginals. The joint PDF of $(Y,N) \sim \mathcal{BTLG}(\beta,p)$, derived in [kozubowski and Panorska, 2006], is

$$f(y,n) = n\beta p e^{-\beta y}[(1-e^{-\beta y})(1-p)]^{n-1}, \; y > 0, \; n = 1,2,\ldots. \tag{9.23}$$

The corresponding distribution function and the survival function are

$$\mathbb{P}(Y \le y, N \le n) = \frac{p}{q}\frac{1-q}{1-p}(1-(1-q)^n), \; y > 0, n = 1,2,\ldots, \tag{9.24}$$

and

$$\mathbb{P}(Y > y, N > n) = (1-p)^n \left[1 - \frac{p}{q}\frac{(1-q)^{n+1}}{(1-p)^{n+1}} \right], \; y > 0, n = 1,2,\ldots, \tag{9.25}$$

respectively, where

$$q = q(p,\beta,y) = p + (1-p)e^{-\beta y} \in (0,1). \tag{9.26}$$

The MGF of $(Y,N) \sim \mathcal{BTLG}(\beta, p)$ can be expressed in terms of the Gaussian hypergeometric function [Gradshteyn et al., 1994]

$$F(\alpha,\beta;\gamma;z) = 1 + \frac{\alpha\cdot\beta}{\gamma\cdot 1}z + \frac{\alpha(\alpha+1)\beta(\beta+1)}{\gamma(\gamma+1)\cdot 1\cdot 2}z^2 + \cdots \tag{9.27}$$

As shown in [Kozubowski and Panorska, 2006], we have

$$M_{Y,N}(t,s) = \mathbb{E}e^{tY+sN} = \frac{pe^s}{1-t/\beta}\cdot F(2,1;2-t/\beta;(1-p)e^s),\ t<\beta, s<-\log(1-p), \tag{9.28}$$

where F is given by (9.27). The function $M_{Y,N}(t,s)$ admits the following integral and series representations:

$$pe^s\int_0^1 \frac{1}{(1-u)^{t/\beta}}\frac{1}{[1-(1-p)ue^s]^2}du = pe^s\sum_{n=1}^{\infty} n[(1-p)e^s]^{n-1}\frac{\Gamma(1-t/\beta)\Gamma(n)}{\Gamma(1+n-t/\beta)}. \tag{9.29}$$

9.3.1. *Marginal distributions*

By definition, the marginal distribution of N is geometric with the PDF (9.2). Since Y is a maximum of a geometric number of exponential variables with parameter β, the CDF of Y can be written as $F_Y(y) = G_N(F(y))$, where $F(y) = 1 - e^{-\beta y}$ and G_N is the probability generating function (PGF) of N [Marshall and Olkin, 1997]. This leads to

$$F_Y(y) = \frac{p(1-e^{-\beta y})}{p+(1-p)e^{-\beta y}} = \frac{p}{q}\frac{1-q}{1-p},\ y\geq 0, \tag{9.30}$$

with q as above. The corresponding pdf is

$$f_Y(y) = \frac{\alpha\beta e^{-\beta y}}{[1-(1-\alpha)e^{-\beta y}]^2},\ y>0, \tag{9.31}$$

with $\alpha = 1/p > 1$. It follows that Y has a logistic distribution [see references numbered 2,30] with the CDF

$$F(y) = \frac{1}{1+\exp\left(\frac{y-\mu}{\sigma}\right)},\ y\in\mathbb{R}, \tag{9.32}$$

and parameters $\mu = \log[(1-p)/p]/\beta \in \mathbb{R}$ and $\sigma = 1/\beta > 0$, truncated below at zero. When $\mu = 0$ ($p = 1/2$), this reduces to a (scaled) *half-logistic* (or *folded logistic*) distribution introduced in [Balakrishnan, 1985] and studied in [see references nnumbered 6-9,52]. Incidentally, the density (9.31) is well-defined for

$\alpha \in (0,1)$ as well, and it corresponds to the *minimum* of N IID $\mathcal{EXP}(\beta)$ variables, where $N \sim \mathcal{GEO}(\alpha)$, [Marshalland Olkin, 1997] for more details.

9.3.2. *Conditional distributions*

Given $N = n$, Y is the maximum of n IID $\mathcal{EXP}(\beta)$ variables. Consequently, the CDF and PDF of Y are

$$F_{Y|N}(y|n) = (1 - e^{-\beta y})^n, \ y > 0, \tag{9.33}$$

and

$$f_{Y|N}(y|n) = n\beta e^{-\beta y}(1 - e^{-\beta y})^{n-1}, \ y > 0, \tag{9.34}$$

respectively. This is a particular case of *generalized exponential* distribution with an integer shape parameter, [see references numbered 25-29,50,51,65] for more details. Next, by dividing the marginal pdf of Y (9.31) into the joint pdf (9.23), we obtain the conditional pdf of N given $Y = y > 0$,

$$f_{N|Y}(n|y) = \mathbb{P}(N = n|Y = y) = nq(1-q)^n, \ n = 1, 2, \ldots, \tag{9.35}$$

where q is given by (9.26) as before. Since the above pdf is of the form $n\mathbb{P}(N_q = n)/\mathbb{E}N_q$, where $N_q \sim \mathcal{GEO}(q)$, this is a *length-biased* (or *tilted*) distribution [Patil et al., 1988] corresponding to the geometric variable N_q. Incidentally, the distribution of N_q is the same as that of N given $Y \leq y$.

9.3.2.1. *The conditional distribution of (Y,N) given $N > n$*

As shown in [Kozubowski and Panorska, 2006], for any real $y > 0$ and positive integers m, n we have

$$\mathbb{P}(Y > y, N > m|N > n) = \begin{cases} (1-p)^{m-n}\left(1 - \frac{p(1-e^{-\beta y})(1-e^{-\beta y})^m}{p+(1-p)e^{-\beta y}}\right) & \text{for } m \geq n > 0 \\ 1 - \frac{p(1-e^{-\beta y})(1-e^{-\beta y})^n}{p+(1-p)e^{-\beta y}} & \text{for } n \geq m > 0. \end{cases}$$

The case $n \geq m > 0$ coincides with the conditional survival function of Y given $N > n$.

9.3.2.2. *The conditional distribution of (Y,N) given $Y > u$*

As shown in [Kozubowski and Panorska, 2006], for any integer $n > 0$ and real $u, y > 0$, the conditional survival function $S(y,n|u) = \mathbb{P}(Y > y, N > n|Y > u)$ is of the form

$$S(y,n|u) = \begin{cases} \frac{p+(1-p)e^{-\beta u}}{e^{-\beta u}}(1-p)^n\left(1 - \frac{p(1-e^{-\beta y})(1-e^{-\beta y})^n}{p+(1-p)e^{-\beta y}}\right) & \text{for } y \geq u > 0 \\ (1-p)^{n+1} + (1-p)^n p e^{\beta u}[1 - (1-e^{-\beta u})^{n+1}] & \text{for } u \geq y > 0. \end{cases}$$

The case $u \geq y > 0$ reduces to the conditional survival function of N given $Y > u$, with the corresponding pdf $\mathbb{P}(N = n|Y > u) = C_{p,r}(1-p)^{n-1}(1-r^n)$, where $r = 1 - e^{-\beta u} \in [0,1)$ and $C_{p,r} = p[1-(1-p)r]/(1-r)$. This further reduces to $\mathcal{GEO}(p)$ when $r = 0$ $(u = 0)$.

9.3.3. *Moments*

Routine calculations lead to series representations for the mixed moments $\mathbb{E}Y^\eta N^\gamma$ of $(Y,N) \sim \mathcal{BTLG}(\beta,p)$. For any real $\eta,\gamma \geq 0$ we have

$$\mathbb{E}Y^\eta N^\gamma = \frac{\Gamma(\eta+1)}{\beta^\eta} \sum_{n=1}^{\infty} n^{\gamma+1} p(1-p)^{n-1} \sum_{k=0}^{n-1} (-1)^k \binom{n-1}{k} \left(\frac{1}{k+1}\right)^{\eta+1}.$$

In particular, as shown in [Kozubowski and Panorska, 2006], we have $\mathbb{E}Y = -\log p/[(1-p)\beta]$, $\mathbb{E}N = p^{-1}$, and $\mathbb{E}YN = (1-\log p)/(\beta p)$. Further, the covariance matrix of (Y,N) is

$$\Sigma = \begin{bmatrix} \frac{pc_p-(\log p)^2}{\beta^2(1-p)^2} & \frac{1-p+p\log p}{\beta p(1-p)} \\ \frac{1-p+p\log p}{\beta p(1-p)} & \frac{1-p}{p^2} \end{bmatrix}, \tag{9.36}$$

where

$$c_p = \int_0^\infty \frac{u^2 e^{-u} du}{[e^{-u} + p/(1-p)]^2}. \tag{9.37}$$

The correlation coefficient of Y and N is positive and of the form

$$\rho = \frac{1-p+p\log p}{\sqrt{1-p}\sqrt{pc_p - (\log p)^2}}.$$

9.3.4. *Representations*

As noted in [Kozubowski and Panorska, 2006], a random variable $(Y,N) \sim \mathcal{BTLG}(\beta,p)$ admits the representation

$$(Y,N) \stackrel{d}{=} \left(\sum_{j=1}^{N} \frac{E_j}{j}, N\right), \tag{9.38}$$

where the $\{E_j\}$ are IID $\mathcal{EXP}(\beta)$ variables independent of $N \sim \mathcal{GEO}(p)$. This is a consequence of the fact that the distribution of the conditional random variable $Y|N = n$ is a convolution of n independent exponential variables with parameters βj, $j = 1,\ldots,n$, [Gupta and Kundu 1999]. The variables under the geometric sum in (9.38) can be thought of as the spacings corresponding to N IID $\mathcal{EXP}(\beta)$ variables.

9.3.5. *Stability properties*

As we noted earlier, the geometric distributions is stable with respect to geometric compounding. On the other hand, the truncated logistic distribution (9.30) is stable with respect to geometric maximum: if N_q is a geometric variable $\mathcal{GEO}(q)$ with $q \in (0,1)$, then the distribution of the random maximum $\bigvee_{i=1}^{N_q} Y_i$ has a truncated logistic distribution if the $\{Y_i\}$ are IID with a truncated logistic distribution, whenever the $\{Y_i\}$ and N_q are independent [Marshall and Olkin, 1997]. To prove an extension of these results to the BTLG model, we first derive the hybrid cumulative distribution function - characteristic function (CDF-ChF) of the BTLG distribution, defined as

$$\phi(r,s) = \mathbb{E}\{\chi(Y \le r)\cdot e^{isN}\}, \; r,s \in \mathbb{R}, \tag{9.39}$$

where $\chi(A)$ is the indicator of the set A. This function uniquely determines the probability distribution, and is particularly useful when dealing with sums and maxima [see references numbered 2,15].

Proposition 3 If $(Y,N) \sim \mathcal{BTLG}(\beta,p)$ then

$$\phi(r,s) = \mathbb{E}\{\chi(Y \le r)\cdot e^{isN}\} = \frac{pF(r)e^{is}}{1-(1-p)F(r)e^{is}}, \; r,s \in \mathbb{R}, \tag{9.40}$$

where $F(r) = 1 - e^{-\beta r}$, $r \ge 0$ (and zero otherwise) is the CDF of exponential distribution with parameter $\beta > 0$.

Proof. Recalling the representation (9.22) and conditioning on N, we have

$$\phi(r,s) = \mathbb{E}\left\{\mathbb{E}\{\chi(Y \le r)\cdot e^{isN}|N\}\right\} = \sum_{n=1}^{\infty} e^{isn}\mathbb{E}\left\{\chi(\bigvee_{j=1}^{n} E_j \le r)\right\} p(1-p)^{n-1}.$$

Since $\mathbb{E}\left\{\chi(\bigvee_{j=1}^{n} E_j \le r)\right\} = [F(r)]^n$, the result follows by summing up the geometric series. □

Note that the quantity $\phi(\infty,s)$ reduces to the ChF of N, while $\phi(r,0)$ coincides with the CDF of Y given in (9.30). We are now ready to establish the following new result.

Proposition 4 Let $\beta > 0$ and $p,q \in (0,1)$. If N_q is a geometric $\mathcal{GEO}(q)$ variable and (Y_i,N_i) are IID $\mathcal{BTLG}(\beta,p)$ variables, independent of N_q, then the distribution of $(\bigvee_{i=1}^{N_q} Y_i, \sum_{i=1}^{N_q} N_i)$ is $\mathcal{BTLG}(\beta,pq)$.

Proof. We proceed by evaluating the CDF-ChF $\psi(r,s)$ of the random vector $(\bigvee_{i=1}^{N_q} Y_i, \sum_{i=1}^{N_q} N_i)$. By conditioning on N_q, we have

$$\psi(r,s) = \sum_{n=1}^{\infty} \mathbb{E}\left\{\chi(\bigvee_{j=1}^{n} Y_j \le r)\cdot e^{is\sum_{i=1}^{n} N_i}\right\} q(1-q)^{n-1}.$$

By independence of the (Y_i, N_i), the expectation under the sum above reduces to $[\phi(r,s)]^n$. Summing up the resulting geometric series, followed by straightforward algebra, leads to

$$\psi(r,s) = \frac{q\phi(r,s)}{1-(1-q)\phi(r,s)} = \frac{pqF(r)e^{is}}{1-(1-pq)F(r)e^{is}},$$

which by Proposition 3 is the CDF-ChF of the $\mathcal{BTLG}(\beta, pq)$ distribution. This completes the argument. □

9.3.6. *Estimation*

Let $(Y_1, N_1), \ldots, (Y_n, N_n)$ be a random sample from a $\mathcal{BTLG}(\beta, p)$ distribution. By replacing the theoretical moments $\mathbb{E}Y$ and $\mathbb{E}N$ in the equations

$$\mathbb{E}Y = -\frac{\log p}{(1-p)\beta} \text{ and } \mathbb{E}N = \frac{1}{p}$$

by their respective sample means and solving for β and p, we obtain a unique pair of moment estimators,

$$\hat{\beta}_n = \frac{\overline{N}_n}{\overline{Y}_n} \frac{\log \overline{N}_n}{\overline{N}_n - 1}, \quad \hat{p}_n = \frac{1}{\overline{N}_n}. \tag{9.41}$$

As shown in [Kozubowski and Panorska (2008)], these are consistent and asymptotically normal.

Proposition 5 The estimators (9.41) are consistent and asymptotically normal: $\sqrt{n}[(\hat{\beta}_n, \hat{p}_n)' - (\beta, p)']$ converges in distribution to a bivariate normal distribution with the (vector) mean zero and the covariance matrix

$$\Sigma_{MME} = \begin{bmatrix} \beta^2 b_p & 0 \\ 0 & p^2(1-p) \end{bmatrix}, \tag{9.42}$$

where

$$b_p = \frac{1}{(\log p)^2} \left[pc_p - \frac{(1-p+p\log p)^2}{1-p} \right] - 1 \tag{9.43}$$

and c_p is given by (9.37).

Next, we summarize maximum likelihood estimation connected with the BTLG model. As shown in [Kozubowski and Panorska, 2006], the log-likelihood function

$$\log L(\beta, p) =$$

$$n\left\{ \frac{1}{n} \sum_{j=1}^{n} \log \left[N_j (1-e^{-\beta Y_j})^{N_j - 1} \right] + \log(\beta p) - \beta \overline{Y}_n + (\overline{N}_n - 1)\log(1-p) \right\} \tag{9.44}$$

can be maximized separately with respect to β and p, leading to a unique pair of MLEs $\hat{\beta}_n$ and $\hat{p}_n$, where $\hat{p}_n = 1/\overline{N}_n$ and $\hat{\beta}_n \in [1/\overline{Y}_n, \overline{N}_n/\overline{Y}_n]$ is the unique solution of the equation

$$\beta = \frac{1}{n\overline{Y}_n}\sum_{j=1}^{n}\left(1+\frac{(N_j-1)Y_j\beta}{e^{\beta Y_j}-1}\right). \tag{9.45}$$

Although here a numerical search is required to find the MLE of β, this is straightforward since the function on the right-hand-side of (9.45) is decreasing in β. When iterations are used, one can use the method of moments estimate of β, which is always in the interval $[1/\overline{Y}_n, \overline{N}_n/\overline{Y}_n]$), as the starting value. As discussed in,[35] these MLEs are consistent, asymptotically normal, and efficient.

Proposition 6 The vector MLE $(\hat{\beta}_n, \hat{p}_n)'$ described above is
(i) consistent;
(ii) asymptotically normal, that is $\sqrt{n}[(\hat{\beta}_n, \hat{p}_n)' - (\beta, p)']$ converges in distribution to a bivariate normal distribution with the (vector) mean zero and the covariance matrix

$$\Sigma_{MLE} = \begin{bmatrix} \frac{\beta^2}{1+d_p} & 0 \\ 0 & p^2(1-p) \end{bmatrix}, \tag{9.46}$$

where

$$d_p = 2p(1-p)\int_0^1 \frac{u(\log u)^2 du}{(1-u)[p+(1-p)u]^3}. \tag{9.47}$$

(iii) asymptotically efficient - the asymptotic covariance matrix (9.46) coincides with the inverse of the Fisher information matrix.

9.4. Extensions

Here we present two generalizations of the BEG and BTLG models, connected with a tri-variate distribution of (X, Y, Z) given by (9.4), where the (X_i, Y_i, N_i) are IID random vectors admitting the stochastic representation (9.1). Our focus will be again the bivariate distributions of (X, Z) and (Y, Z). Most results connected with the former are taken from [Kozubowski et al., 2008], while those related to the latter are all new.

9.4.1. *The BGNB model*

With (X, Y, Z) defined in (9.4), the distribution of (X, Z) is the same as that of the sum of n IID BEG random vectors, and it coincides with the marginal distribution at $t = n$ of a bivariate Lévy process $\{(X(t),Z(t)),t \geq 0\}$, where $(X(1),Z(1))$ is given by (9.5). This process can be represented as

$$\{(X(t),Z(t)),t \geq 0\} \stackrel{d}{=} \left\{ \left(\sum_{i=1}^{NB(t)} E_i + G(t), NB(t)+t \right), t \geq 0 \right\}, \qquad (9.48)$$

where the $\{E_i\}$ are, as before, IID exponential variables with parameter β, $\{G(t),t \geq 0\}$ is a gamma Lévy process starting at zero, based on the exponential distribution (9.3), and $\{NB(t),t \geq 0\}$ is a negative binomial (NB) Lévy process starting at zero, with the ChF

$$\mathbb{E}e^{isNB(t)} = \left(\frac{p}{1-(1-p)e^{is}} \right)^t, \ s \in \mathbb{R}, \qquad (9.49)$$

studied in [see references numbered 40,41]. This continuous time model, discussed in [Kozubowski et al., 2008] along with three other related processes obtained by replacing either t or $G(t)$ (or both) on the right-hand-side of (9.48) by zero, has high potential use in stochastic modeling involving negative binomial sums of independent random quantities. Here we shall focus on the bivariate distribution of the process (9.48) with deleted t, which for $t = 1$ coincides with the $\mathcal{BEG}(\beta, p)$ distribution shifted by $(0,-1)$. As shown in [Kozubowski et al., 2008], the joint pdf of this model is of the form

$$f(x,k) = \frac{\beta^{t+k}}{k!\Gamma(t)} x^{k+t-1} e^{-\beta x} p^t (1-p)^k, \ x > 0, k = 0,1,2,\dots. \qquad (9.50)$$

Following [Kozubowski et al., 2008], we say that a random vector (X,N) with the pdf (9.50) has a BGNB distribution with parameters $t > 0$, $\beta > 0$ and $p \in (0,1)$, denoted by $\mathcal{BGNB}(t,\beta,p)$. The name reflects the marginal distributions: X has a gamma distribution with shape parameter t and scale $p\beta$, while N has a negative binomial distribution supported on non-negative integers, given by the pdf

$$\mathbb{P}(N = k) = \frac{\Gamma(k+t)}{k!\Gamma(t)} p^t (1-p)^k, \ k = 0,1,2,\dots. \qquad (9.51)$$

As shown in [Kozubowski et al., 2008], the CDF and the survival function of $(X,N) \sim \mathcal{BGNB}(t,\beta,p)$ are

$$P(X \leq x, N \leq y) = \frac{p^t}{\Gamma(t)} \sum_{j=0}^{[y]} \frac{(1-p)^j}{j!} \{\Gamma(j+t) - \Gamma(j+t, \beta x)\} \qquad (9.52)$$

and

$$P(X > x, N > y) = \frac{\Gamma(t, p\beta x)}{\Gamma(t)} - \frac{p^t}{\Gamma(t)} \sum_{j=0}^{[y]} \frac{(1-p)^j}{j!} \Gamma(j+t, \beta x), \qquad (9.53)$$

respectively, where $x, y \geq 0$, $[y]$ is the integer part of y, and the quantity

$$\Gamma(\alpha, x) = \int_x^\infty w^{\alpha-1} e^{-w} dw \tag{9.54}$$

is the incomplete gamma function.

9.4.1.1. *Conditional distributions*

The conditional distribution of X given $N = k$, $k = 0, 1, 2\ldots$, is gamma with shape parameter $t + k$ and scale β, so that the conditional pdf is

$$f_{X|N=k}(x) = \frac{\beta^{k+t}}{\Gamma(t+k)} x^{k+t-1} e^{-\beta x}, \ x > 0. \tag{9.55}$$

The conditional PDF of N given $X = x$ is given by

$$f_{N|X=x}(k) = \frac{[\beta(1-p)x]^k e^{-\beta(1-p)x}}{k!}, \ k = 0, 1, 2, \ldots, \tag{9.56}$$

which is a Poisson distribution with mean $\beta(1-p)x$. Further, for any real $x > 0$ and integers $m, n \geq 0$ we have

$$\mathbb{P}(X > x, N > m | N > n) = \frac{1}{c_n} \left\{ \frac{\Gamma(t, p\beta x)}{\Gamma(t)} - \frac{p^t}{\Gamma(t)} \sum_{j=0}^{m \vee n} \frac{(1-p)^j}{j!} \Gamma(j+t, \beta x) \right\},$$

where

$$c_n = \mathbb{P}(N > n) = 1 - \sum_{j=0}^{n} \frac{\Gamma(j+t)}{j!\Gamma(t)} p^t (1-p)^j. \tag{9.57}$$

For $0 \leq m \leq n$ this gives the survival function of X given $N > n$. Similarly, for any integer $n \geq 0$ and real $x, u > 0$ we have

$$\mathbb{P}(X > x, N > n | X > u) = \frac{\Gamma(t, p\beta(x \vee u))}{\Gamma(t, p\beta u)} - \frac{p^t}{\Gamma(t, p\beta u)} \sum_{j=0}^{n} \frac{(1-p)^j}{j!} \Gamma(j+t, \beta(x \vee u)),$$

which coincides with the survival function of N given $X > u$ when $u \geq x > 0$.

9.4.1.2. *Moments and related parameters*

As shown in [Kozubowski et al., 2008], the joint moments of $(X, N) \sim \mathcal{BGNB}(t, \beta, p)$ are given by

$$\mathbb{E}X^\eta N^\gamma = \frac{\Gamma(\eta+t)}{\Gamma(t)} \left(\frac{1}{p\beta} \right)^\eta \mu_\gamma, \tag{9.58}$$

where $\mu_\gamma = \mathbb{E}W^\gamma$ and W has a NB distribution with parameters $\eta + t > 0$ and $p \in (0,1)$. In particular, $\mathbb{E}X = t(\beta p)^{-1}$, $\mathbb{E}N = t(1-p)p^{-1}$, $\mathbb{E}XN = (1+t^2)(1-p)/(\beta p^2)$, and the covariance matrix of (X,N) is

$$\Sigma = t \cdot \begin{bmatrix} \frac{1}{\beta^2 p^2} & \frac{1-p}{\beta p^2} \\ \frac{1-p}{\beta p^2} & \frac{1-p}{p^2} \end{bmatrix}. \tag{9.59}$$

The correlation coefficient of X and N is the same as that of the BEG model, $\rho = \sqrt{1-p}$.

9.4.1.3. *Representations*

By definition, a BGNB random vector with parameters $t > 0$, $\beta > 0$ and $p \in (0,1)$ admits the stochastic representation

$$(X,N) \stackrel{d}{=} \left(\sum_{i=1}^{N} E_i + G, N \right), \tag{9.60}$$

where all the variables on the right-hand-side of (9.60) are mutually independent, the $\{E_i\}$ are IID exponential variables with pdf (9.3), G has a gamma distribution with shape parameter t and scale β, and N is a NB variable with the pdf (9.51). In addition, we have two related representations in terms of randomly stopped gamma and Poisson processes, which follow from more general results established in [Kozubowski et al., 2008]. The first representation of $(X,N) \sim \mathcal{BGNB}(t,\beta,p)$ is

$$(X,N) \stackrel{d}{=} (G(N+t), N). \tag{9.61}$$

Here, $\{G(t), t \geq 0\}$ is a gamma Lévy process, where $G(1)$ has the exponential distribution with PDF (9.3), and N is an independent negative binomial variable with pdf (9.51). The second representation is

$$(X,N) \stackrel{d}{=} (X, N(X)), \tag{9.62}$$

where $\{N(t),\, t \geq 0\}$ is a Poisson process with parameter $\lambda = \gamma(1-p)/p$ while X is an independent gamma variable with parameter $p\beta$.

9.4.2. *The BGTLNB model*

With (X,Y,Z) defined in (9.4), we have

$$(Y,Z) \stackrel{d}{=} \left(\bigvee_{i=1}^{n} Y_i, \sum_{i=1}^{n} N_i \right), \tag{9.63}$$

where the $\{Y_i, N_i\}$ are IID variables following the $\mathcal{BTLG}(\beta, p)$ distribution. Since the $\{N_i\}$ are IID geometric variables with pdf (9.2), their sum in (9.63) has the same distribution as $N+n$, where N has a negative binomial distribution with pdf (9.51). Using the fact that each Y_i is the maximum of N_i IID exponential variables with pdf (9.3), we obtain

$$Y \stackrel{d}{=} \bigvee_{i=1}^{n} Y_i \stackrel{d}{=} \bigvee_{i=1}^{N_1+\cdots+N_n} E_i \stackrel{d}{=} \bigvee_{i=1}^{N+n} E_i \stackrel{d}{=} \bigvee_{i=1}^{N} E_i \vee R_n, \tag{9.64}$$

where R_n has a generalized exponential distribution with the PDF (9.34), being the maximum of n IID exponential variables with parameter β. Consequently, we arrive at the representation

$$(Y, Z) \stackrel{d}{=} \left(\bigvee_{i=1}^{N} E_i \vee R_n, N+n \right), \tag{9.65}$$

with all variables on the right-hand-side being mutually independent. More generally, we can define a stochastic process

$$\{(Y(t), Z(t)), t \geq 0\} \stackrel{d}{=} \left\{ \left(\bigvee_{i=1}^{NB(t)} E_i \vee R_t, NB(t)+t \right), t \geq 0 \right\}, \tag{9.66}$$

where, again, $\{NB(t), t \geq 0\}$ is a negative binomial Lévy process with the ChF (9.49) while $\{R_t, t \geq 0\}$ is an extremal process defined via the CDF

$$\mathbb{P}(R_t \leq x) = (1 - e^{-\beta x})^t. \tag{9.67}$$

When $t = 1$, the marginal distribution of this process reduces to the $\mathcal{BTLG}(\beta, p)$ model discussed in Section 9.3. Moreover, similarly to the BGNB models discussed above, here we can also define three other processes, by replacing either t or R_t (or both) on the right-hand-side of (9.66) by zero. As in the BGNB case, we shall consider the model resulting from deleting t, focusing our attention on bivariate marginal distributions of this process. As we show below, each of these distributions has *generalized truncated logistic* and *negative binomial* marginal distributions, so we shall refer to this model as the BGTLNB model, which stands for a **bi**variate distribution with **g**eneralized **t**runcated **l**ogistic and **n**egative **b**inomial marginals.

Definition A random vector (Y, N) with the stochastic representation

$$(Y, N) \stackrel{d}{=} \left(\bigvee_{i=1}^{N} E_i \vee R, N \right), \tag{9.68}$$

where the $\{E_i\}$ are IID exponential variables (9.3), R is a generalized exponential variable with the CDF (9.67), and N is a NB variable (9.51), with all the variables mutually independent, is said to have a BGTLNB distribution with parameters $t > 0$, $\beta > 0$ and $p \in (0,1)$. This distribution is denoted by $\mathcal{BGTLNB}(t,\beta,p)$.

The joint pdf of $(Y,N) \sim \mathcal{BGTLNB}(t,\beta,p)$ can be derived through a conditioning argument. Given $N = n$, Y is the maximum of n IID $\mathcal{EXP}(\beta)$ variables and a generalized exponential variable with CDF (9.67), so that the CDF and the PDF of Y are

$$F_{Y|N}(y|n) = (1-e^{-\beta y})^{n+t}, \; y > 0, \tag{9.69}$$

and

$$f_{Y|N}(y|n) = (n+t)\beta e^{-\beta y}(1-e^{-\beta y})^{n+t-1}, \; y > 0, \tag{9.70}$$

respectively. Since N is a NB variable with the pdf (9.51), the joint PDF of this model is of the form

$$f(y,n) = \frac{\beta(n+t)\Gamma(n+t)}{n!\Gamma(t)} e^{-\beta y}(1-e^{-\beta y})^{t+n-1} p^t (1-p)^n, \; y > 0, n = 0,1,2,\ldots. \tag{9.71}$$

Note that when $t = 1$ this reduces to the pdf of the $\mathcal{BTLG}(\beta,p)$ distribution shifted by $(0,-1)$. Similar conditioning leads to the CDF of Y:

$$\mathbb{P}(Y \le y) = \sum_{n=0}^{\infty} \mathbb{P}(\bigvee_{j=1}^{n} E_j \vee R \le y)\mathbb{P}(N = n) = (F(y))^t G_N(F(y)),$$

where $F(\cdot)$ is the CDF of the $\{E_i\}$ and $G_N(\cdot)$ is the generating function of N. After further simplifications we obtain

$$F_Y(y) = \left(\frac{p(1-e^{-\beta y})}{p+(1-p)e^{-\beta y}}\right)^t = \left(\frac{p}{q}\frac{1-q}{1-p}\right)^t, \; y \ge 0, \tag{9.72}$$

with q defined in (9.26). Since this is a power of the truncated logistic CDF (9.30), in analogy to the generalized exponential distribution [Gupta and Kundu 1999], we shall refer to this as a *generalized truncated logistic distribution* (GTL) with shape parameter $t > 0$ and scale parameter $\beta > 0$.

To obtain the joint CDF of the BGTLNB model, we start by writing

$$\mathbb{P}(Y \le y, N \le n) = \sum_{k=0}^{n} \frac{\Gamma(k+t)}{k!\Gamma(t)} p^t (1-p)^k \int_0^y \beta(k+t)e^{-\beta x}(1-e^{-\beta x})^{t+k-1} dx$$

for any $y > 0$ and $n = 0,1,2,\ldots$. Since the integral above simplifies to $(1-e^{-\beta y})^{k+t}$, after further simplifications we obtain

$$\mathbb{P}(Y \le y, N \le n) = \left(\frac{p}{q}\frac{1-q}{1-p}\right)^t \sum_{k=0}^{n} \frac{\Gamma(k+t)}{k!\Gamma(t)} q^t (1-q)^k, \; y > 0, n = 0,1,2,\ldots, \tag{9.73}$$

with q given by (9.26) as before. Note that the summation above is the same as the probability $\mathbb{P}(N_q \le n)$, where N_q is a NB variable with parameters $t > 0$ and $q \in (0,1)$. To obtain the corresponding survival function $S(y,n) = \mathbb{P}(Y > y, N > n)$, we use the relation

$$\mathbb{P}(Y \le y, N \le n) = 1 - \mathbb{P}(Y > y) - \mathbb{P}(N > n) + \mathbb{P}(Y > y, N > n)$$

along with (9.72) and (9.57), leading to the following formula valid for $y \ge 0$ and $n = 0, 1, 2, \ldots$:

$$S(y,n) = 1 - \left(\frac{p}{q}\frac{1-q}{1-p}\right)^t \left\{1 - \sum_{k=0}^{n} \frac{\Gamma(k+t)}{k!\Gamma(t)} q^t (1-q)^k \left(1 - \left(\frac{1-p}{1-q}\right)^{t+k}\right)\right\}. \tag{9.74}$$

9.4.2.1. *Conditional distributions*

As seen above, the conditional distribution of Y given $N = n$ is generalized exponential with the CDF and the pdf given by (9.69) and (9.70), respectively. Dividing the joint pdf (9.71) by the marginal pdf of Y, obtained by differentiating the CDF (9.72), leads to the following expression for the conditional PDF of N given $Y = y$:

$$f_{N|Y}(n|y) = \mathbb{P}(N = n | Y = y) = \frac{(n+t)\Gamma(n+t)}{n!\Gamma(t)t} q^{t+1}(1-q)^n, \ n = 0, 1, 2, \ldots, y > 0, \tag{9.75}$$

with q given by (9.26). Since the above PDF is of the form $(n+t)\mathbb{P}(N_q = n)/\mathbb{E}(N_q + t)$, where N_q is a NB variable with parameters $t > 0$ and $q \in (0,1)$, this is a *weighted* (*tilted*) NB distribution. Further, for any real $y > 0$ and integers $m, n \ge 0$, the conditional probability $\mathbb{P}(Y > y, N > m | N > n)$ is equal to

$$\frac{1}{c_n}\left\{1 - \left(\frac{p}{q}\frac{1-q}{1-p}\right)^t \left\{1 - \sum_{k=0}^{m \vee n} \frac{\Gamma(k+t)}{k!\Gamma(t)} q^t (1-q)^k \left(1 - \left(\frac{1-p}{1-q}\right)^{t+k}\right)\right\}\right\},$$

where q and c_n are given by (9.26) and (9.57), respectively. For $0 \le m \le n$ this gives the survival function of Y given $N > n$. Similarly, for any integer $n \ge 0$ and real $y, u > 0$, the conditional probability $\mathbb{P}(Y > y, N > n | Y > u)$ is equal to

$$\frac{1}{c}\left\{1 - d\left\{1 - \sum_{k=0}^{n} \frac{\Gamma(k+t)}{k!\Gamma(t)} q_{max}^t (1-q_{max})^k \left(1 - \left(\frac{1-p}{1-q_{max}}\right)^{t+k}\right)\right\}\right\}, \tag{9.76}$$

where

$$c = \mathbb{P}(Y > u) = 1 - \left(\frac{p}{q_u}\frac{1-q_u}{1-p}\right)^t, \ d = \left(\frac{p}{q_{max}}\frac{1-q_{max}}{1-p}\right)^t,$$

and $q_{max} = q(p, \beta, y \vee u)$, $q_u = q(p, \beta, u)$, with the quantity $q(\cdot,\cdot,\cdot)$ given by (9.26). The function in (9.76) reduces to the survival function of N given $Y > u$ when $u \geq y > 0$.

9.4.2.2. *Moments and related parameters*

Routine calculations lead to the following integral-series representations for the mixed moments $\mathbb{E}Y^{\eta}N^{\gamma}$ of $(Y,N) \sim \mathcal{BGTLNB}(t,\beta,p)$:

$$\mathbb{E}Y^{\eta}N^{\gamma} = \frac{p^t}{\beta^{\eta}} \sum_{n=0}^{\infty} \binom{n+t+1}{n} n^{\gamma}(1-p)^n (n+t) A_n(t,\eta), \ \ \eta,\gamma \geq 0,$$

where

$$A_n(t,\eta) = \int_0^1 (-\log u)^{\eta}(1-u)^{t+n-1} du. \tag{9.77}$$

When $\eta = 1$, the integral in (9.77) reduces to $(\psi(t+n+1)+C)/(t+n)$, where ψ is the Digamma function and $C = 0.5772156\ldots$ is the Euler's constant [Gradshetyn and Ryzhik, 1994], formula 1 in 4.253, p.570). Using the relation $\psi(t+n+1) = \psi(t) + \sum_{k=0}^{n}(t+k)^{-1}$, after some algebra we obtain the following expression for the mean of Y $(\eta = 1, \gamma = 0)$:

$$\mathbb{E}Y = \frac{1}{\beta}(C + \psi(t) + a_{t,p})$$

where

$$a_{t,p} = \mathbb{E}\sum_{k=0}^{N} \frac{1}{t+k}$$

and N has a negative binomial distribution with pdf (9.51). Note that for $t = 1$ this reduces to $\mathbb{E}Y = -\log p/[(1-p)\beta]$ (cf. Subsection 9.3.3). Similar calculations incorporating the fact that $\mathbb{E}N = t(1-p)/p$, show that

$$\mathbb{E}[YN] = \frac{1}{\beta}\left\{t\frac{1-p}{p}(C+\psi(t)) + b_{t,p}\right\}$$

where

$$b_{t,p} = \mathbb{E}\left\{N\sum_{k=0}^{N} \frac{1}{t+k}\right\}.$$

In case $\eta = 2$, the integral in (9.77) reduces to $\{\psi(t+n+1)+C)^2 + \pi^2/6 - \psi'(t+n+1)\}/(t+n)$, where ψ' is the Trigamma function [Gradshetyn and Ryzhik,

1994], formula 17 in 4.261, p. 573). This produces

$$\mathbb{E}[Y^2] = \frac{1}{\beta^2}\left(\mathbb{E}\left\{\left(\psi(t)+C+\sum_{k=0}^{N}\frac{1}{t+k}\right)^2 - \psi'(t+n+1)\right\}+\frac{\pi^2}{6}\right),$$

and further leads to the variance of Y, incorporated in the covariance matrix Σ of (Y,N), where

$$\Sigma = \begin{bmatrix} \frac{1}{\beta^2}\{c_{t,p}-d_{t,p}+\pi^2/6\} & \frac{(1-p)t}{\beta p}\{b_{t,p}-a_{t,p}\} \\ \frac{(1-p)t}{\beta p}\{b_{t,p}-a_{t,p}\} & \frac{1-p}{p^2} \end{bmatrix}. \tag{9.78}$$

Here, the quantities $a_{t,p}$ and $b_{t,p}$ are as before, while

$$c_{t,p} = Var\left\{\sum_{k=0}^{N}\frac{1}{t+k}\right\},\ d_{t,p} = \mathbb{E}[\psi'(t+N+1)], \tag{9.79}$$

with N having the negative binomial distribution (9.51). The correlation coefficient of Y and N is of the form

$$\rho = \frac{t\sqrt{1-p}(b_{t,p}-a_{t,p})}{\sqrt{c_{t,p}-d_{t,p}+\pi^2/6}}.$$

9.4.2.3. *Representations*

In addition to the basic representation (9.68), a random vector $(Y,N) \sim \mathcal{BGTLNG}(t,\beta,p)$ admits a representation analogous to (9.38). Indeed, since given $N=n$, Y is a generalized exponential variable with shape parameter $n+t$ and scale β, by the results of Gupta and Kundu [1999], its distribution is the same as that of the sum

$$\sum_{j=1}^{n+k}\frac{E_j}{j+\alpha}+R_\alpha,$$

where $t=k+\alpha$ with $k=0,1,2,\ldots$ and $\alpha\in(0,1]$, the $\{E_j\}$ are IID exponential variables with parameter β, and R_α has a generalized exponential distribution with shape parameter α and scale β. It follows that

$$(Y,N) \stackrel{d}{=} \left(\sum_{j=1}^{N+k}\frac{E_j}{j+\alpha}+R_\alpha, N\right), \tag{9.80}$$

with the $\{E_j\}$ and R_α as above, NB variable N, and all the variables on the right-hand-side being mutually independent. Note that when $t=1$, in which case $k=0$

and $\alpha = 1$, the variable R_1 is also exponential with parameter β, and the right-hand-side of (9.80) reduces to

$$\left(\sum_{j=1}^{N}\frac{E_j}{j+1}+R_1,N\right)\stackrel{d}{=}\left(\sum_{j=0}^{N}\frac{E_j}{j+1},N\right)\stackrel{d}{=}\left(\sum_{j=1}^{N+1}\frac{E_j}{j},N+1\right)-(0,1).$$

Since $N+1 \sim \mathcal{GEO}(p)$ and the $\mathcal{BGTLNG}(t,\beta,p)$ distribution with $t=1$ coincides with the $\mathcal{BTLG}(\beta,p)$ distribution shifted by $(0,-1)$, we see that (9.80) is in agreement with (9.38).

9.5. Acknowledgements

Kozubowski's research was partially supported by NSF grant ATM-0503722; Panorska's research was partially supported by NSF grants ATM-0236898 and ATM-0503722; Biondi's research was partially supported by NSF grants ATM-0503722 and ATM-CAREER-0132631.

References

1. Amaral, L.A.N., Buldyrev, S.V., Havlin, S., Salinger, M.A. and Stanley, H.E. (1998). Power law scaling for a system of interacting units with complex internal structure, *Physical Rev. Lett.* **80**(7), 1385-1388.
2. Anderson, C.W. and Turkman, K.F. (1995). Sums and maxima of stationary sequences with heavy tailed distributions, *Sankhyā* **57**(1), 1-10.
3. Arnold, B.C., (1973). Some characterizations of the exponential distribution by geometric compounding, *SIAM J. Appl. Math.* **24**, 242-244.
4. Balakrishnan, N. (1985). Order statistics from the half logistic distribution, *J. Statist. Comput. Simulation* **20**, 287-309.
5. Balakrishnan, N. (Ed.) (1992). *Handbook of the Logistic Distribution*, Marcel Dekker, New York.
6. Balakrishnan, N. and Chan, P.S. (1992). Estimation for the scaled half logistic distribution under Type II censoring, *Comput. Statist. Data Anal.* **13**, 123-141.
7. Balakrishnan, N. and Puthenpura, S. (1986). Best linear unbiased estimators of location and scale parameters of the half logistic distribution, *J. Statist. Comput. Simulation* **25**, 193-204.
8. Balakrishnan, N. and Wong, K.H.T. (1991). Approximate MLE's for the location and scale parameters of the half logistic distribution with Type II censoring, *IEEE Trans. Reliab.* **40**, 140-145.
9. Balakrishnan, N. and Wong, K.H.T. (1994). Best linear unbiased estimation of location and scale parameters of the half-logistic distribution based on Type II censored samples, *Amer. J. Math. Management Sci.* **14**(1-2), 53-101.
10. Balkema, A.A. and de Haan, L. (1974). Residual life time at great age, *Ann. Probab.* **2**(5), 792-804.

11. Biondi, F., Kozubowski, T.J. and Panorska, A.K. (2002). Stochastic modeling of regime shifts *Climate Research* **23**, 23-30.
12. Biondi, F., Kozubowski, T.J. and Panorska, A.K. (2005). A new model for quantifying climate episodes, *Intern. J. Climatology* **25**, 1253-1264.
13. Biondi, F., Kozubowski, T.J., Panorska, A.K. and Saito, L. (2008). A new stochastic model of episode peak and duration for eco-hydro-climatic applications, *Ecological Modelling* **211**, 383-395.
14. Buldyrev, S.V., Amaral, L.A.N., Havlin, S., Leschhorn, H., Maass, P., Salinger, M.A., Stanley, H.E. and Stanley, M.H.R. (1997). Scaling behavior in economics. II. Modeling of company growth, *J. Phys. I* (France) **7**(4), 635-650.
15. Chow, T.L. and Teugels, J. (1979). The sum and the maximum of i.i.d. random variables, in: *Proceedings of the Second Prague Symposium on Asymptotic Statistics (Hradec Králové, 1978)*, pp. 81-92, North-Holland, Amsterdam-New York.
16. Cook, E.R. and Krusic, P.J. (2003). The North American Drought Atlas, *EOS Transactions of the American Geophysical Union* **84**, Abstract GC52A-01.
17. Correira, F.N. (1987). Multivariate partial duration series in flood risk analysis, in: V.P. Singh (Ed.), *Hydrologic Frequency Modelling*, pp. 541-554, Reidel, Dordrecht, The Netherlands.
18. Davison, A.C. and Smith, R.L. (1990). Models for exceedances over high thresholds, *J. Roy. Statist. Soc. Ser. B* **52**, 393-442.
19. Feller, W. (1957). *An Introduction to Probability Theory and Its Applications*, Vol. I, 2nd edition, Wiley, New York.
20. Fernández, B. and Salas, J.D. (1999). Return period and risk of hydrologic events. I: Mathematical formulation, *Journal of Hydrologic Engineering* **4**, 297-307.
21. Gnedenko, B.V. and Korolev, V.Yu. (1996). *Random Summation: Limit Theorems and Applications*, CRC Press, Boca Raton.
22. Goel, N.K., Seth, S.C. and Chandra, S. (1998). Multivariate modeling of flood flows, *Journal of Hydraulic Engineering* **124**, 146-155.
23. González, J., Valdés, J.B. (2003). Bivariate drought recurrence analysis using tree-ring reconstructions, *Journal of Hydrologic Engineering* **8**, 247-258.
24. Gradshteyn, I.S. and Ryzhik, I.M. (1994). *Table of Integrals, Series, and Products*, 5th edition (edited by A. Jeffrey), Academic Press, San Diego.
25. Gupta, R.D. and Kundu, D. (1999). Generalized exponential distributions, *Austral. & New Zealand J. Statist.* **41**(2), 173-188.
26. Gupta, R.D. and Kundu, D. (2001). Generalized exponential distribution: Different methods of estimations, *J. Statist. Comput. Simul.* **69**(4), 315-338.
27. Gupta, R.D. and Kundu, D. (2001). Exponentiated exponential distribution: An alternative to gamma and Weibull distributions, *Biometrical J.* **43**(1), 117-130.
28. Gupta, R.D. and Kundu, D. (2002). Generalized exponential distribution: Statistical inferences, *J. Statist. Theory Appl.* **1**, 101-118.
29. Gupta, R.D. and Kundu, D. (2005). Estimation of $P(Y < X)$ for generalized exponential distribution, *Metrika* **61**, 291-308.
30. Johnson, N.L., Kotz, S. and Balakrishnan, N. (1995). *Continuous Univariate Distributions*, Vol. 2, 2nd edition. Wiley, New York.
31. Kim, T.-W., Valdés, J.B. and Yoo, C. (2003). Nonparametric approach for estimating return periods of droughts in arid regions, *Journal of Hydrologic Engineering* **8**, 237-

246.
32. Klebanov, L.B., Kozubowski, T.J. and Rachev, S.T. (2007). *Ill-Posed Problems in Probability and Stability of Random Sums*, Nova Science Publishers, New York.
33. Klebanov, L.B., Maniya, G.M., Melamed, I.A. (1984). A problem of Zolotarev and analogs of infinitely divisible and stable distributions in a scheme for summing a random number of random variables, *Theory Probab. Appl.* **29**, 791-794.
34. Kozubowski, T.J. and Panorska, A.K. (2005). A mixed bivariate distribution with exponential and geometric marginals, *J. Statist. Plann. Inference* **134**, 501-520.
35. Kozubowski, T.J. and Panorska, A.K. (2008). A mixed bivariate distribution connected with geometric maxima of exponential variables, *Comm. Statist. Theory Methods* **37**, 2903-2923.
36. Kozubowski, T.J., Panorska, A.K. and Podgórski, K. (2008). A bivariate Lévy process with negative binomial and gamma marginals, *J. Multivariate Anal.* **99**, 1418-1437.
37. Kozubowski, T.J. and Podgórski, K. (2000). Asymmetric Laplace distributions, *Math. Sci.* **25**, 37-46.
38. Kozubowski, T.J. and Podgórski, K. (2001). Asymmetric Laplace laws and modeling financial data, *Math. Comput. Modelling* **34**, 1003-1021.
39. Kozubowski, T.J. and Podgórski, K. (2003). A log-Laplace growth-rate model, *Math. Sci.* **28**, 49-60.
40. Kozubowski, T.J. and Podgórski, K. (2005). Distributional properties of the negative binomial Lévy process, *Probability and Mathematical Statistics* (in press).
41. Kozubowski, T.J. and Podgórski, K. (2007). Invariance properties of the negative binomial Lévy process and stochastic self-similarity, *Intern. Math. Forum* **2**(30), 1457-1468.
42. Krstanovic, P.F., and Singh, V.P. (1987). A multivariate stochastic flood analysis using entropy, in: V.P. Singh (Ed.), *Hydrologic Frequency Modelling*, pp. 515-539. Reidel, Dordrecht, The Netherlands.
43. Lee, Y., Amaral, L.A.N., Canning, D., Meyer, M., Stanley, H.E. (1998). Universal features in the growth dynamics of complex organizations, *Physical Rev. Lett.* **81**(15), 3275-3278.
44. Madan, D.B., Carr, P.P. and Chang, E.C. (1998). The variance gamma process and option pricing, *European Finance Rev.* **2**, 79-105.
45. Marshall, A.W. and Olkin, I. (1997). A new method for adding a parameter to a family of distributions with application to the exponential and Weibull families, *Biometrika* **84**(3), 641-652.
46. Mathier, L., Perreault, L., Bobée, B. and Ashkar, F. (1992). The use of geometric and gamma-related distributions for frequency analysis of water deficit, *Stochastic Hydrology and Hydraulics* **6**, 239-254.
47. Nolan, J.P. (2001). Maximum likelihood estimation and diagnostics for stable distributions, in: O.E. Barndorff-Nielsen, T. Mikosh, and S. Resnick, Eds., *Lévy Processes*, Birkhäuser, Boston, 379-400.
48. Patil, G.P., Rao, C.R. and Zelen, M. (1988). Weighted distributions, in *Encyclopedia of Statistical Sciences, Vol. 9* (Eds., S. Kotz, N.L. Johnson and C.B. Read), pp. 565-571, Wiley, New York.
49. Pickands, J. (1975). Statistical Inference using extreme order statistics, *Ann. Statist.* **3**, 119-131.

50. Raqab, M.Z. (2002). Inference for generalized exponential distribution based on record statistics, *J. Statist. Plann. Inference* **104**, 339-350.
51. Raqab, M.Z. and Ahsanullah, M. (2001). Estimation of the location and scale parameters of the generalized exponential distribution based on order statistics, *J. Statist. Comput. Simul.* **69**, 109-124.
52. Ratnam, R.R.L., Rosaiah, K., and Anjaneyulu, M.S.R. (2000). Estimation of reliability in multicomponent stress-strength model: half logistic distribution, *IAPQR Trans.* **25**(2), 43-52.
53. Reed, W.J. (2001). The Pareto, Zipf and other power laws, *Economics Letters* **74**, 15-19.
54. Sackl, B., and Bergmann, H. (1987). A bivariate flood model and its application, in: V.P. Singh (Ed.), *Hydrology Frequency Modelling*, pp. 571-582, Reidel, Dordrecht, The Netherlands.
55. Sen, Z. (1976). Wet and dry periods for annual flow series, *Journal of Hydraulic Engineering Division, ASCE* **102**, 1503-1514.
56. Sen, Z. (1980). Statistical analysis of hydrologic critical droughts *Journal of Hydraulic Engineering Division, ASCE* **106**, 99-115.
57. Shiau, J.-T. (2003). Return period of bivariate distributed extreme hydrological events, *Stochastic Environmental Research and Risk Assessment* **17**, 42-57.
58. Shiau, J.-T. and Shen, H.W. (2001). Recurrence analysis of hydrologic droughts of differing severity, *Journal of Water Resources Planning and Management* **127**, 30-40.
59. Stanley, M.H.R., Amaral, L.A.N., Buldyrev, S.V., Havlin, S., Leschorn, H., Maass, P, Salinger, M.A., Stanley, H.E. (1996). Scaling behavior in the growth of companies, *Nature* **379**, 804.
60. Takayasu, H. and Okuyama, K. (1998). Country dependence on company size distributions and a numerical model based on competition and cooperation, *Fractals* **6**(1), 67-69.
61. Todorovic, P. and Woolhiser, D.A. (1976). Stochastic structure of the local pattern of precipitation, in: *Stochastic Approaches to Water Resources*, Vol. 2 (Ed., H.W. Shen), Colorado State University, Fort Collins, Colorado.
62. Yue, S. (2001). A statistical measure of severity of El Nio events, *Stochastic Environmental Research and Risk Assessment* **15**, 153-172.
63. Yue, S., Ouarda, T.B.M.J., Bobe, B., Legendre, P. and Bruneau, P. (1999). The Gumbel mixed model for flood frequency analysis, *Journal of Hydrology* **226**, 88-100.
64. Zelenhasić, E. and Salvai, A. (1987). A method of streamflow drought analysis, *Water Resour. Res.* **23**(1), 156-168.
65. Zheng, G. (2002). On the Fisher information matrix in type-II censored data from the exponentiated exponential family, *Biometrical J.* **44**, 353-357.

Chapter 10

Estimation of the Multivariate Box-Cox Transformation Parameters

Mezbahur Rahman[1] and Larry M. Pearson[2]
Minnesota State University, Mankato, MN 56001, USA
E-mail:mezbahur.rahman@mnsu.edu; larry.pearson@mnsu.edu

The Box-Cox transformation is a well known family of power transformations that brings a set of data closer into agreement with the normality assumption of the residuals and, hence, the response variables of a postulated model in regression analysis. This paper implements the Newton-Raphson method in estimating the multivariate Box-Cox transformation parameters and gives a new method of estimation of the parameters by maximizing the multivariate Shapiro-Wilk statistic. Simulation is performed to compare the two methods for bivariate transformations.

Contents

10.1. Introduction

In regression analysis, often the key assumption regarding normality of the response variables is violated. The commonly used remedy is the Box-Cox family of power transformations (Box and Cox (1964)). The process is to select a parameter in the Box-Cox transformation which maximizes the normal likelihood using the data at hand and then apply regression analysis on the transformed response variables. There is no role of the estimates of the location and the scale parameters which were derived in the process of estimating the power transformation parameters in the analysis. The model parameters are usually estimated seperately after the necessary Box-Cox power transformation parameters are selected.

In the literature, the estimation procedures of the multivariate Box-Cox power transformation parameters have not received as much attention as in the univariate case. The univariate transformation parameter is usually estimated using the maximization of the normal likelihood function as suggested by Box and Cox (1964), the robustified version of the normal likelihood method of Carroll (1980) and of Bickel and Doksum (1981), the transformation to symmetry method of Hinkley (1975), the quick estimate of Hinkley (1977) and of Taylor (1985). Lin and Vonesh (1989) constructed a nonlinear regression model which is used to estimate the transformation parameter such that the normal probability plot of the data on the transformed scale is as close to linearity as possible. Following Box and Cox (1982) and Lin and Vonesh (1989), Halawa (1996) considered the power transformation parameter estimation procedure using an artificial regression model which gives estimates with very small variabilities compared to the normal likelihood procedure. Halawa (1996) conducted an exhaustive comparative study with the normal likelihood procedure. In that study, he also considered estimation procedures of the location and the scale parameters in the likelihood. Most recently, Rahman (1999) introduced a method of estimating the Box-Cox power transformation parameter using maximization of the Shapiro-Wilk W (Shapiro and Wilk (1965)) statistic along with a comparative study of the normal likelihood method (Carroll (1980)), and of the artificial regression model method (Halawa (1996)).

In this paper, the estimation procedure for the multivariate Box-Cox power transformation parameters is considered using maximization of the normal likelihood along with the multivariate Newton-Raphson algorithm. In addition, the maximization of the multivariate Shapiro-Wilk W statistic method (Rahman (1999)) is implemented in the multivariate case. Andrews et al. (1971) considered both the marginal and the joint transformations and noted that for most purposes the marginal transformation is sufficient to achieve the goal. Here, we will also consider the marginal transformation.

10.2. Box-Cox Transformation

Let $\mathbf{Y}_1, \mathbf{Y}_2, \cdots, \mathbf{Y}_n$ be a random sample of p-variate vectors from a population whose functional form is unknown. The multivariate version of the Box and Cox (1964) transformation suggested by Velilla (1993) is given by

$$\mathbf{X}^{(\Lambda)} = \left(X_1^{(\lambda_1)}, X_2^{(\lambda_2)}, \ldots, X_i^{(\lambda_i)}, \ldots, X_p^{(\lambda_p)}\right)' \sim N_p(\mu, \Sigma) \tag{10.1}$$

where

$$X_i^{(\lambda_i)} = \begin{cases} \frac{Y_i^{\lambda_i}-1}{\lambda_i}, & \lambda_i \neq 0 \\ \ln(Y_i), & \lambda_i = 0 \end{cases},$$

$\mu = (\mu_1, \mu_2, \ldots, \mu_p)'$ and $\Sigma = (\sigma_{ik})_{p \times p}$.

10.3. Maximum Likelihood Estimation Using The Newton-Raphson Method

After applying the transformation identified in equation (10.2.1), the likelihood function of the data can be written as

$$L(\mathbf{y};\mu,\Sigma,\Lambda) = |2\pi\Sigma|^{-n/2} e^{\mathrm{tr}[-\frac{1}{2}\Sigma^{-1}V^{(\Lambda)}]} \cdot J, \tag{10.2}$$

where $J = \prod_{i=1}^{p}\prod_{j=1}^{n} y_{ij}^{\lambda_i - 1}$ is the Jacobian of the transformations and $V^{(\Lambda)} = (v_{ik})_{p \times p}$, $v_{ik} = \sum_{j=1}^{n}(x_{ij}^{(\lambda_i)} - \mu_i)(x_{kj}^{(\lambda_k)} - \mu_k)$.
For a given Λ, the maximum likelihood estimates (MLE's) of μ and Σ are given by $\hat{\mu} = \bar{\mathbf{X}}^{(\Lambda)}$ and $\hat{\Sigma} = (S_{ik}^{(\Lambda)})_{p \times p}$ where $\bar{\mathbf{X}}^{(\Lambda)} = (\bar{X}_1^{(\lambda_1)}, \bar{X}_2^{(\lambda_2)}, \ldots, \bar{\mathbf{X}}_p^{(\lambda_p)})'$ and $S_{ik}^{(\Lambda)} = \frac{1}{n}\sum_{j=1}^{n}(X_{ij}^{(\lambda_i)} - \bar{X}_i^{(\lambda_i)})(X_{kj}^{(\lambda_k)} - \bar{X}_k^{(\lambda_k)})$, and hence

$$\ell_{\max}(\Lambda) \equiv L_{\max}(\Lambda) = -\frac{np}{2}\log(2\pi) - \frac{n}{2}\log|\hat{\Sigma}| + \frac{np}{2}$$

$$+\left\{\sum_{i=1}^{p}(\lambda_i - 1)\sum_{j=1}^{n}\log y_{ij}\right\}, \tag{10.3}$$

where $\ell_{\max}(\Lambda)$ and $L_{\max}(\Lambda)$ are the logarithm of the likelihood function 10.2. According to Harville (1999, p.309 (8.6)), the likelihood equations are

$$\frac{\partial \ell_{\max}(\Lambda)}{\partial \lambda_i} = \frac{\partial - \frac{n}{2}\log|\hat{\Sigma}|}{\partial \lambda_i} + \sum_{j=1}^{n}\log Y_{ij} = -\frac{n}{2}\mathrm{tr}\left(\hat{\Sigma}^{-1}\frac{\partial \hat{\Sigma}}{\partial \lambda_i}\right) + \sum_{j=1}^{n}\log Y_{ij} = 0, \tag{10.4}$$

for $i = 1, 2, \ldots, p$, with $\frac{\partial \hat{\Sigma}}{\partial \lambda_i} = (D_{mq})_{p \times p}$, where, for $i' \neq i$,

$$D_{ii'} = D_{i'i} = \frac{1}{n}\sum_{j=1}^{n}\left(frac\lambda_i Y_{ij}^{\lambda_i}\log Y_{ij} - (Y_{ij}^{\lambda_i} - 1)\lambda_i^2 - \frac{1}{n}\sum_{l=1}^{n}\frac{\lambda_i Y_{il}^{\lambda_i}\log Y_{il} - (Y_{il}^{\lambda_i} - 1)}{\lambda_i^2}\right)$$

$$\cdot\left(\frac{Y_{i'j}^{\lambda_{i'}} - 1}{\lambda_{i'}} - \frac{1}{n}\sum_{l=1}^{n}\frac{Y_{i'l}^{\lambda_{i'}} - 1}{\lambda_{i'}}\right) \tag{10.5}$$

$$D_{ii} = \frac{2}{n}\sum_{j=1}^{n}\left(\frac{\lambda_i Y_{ij}^{\lambda_i}\log Y_{ij} - (Y_{ij}^{\lambda_i}-1)}{\lambda_i^2} - \frac{1}{n}\sum_{l=1}^{n}\frac{\lambda_i Y_{il}^{\lambda_i}\log Y_{il} - (Y_{il}^{\lambda_i}-1)}{\lambda_i^2}\right)$$

$$\cdot\left(\frac{Y_{ij}^{\lambda_i}-1}{\lambda_i} - \frac{1}{n}\sum_{l=1}^{n}\frac{Y_{il}^{\lambda_i}-1}{\lambda_i}\right) \tag{10.6}$$

and $D_{mq} = 0$ for $m \neq i$ and $q \neq i$.
To obtain $\hat{\Lambda}$, we solve the equation (4) using the Newton-Raphson method as

$$\hat{\Lambda}^{(t+1)} = \hat{\Lambda}^{(t)} - \left(H(\hat{\Lambda}^{(t)})\right)^{-1} D(\hat{\Lambda}^{(t)}) \tag{10.7}$$

where $D^{(t)}$ is computed using 10.5 and 10.6 and the derivative of $D^{(t)}$, denoted by $H^{(t)}$, is computed using 10.3, 10.3, and 10.3. For the initial value of $\hat{\Lambda}^{(t)}$, $\hat{\Lambda}^{(t)} = \mathbf{1}$ is an obvious choice.
Note that,

$$\frac{\partial^2 \log|\hat{\Sigma}|}{\partial\lambda_i\partial\lambda_k} = \mathrm{tr}\left(\hat{\Sigma}^{-1}\frac{\partial^2\hat{\Sigma}}{\partial\lambda_i\partial\lambda_k}\right) - \mathrm{tr}\left(\hat{\Sigma}^{-1}\frac{\partial\hat{\Sigma}}{\partial\lambda_i}\hat{\Sigma}^{-1}\frac{\partial\hat{\Sigma}}{\partial\lambda_k}\right),$$

where for $i \neq k$,

$$\frac{\partial^2\hat{\Sigma}}{\partial\lambda_i\partial\lambda_k} = (H_{mq})_{p\times p}$$

with

$$H_{ik} = H_{ki}$$

$$= \frac{1}{n}\sum_{j=1}^{n}\left(\frac{\lambda_i Y_{ij}^{\lambda_i}\log(Y_{ij}) - (Y_{ij}^{\lambda_i}-1)}{\lambda_i^2} - \frac{1}{n}\sum_{l=1}^{n}\frac{\lambda_i Y_{il}^{\lambda_i}\log(Y_{il}) - (Y_{il}^{\lambda_i}-1)}{\lambda_i^2}\right)$$

$$\cdot\left(\frac{\lambda_k Y_{kj}^{\lambda_k}\log(Y_{kj}) - (Y_{kj}^{\lambda_k}-1)}{\lambda_k^2} - \frac{1}{n}\sum_{l=1}^{n}\frac{\lambda_k Y_{kl}^{\lambda_k}\log(Y_{kl}) - (Y_{kl}^{\lambda_k}-1)}{\lambda_k^2}\right) \tag{10.8}$$

and $H_{mq} = 0$ for $(m,q) \neq (k,i)$ or (i,k).
For $i = k$,

$$\frac{\partial^2\hat{\Sigma}}{\partial\lambda_i^2} = (G_{mq})_{p\times p}$$

where, for $i' \neq i$,

$$G_{ii'} = G_{i'i} = \frac{1}{n}\sum_{j=1}^{n}\left(\frac{Y_{ij}^{\lambda_i}(\log Y_{ij})^2}{\lambda_i} - \frac{2Y_{ij}^{\lambda_i}\log Y_{ij}}{\lambda_i^2} + \frac{2(Y_{ij}^{\lambda_i}-1)}{\lambda_i^3}\right.$$

$$\left. - \frac{1}{n}\sum_{l=1}^{n}\frac{Y_{il}^{\lambda_i}(\log Y_{il})^2}{\lambda_i} - \frac{2Y_{il}^{\lambda_i}\log Y_{il}}{\lambda_i^2} + \frac{2(Y_{il}^{\lambda_i}-1)}{\lambda_i^3}\right)\left(\frac{Y_{i'j}^{\lambda_{i'}}-1}{\lambda_{i'}} - \frac{1}{n}\sum_{l=1}^{n}\frac{Y_{i'l}^{\lambda_{i'}}-1}{\lambda_{i'}}\right), \tag{10.9}$$

$$G_{ii} = \frac{2}{n}\sum_{j=1}^{n}\left(\left(\frac{\lambda_i Y_{ij}^{\lambda_i}\log Y_{ij} - (Y_{ij}^{\lambda_i}-1)}{\lambda_i^2} - \frac{1}{n}\sum_{l=1}^{n}\frac{\lambda_i Y_{il}^{\lambda_i}\log Y_{il} - (Y_{il}^{\lambda_i}-1)}{\lambda_i^2}\right)^2\right.$$

$$+\left(\frac{Y_{ij}^{\lambda_i}(\log Y_{ij})^2}{\lambda_i} - \frac{2Y_{ij}^{\lambda_i}\log Y_{ij}}{\lambda_i^2} + \frac{2(Y_{ij}^{\lambda_i}-1)}{\lambda_i^3}\right.$$

$$\left.\left. - \frac{1}{n}\sum_{l=1}^{n}\frac{Y_{il}^{\lambda_i}(\log Y_{il})^2}{\lambda_i} - \frac{2Y_{il}^{\lambda_i}\log Y_{il}}{\lambda_i^2} + \frac{2(Y_{il}^{\lambda_i}-1)}{\lambda_i^3}\right)\cdot\left(\frac{Y_{ij}^{\lambda_i}-1}{\lambda_i} - \frac{1}{n}\sum_{l=1}^{n}\frac{Y_{il}^{\lambda_i}-1}{\lambda_i}\right)\right) \tag{10.10}$$

and $G_{mq} = 0$ for $m \neq i$ and $q \neq i$.

10.4. Maximization of the Multivariate Shapiro-Wilk *W* Statistic

Malkovich and Afifi (1973) suggested a test for multivariate normality by introducing the multivariate form of the Shapiro and Wilk (1965) W statistic. An exhaustive reference of tests for multivariate normality is given by Mecklin and Mundfrom (2004).

Rahman (1999) showed that by maximizing the W statistic, the Box-Cox transformation parameter also can be estimated successfully with high precision. Here we maximize the Malkovich and Afifi W^* statistic to obtain the multivariate Box-Cox transformation parameters. Now,

$$W^* = \frac{1}{n}\frac{[\sum_{j=1}^{n} a_j U_{(j)}]^2}{\left(\mathbf{X}_m^{(\Lambda)} - \bar{\mathbf{X}}^{(\Lambda)}\right)'\hat{\Sigma}^{-1}\left(\mathbf{X}_m^{(\Lambda)} - \bar{\mathbf{X}}^{(\Lambda)}\right)}. \tag{10.11}$$

where $\mathbf{X}^{(\Lambda)}$, $\bar{\mathbf{X}}^{(\Lambda)}$, and $\hat{\Sigma}$ are as defined earlier, the vector $\mathbf{a} = \frac{\mathbf{M}'V^{-1}}{(\mathbf{M}'\mathbf{V}^{-1}\mathbf{V}^{-1}\mathbf{M})^{1/2}}$, $\mathbf{M}$ is the mean vector and $\mathbf{V}$ is the variance covariance matrix of the standard normal order statistics, the $U_{(j)}$ are the ordered

$$U_j = \left(\mathbf{X}_m^{(\Lambda)} - \bar{\mathbf{X}}^{(\Lambda)}\right)' \hat{\Sigma}^{-1} \left(\mathbf{X}_j^{(\Lambda)} - \bar{\mathbf{X}}^{(\Lambda)}\right) = \mathbf{Z}_m^{(\Lambda)\prime}\hat{\Sigma}^{-1}\mathbf{Z}_j^{(\Lambda)}, \text{ for}$$

$j = 1, 2, \ldots, n$, and

$$\mathbf{Z}_m^{(\Lambda)\prime}\hat{\Sigma}^{-1}\mathbf{Z}_m^{(\Lambda)} = \max_{1 \le j \le n} \mathbf{Z}_j^{(\Lambda)\prime}\hat{\Sigma}^{-1}\mathbf{Z}_j^{(\Lambda)}.$$

Now,

$$W^* = \frac{1}{n}\frac{[\sum_{j=1}^n a_j U_{(j)}]^2}{\mathbf{Z}_m^{(\Lambda)\prime}\hat{\Sigma}^{-1}\mathbf{Z}_m^{(\Lambda)}} \quad \text{and} \quad U_{(j)} = \mathbf{Z}_m^{(\Lambda)\prime}\hat{\Sigma}^{-1}\mathbf{Z}_{(j*)}^{(\Lambda)},$$

where $\mathbf{Z}_{(j*)}^{(\Lambda)}$ corresponds to $U_{(j)}$. Then W^* can be maximized by solving the equations $\frac{\partial W^*}{\partial \lambda_i} = 0$ for $i = 1, 2, \ldots, p$ where

$$\frac{\partial W^*}{\partial \lambda_i} = \frac{1}{n}\frac{\mathbf{Z}_m^{(\Lambda)\prime}\hat{\Sigma}^{-1}\mathbf{Z}_m^{(\Lambda)} 2\sum_{j=1}^n a_j U_{(j)} \sum_{j=1}^n a_j \frac{\partial U_{(j)}}{\partial \lambda_i} - \left(\sum_{j=1}^n a_j U_{(j)}\right)^2 \frac{\partial \mathbf{Z}_m^{(\Lambda)\prime}\hat{\Sigma}^{-1}\mathbf{Z}_m^{(\Lambda)}}{\partial \lambda_i}}{\left\{\mathbf{Z}_m^{(\Lambda)\prime}\hat{\Sigma}^{-1}\mathbf{Z}_m^{(\Lambda)}\right\}^2}.$$

Solving $\frac{\partial W^*}{\partial \lambda_i} = 0$ is equivalent to solving

$$\mathbf{Z}_m^{(\Lambda)\prime}\hat{\Sigma}^{-1}\mathbf{Z}_m^{(\Lambda)} 2\sum_{j=1}^n a_j U_{(j)} \sum_{j=1}^n a_j \frac{\partial U_{(j)}}{\partial \lambda_i} - \left(\sum_{j=1}^n a_j U_{(j)}\right)^2 \frac{\partial \mathbf{Z}_m^{(\Lambda)\prime}\hat{\Sigma}^{-1}\mathbf{Z}_m^{(\Lambda)}}{\partial \lambda_i} = 0, \tag{10.12}$$

where

$$\frac{\partial U_{(j)}}{\partial \lambda_i} = \frac{\partial \mathbf{Z}_m^{(\Lambda)\prime}\hat{\Sigma}^{-1}\mathbf{Z}_{(j*)}^{(\Lambda)}}{\partial \lambda_i} = \mathbf{Z}_m^{(\Lambda)\prime}\hat{\Sigma}^{-1}\frac{\partial \mathbf{Z}_{(j*)}^{(\Lambda)}}{\partial \lambda_i} + \frac{\partial \mathbf{Z}_m^{(\Lambda)\prime}}{\partial \lambda_i}\hat{\Sigma}^{-1}\mathbf{Z}_{(j*)}^{(\Lambda)} - \mathbf{Z}_m^{(\Lambda)\prime}\hat{\Sigma}^{-1}\frac{\partial \hat{\Sigma}}{\partial \lambda_i}\hat{\Sigma}^{-1}\mathbf{Z}_{(j*)}^{(\Lambda)}$$

(using Harville (1999, p.307 (8.15)), $\frac{\partial \mathbf{Z}_{(j*)}^{(\Lambda)}}{\partial \lambda_i}$ is a vector of zeros except for the ith element which is

$$d_{j*i} = \frac{\lambda_i Y_{ij*}^{\lambda_i} \log Y_{ij*} - (Y_{ij*}^{\lambda_i} - 1)}{\lambda_i^2} - \frac{1}{n}\sum_{l=1}^n \frac{\lambda_i Y_{il}^{\lambda_i} \log Y_{il} - (Y_{il}^{\lambda_i} - 1)}{\lambda_i^2},$$

$\frac{\partial \mathbf{Z}_m^{(\Lambda)\prime}}{\partial \lambda_i}$ is a vector of zeros except for the ith element which is

$$d_{mi} = \frac{\lambda_i Y_{im}^{\lambda_i} \log Y_{im} - (Y_{im}^{\lambda_i} - 1)}{\lambda_i^2} - \frac{1}{n}\sum_{l=1}^n \frac{\lambda_i Y_{il}^{\lambda_i} \log Y_{il} - (Y_{il}^{\lambda_i} - 1)}{\lambda_i^2},$$

$\frac{\partial \hat{\Sigma}}{\partial \lambda_i}$ is defined in Section 10.3, and

$$\frac{\partial \mathbf{Z}_m^{(\Lambda)\prime}\hat{\Sigma}^{-1}\mathbf{Z}_m^{(\Lambda)}}{\partial \lambda_i} = 2\frac{\partial \mathbf{Z}_m^{(\Lambda)\prime}}{\partial \lambda_i}\hat{\Sigma}^{-1}\mathbf{Z}_m^{(\Lambda)} - \mathbf{Z}_m^{(\Lambda)\prime}\hat{\Sigma}^{-1}\frac{\partial \hat{\Sigma}}{\partial \lambda_i}\hat{\Sigma}^{-1}\mathbf{Z}_m^{(\Lambda)}.$$

To solve (10.12) using the Newton-Raphson method let

$$\tilde{\Lambda}^{(t+1)} = \tilde{\Lambda}^{(t)} - \left(h(\tilde{\Lambda}^{(t)})\right)^{-1} g(\tilde{\Lambda}^{(t)}), \tag{10.13}$$

where the first derivative of $g(\Lambda)$ is denoted as

$$h(\Lambda) = \frac{\partial}{\partial \lambda_k}\left(\mathbf{Z}_m^{(\Lambda)\prime}\hat{\Sigma}^{-1}\mathbf{Z}_m^{(\Lambda)}\right) 2\sum_{j=1}^{n} a_j U_{(j)} \sum_{j=1}^{n} a_j \frac{\partial U_{(j)}}{\partial \lambda_i} + \mathbf{Z}_m^{(\Lambda)\prime}\hat{\Sigma}^{-1}$$

$$\mathbf{Z}_m^{(\Lambda)} 2\sum_{j=1}^{n} a_j \frac{\partial U_{(j)}}{\partial \lambda_k} \sum_{j=1}^{n} a_j \frac{\partial U_{(j)}}{\partial \lambda_i} + \mathbf{Z}_m^{(\Lambda)\prime}\hat{\Sigma}^{-1}\mathbf{Z}_m^{(\Lambda)} 2\sum_{j=1}^{n} a_j U_{(j)} \sum_{j=1}^{n} a_j \frac{\partial}{\partial \lambda_k}\left(\frac{\partial U_{(j)}}{\partial \lambda_i}\right)$$

$$-2\sum_{j=1}^{n} a_j U_{(j)} \sum_{j=1}^{n} a_j \frac{\partial U_{(j)}}{\partial \lambda_k} \frac{\partial \mathbf{Z}_m^{(\Lambda)\prime}\hat{\Sigma}^{-1}\mathbf{Z}_m^{(\Lambda)}}{\partial \lambda_i} - \left(\sum_{j=1}^{n} a_j U_{(j)}\right)^2 \frac{\partial}{\partial \lambda_k}\left(\frac{\partial \mathbf{Z}_m^{(\Lambda)\prime}\hat{\Sigma}^{-1}\mathbf{Z}_m^{(\Lambda)}}{\partial \lambda_i}\right)$$

where for $k \neq i$,

$$\frac{\partial}{\partial \lambda_k}\left(\frac{\partial U_{(j)}}{\partial \lambda_i}\right) = \frac{\partial}{\partial \lambda_k}\left(\frac{\partial \mathbf{Z}_m^{(\Lambda)\prime}\hat{\Sigma}^{-1}\partial \mathbf{Z}_{(j*)}^{(\Lambda)}}{\partial \lambda_i}\right)$$

$$= \left[\left(\frac{\partial \mathbf{Z}^{(\Lambda)_m}}{\partial \lambda_k}\right)' - \mathbf{Z}_m^{(\Lambda)\prime}\hat{\Sigma}^{-1}\left(\frac{\partial \hat{\Sigma}}{\partial \lambda_k}\right)\right]\hat{\Sigma}^{-1}\left(\frac{\partial \mathbf{Z}_{(j*)}^{(\Lambda)}}{\partial \lambda_i}\right) + \mathbf{Z}_m^{(\Lambda)\prime}\hat{\Sigma}^{-1}\frac{\partial}{\partial \lambda_k}\frac{\partial \mathbf{Z}_{(j*)}^{(\Lambda)}}{\partial \lambda_i}$$

$$+\left[\frac{\partial}{\partial \lambda_k}\left(\frac{\partial \mathbf{Z}^{(\Lambda)_m}}{\partial \lambda_i}\right)' - \frac{\partial \mathbf{Z}_m^{(\Lambda)\prime}}{\partial \lambda_i}\left(\hat{\Sigma}^{-1}\frac{\partial \hat{\Sigma}}{\partial \lambda_k}\right)\right]\hat{\Sigma}^{-1}\mathbf{Z}_{(j*)}^{(\Lambda)} + \frac{\partial \mathbf{Z}_m^{(\Lambda)\prime}}{\partial \lambda_i}\hat{\Sigma}^{-1}\frac{\partial \mathbf{Z}_{(j*)}^{(\Lambda)}}{\partial \lambda_k}$$

$$-\frac{\partial \mathbf{Z}_m^{(\Lambda)\prime}}{\partial \lambda_k}\left(\hat{\Sigma}^{-1}\frac{\partial \hat{\Sigma}}{\partial \lambda_i}\hat{\Sigma}^{-1}\mathbf{Z}_{(j*)}^{(\Lambda)}\right) + \mathbf{Z}_m^{(\Lambda)\prime}\hat{\Sigma}^{-1}\frac{\partial \hat{\Sigma}}{\partial \lambda_k}\hat{\Sigma}^{-1}\frac{\partial \hat{\Sigma}}{\partial \lambda_i}\hat{\Sigma}^{-1}\mathbf{Z}_{(j*)}^{(\Lambda)}$$

$$-\mathbf{Z}_m^{(\Lambda)\prime}\hat{\Sigma}^{-1}\frac{\partial}{\partial \lambda_k}\left(\frac{\partial \hat{\Sigma}}{\partial \lambda_i}\right)\hat{\Sigma}^{-1}\mathbf{Z}_{(j*)}^{(\Lambda)} + \mathbf{Z}_m^{(\Lambda)\prime}\hat{\Sigma}^{-1}\frac{\partial \hat{\Sigma}}{\partial \lambda_i}\hat{\Sigma}^{-1}\frac{\partial \hat{\Sigma}}{\partial \lambda_k}\hat{\Sigma}^{-1}\mathbf{Z}_{(j*)}^{(\Lambda)}$$

$$-\mathbf{Z}_m^{(\Lambda)\prime}\hat{\Sigma}^{-1}\frac{\partial \hat{\Sigma}}{\partial \lambda_i}\hat{\Sigma}^{-1}\frac{\partial \mathbf{Z}_{(j*)}^{(\Lambda)}}{\partial \lambda_k},$$

for $k = i$,

$$\frac{\partial^2 U_{(j)}}{\partial\lambda_i^2} = \mathbf{Z}_m^{(\Lambda)\prime}\hat{\Sigma}^{-1}\frac{\partial^2 \mathbf{Z}_{(j*)}^{(\Lambda)}}{\partial\lambda_i^2} - 2\mathbf{Z}_m^{(\Lambda)\prime}\hat{\Sigma}^{-1}\frac{\partial\hat{\Sigma}}{\partial\lambda_i}\hat{\Sigma}^{-1}\frac{\partial \mathbf{Z}_{(j*)}^{(\Lambda)}}{\partial\lambda_i} + 2\frac{\partial \mathbf{Z}_m^{(\Lambda)\prime}}{\partial\lambda_i}\hat{\Sigma}^{-1}\frac{\partial \mathbf{Z}_{(j*)}^{(\Lambda)}}{\partial\lambda_i}$$

$$-2\frac{\partial \mathbf{Z}_m^{(\Lambda)\prime}}{\partial\lambda_i}\hat{\Sigma}^{-1}\frac{\partial\hat{\Sigma}}{\partial\lambda_i}\hat{\Sigma}^{-1}\mathbf{Z}_{(j*)}^{(\Lambda)} + \frac{\partial^2 \mathbf{Z}_m^{(\Lambda)\prime}}{\partial\lambda_i^2}\hat{\Sigma}^{-1}\mathbf{Z}_{(j*)}^{(\Lambda)} + 2\mathbf{Z}_m^{(\Lambda)\prime}\hat{\Sigma}^{-1}\frac{\partial\hat{\Sigma}}{\partial\lambda_i}\hat{\Sigma}^{-1}\frac{\partial\hat{\Sigma}}{\partial\lambda_i}\hat{\Sigma}^{-1}\mathbf{Z}_{(j*)}^{(\Lambda)}$$

$$-\mathbf{Z}_m^{(\Lambda)\prime}\hat{\Sigma}^{-1}\frac{\partial^2\hat{\Sigma}}{\partial\lambda_i^2}\hat{\Sigma}^{-1}\mathbf{Z}_{(j*)}^{(\Lambda)},$$

$$\frac{\partial^2 \mathbf{Z}_m^{(\Lambda)\prime}\hat{\Sigma}^{-1}\mathbf{Z}_m^{(\Lambda)}}{\partial\lambda_i^2} = 2\frac{\partial \mathbf{Z}_m^{(\Lambda)\prime}}{\partial\lambda_i}\hat{\Sigma}^{-1}\frac{\partial \mathbf{Z}_m^{(\Lambda)}}{\partial\lambda_i} - 4\frac{\partial \mathbf{Z}_m^{(\Lambda)\prime}}{\partial\lambda_i}\hat{\Sigma}^{-1}\frac{\partial\hat{\Sigma}}{\partial\lambda_i}\hat{\Sigma}^{-1}\mathbf{Z}_m^{(\Lambda)}$$

$$+2\frac{\partial^2 \mathbf{Z}_m^{(\Lambda)\prime}}{\partial\lambda_i^2}\hat{\Sigma}^{-1}\mathbf{Z}_m^{(\Lambda)} + 2\mathbf{Z}_m^{(\Lambda)}\hat{\Sigma}^{-1}\frac{\partial\hat{\Sigma}}{\partial\lambda_i}\hat{\Sigma}^{-1}\frac{\partial\hat{\Sigma}}{\partial\lambda_i}\hat{\Sigma}^{-1}\mathbf{Z}_{(j*)}^{(\Lambda)} - \mathbf{Z}_m^{(\Lambda)}\hat{\Sigma}^{-1}\frac{\partial^2\hat{\Sigma}}{\partial\lambda_i^2}\hat{\Sigma}^{-1}\mathbf{Z}_{(j*)}^{(\Lambda)},$$

$\frac{\partial^2 \mathbf{Z}_{(j*)}^{(\Lambda)}}{\partial\lambda_i^2}$ is a vector of zeros except for the ith element which is

$$d_{j*i}^{(2)} = \left(\frac{\lambda_i^3(\log Y_{ij*})^2 Y_{ij*}^{\lambda_i} - 2\lambda_i^2(\log Y_{ij*})Y_{ij*}^{\lambda_i} + 2\lambda_i Y_{ij*}^{\lambda_i} - 2\lambda_i}{\lambda_i^4}\right)$$

$$-\frac{1}{n}\sum_{l=1}^{n}\left(\frac{\lambda_i^3(\log Y_{il})^2 Y_{il}^{\lambda_i} - 2\lambda_i^2(\log Y_{il})Y_{il}^{\lambda_i} + 2\lambda_i Y_{il}^{\lambda_i} - 2\lambda_i}{\lambda_i^4}\right)$$

and $\frac{\partial^2 \mathbf{Z}_m^{(\Lambda)\prime}}{\partial\lambda_i^2}$ is a vector of zeros except for the ith element which is

$$d_{mi}^{(2)} = \left(\frac{\lambda_i^3(\log Y_{im})^2 Y_{im}^{\lambda_i} - 2\lambda_i^2(\log Y_{im})Y_{im}^{\lambda_i} + 2\lambda_i Y_{im}^{\lambda_i} - 2\lambda_i}{\lambda_i^4}\right)$$

$$-\frac{1}{n}\sum_{l=1}^{n}\left(\frac{\lambda_i^3(\log Y_{il})^2 Y_{il}^{\lambda_i} - 2\lambda_i^2(\log Y_{il})Y_{il}^{\lambda_i} + 2\lambda_i Y_{il}^{\lambda_i} - 2\lambda_i}{\lambda_i^4}\right)$$

and $\frac{\partial^2\hat{\Sigma}^2}{\partial\lambda_i^2}$ is defined in Section 10.3.

10.5. Simulation Study

One thousand samples of size 50 were selected from a bivariate normal population for each parameter combination given in the table below. The means and the standard deviations were kept fixed. In the table, W_0 indicates the multivariate Shapiro-Wilk statistic for the generated sample, $W(\hat{\Lambda}_L)$ is the multivariate Shapiro-Wilk statistic for the transformed data using the maximum likelihood method, and $W(\hat{\Lambda}_W)$ is the multivariate Shapiro-Wilk statistic for the transformed data using the maximization of the multivariate Shapiro-Wilk statistic method. For the maximum likelihood method, the Newton-Raphson algorithm is implemented as described in section 10.3. When applying the Newton-Raphson algorithm to implement the maximization of the multivariate Shapiro-Wilk statistic method, different starting values yielded various local maximum values and as a result the Newton-Raphson algorithm was not used for this method. Thus, a grid search using a three standard deviation range of the maximum likelihood estimate is implemented in the maximization of the multivariate Shapiro-Wilk statistic method. The grid search procedure for the maximization of the likelihood method yielded similar estimates as the Newton-Raphson method.

The coefficients a_i's in the Shapiro-Wilk W statistic were obtained from Parish (1992a and 1992b) as the most accurate values available. Means (m) and standard deviations (s) are given for the estimates of the Box-Cox transformation parameters and their corresponding W statistics are displayed in Table 1.

The means of the W statistics are consistently higher with lower standard errors for the maximization of the W statistic method in comparison to the maximum likelihood method.

The biases in estimating the Box-Cox parameters are lower for the maximization of the W statistic method but the standard errors of the estimates are higher which leads to higher mean squared errors.

References

1. Andrews, D. F., R. Gnanadesikan, and J. L. Warner (1971). *Transformations of Multivariate Data.* Biometrics, 27, 825-840.
2. Bickel, P. J. and K. A. Doksum (1981). *An Analysis of Transformations Revisited.* Journal of the American Statistical Association, 76, 296-311.
3. Box, G. E. P. and D. R. Cox (1964). *An Analysis of Transformations.* Journal of the Royal Statistical Society, Series B.,26, 211-252.
4. Box, G. E. P. and D. R. Cox (1982). *An Analysis of Transformations Revisited (Rebutted).* Journal of the American Statistical Association, 77, 209-210.

Table 10.1. Simulation Results Displaying the Means and the Standard Errors of the Estimates

	W_0	$W(\hat{\Lambda}_L)$	$W(\hat{\Lambda}_W)$	$\hat{\lambda}_{1L}$	$\hat{\lambda}_{2L}$	$\hat{\lambda}_{1W}$	$\hat{\lambda}_{2W}$
	Normal: $\mu_1 = -5$, $\mu_2 = 10$, σ_1=1, σ_2=2, $\rho = -0.7$, λ_1=-2, λ_2=1						
m	0.9311	0.9775	0.9870	-1.9075	0.9455	-2.1225	1.0231
s	0.0555	0.0117	0.0057	1.1517	0.5772	1.8002	0.8968
	Normal: $\mu_1 = -5$, $\mu_2 = 10$, σ_1=1, σ_2=2, $\rho = -0.3$, λ_1=-2, λ_2=1						
m	0.9393	0.9782	0.9866	-1.8526	0.9260	-2.1015	1.0700
s	0.0494	0.0109	0.0058	1.2970	0.6781	1.9349	1.0258
	Normal: $\mu_1 = -5$, $\mu_2 = 10$, σ_1=1, σ_2=2, $\rho = 0$, λ_1=-2, λ_2=1						
m	0.9410	0.9785	0.9867	-1.9039	0.9189	-2.0395	1.0229
s	0.0464	0.0104	0.0058	1.2688	0.6549	1.9481	0.9905
	Normal: $\mu_1 = -5$, $\mu_2 = 10$, σ_1=1, σ_2=2, $\rho = 0.3$, λ_1=-2, λ_2=1						
m	0.9402	0.9785	0.9866	-1.8439	0.9734	-1.9180	0.9997
s	0.0497	0.0107	0.0059	1.3060	0.6535	2.0447	0.9904
	Normal: $\mu_1 = -5$, $\mu_2 = 10$, σ_1=1, σ_2=2, $\rho = 0.7$, λ_1=-2, λ_2=1						
m	0.9356	0.9777	0.9870	-1.8420	0.9457	-1.9706	1.0029
s	0.0504	0.0114	0.0054	1.2036	0.5866	1.8054	0.8908
	Normal: $\mu_1 = -5$, $\mu_2 = 10$, σ_1=1, σ_2=2, $\rho = -0.7$, λ_1=-1, λ_2=2						
m	0.9145	0.9774	0.9871	-0.9642	1.8857	-1.0607	1.9823
s	0.0641	0.0100	0.0057	0.6570	1.1128	0.9998	1.7320
	Normal: $\mu_1 = -5$, $\mu_2 = 10$, σ_1=1, σ_2=2, $\rho = -0.3$, λ_1=-1, λ_2=2						
m	0.9234	0.9781	0.9865	-0.9437	1.9155	-0.9892	2.0127
s	0.0621	0.0108	0.0060	0.7017	1.2119	1.0527	1.9181
	Normal: $\mu_1 = -5$, $\mu_2 = 10$, σ_1=1, σ_2=2, $\rho = 0$, λ_1=-1, λ_2=2						
m	0.9273	0.9786	0.9865	-0.9279	1.9002	-1.0350	1.9636
s	0.0547	0.0114	0.0055	0.7550	1.2833	1.1268	1.9926
	Normal: $\mu_1 = -5$, $\mu_2 = 10$, σ_1=1, σ_2=2, $\rho = 0.3$, λ_1=-1, λ_2=2						
m	0.9266	0.9783	0.9865	-0.9056	1.8847	-1.0311	1.9632
s	0.0571	0.0105	0.0062	0.7332	1.2594	1.0787	1.9684
	Normal: $\mu_1 = -5$, $\mu_2 = 10$, σ_1=1, σ_2=2, $\rho = 0.7$, λ_1=-1, λ_2=2						
m	0.9125	0.9781	0.9871	-0.9410	1.8121	-1.0675	1.9703
s	0.0681	0.0106	0.0059	0.6863	1.1727	1.0538	1.7423

5. Carroll, R. J. (1980). *A Robust Method for Testing Transformations to Achieve Approximate Normality.* Journal of the Royal Statistical Society, Series B., 42, 71-78.
6. Halawa, Adel M. (1996). *Estimating the Box-Cox Transformation via an Artificial Regression Model.* Communications in Statistics Simulation and Computation, 25(2), 331-350.
7. Harville, David A. (1999). *Matrix Algebra From A Statistician's Perspective.* Springer, New York.
8. Hinkley, D. V. (1975). *On Power Transformation to Symmetry.* Biometrika, 62, 101-111.
9. Hinkley, D. V. (1977). *On Quick Choice of Power Transformation.* Applied Statistics, 26, 67-68.
10. Lin, L. I. and E. F. Vonesh (1989). *An Empirical Nonlinear Data-Fitting Approach for Transforming Data to Normality.* American Statistician, 43, 237-243.

11. Malkovich, J. F. and Afifi, A. A. (1973). *On Tests for Multivariate Normality.* Journal of the American Statistical Association, 68(341), 176-179.
12. Mecklin, Christopher J. and Mundfrom, Daniel J. (2004). *An Appraisal and Bibliography of Tests for Multivariate Normality,* International Statistical Review, 72(1), 123-138.
13. Parish, Rudolph S. (1992a). *Computing Expected Values of Normal Order Statistics.* Communications in Statistics - Simulation and Computation, 21(1), 57-70.
14. Parish, Rudolph S. (1992b). *Computing Variances and Covariances of Normal Order Statistics.* Communications in Statistics - Simulation and Computation, 21(1), 71-101.
15. Rahman, Mezbahur (1999). *Estimating the Box-Cox Transformation via Shapiro-Wilk W Statistic.* Communications in Statistics - Simulation and Computation, 28(1), 223-241.
16. Shapiro, S. S. and M. B. Wilk (1965). *An Analysis of Variance Test for Normality.* Biometrika, 52, 3 and 4, 591-611.
17. Taylor, J. M. G. (1985). *Power Transformations to Symmetry.* Annals of Mathematical Statistics, 33, 1-67.
18. Velilla, Santiago (1993). *A Note on the Multivariate Box-Cox Transformation.* Statistics & Probability Letters, 17, 259-263.

Chapter 11

Generation of Multivariate Densities

R. N. Rattihali and A. N. Basugade

Department of Statistics, Shivaji University, Kolhapur India- 416004,
E-mail: rnr5@rediffmail.com; anb9@rediffmail.com

We introduced the concept of Contour transformation and use it to generate a class of multivariate densities. We consider a p-variate C-contoured unimodal probability density function (p.d.f.) f with modal value 0 and by using contour transformation a new family of $C_\delta{}^*$ - contoured density functions $\{f^*(x,\delta) : \delta \in \Delta$, a suitable set of parameters$\}$ is obtained. The density f is a member of this family. Some properties of $f^*(x,\delta)$ are studied. Further location and scale parameters can be introduced.

Contents

11.1. Introduction

A new class of densities can be generated by conditioning on the variables or by introducing a new parameter. For example skew normal distribution of Azzalini (1985) and Multivariate skew normal distribution of Azzalini and Dalla Valle (1996) can be obtained by conditioning and epsilon-skew normal distribution of Mudholkar and Hutson(2000) is obtained by introducing new parameter, epsilon. In this article we generate a class of densities using the concept of contour transformation. Let $f(x)$ be p-variate density function with modal value 0. Then for $0 < u \leq f(0)$, the set $C_f(u) = \{x : f(x) \geq u\}$ be called the u-contour of f. A function f uniquely determines the class $\mathbf{C}_f = \{C_f(u) : u \in Range(f)\}$ and conversely. It is a non-increasing class of contours. Further a non-increasing class $\{C(t), for\, r \leq t \leq \infty\}$ of sub sets of R^p uniquely determines the function g with domain $C(r)$ given by

$$g(x) = sup\{u : x \in C(u)\}$$

If f is unimodal, symmetric about 0, $(f(x) = f(-x))$ then for $0 < u \leq f(0)$, the set $C_f(u)$ is symmetric about 0.

Contour Transformation:
For $0 < u \leq f(0)$ consider a transformation of contours $C_f(u)$ to $C^*(u)$ such that

i) class $\{C^*(u) : 0 \leq u \leq f(0)\}$ is non-increasing and
ii) $\Lambda(C_f(u)) = \Lambda(C^*(u)), 0 \leq u \leq f(0)$

where Λ is the Lebesgue measure on R^p. Note that corresponding to such a class

$$C^* = \{C^*(u) : 0 \leq u \leq f(0)\}$$

there exists a function $f^*(say)$ given by $f^*(x) = sup\{u : x \in C^*(u)\}$ such that

$$C^*(u) = C_{f^*}(u)$$

In the above transformation requirement i) is major constraint. An arbitrary transformed class of contours may not correspond to a function.
Definition: A 'Contour Transformation' (CT) transforms each member $C_f(u)$ of the class $\mathbf{C}_f$ to $C^*(u)$, so that

$$\Lambda(C_f(u)) = \Lambda(C^*(u)) \;\; for \;\; 0 \leq u \leq f(0)$$

and empty set is transformed to empty set itself, so that $C^*(u) = C_{f^*}(u)$, for some density f^*. The p.d.f. f^* is said to be obtained by a CT of the p.d.f. f.
Let $A \subset R^p$, $\theta \in R^p$ and M be a $p \times p$ matrix and $MA + \theta = \{Mx + \theta : x \in A\}$. Let $f(x)$ be an unimodal p.d.f. and $C_f(u) = \{x \in R^p : f(x) \geq u\}$. Then it is clear that $\{C_f(u), o < u \leq f(0)\}$ is decreasing class of convex contours. We note that $\Lambda(MA + \theta) = |M|\Lambda(A)$, where $|M|$ denotes the determinant of the matrix M.
It is to be noted that if a set A is convex then $MA + \theta$ is also convex. If $C^*(u) = MC(u) + \theta$ with $|M| = 1$ then $\{C^*(u), 0 < u \leq f(0)\}$ satisfies the required conditions and $f^*(x) = sup\{u : x \in C^*(u)\}$. In fact $f^*(x) = f(M^{-1}x - \theta)$.
Let C_0 be a subset of R^p containing 0. A p.d.f. f is said to be C_0-*contoured density function*, if every contour of f is of the form $kC_0 = \{kx : x \in C_0\}$ where, $k \leq 0$. Consider the transformation which transforms C_0 to $C_0{}^*$, so that $\Lambda(C_0) = \Lambda(C_0{}^*)$ (as a desired property $C_0{}^*$ can be chosen to be a convex set in R^p). This induces a transformation of $C = kC_0$ to $C^* = kC_0{}^*$. Then f^* the p.d.f. corresponding to

C^* is a $C_0{}^*$-contoured density function. The transformations $C_0{}^* = C_0 + \theta$ and $C_0{}^* = PC_0$, where P is $p \times p$ orthogonal matrix respectively correspond to the location transformation and rotational transformation. If P is not orthogonal but with $|P| = 1$ and $C_0{}^* = PC_0$, then $\Lambda(C_0{}^*) = \Lambda(C_0)$, here the shape need not remain same.

11.2. Generation of Densities by Contour Transformation

In section 11.2.1 we generate a bivariate density by using a density corresponding to two i.i.d. standard Normal variates. In section 11.2.2 we illustrate how this technique can be used for any arbitrary circular contour density. In section 11.2.3 we generalize the technique for multivariate case.

11.2.1. Bivariate densities obtained by two i.i.d. standard normal variates

Let $f(x,y) = (2\pi)^{-1} exp\{-(x^2+y^2)/2\}, (x,y) \in R^2$ and for simplicity $C_f(u)$ be denoted by $C(u)$. Then $C(u) = \{(x,y) : (x^2+y^2) \leq -2log(2\pi u)\}$. If $u_1 = (2\pi)^{-1} exp(-1/2\pi)$, then

$$C(u_1) = C_0 = \{(x,y) : (x^2+y^2) \leq \pi^{-1}\}$$

is a circular disc with unit area. For $\delta < \pi^{-1}$ and $k > 0$ transform the contours $kC_0\, to\, kC_\delta{}^*$, where

$$C_\delta{}^* = \{(x,y) : (x^2+y^2) \leq \pi^{-1} - \delta cos(\theta)\} \tag{11.1}$$

$cos(\theta) = x/(x^2+y^2)^{1/2}$ and is of unit area. It can be analytically shown that the set $C_\delta{}^*$ is convex for $\pi\delta \leq (2/3)$ and is not convex for $\pi\delta > (2/3)$. The corresponding density function $f^*(x,y)$ is given by

$$f^*(x,y) = (2\pi)^{-1} exp\{-(x^2+y^2)/2(1-\pi\delta x/(x^2+y2)^{1/2})\} \tag{11.2}$$

It is to be noted that $f^*(x,y)$ is a $C_\delta{}^*$-contoured density function. For $\delta = 0$ we get the circular contours and the graphs of p.d.f. $f^*(x,y)$ and their contours for $\pi\delta = 0.6$ and $\pi\delta = 0.85$ are given in Figure 11.1 to Figure 11.4. If $x = rcos(\theta)$ and $y = rsin(\theta)$ then the joint density function of (r,θ) is given by

$$g^*(r,\theta) = (2\pi)^{-1} r\, exp\{-(r^2)/2(1-\pi\delta cos(\theta))\}, 0 < r < \infty, 0 < \theta \leq 2\pi.$$

Properties of the p.d.f. f^*:
P-1: *If* $\delta = 0$ *then* $f = f^*$.

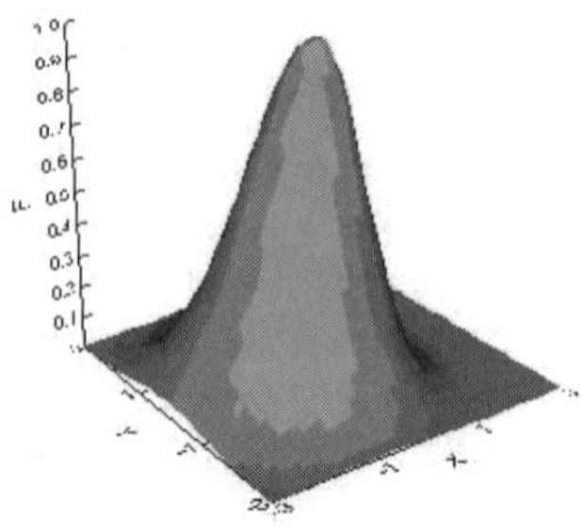

Fig. 11.1. *Graph of p.d.f.* $f^*(x,y)$ *for* $\pi\delta = 0.6$

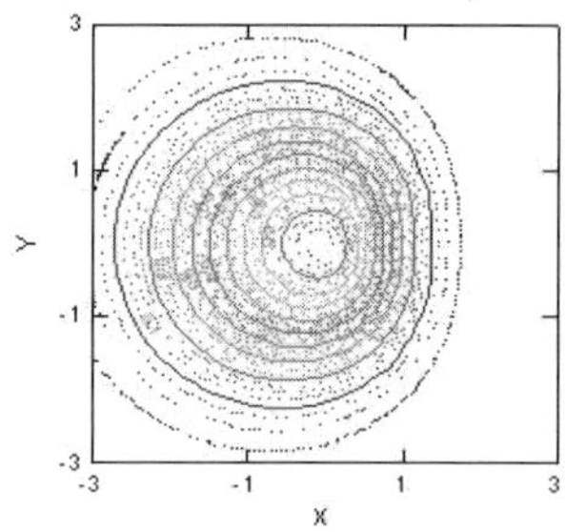

Fig. 11.2. *Graph of Contours of* $f^*(x,y)$ *for* $\pi\delta = 0.6$

P-2: *The marginal density function of* θ *is*

$$f_1(\theta) = (1/2\pi)[1 - \pi\delta cos\theta],\ 0 < \theta \leq 2\theta.$$

P-3: *The marginal density function of r is*

$$f_2(r) = (r)\sum_{i=0}^{\infty} \frac{(-r^2/2)^i}{i!}\ {}_2F_1(i/2,(i+1)/2;1;\pi^2\delta^2)$$

Note that

$$f_2(r) = (r/2\pi)\int_0^{2\pi} exp\{-r^2/2[1-\pi\delta cos\theta]\}d\theta$$

$$exp\left\{\left[\frac{-r^2}{2[1-\pi\delta cos\theta]}\right]\right\} = \sum_{i=0}^{\infty} \frac{(-1)^i}{i!}\left[\frac{r^2}{2[1-\pi\delta cos\theta]}\right]^i$$

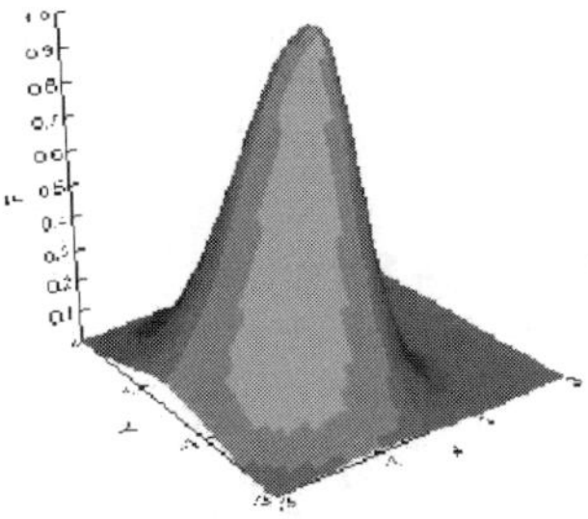

Fig. 11.3. *Graph of p.d.f.* $f^*(x,y)$ *for* $\pi\delta = 0.85$

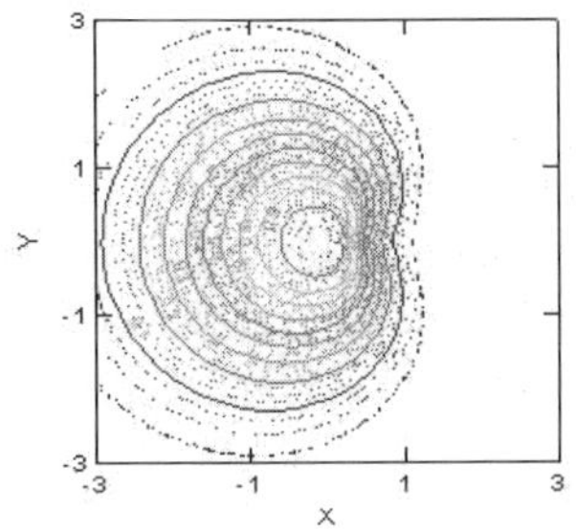

Fig. 11.4. *Graph of Contours of* $f^*(x,y)$ *for* $\pi\delta = 0.85$

$$= \sum_{i=0}^{\infty} \frac{(-r^2/2)^i}{i!} [1 - \pi\delta cos\theta]^{-i}$$

$$f_2(r) = (r/2\pi) \sum_{i=0}^{\infty} \frac{(-r^2/2)^i}{i!} \sum_{j=0}^{\infty} (-\pi\delta)^j \ {}^{-i}C_j \int_0^{2\pi} cos^j\theta \, d\theta$$

$$= (r/2\pi) \sum_{i=0}^{\infty} \frac{(-r^2/2)^i}{i!} \sum_{j\ even}^{\infty} (-\pi\delta)^j \ {}^{-i}C_j \left[\frac{(j-1)(j-3)\dots 3}{j(j-2)(j-4)\dots 2} 2\pi \right]$$

for j odd the integral is zero and hence setting $j = 2t$, we get

$$= (r/2\pi) \sum_{i=0}^{\infty} \frac{(-r^2/2)^i}{i!} \sum_{j=0}^{\infty} (-\pi\delta)^{2t} \ {}^{-i}C_{2t} \left[\frac{(2t-1)(2t-3)\dots 3}{2t(2t-2)(2t-4)\dots 2} 2\pi \right]$$

$$= (r/2\pi)\sum_{i=0}^{\infty}\frac{(-r^2/2)^i}{i!}\sum_{j=0}^{\infty}(-\pi\delta)^{2t}\,\frac{(-1)^{2t}(i)_{2t}}{(2t)!}\left[\frac{(2t-1)(2t-3)\ldots 3}{2t(2t-2)(2t-4)\ldots 2}\,2\pi\right]$$

$$f_2(r) = (r/2\pi)\sum_{i=0}^{\infty}\frac{(-r^2/2)^i}{i!}2\pi\sum_{t=0}^{\infty}(\pi\delta)^{2t}\left[\frac{i(i+1)(i+2)\ldots(i+2t-1)}{2^{2t}\,t!\,t!}\right]$$

$$f_2(r) = (r/2\pi)\sum_{i=0}^{\infty}\frac{(-r^2/2)^i}{i!}\;2\pi\;\sum_{t=0}^{\infty}\frac{(\pi^2\delta^2)^t}{t!}\left[\frac{(\frac{i}{2})_t(\frac{i+1}{2})_t}{(1)_t}\right]$$

$$f_2(r) = (r)\sum_{i=0}^{\infty}\frac{(-r^2/2)^i}{i!}\;{}_2F_1\left(i/2,\,(i+1)/2;1;\pi^2\delta^2\right)$$

where, ${}_2F_1(a,b;c;z)$ is a Gauss Hypergeometric function given in Harry Bateman (1953).

P-4: $f^*(x,y) \le (1+\pi\delta)\,g(x,y)$ *where*, $g(x,y) = (2\pi)^{-1}(1+\pi\delta)^{-1}exp\{-(x^2+y2)/2(1+\pi\delta)\}$, which corresponds to the joint density function of *i.i.d.* normal variates with mean 0 and variance $(1+\pi\delta)$.

Since $(-cos\theta) < 1$, we note that $(x^2+y^2)/2(1-\pi\delta cos\theta) > (x^2+y^2)/2(1+\pi\delta)$. Thus we have

$$f^*(x,y) = (2\pi)^{-1}exp\{-(x^2+y^2)/2(1-\pi\delta cos\theta)\}$$

$$\le [(1+\pi\delta)/2\pi][1/(1+\pi\delta)]exp\{-(x^2+y^2)/2(1+\pi\delta)\}$$

$$= (1+\pi\delta)\,g(x,y)$$

P-5: $P[X \le 0] = [(1/2)-\delta]$

Note that

$$P[X>0] = \int_0^{1/2\pi}\Lambda(A^*(u))du \quad and \quad P[X<0] = \int_0^{1/2\pi}\Lambda(B^*(u))du \tag{11.3}$$

where,

$$A^*(u) = \{(x,y) : x \ge 0 \quad f^*(x,y) \ge u\} \quad and$$

$$B^*(u) = \{(x,y) : x < 0 \quad f^*(x,y) \ge u\}$$

Let $D^* = \{f(x,y) : (x^2+y^2) \le \pi^{-1} - \delta\cos(\theta), x \ge 0\}$ *and*

$$E^* = \{(x,y) : (x^2+y^2) \le \pi^{-1} - \delta cos(\theta), x < 0\}$$

But for each u, $0 < u < (1/2\pi)$, each contour $C_{f^*}(u) = A^*(u) \cup B^*(u)$ is of the form $kC_\delta{}^*$ and $C_\delta{}^* = D^* \cup E^*$ $\frac{\Lambda(A^*(u))}{\Lambda(B^*(u))}$ is same as that of $\frac{\Lambda(D^*)}{\Lambda(E^*)}$
But

$$\frac{\Lambda(D^*)}{\Lambda(E^*)} = \frac{\int_{-\pi/2}^{\pi/2}(1-\pi\delta cos\theta)\,d\theta}{\int_{\pi/2}^{3\pi/2}(1-\pi\delta cos\theta)\,d\theta} = \frac{1-2\delta}{1+2\delta}.$$

Hence

$$\Lambda(A^*(u)) = \frac{1-2\delta}{1+2\delta}\Lambda(B^*(u)) \tag{11.4}$$

From 11.3 and 11.4 we have

$$P[X \geq 0] = \int_0^{1/2\pi} \Lambda(A^*(u))du = \int_0^{1/2\pi} \frac{(1-2\delta)}{(1+2\delta)}\Lambda(B^*(u))du = \frac{(1-2\delta)}{(1+2\delta)}P[X<0]$$

$$= \frac{(1-2\delta)}{(1+2\delta)}\{1-P[X \geq 0]\}$$

$$P[X \geq 0] = \frac{(1-2\delta)}{2} = \frac{1}{2} - \delta > \frac{1}{2} - \frac{1}{\pi} = \frac{\pi-2}{\pi} > 0 \;\; (Since\,\delta < \frac{1}{\pi})$$

Remarks: To simulate the observations from $f^*(x,y)$ one can use the marginal of θ and the conditional distribution of r given θ and then transform (r,θ) to (x,y). Alternatively by using a suitable upper bound for $f^*(x,y)$ in terms of a scale multiple of a suitable density (as given in P-4) one can generate observations from $f^*(x,y)$ by using Rejection Acceptance method. By using P-5, one can obtain trivial estimate of δ by equating $P[X \geq 0] = (1/2) - \delta$ to the sample proportion.

11.2.2. General form of densities

2.2. a). In general the form of circular bivariate density function is $f(x,y) = h(r^2)$ and the transformed density is given by

$$f^*(x,y) = h((r^2)/(1-\pi\delta cos\theta)) \tag{11.5}$$

and the corresponding density function of (r,θ) is given by

$$g(r,\theta) = r\,h(r^2/(1-\pi\delta cos\theta))$$

The marginal density of θ, yield circular models that depend on $h(.)$ that include uniform density when $\delta = 0$. The above can be extended to multivariate cases also. Note that $h(x,y) = h(x,-y)$.

2.2. b). In the above C_0 was taken to be a circular disc of unit area and was transformed to $C_0{}^*$ of unit area. However one can choose C_0 to be of unit radius which in this case is $\{(x,y) : h(x^2+y^2) \geq h(1)\} = \{(x,y) : x^2+y^2 \leq 1\}$.
Let $g(\theta)$ be a circular density over $(0, 2\pi)$ and define the function

$$f^*(x,y) = h(r^2/(2\pi g(\theta)))$$

Note that $C^*(h(1)) = \{(x,y) : h((x^2+y^2)/2\pi g(\theta)) \geq h(1)\} = \{(x,y) : (x^2+y^2) \leq 2\pi g(\theta))\}$ and its area is $2^{-1} \int r^2(\theta) d\theta = \pi \int g(\theta) d\theta = \pi$. With $g(\theta) = (2\pi)^{-1}(1 - \pi\delta cos\theta)$, we get the density given in 11.5.

11.2.3. Multivariate models obtained from circular contoured densities

Let $f(x) = h(x'x), x \in R^p$ be a circular contoured density with h(.) decreasing on $[0, \infty)$. For example $h(t) = (2\pi)^{-p/2} exp\{-t/2\}$ and $h(t) = C_n\{1 + t't/n\}^{-\frac{n+p}{2}}$ respectively correspond to multivariate normal and multivariate t density functions. For $x = (x_1, \ldots, x_p) \in R^p$, let $t = x'x$ and $cos(\theta_i) = x_i t^{-1/2}$, for $i = 1, 2, \ldots, p-1$. Let $g(\theta_1, \theta_2, \ldots, \theta_{p-1})$ be spherical density function defined on the surface of unit sphere in R^p. Consider the function

$$f^*(x,g) = h(x'x/ [kV_p g(\theta_1, \theta_2, \ldots, \theta_{p-1})]^{2/p}) \tag{11.6}$$

For $u = h(1)$ the u-contours $C(u)$ and $C^*(u)$ of f and f^* are given by

$$C(u) = \{x : h(x'x) \geq h(1), x \in R^p\} = \{x : x'x \leq 1, x \in R^p\}$$

and

$$C^*(u) = \{x : h(x'x/[pV_p g(\theta_1, \theta_2, \ldots \theta_{p-1})]^{2/p}) \geq h(1), x \in R^p\}$$

$$= \{x : x'x \leq [pV_p g(\theta_1, \theta_2, \ldots \theta_{p-1})]^{2/p} = r^2(\theta)\ (say) : x \in R^p\}$$

Then volume of $C^*(u)$ is $p^{-1} \int r^p(\theta) d\theta = V_p \int g(\theta) d\theta = V_p$, which is the volume of $C(u)$. Thus $f^*(x,g)$ is a multivariate density function.

11.3. Acknowledgement

Thanks to Professor P. N. Rathie, University of Brasilia, Brasilia, Brazil for bringing to our notice the use of Gauss Hypergeometric functions in evaluating the marginal density of r.

References

1. Azzalini, A. (1985). *A class of distributions which includes the normal ones.* Scandinavian Journal of Statistics, 12 (171-178).
2. Azzalini, A. and Dalla Valle A. (1996). *The Multivariate Skew-normal distribution.* Biometrica, 84/4, (715-726).
3. Bateman H.(1953). *Higher Transcendental Functions*, Vol. 1.McGraw-Hill, New York.
4. Mudholkar G. S. and Hutson A. D. (2000). *The eplison-skew-normal distribution for analyzing near-normal data.* Journal of Statistical Planning and Inference, 83, (292-309).

Chapter 12

Smooth Estimation of Multivariate Distribution and Density Functions

Yogendra P. Chaubey

Department of Mathematics and Statistics
Concordia University
Montreal, Canada H3G 1M8
chaubey@alcor.concordia.ca

Contents

12.1. Intoduction

Estimating a multivariate density, in general, has become an important area of data analysis in modern era of statistical development (see Scott[48]). The more general problem of estimating functionals of the distribution function (DF), especially in the non-parametric setup, has also acquired a significant portion of research endeavours in the modern statistical literature (see Prakasa Rao[43] and Sylvapulle and Sen[50]).

For the DF belonging to a specific parametric family, involving a finite number of (unknown) parameters, statistical inference about them can be made exclusively via these parameters. Such parametric inference procedures are efficient as long as the assumed and true distributions are the same. However, model departures, *i.e.,* any discrepancy between the assumed and true distributions, may grossly affect parametric procedures, and thereby they are notoriously non-robust (and even inefficient to inconsistent). For example, if the underlying pdf is gamma with shape parameter $p(>> 1)$ while we assume it to be exponential (for which $p = 1$), then the true and assumed pdf differ drastically at the lower end point (0) as well as in the upper tail $(x \to \infty)$. This is one of the primary reasons for non-parametric density estimation and its functionals.

Scott[48] provides a brief chronology of developments in nonparametric density estimation. Until the 1950s,"the histogram stood as the only non-parametric density estimator" when "substantial and simultaneous progress was made in density estimation and spectral density estimation." The next decade saw "several general algorithms and alternative theoretical modes were introduced by" Rosenblatt,[44] Parzen[40] and Çencov.[9] "There followed a second wave of primarily theoretical papers by" Watson and Leadbetter,[62] Loftsgaarden and Quesenberry,[38] Schwarz,[46] Epanechnikov,[30] Tarter and Kronmal,[55] and Wahba.[59] "The natural multivariate generalization was introduced by Cacoullos".[8] "Finally, in 1970s came the first papers focusing on the practical application of these methods:" Scott et al.[47] and Silverman.[51]

The computing revolution brought a surge into these and later applications in the recent years (see Härdle[35]). The basic kernel method of Rosenblatt,[44] in its various forms still remains to be one of the most popular estimators for density estimation.

Based on a random sample $(X_1,X_2,...,X_n)$, from a univariate distribution with density $f(.)$, the kernel estimator (Rosenblatt;[44] Parzen[40]) of f is generally taken in the form

$$\hat{f}_n(x) = (nh_n)^{-1} \sum_{i=1}^{n} k((X_i - x)/h_n), \tag{12.1}$$

where $h_n(> 0)$, known as the band-width is so chosen that

$$h_n \to 0 \text{ but } nh_n \to \infty, \quad \text{as } n \to \infty; \tag{12.2}$$

$k(.)$ is termed, the kernel function, and it is typically taken to be symmetric.

It can be motivated by several techniques as shown in Scott.[48] Furthermore, it has been demonstrated by Walter and Blum[60] and proved rigourously by Terell and Scott[56] that virtually all nonparametric algorithms for density estimation are asymptotically kernel methods.

Here I will demonstrate yet another motivation for the kernel method which yields new nonparametric estimators of density but alleviates the problems by classical kernel estimator, such as providing inconsistent estimator at the boundaries as well as assigning positive mass in the non-positive region. The basic motivation comes from the paper by Chaubey and Sen[14] which considered estimating the density and survival function for non-negative data. The details for the univariate case are given in the next section along with a generalization. Some applications of this technique are outlined in Section 3. Section 4 provides the basic generalization of Chaubey and Sen[14] to the multivariate case for non-negative data as given in Chaubey and Sen[18] and the final section gives some recent developments using the new estimators.

12.2. Hille's Theorem: Univariate Case

12.2.1. *Smooth estimation of Survival and Density Functions*

Silverman[52] noted the inadequacy of the kernel density estimator in assigning positive mass to some $x \in (-\infty, 0)$, while illustrating the method where the random variable takes only positive values. The problem was also addressed by Bagai and Prakasa Rao,[5] by proposing to replace the kernel k by a non-negative density function k^*, such that

$$\int_0^\infty x^2 k^*(x)dx < \infty.$$

They show that the resulting estimator has similar asymptotic properties as the usual kernel estimator under some regularity conditions. This alleviates the problem of positive probability in the negative region. However, as noted by them, in estimating $f(x)$ for $X_{(r)} < x \leq X_{(r+1)}$, $X_{(i)}$ denoting the ith order statistic, only the first r order statistics contribute to the value of the modified estimator. Chaubey and Sen[14] considered the problem of estimating the survival function and the corresponding density for survival data where the density is typically supported on the non-negative half of the real line. The approach taken in this paper is to smooth the distribution (survival) function based on Poisson weights, and consider the density estimator as its derivative. According to this approach, essentially, a smooth estimator of $F(x)$ is given by

$$\tilde{F}_n(x) = \sum_{k=0}^{\infty} F_n\left(\frac{k}{\lambda_n}\right) p_k(\lambda_n x) \tag{12.3}$$

where

$$p_k(\mu) = e^{-\mu}\frac{\mu^k}{k!}, k = 0, 1, 2, \ldots,$$

where $\lambda_n \to \infty$ as $n \to \infty$. And the derived smooth estimator of $f(x)$ is given by

$$\tilde{f}_n(x) = \lambda_n \sum_{k=0}^{\infty} \left[F_n\left(\frac{k+1}{\lambda_n}\right) - F_n\left(\frac{k}{\lambda_n}\right) \right] p_k(\lambda_n x) \tag{12.4}$$

The above proposal, in contrast to that of Bagai and Prakasa Rao,[5] uses the whole data and naturally provides consistent estimate of $f(0)$, in case $f(0) > 0$. .
Figure 12.1 gives kernel estimators for different choices of bandwidths using the Gaussian kernel for the *Suicide Data* given in Silverman (page 8)[52] consisting of lengths of 86 spells of psychiatric treatment undergone by patients used as controls in a study of the relationship between suicide risk and time under treatment. The reader may refer to Venebles and Ripley[57](§5.5) for a description of these choices. The bandwidth choices given by the *default* option and BCV criterion produce smoother estimates as compared to the other two choices. However, all the bandwidth choices produce non-negative estimates of the density below zero where it should be definitely zero.

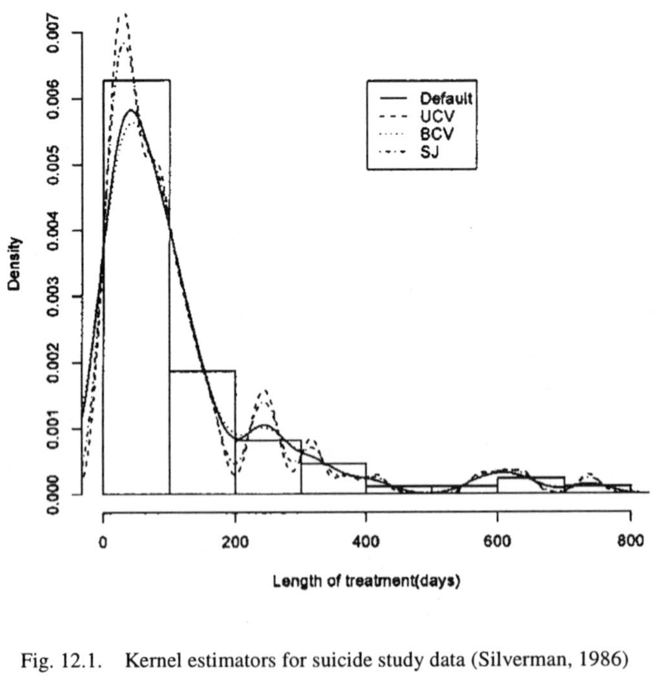

Fig. 12.1. Kernel estimators for suicide study data (Silverman, 1986)

In Figure 12.2, we produce the estimators due to Bagai and Prakasa Rao[5] using a standard exponential kernel with the same band widths as used for producing Figure 12.1. Here, we see that the density is necessarily zero at zero which may not be desirable. We may also note that some boundary correction methods (see Karunamuni and Alberts[37]) produce density estimators for this data with positive mass at zero.

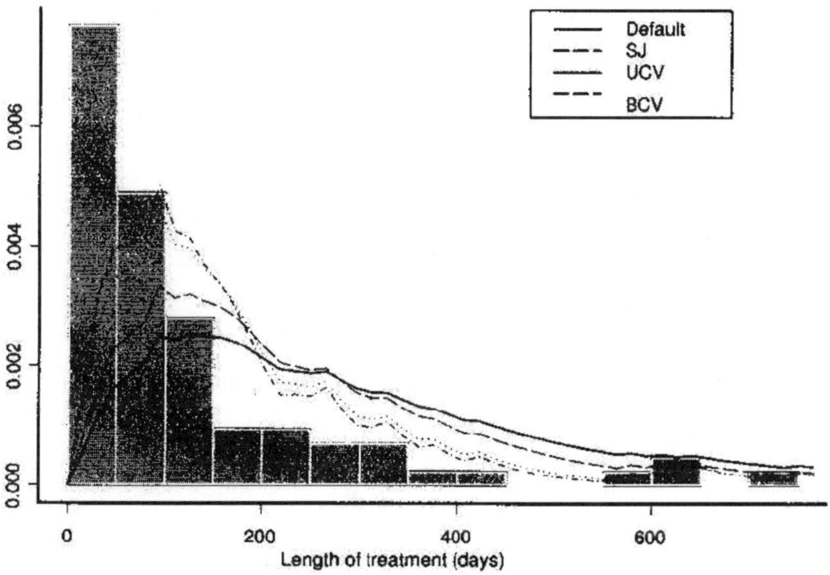

Fig. 12.2. Bagai-Parakasa Rao estimators for suicide study data (Silverman, 1986)

The method given in Chaubey and Sen[14] is actually a truncated version of the above estimator, which is noted to be inappropriate for estimating mean residual life function (see Chaubey and Sen[17]), hence the un-truncated version may be preferred in practice. The value of λ_n is to be chosen adaptive to the data. They recommend the choice $\lambda_n = cn/X_{(n)}$, where c has to be chosen data adaptively. Figure 12.3 gives smooth density estimators for the same data as in Figure 12.1 for different values of c, obtained from the expression in Eq. (12.4) based on the Poisson weights. Here, we note that a quite different picture emerges.
The motivation for the estimator proposed in Chaubey and Sen[14] is based on the smoothing lemma of Hille.[36]

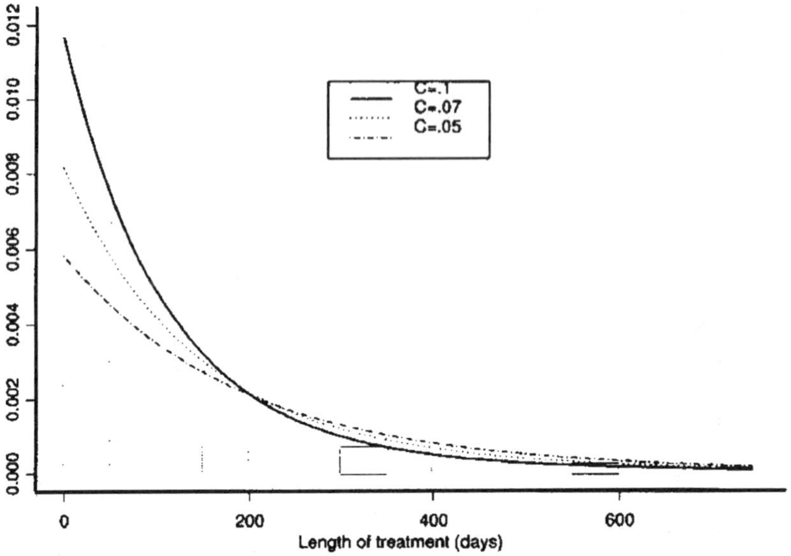

Fig. 12.3. New estimators for suicide study data (Silverman, 1986)

Lemma 12.1. *Let u(x) be bounded, continuous function on* $\mathbf{R}^+$. *Then*

$$e^{-\lambda x} \sum_{k \geq 0} u(k/\lambda)(\lambda x)^k/k! \to u(x), \ \text{ as } \lambda \to \infty, \tag{12.5}$$

uniformly in any finite interval $\mathbf{J}$ *contained in* $\mathbf{R}^+$.

Adapting this lemma for estimation of $F(x)$, replacing $u(x)$ by the empirical distribution function $F_n(x)$ provides the estimator given in equation (12.3).
They established the following asymptotic properties for the smooth estimators of survival and density functions.

Theorem 12.1. *If* $S(t)$ *is continuous (a.e.),* $\lambda_n \to \infty$ *and* $n^{-1}\lambda_n \to 0$ *then*

$$||\tilde{S}_n - S|| = \sup\{|\tilde{S}_n(t) - S(t)| : t \in \mathbf{R}^+\} \to 0 \ a.s., \ as \ n \to \infty. \tag{12.6}$$

The closeness of the smooth estimator of survival function to the true survival function as given in the following theorem implies also the same asymptotic normality for the smooth as well as the empirical estimator of the survival function.

Theorem 12.2. *Under the hypothesis on* λ_n *in the previous theorem, whenever* $f(t)$ *is absolutely continuous with a bounded derivative* $f'(\cdot)$ a.e. *on* $\mathbf{R}^+$,

$$||\tilde{S}_n - S_n|| = O(n^{-3/4}(\log n)^{1+\delta}) \ \text{a.s., as } n \to \infty, \tag{12.7}$$

where $\delta(> 0)$ *is arbitrary.*

Asymptotic properties for the derived density functions were also established as given in the following theorems.

Theorem 12.3. *Under the hypothesis of Theorem 2.2, if* $\lambda_n = O(n^{\alpha})$ *for some* $\alpha < 3/4$,

$$\|\tilde{f}_n - f\| \rightarrow 0 \ a.s. \ as \ n \rightarrow \infty. \tag{12.8}$$

Theorem 12.4. *Consider a nonstochastic, nondecreasing sequence* $\{\lambda_n\}$ *of positive numbers such that* $\lambda_n = O(n^{2/5})$, *and assume that*

$$|f'(t) - f'(s)| \leq K|t-s|^{\alpha}, \text{ for every } t, s \in \mathbf{R}^+. \tag{12.9}$$

Then, for every fixed $t \in \mathbf{R}^+$,

$$n^{2/5}(\tilde{f}_n(t) - f(t)) - \frac{1}{2}\delta^{-2}f'(t) \xrightarrow{D} \mathcal{N}(0, \gamma_t^2), \ as \ n \rightarrow \infty, \tag{12.10}$$

where

$$\gamma_t^2 = \frac{1}{2}(\pi t)^{-1/2} f(t)\delta; \ \delta = \lim_{n\rightarrow\infty} \left(n^{-1/5}\lambda_n^{1/2}\right). \tag{12.11}$$

Remark 12.1. The regularity assumption on f (and f') here in Theorem 12.2 are isomorphic to those in Theorem 1 on p. 765) of Shorack and Wellner;[49] but compared to their rate of $n^{-1/6}$ here we have a better rate of $n^{-1/4}(\log n)^{1+\delta}$ a.s.. In recent years, better rates have been obtained under more stringent smoothness conditions on f (such that $\|f''\| < \infty$ or boundedness of the 4th derivative *etc.*). With the proposed smoothing method, even without imposing a second derivative condition on f, their rate may be matched and better rates under such extra conditions can be obtained.

Remark 12.2. Chaubey and Sen[14] also note that for every $t \in \mathbf{R}^+$,

$$\begin{aligned} E\{(\tilde{S}_n(t) - S_n(t))^2\} &\leq c^2 n^{-3/2} + O(n^{-7/4}) \\ &= O(n^{-3/2}). \end{aligned} \tag{12.12}$$

The kernel method, on the contrary, yields the corresponding order as $O(n^{-4/3})$ (*viz.*, Azzalini[1]), so that in this respect too, the proposed smooth estimator fares better.

12.3. Smooth Estimators of Other Functionals

The paper by Chaubey and Sen[15] investigated the asymptotic properties of the derived functionals of *distribution function* (d.f.) $F(\cdot)$,such as the *hazard function*

$h(t) = f(t)/S(t) = -(d/dt)\log S(t)$ and, the *cumulative hazard function* (c.h.f.) $H(t)$

$$H(t) = \int_0^t h(y)dy = -\log S(t),\ t \geq 0. \tag{12.13}$$

Note that all the above papers considered the truncated version of the estimator given in Eq. (12.3), *i.e*

$$F_n^T(t) = \sum_{k=0}^{\infty} F_n\left(\frac{k}{\lambda_n}\right) w_k(\lambda_n x), \tag{12.14}$$

where

$$w_k(\mu) = \frac{p_k(\mu)}{\sum_{j=0}^{k} p_j(\mu)}.$$

This form was found to be inadequate for the estimating the *Mean Residual Life Function* $m(t) = \int_t^\infty S(u)du/S(t)$, however the estimator given in Eq. (12.3) is found to be adequate. This choice was investigated in Chaubey and Sen,[17] establishing almost sure convergence and asymptotic normality of the resulting smooth estimator of $m(t)$, as given by

$$\tilde{m}_n(t) = \frac{1}{\lambda_n} \frac{\sum_{k=0}^{n} \sum_{r=0}^{k} \frac{(t\lambda_n)^{k-r}}{(k-r)!} S_n(\frac{k}{\lambda_n})}{\sum_{k=0}^{n} \frac{(t\lambda_n)^k}{k!} S_n(\frac{k}{\lambda_n})}. \tag{12.15}$$

12.3.1. *Censored Data*

Chaubey and Sen[16] investigated the properties of the smooth estimator of survival, density, hazard and cumulative hazard functions in the case of randomly censored data establishing similar results. In this case, we do not have the complete sample, rather we observe the censored life times $Z_i = \min(X_i, Y_i)$ and the indicator variable $\delta_i = I(X_i, Y_i)$, where $Y_1, Y_2, \ldots$ is an independent sequence of non-negative independent random variables representing random censoring times. For notational convenience, assume that all the random variables are defined a probability space $(\Omega, \mathcal{B}, P)$. The role of $S_n = 1 - F_n$ is played by the Kaplan-Meier product-limit estimator (PLE) defined by

$$\hat{S}_n(t) = 1 - \hat{F}_n(t) = \prod_{i=1}^{n} \left(1 - \frac{\delta_{[i:n]}}{n-i+1}\right)^{I_{[Z_{i:n} \leq t]}} \tag{12.16}$$

where $(Z_{n:0} = 0) < Z_{n:1} < \ldots < Z_{n:n}(Z_{n:n+1} < +\infty)$ denote the ordered statistics corresponding to the observations Z_i, $i \leq n$ and $\delta_{[i:n]}$ is the value of δ corresponding to $Z_{i:n}$.

These results have been recently extended in Chaubey and Sen[13] in the context of estimating the MRL function, where the expression for the asymptotic variance in the uncensored case is also corrected.

12.3.2. *Other Applications*

Chaubey and Sen[18] noted that if $u(x)$ is monotone then so is $\tilde{u}(x)$. This fact is used in proposing nonparametric smooth estimators of monotone hazard rate, monotone density and monotone mean residual life functions. Chaubey and Kochar[10,11] have also used this fact for estimating the survival function subject to stochastic ordering and uniform stochastic ordering constraints. More recently, Chaubey and Xu[13] have considered the use of smoothing given by the Hille's theorem for estimating survival functions under mean residual life ordering.

12.3.3. *Generalized Smoothing Lemma and Applications*

Recently, Chaubey, Sen and Sen[20] have used the following generalization of Hille's theorem in proposing alternative estimators of density and distribution functions.

Lemma 12.2. (Lemma 1, §VII.1, Feller[32]) *Let u be any bounded and continuous function. Let $G_{x,n}, n = 1,2,...$ be a family of distributions with mean x and variance $h_n^2(x)$ then we have for $h_n(x) \to 0$*

$$\tilde{u}(x) = \int_{-\infty}^{\infty} u(t)dG_{x,n}(t) \to u(x). \tag{12.17}$$

The convergence is uniform in every subinterval in which $h_n(x) \to 0$ and u is uniformly continuous.

Hille's lemma is obtained by choosing $G_{x,n}$ generated by attaching probabilities $p_k(\lambda_n x)$ to (k/λ_n), thus $G_{x,n}$ having mean x and variance $h_n^2(x) = x/\lambda_n$, in case the support of F is $[0,\infty)$. This generalization may be easily adapted for smooth estimation of the distribution(survival) function as given below ;

$$\tilde{F}_n(x) = \int_{-\infty}^{\infty} F_n(t)dG_{x,n}(t) \tag{12.18}$$

Strong convergence of $\tilde{F}_n(x)$ parallels to that of the strong convergence of the empirical distribution function as stated in the following theorem.

Theorem 12.5. *If $h \equiv h_n(x) \to 0$ for every fixed x as $n \to \infty$ we have*

$$\sup_x |\tilde{F}_n(x) - F(x)| \stackrel{a.s.}{\to} 0 \tag{12.19}$$

as $n \to \infty$.

Technically, $G_{x,n}$ can have any support but it may be prudent to choose it so that it has the same support as the random variable under consideration; because this will get rid of the problem of the estimator assigning positive mass to undesired region.
For $\tilde{F}_n(x)$ to be a proper distribution function, $G_{x,n}(t)$ must be decreasing function of x. This can be shown using an alternative form of $\tilde{F}_n(x)$. In addition to being computationally attractive, this form provides an insight into the usual kernel estimator for distributions with infinite support.
We can write

$$\tilde{F}_n(x) = \sum_{i=1}^{n} \int_{X_{(i)}}^{X_{(i+1)}} \frac{i}{n} dG_{x,n}(t) \tag{12.20}$$

$$= \sum_{i=1}^{n} a_i B_i \tag{12.21}$$

where $X_{(1)} \leq X_{(2)} \leq ... \leq X_{(n)}$ represent the ordered values of the sample, $a_i = G_{x,n}(X_{(i+1)}) - G_{x,n}(X_{(i)})$ and $B_i = i/n$.We can write $\sum_{i=1}^{n} a_i B_i = \sum_{i=1}^{n} b_i A_i$ where $b_i = B_i - B_{i-1}$ and $A_i = a_i + ... + a_n$. Thus we have

$$\tilde{F}_n(x) = 1 - \frac{1}{n} \sum_{i=1}^{n} G_{x,n}(X_i). \tag{12.22}$$

This also leads us to propose a smooth estimator of the density as

$$\tilde{f}_n(x) = \frac{d\tilde{F}_n(x)}{dx} = -\frac{1}{n} \sum_{i=1}^{n} \frac{d}{dx} G_{x,n}(X_i), \tag{12.23}$$

The representation given by Eq. (12.23) can also be used to derive the kernel estimator given by Eq. (12.1) as follows. Let $G_{x,n}(.)$ be given by

$$G_{x,n}(t) = K\left(\frac{t-x}{h}\right),$$

with has mean x and variance h^2, where $K(.)$ is a distribution function with mean zero and variance 1. Note also that symmetry of K is not required as is usually required in kernel estimation. The condition that $h \equiv h_n(x) \to 0$, which is enough to guarantee the almost sure convergence of $\tilde{F}_n(x)$, may not be enough for consistency of $f_n(x)$.
The general smoothing lemma may also be used directly for proposing motivating the kernel estimator. Replacing $u(x)$ by $f(x)$, we see that a close approximation to $f(x)$ is given by

$$\tilde{f}(x) = E(g_{x,n}(X)) \tag{12.24}$$

which can be estimated for an *i.i.d.* sample by

$$\tilde{f}_n(x) = \frac{1}{n}\sum_{i=1}^{n} g_{x,n}(X_i). \tag{12.25}$$

Choosing $g_{x,n}$ as an appropriate symmetric density gives the usual kernel estimator; however, it also suggests alternative estimators in the case of restricted support for the distribution. Chen[23] chose Gamma kernels for non-negative random variables and Beta kernels for distributions centered on the unit interval, which can be motivated from the above approach, although no such motivation is provided in Chen.[23] Similar estimators for the density function are also given by Scaillet[45] using inverse Gaussian and reciprocal inverse Gaussian kernels. One of the major drawback of these estimators is that the resulting estimate of the density is always zero at $x = 0$. Also note that in general the two estimators given by Eqs. (12.23) and (12.25) may be different. Bouezmarni and Scaillet[7] have considered the use of Chaubey and Sen[14] and Chen[23] estimators for estimation of the density of a time series with $\alpha-$ mixing.
For the data concentrated on $[0,1]$, the use of Lemma 12.2 provides Bernstein-polynomial estimator for the density function by using an appropriate binomial distribution. Such an estimator was originally proposed by Vitale[58] and it was later investigated by Babu, Canty and Chaubey[2] from the point of view of the Hille's theorem.

12.4. Multivariate Generalization of Hille's Lemma

Let **X** denote a d-dimensional vector random variable defined on $\mathbf{R}^{+d}$, with a distribution function denoted by $F(\mathbf{x})$ and a continuous density $f(\cdot)$, so that

$$F(\mathbf{x}) = \int_{\mathbf{t}\leq\mathbf{x}} f(\mathbf{t})d\mathbf{t}. \tag{12.26}$$

The corresponding multivariate survival function is defined as

$$S(\mathbf{x}) = \int_{\mathbf{t}>\mathbf{x}} f(\mathbf{t})d\mathbf{t}. \tag{12.27}$$

(The ordered relations $\leq$ and $>$ are taken to be coordinate wise.) These distributions are quite common in biomedical and industrial studies where each component of the random vector **X** represents failure time of a unit of a system (see Cox and Oaks[24]). Note that in the univariate case $F(x)+S(x) = 1$ for all **x**, however, this may not be true in the multivariate case. Thus, estimation of $S(\mathbf{x})$ does not follow directly from that of $F(\mathbf{x})$ in general for $d > 1$ as opposed to the case

$d = 1$. In this article, we shall implicitly consider the case $d > 1$. Based on a random sample $(\mathbf{X}_1, \mathbf{X}_2, ..., \mathbf{X}_n)$ of n failure times of the system, $F(\mathbf{x})$ is, generally, estimated by the empirical distribution function, $F_n(\mathbf{x})$, defined as

$$F_n(\mathbf{x}) = \frac{1}{n}\sum_{i=1}^{n}\prod_{j=1}^{d} I(X_{ij} \leq x_i) = \frac{1}{n}\sum_{i=1}^{n} I(X_{i1} \leq x_1, ...X_{id} \leq x_d), \tag{12.28}$$

where X_{ij} denotes the jth component of the ith sample observation. The corresponding estimator of the survival function $S(x)$ is given by

$$S_n(\mathbf{x}) = \frac{1}{n}\sum_{i=1}^{n} I(X_{i1} > x_1, ...X_{id} > x_d). \tag{12.29}$$

The discontinuities in $F_n(\mathbf{x})$ and $S_n(\mathbf{x})$ render these estimators useless as "graphical tools" and otherwise in providing smooth estimators of the density function. As a result a lot of effort has been afforded to smooth estimation of multivariate density functions (see Scott[48]).

For $d = 1$, the pioneering work of Rosenblatt[44]and Parzen[40] created a tremendous scope for smooth estimation of density and related functionals. The offshoot of this approach in the area of nonparametric regression has rendered this area an enormous vitality due to its wide ranging applications in health sciences and communications. We may refer to some recent excellent monographs highlighting the details of these methods, *e.g.* Fan and Gijbels,[31] Wand and Jones,[61] Green and Silverman[34] and Härdle[35]). Some of the univariate methods have been extended to the multivariate case (see for example, Cacoullos,[8] Deheuvels,[29] Loftsgaarden and Quesenberry[38] and Murthy[39]). Some of these methods can be treated under a unified manner using the so called "delta-sequence" (see Bosq,[6] Földes and Révész[33] and Walter and Blum[60]). Stone's[53] paper can be also put in this category with respect to the nonparametric regression. Susarla and Walter[54] extended it to the multivariate case.

Chaubey and Sen[19] considered the following generalization of Hille's theorem in proposing smooth estimators of multivariate survival and density functions.

Lemma 12.3. *Consider a sequence* $\{\Phi_n(\mathbf{y}, \mathbf{t})\}_{n=1}^{\infty}$ *of distribution functions in* $\mathbf{R^d}$ *for every fixed* $\mathbf{t} \in \mathbf{R^d}$, *such that for* $\mathbf{Y}_n \sim \Phi_n(\cdot, \mathbf{t})$

(i) $E\mathbf{Y}_n = \mathbf{t}$

(ii) $v_n(\mathbf{t}) = \max_{1\leq i\leq d} Var(Y_{in}) \to 0$ *as* $n \to \infty$, *for every fixed* $\mathbf{t}$.

(iii) $\Phi_n(\mathbf{y}, \mathbf{t})$ *is continuous in* $\mathbf{t}$.

Define for any bounded continuous multivariate function $u(\mathbf{t})$

$$\tilde{u}_n(\mathbf{t}) = \int_{\mathbf{R^d}} u(\mathbf{x}) d\Phi_n(\mathbf{x}, \mathbf{t}). \tag{12.30}$$

Then $\tilde{u}_n(\mathbf{t}) \to u(\mathbf{t})$ as $n \to \infty$ for $\mathbf{t}$ in any compact subset of $\mathbf{R^d}$, the convergence being uniform over any subset over which $u(\mathbf{t})$ is uniformly continuous. Furthermore, if the function $u(\mathbf{t})$ is monotone, the convergence holds uniformly over entire $\mathbf{R^d}$.

The multivariate Chaubey-Sen estimator. The generalization of the the estimator of Chaubey and Sen[14] to the multivariate case (here we consider the untruncated case only) is described below. The smooth estimator of the survival function $S(\mathbf{x})$ is given by

$$\tilde{S}_n(\mathbf{x}) = \sum_{j_1=0}^{\infty} \sum_{j_2=0}^{\infty} \cdots \sum_{j_d=0}^{\infty} w_{j_1,\ldots,j_d}(x_1\lambda_{1n},\ldots,x_d\lambda_{dn}) S_n\left(\frac{j_1}{\lambda_{1n}},\ldots,\frac{j_d}{\lambda_{dn}}\right) \tag{12.31}$$

where the weights $w_{j_1,j_2,\ldots,j_d}(t_1,t_2,\ldots,t_d)$ being defined as

$$w_{j_1,j_2,\ldots,j_d}(t_1,t_2,\ldots,t_d) = e^{-\sum_{i=1}^{d} t_i} \prod_{i=1}^{d} \frac{(t_i)^{j_i}}{j_i!}. \tag{12.32}$$

λ_{in} $i = 1,2,\ldots,d$ may be chosen as

$$\lambda_{in} = \frac{n}{\max(X_{1i},\ldots,X_{ni})},\ i = 1,2,\ldots,d. \tag{12.33}$$

The above choice of the constant λ_{jn} is obviously stochastic and is motivated from the fact that there is no observed mass beyond the point $\max(X_{1j},\ldots,X_{nj})$ for the variable X_j. The deterministic choice has to be such that $\lambda_{jn} \to \infty$ but $n^{-1}\lambda_{jn} \to 0$ (see Chaubey and Sen[14] for discussion on other choices.)

In the univariate case the positive half of the real line is partitioned into a lattice of points $0, 1, 2, \ldots.$ and a suitable Poisson distribution is superimposed on the empirical survival function to provide the smooth estimator. In the multivariate case, we consider a lattice of hypercubes and generate the weights by product of proper Poisson weights. The corresponding estimator of the distribution function is given as

$$\tilde{F}_n(\mathbf{x}) = \sum_{j_1=0}^{\infty} \sum_{j_2=0}^{\infty} \cdots \sum_{j_d=0}^{\infty} w_{j_1,\ldots,j_d}(x_1\lambda_{1n},\ldots,x_d\lambda_{dn}) F_n\left(\frac{j_1}{\lambda_{1n}},\ldots,\frac{j_d}{\lambda_{dn}}\right). \tag{12.34}$$

The smooth estimator of the density function may be readily obtained by considering the derivative of $\tilde{F}_n(x)$ which is given by

$$
\begin{aligned}
\tilde{f}_n(\mathbf{x}) &= \sum_{j_1=0}^{\infty}\sum_{j_2=0}^{\infty}\cdots\sum_{j_d=0}^{\infty} \frac{\partial^d}{\partial x_1 \partial x_2 \ldots \partial x_d} w_{j_1,\ldots,j_d}(x_1\lambda_{1n},\ldots,x_d\lambda_{dn}) F_n\left(\frac{j_1}{\lambda_{1n}},\ldots,\frac{j_d}{\lambda_{dn}}\right) \\
&= \sum_{j_1=0}^{\infty}\sum_{j_2=0}^{\infty}\cdots\sum_{j_d=0}^{\infty} w_{j_1,\ldots,j_d}(x_1\lambda_{1n},\ldots x_d\lambda_{dn}) \\
&\qquad \left(\prod_{i=1}^{d}\lambda_{in}\left(\frac{j_i}{\lambda_{in}x_i}-1\right)\right) F_n\left(\frac{j_1}{\lambda_{1n}},\ldots,\frac{j_d}{\lambda_{dn}}\right) \qquad (12.35)
\end{aligned}
$$

Some asymptotic properties of the estimators $\tilde{F}_n$ and $\tilde{f}_n$ similar to the univariate case have been established in Chaubey and Sen[19] as given below.

Theorem 12.6. *Suppose that for each i, $\lambda_{in} \to \infty, n^{-1}\lambda_{in} \to 0$ as $n \to \infty$* and $f(\mathbf{x})$ *has bounded first derivatives then*

$$\|\tilde{F}_n - F\| = \sup_{\mathbf{x}\in\mathbf{R}^{+\mathbf{d}}} |\tilde{F}_n(\mathbf{x}) - F(\mathbf{x})| \;\to\; 0 \;\; a.s. \text{ as } n \to \infty.$$

Theorem 12.7. *Suppose that F is absolutely continuous and for each i, $\lambda_{in} \to \infty$ and $n^{-1}\lambda_{in} \to 0$ then we have for any $\eta > 0$*

$$\sup_{\mathbf{x}\in\mathbf{R}^{+d}} |\tilde{F}_n(\mathbf{x}) - F_n(\mathbf{x})| = O(n^{-\frac{3}{4}}(\log n)^{1+\eta}) \text{ a.s.}.$$

Theorem 12.8. *Assuming that the density function f is bounded with bounded first derivatives and $\lambda_{in} = O(n^{\alpha/d})$ for some $\alpha < 3/4$, for $i = 1,2,\ldots,d$, we have*

$$\sup_{\mathbf{x}\in\mathbf{R}^{+d}} |\tilde{f}_n(\mathbf{x}) - f(\mathbf{x})| \to 0 \; a.s. \text{ as } n \to \infty.$$

Theorem 12.9. *Consider a non-stochastic sequence $\{\lambda_{in}\}_{n=1}^{\infty}, i = 1,2,\ldots,d$ such that $\lambda_{in} = O(n^{2/(4+d)})$.* Further assume that the first order partial derivatives are Lipschitz continuous, then for every fixed $\mathbf{x} \in \mathbf{R}^{+d}$,

$$n^{2/(4+d)}(\tilde{f}_n(\mathbf{x}) - f(\mathbf{x})) - (1/2)b(\mathbf{x}) \xrightarrow{D} \mathcal{N}(0,\sigma^2(\mathbf{x})), \;\; \text{as } n \to \infty.$$

where

$$b(\mathbf{x}) = \lim_{n\to\infty} n^{2/(4+d)} \sum_{i=1}^{d} \lambda_{in}^{-1} f'_{x_i}(\mathbf{x}) \qquad (12.36)$$

$$\text{and } \sigma^2(\mathbf{x}) = \lim_{n\to\infty} \frac{1}{2^d \pi^{d/2} \prod x_i^{1/2}} n^{-\frac{d}{4+d}} \prod \lambda_{in}^{\frac{1}{2}} f(\mathbf{x}). \qquad (12.37)$$

This generalization is further used in proposing an smooth estimator of the regression function for non-negative random variables by Chaubey, Sen and Zhou[21] with an application in finite population sampling.

An alternative estimator. An estimator similar to that of Chen,[23] which in fact appears to be somewhat simpler than the one above, can be obtained using the motivation outlined after Eq.(12.23). As noted in Eq.(12.24), an approximation to $f(x)$ is given by

$$\tilde{f}(\mathbf{x}) = \int f(\mathbf{t}) g_{\mathbf{x},n}(\mathbf{t}) d\mathbf{t} = E(g_{\mathbf{x},n}(\mathbf{X})) \tag{12.38}$$

which can be estimated for an *i.i.d.* sample by

$$\tilde{f}_n(\mathbf{x}) = \frac{1}{n} \sum_{i=1}^{n} g_{\mathbf{x},n}(\mathbf{X}_i). \tag{12.39}$$

We could choose $g_{\mathbf{x},n}(\cdot)$, for instance, as a product-density $g_{\mathbf{x},n}(\mathbf{t}) = \prod_{i=1}^{d} g_{x_i,n}(t_i)$ where $g_{x_i,n}(\cdot)$ is a univariate density with $\mu(g_{x_i,n}) \to x_i$ and $\sigma(g_{x_i,n}) \to 0$.

The multivariate generalization of the Bernstein estimator is also straight forward and it has been investigated in the case of dependent data (see Babu and Chaubey[3]). This generalization has been used for studying the regression function in the context of filtering and Entropy estimation of a chaotic map from noisy data (see Babu et al.[4]).

12.5. Further Developments

It is of interest to consider the generalization of the estimator in Eq. (12.23) to the multivariate case, especially to the case of restricted support and/or dependent observations. This has been recently explored by Chaubey, Läib and Sen.[12] Another direction of application is multivariate censored survival data.

A natural interest is also to explore the practical case of censored survival data. Since the components of the random vector $\mathbf{X}$ are generally not independent, the mechanism of censoring may be much more complex than in the univariate case. A censoring may be with respect to one response variable only; for example, X_1 relates to the primary end point, while the remaining d-1 coordinates refer to secondary or surrogate end points, a censoring may specifically relate to X_1 alone. In such a case, under noninformative and random censoring scheme, we may conceive of a nonnegative random variable C which is independent of X_1 (and does not depend on the other auxiliary or concomitant variables), such that the observable random elements are (i) $X_1^o = \min(X_1, C)$, (ii) $\delta = I(X_1^o = X_1)$, and (iii)

$X_j, j=2,...,d$, along with other auxiliary variables. In the more likely case, a censoring may be due to withdrawal or drop-out, and if so, it automatically removes the observation vector as a whole from further study. In that case, we have observed multivariate survival time $\mathbf{X^o} = (\min(X_j, c), j = 1, \ldots, d)$. The well known Kaplan-Meier product limit estimator of the survival function in the univariate case has been extended to specific multivariate models under specific regularity assumptions (see Dabrowska,[25–27] Dabrowska *et al.*[28] and Pedroso de Lima and Sen[41,42] for some discussion). Given these developments, one can incorporate the univariate results of Chaubey and Sen[16] on smoothing of the Product Limit estimator, and formulate their multivariate counterparts. For intended brevity of presentation, the details are not given here.

12.5.1. *Quantile Estimation*

We may define the usual quantile process $\mathbf{Q}_n = \{Q_n(t) : 0 \leq t \leq 1\}$ by letting

$$Q_n(t) = \sup\{x : S_n(x) \geq 1 - t\}, \ 0 \leq t \leq 1.$$

Side by side, we introduce the smoothed quantile process $\tilde{\mathbf{Q}}_n = \{\tilde{Q}_n(t) : 0 \leq t \leq 1\}$ by letting

$$\tilde{Q}_n(t) = \sup\{x : \tilde{S}_n(x) \geq 1 - t\}, \ 0 \leq t \leq 1. \tag{12.40}$$

Then, for every $t : 0 \leq t \leq 1$, the improved order of approximations discussed in the preceding two remarks also pertain to $\tilde{Q}_n$. It may be noted that $\tilde{S}_n(\cdot)$, being differentiable, is smooth, and hence in Eq. (12.40) we may as well replace "$\geq$" by "$=$". Deeper smoothness properties of $\tilde{Q}_n(t)$ depend on more specific choice of $\{\lambda_n\}$.

12.5.2. *Other Problems*

Another area is to explore the conditional distributions and quantile functions on restricted support. The multivariate generalization paves the way for discussing such problems. Another direction for generalization of these results is towards the study of asymptotic properties of the resulting estimators of the density for dependent data and other kinds of stochastic orderings.

It may be of interest to combine the non-parametric estimator with a member of some specific class, so that the data can decide which estimator is to be given higher weight. A comparative study on the properties of such estimators along with those based on other smoothing methods such as wavelets, local linear polynomials, splines and nearest neighbour methods and so on may also be important.

Acknowledgments

The author would like to thank the referee and Ashis Sengupta for their careful reading of the earlier manuscript. This research was partially supported by NSERC Canada through a Discovery Grant held by the author.

References

1. Azzalini, A. *A note on the estimation of a distribution function and quantiles by a kernel method*, Biometrika, vol. 68, pp. 326–328, 1981.
2. Babu, G. J., A. J. Canty and Y. P. Chaubey, *Application of Bernstein polynomials for smooth estimation of distribution and density function*, Journal of Statistical Planning and Inference, vol. 105, pp. 377–392, 2002.
3. Babu, G. J. and Y. P. Chaubey, *Smooth estimation of a distribution and density function on a hypercube using Bernstein polynomials for dependent random vectors*, Statistics and Probability Letters, vol. 76, pp. 377–392, 2006.
4. Babu, G. J., A. Boyarsky, Y. P. Chaubey and P. Gora, *A new statistical method for filtering and Entropy estimation of a chaotic map from noisy data*, Bifurcation and Chaos, vol. 14, pp. 3989–3994, 2004.
5. Bagai, I. and B. L. S. Prakasa Rao, *Kernel type density estimates for positive valued random variables*, Sankhya, vol. A57, pp. 56–67, 1996.
6. Bosq, D. *Study of a class of density estimators*, Recent Developments in Statistics, Eds.: J. F. Barra, F. Brodeau, G. Romier and B. van Kutsem, North Holland, Amsterdam, 1977.
7. Bouezmarni, T. and O. Scaillet, *Consistency of asymmetric kernel density estimators and smoothed histograms with application to income data*, Econometric Theory, vol. 21, pp. 390–412, 2005.
8. Cacoullos, T. *Estimation of multivariate density*, Annals of Institute of Statistical Mathematics, vol. 18, pp. 179–189, 1966.
9. Çencov, N. N. *Evaluation of an unknown density from observations*, Soviet Mathematics, vol. 3, pp. 1559–1562, 1962.
10. Chaubey, Y. P. and S. C. Kochar, *Smooth estimation of stochastically ordered survival functions*, Journal of Indian Statistical Association, vol. 38, pp. 209–225, 2000.
11. Chaubey, Y. P. and S. C. Kochar, *Smooth estimation of uniformly stochastically ordered survival functions*, Journal of Combinatorics, Information and System Sciences, vol. 31, pp. 1-13, 2006.
12. Chaubey, Y. P., N. Läib and A. Sen, *A Smooth Estimator of Regression Function for Non-Negative Dependent Random Variables*, Technical Report No.02/08, Department of Mathematics and Statistics, Concordia University, Montreal, Canada, 2008.
13. Chaubey, Y. P. and A. Sen, *Smooth estimation of mean residual life under random censoring*, IMS Collections, vol. 1, Eds.: N. Balakrishnan, Edsel Pena and Mervyn J. Silvapulle, Institute of Mathematical Statistics, Maryland, 2007.
14. Chaubey, Y. P. and P. K. Sen, On smooth estimation of survival and density functions. Statistics & Decisions, vol. 14, pp. 1–22, 1996.
15. Chaubey, Y. P. and P. K. Sen, *On smooth estimation of hazard and cumulative hazard*

functions, Frontiers of Probability and Statistics, Ed.: S. P. Mukherjee et al., Narosa Publishing House, New Delhi, pp. 91–99, 1997.
16. Chaubey, Y. P. and P. K. Sen, *On smoothed functional estimation under random censoring*, Frontiers in Reliability, Eds.: A. P. Basu, S. K. Basu and S. Mukhopadhyay, World Scientific, Singapore, 1998.
17. Chaubey, Y. P. and P. K. Sen, *On smooth estimation of mean residual life*, Journal of Statistical Planning and Inference, vol. 75, pp. 223–236, 1999.
18. Chaubey, Y. P. and P. K. Sen, *Smooth isotonic estimation of density, hazard and MRL functions*, Calcutta Statistical Association Bulletin, vol. 52, pp. 99–116, 2002.
19. Chaubey, Y. P. and P. K. Sen, *Smooth estimation of Multivariate Survival and Density Functions*, Journal of Statistical Planning and Inference, pp. vol. 103, pp. 361–376, 2002.
20. Chaubey, Y. P., A. Sen, and P. K. Sen, *A New Smooth Density Estimator for Non-negative Random Variables*, Technical Report No. 01/07, Department of Mathematics and Statistics, Concordia University, Montreal, Canada, 2007.
21. Chaubey, Y. P., P. K. Sen and X. Zhou, *A nonparametric regression smoother for nonnegative data*, Proceedings of International Conference on Recent Advances in Survey Sampling, ICRASS, Carleton University, Ottawa, ONT, July 10-13, 2002, CDROM, LRSP, Carleton University, Ottawa, Canada, 2002.
22. Chaubey, Y. P. and H. Xu, *Smooth estimation of survival functions under mean residual life ordering*, Journal of Statistical Planning and Inference, vol. 137, pp. 3303–3316, 2007.
23. Chen, S. X. *Probability density function estimation using Gamma kernels*, Annals of Institute of Statistical Mathematics, vol. 52, pp. 471-480, 2000.
24. Cox, D. R. and D. Oakes, *Analysis of Survival Data,* Chapman and Hall, London, 1984.
25. Dabrowska, D. M. *Kaplan-Meier estimate on the plane*, Annals of Statistics, vol. 16, pp. 1475–189, 1988.
26. Dabrowska, D. M. *Rank tests for matched pair experiments with censored data*, Journal of Multivariate Analysis, vol. 28, pp. 88–114, 1989.
27. Dabrowska, D. M. *Kaplan-Meier estimate on the plane: weak convergence, LIL, and the bootstrap*, Journal of Multivariate Analysis, vol. 29, pp. 308–325, 1989.
28. Dabrowska, D. M., D. L. Duffy and Z. D. Zhang, *Hazard and density estimation from bivariate censored data*, Journal of Nonparametric Statistics, vol. 10, pp. 67–93, 1999.
29. Deheuvels, P. *Estimation non paramétrique de la densité par histogrammes généralisés (II)*, Publications de l'Institut de Statistique de l'Université de Paris, vol. 22, pp. 1–23, 1977.
30. Epanechnikov, V. K. *Non-parametric estimation of a multivariate density*, Theory of Probability and Applications, vol. 14, pp. 153–158, 1969.
31. Fan, J. and I. Gijbels, *Local polynomial modelling and its applications*, Chapman and Hall, London, 1996.
32. Feller, W. *An Introduction to Probability Theory and Its Applications, Vol. II*, John Wiley and Sons, New York, 1965.
33. Földes, A. and P. Révész, *A general method for density estimation*, Studia Scientiarum Mathematicarum Hungarica, vol. 9, pp. 443–452, 1974.

34. Green, P. J. and B. W. Silverman, *Nonparametric Regression and Generalized Linear Models: A Roughness Penalty Approach,* Chapman and Hall, London, 1994.
35. Härdle, W. *Smoothing Techniques With Implementation in S*, Springer-Verlag, New York, 1991.
36. Hille, E. *Functional Analysis and Semigroups*. American Mathematical Society Colloquium Publication, vol. 31, American Mathematical Society, New York, 1948.
37. Karunamuni, R. J. and T. Alberts, *On boundary correction in kernel density estimation*, Statistical Methodology, vol. 2, pp. 191–212, 2005.
38. Loftsgaarden, D. G. and C. P. Quesenberry, *A non-parametric estimate of multivariate density function*, Annals of Mathematical Statistics, vol. 36, pp. 1049–1051, 1965.
39. Murthy, V. K. *Nonparametric estimation of multivariate densities with applications*, Multivariate Analysis (Proc. Internat. Sympos., Dayton, Ohio, 1965), Ed.: P. R. Krishnaiah, pp. 43-56, Academic Press, New York, 1966.
40. Parzen, E. *On estimation of probability density and mode*, Annals of Mathematical Statistics, vol. 33, pp. 1065–1070, 1962.
41. Pedroso de Lima, A. C. and P. K. Sen, *A matrix-valued counting process with first-order interactive intensities*, Annals of Applied Probability, vol. 7, pp. 494–507, 1997.
42. Pedroso de Lima, A. C. and P. K. Sen, *Time-dependent coefficients in a multi-event model for survival analysis*, Journal of Statistical Planning and Inference, vol. 75, pp. 393–413, 1999.
43. Prakasa Rao, B. L. S. *Nonparametric Functional Estimation*, Academic Press, 1983.
44. Rosenblatt, M. *Remarks on some nonparametric estimates of density functions*, Annals of Mathematical Statistics, vol. 27, pp. 832–837, 1956.
45. Scaillet, O. *Density estimation using inverse Gaussian and reciprocal inverse Gaussian kernels*, Journal of Nonparametric Statistics, vol. 16, pp. 217–226, 2003.
46. Schwarz, S. C. *Estimation of a probability density by an orthogonal series*, Annals of Mathematical Statistics, vol. 38, pp. 1262–1265, 1967.
47. Scott, D. W., A. M. Gotto, J. S. Cole and G. A. Gorry, *Plasma lipids as collateral risk factors in coronary artery disease: A study of 371 males with chest pain*, Journal of Chronic Deseases, vol. 31, pp. 337–345, 1978.
48. Scott, D. W. *Multivariate Density Estimation: Theory, Practice and Visualizations*, John Wiley and Sons, New York, 1992.
49. Shorack, G. R. and J. A. Wellner, *Empirical Processes with Applications to Statistics*, John Wiley and Sons, New York, 1986.
50. Silvapulle, M. J. and P. K. Sen, *Constrained Statistical Inference: Order, Inequality, and Shape Constraints*, John Wiley and Sons, New York, 2004.
51. Silverman, B. W. *Choosing the window width when estimating a density*, Biometrika, vol. 65, pp. 1–11, 1978.
52. Silverman, B. W. *Density Estimation for Statistics and Data Analysis*, Chapman and Hall, London, 1986.
53. Stone, C. *Consistent nonparametric regression*, Annals of Statististics, vol. 5, pp. 595-645, 1977.
54. Susarla, V. and G. Walter, *Estimation of multivariate density function using delta sequences*, Annals of Statistics. vol. 9, pp. 347–355, 1981.
55. Tarter, M. E. and R. A. Kronmal, *On multivariate density estimates based on orthog-*

onal expansions, Annals of Mathematical Statistics, vol. 41, pp. 718–722, 1970.
56. Terrell, G. R. and D. W. Scott, *Variable kernel density estimation*, Annals of Statistics, vol. 20, pp. 1236–1265, 1992.
57. Venable, W. N. and B. D. Ripley, *Modern Applied Statistics with S-Plus*, Springer - Verlag, New York, 1999
58. Vitale, R. A. *A Bernstein polynomial approach to density estimation*, In Statistical Inference and Related Topics, vol. 2, Ed.: M.L. Puri, Academic Press, New York, 1975.
59. Wahba, G. *A polynomial algorithm for density estimation*, Annals of Mathematical Statististics, vol. 42, pp. 1870–1886, 1971.
60. Walter, G. and J. Blum, *Probability density estimation using delta sequences*, Annals of Statististics, vol. 7, pp. 328–340, 1979.
61. Wand, M. P. and M. C. Jones, *Kernel Smoothing*, Chapman and Hall, London, 1995.
62. Watson, G. S. and R. Leadbetter, *On the estimation of the probability density I*, Annals of Mathematical Statististics, vol. 34, pp. 480–491, 1963.

Chapter 13

Estimation Using Quantile Function Structure with Emphasis on Weibull Distribution

Georgia D. Kollia, Govind S. Mudholkar and Deo Kumar Srivastava*

Bristol-Myers Squibb, Princeton, NJ 08540
University of Rochester, Rochester, NY 14627
St. Jude Children's Research Hospital, Memphis, TN 38105-2794
E-mail:kumar.srivastava@stjude.org

An approach to estimation of parameters of quantile function models is proposed and applied to constructing simple L-estimators of the parameters of the Weibull models. The general approach to construction involves starting with convenient pivotal statistics and refining them using the jackknife method to obtain simple closed form estimators. Empirical studies of these closed form L-estimators for the Weibull parameters so obtained show that they are essentially bias free and in terms of MSE compare very well with the MLE and other competitors.

Contents

*Corresponding Author: Department of Biostatistics, St. Jude Children's Research Hospital, Memphis, TN 38105-2794.

13.1. Introduction

In this paper we propose a general heuristic approach, based on idea of using the natural plug-in estimators, to the estimation of parameters when the population is specified in terms of quantile function. The approach consists of two stages. In the first stage, a quantile function analogue of the classical method of moments is used to construct convenient functions of order statistics. In the second stage, these functions termed as pivots are refined using results related to jackknife theory derived for the purpose of constructing the closed form estimators. Applications of the general method to the Weibull case with unknown shape and scale parameters, yields simple linear combination of order statistics as the pivotal statistics, which then yield closed form estimators of the shape and scale parameters as well as an estimate of their covariance matrix.

The general approach is outlined in Section 13.2, by outlining the pivotal statistics and then processing these to derive the final estimators. In Section 13.3, the approach is applied to obtain simple closed form estimators of Weibull parameters. In Section 13.4, these estimators are studied intrinsically and in comparison with the MLE and some other competitors. Section 13.5, contains some conclusions and miscellaneous remarks.

13.2. The Approach in the General Setting

Let $X_1, X_2, \cdots, X_n$ be a random sample from a continuous population c.d.f. $F(x,\theta)$ and the quantile function $Q(u,\theta) = F^{-1}(u,\theta)$, with k-dimensional parameter θ. The common classical approach to estimation of θ involves the simpler method of moments, which often yields suboptimal estimators, or the method of maximum likelihood, which, in general, yields asymptotically efficient estimates but not necessarily closed form estimators, and often requires iterative schemes for solving nonlinear equations. In this section we outline the use of the quantile function structure of the population for constructing closed form alternative estimators.

13.2.1. *Construction of the pivotal statistics*

This step may be compared to the method of moments which involves equating the first k population moments with their sample estimates to obtain estimators of the unknown k-vector parameter θ. It is easy to see that the crude moments of the population specified in terms of the quantile function are given by,

$$E(X^k) = \int_0^1 Q(u;\theta)^k du. \tag{13.1}$$

and more generally, for any real valued function h and a function g defined on $(0,1)$,

$$\phi = \int g(u)h(Q(u))du, \tag{13.2}$$

is a function of θ, $\phi(\theta)$. This stage of the construction involves choosing k functions g_j and h_j in order to obtain k convenient parametric functions $\phi_j(\theta)$ the estimates of which serve as pivotal statistics. For a general discussion on the construction of pivotal quantities one can refer to [Cox and Hinkley 1974] or [Serfling 1980]. We propose estimating ϕ_j using the simple logic in the approximation,

$$\int_0^1 g_j(u)h_j(Q(u))du = \sum_{i=1}^{n} \int_{(i-1)/n}^{i/n} g_j(u)h_j(Q(u))du \cong \frac{1}{n}\sum_{i=1}^{n} g_j(u_i^*)Y_{(i)}, \tag{13.3}$$

where $Y_{(i)} = h_j(X_{(i)})$ and $u_i^* = (i-0.5)/n$, and obtaining a set of equations,

$$\hat{\phi}_j = \frac{1}{n}\sum_{i=1}^{n} g_j((i-0.5)/n)Y_{(i)},\ j = 1,2,\cdots,k. \tag{13.4}$$

If the function h is chosen to be monotone, then $Y_{(i)}$ are the order statistics of $Y_i = h(X_i), i = 1,2,\cdots,n$. Furthermore, it may be possible, as in the Weibull population case, to choose h such that $\hat{\phi}(\theta)$ are functions of some simple transformations θ_i^* of $\theta_i, i = 1,2,\cdots,k$, but the equations (13.2.4) reduce to an easily solvable set of equations,

$$\phi_i(\theta^*) = \hat{\phi}_i(y_1,y_2,\cdots,y_n),\ i = 1,2,\cdots,k. \tag{13.5}$$

13.2.2. *The Jackknifed L-statistic and its variance estimate*

Let $Y_1,Y_2,\cdots,Y_n$ be a random sample of size n and $T_{ni} = \hat{\theta}_i^*$ an estimator of the i-th component θ_i^* of the vector θ^*, after reparameterization of the original θ's. Let $\mathbf{T_n} = \hat{\theta}^*$, and assume that $\mathbf{T_n}$ is a vector of L-statistic in the Y_i's. The problem of jackknifing an L-statistic and estimating its variance has been considered by many including [Gardiner and Sen 1979], [Cheng 1982], [Parr and Schucany

1982], [Shao and Tu, 1996],[Huang, 1991]. In this section we give explicit expressions for jackknifed L-statistic and its estimated variance.
Now consider the case of single parameter, i.e. $k = 1$, and let T_{n-1}^{l} be the estimator of T_n computed after deleting the lth observation from the sample, then the jackknife estimator $\check{T}_n$ obtained from T_n is given by

$$\check{T}_n = nT_n - \frac{n-1}{n}\sum_{l=1}^{n} T_{n-1}^{(l)}. \tag{13.6}$$

The jackknife estimator in (13.6) can also be written as the average of the pseudovalues P_l defined by

$$P_l = nT_n - (n-1)T_{n-1}^{l},\ l = 1,2,\cdots,n. \tag{13.7}$$

Furthermore, despite some shortcomings, see Efron and Stein (1981), the sample variance of the pseudovalues P_l also provides a simple estimator of the variance of $\check{T}_n$, namely

$$\hat{Var}(\check{T}_n) \cong \frac{1}{n(n-1)}\left[\sum P_l^2 - \frac{(\Sigma P_l)^2}{n}\right]. \tag{13.8}$$

We now consider the case of $\hat{\theta}^* = T_n$ and L-statistic given by

$$\hat{\theta}^* = T_n = \frac{1}{n}\sum_{i=1}^{n} c_{i:n}Y_{(i)}\ , \tag{13.9}$$

where $c_i = c_{i:n}$ are fixed constants.
It is obvious that the jackknife version of an L-statistic is also an L-statistic. Specifically, the jackknife estimator $\check{T}_n$ of an L-estimator T_n is now given in an explicit form in the following.
Proposition 2.1. The jackknife estimator $\check{T}_n$ corresponding to an L-statistic $T_n = \sum_{i=1}^{n} c_i Y_{(i)}/n$ is given by

$$\check{T}_n = \sum_{i=1}^{n}\left[c_{i:n} - \frac{(i-1)}{n}c_{i-1:n-1} - \frac{(n-1)}{n}c_{i:n-1}\right]Y_{(i)}\ , \tag{13.10}$$

where $c_{i:n-1}$ is the coefficient of $Y_{(i)}$ in $T_{n-1}^{(l)}$, and $c_{0,n-1} = 0,\ c_{n:n-1} = 1$.

Proof. [Proof of the Proposition] Let $(R_1, R_2, \cdots, R_n)$ be the vector of ranks pertaining to $Y_1, Y_2, \cdots, Y_n$ so that $Y_{R_l:n} = Y_l$ for $1 \le l \le n$. Then, we have

$$\{Y_1, Y_2, \cdots, Y_{l-1}, Y_{l+1}, \cdots, Y_n\} = \{Y_1, Y_2, \cdots, Y_n\}\backslash\{Y_{R_l:n}\} \tag{13.11}$$

so that the order statistics of $Y_1, Y_2, \cdots, Y_{l-1}, Y_{l+1}, \cdots, Y_n$are given by $\{Y_{1:n}, Y_{2:n},, \cdots, Y_{n:n}\} \backslash \{Y_{R_l:n}\}$.
Consequently,, we have

$$T_{n-1}^{(l)} = \frac{1}{n-1}\left[\sum_{i=1}^{R_l-1} c_{i:n-1}Y_{(i)} + \sum_{i=R_l+1}^{n} c_{i-1:n-1}Y_{(i)}\right]. \tag{13.12}$$

Hence, noting that $R_1, R_2, \cdots, R_n$ is a random permutation of $1, 2, \cdots, n$, we have

$$\begin{aligned}
\breve{T}_n &= \sum_{i=1}^{n} c_{i:n}Y_{(i)} - \frac{n-1}{n}\sum_{l=1}^{n}\frac{1}{n-1}\left\{\sum_{l=1}^{R_l-1} c_{i:n-1}Y_{(i)} + \sum_{i=R_l+1}^{n} c_{i-1:n-1}Y_{(i)}\right\} \\
&= \sum_{i=1}^{n} c_{i:n}Y_{(i)} - \frac{1}{n}\sum_{l=1}^{n}\sum_{i=1}^{l-1} c_{i:n-1}Y_{(i)} - \frac{1}{n}\sum_{l=1}^{n}\sum_{i=l+1}^{n} c_{i-1:n-1}Y_{(i)} \\
&= \sum_{i=1}^{n} c_{i:n}Y_{(i)} - \frac{1}{n}\sum_{i=1}^{n}(n-i)c_{i:n-1}Y_{(i)} - \frac{1}{n}\sum_{i=1}^{n}(i-1)c_{i-1:n-1}Y_{(i)},
\end{aligned}$$

which is equation (13.10) and proves the proposition. □

It may be noted that the estimate of the variance of $\breve{T}_n$ given in (13.8) can also be expressed as

$$\hat{Var}(\breve{T}_n) \cong \frac{n-1}{n}\left[\sum\left(T_{n-1}^{(l)}\right)^2 - \frac{\left(\sum T_{n-1}^{(l)}\right)^2}{n}\right], \tag{13.13}$$

where

$$\begin{aligned}
\sum_{l=1}^{n}(T_{n-1}^{(l)})^2 = 2\sum\sum[&(l-1)c_{l-1:n-1}c_{j-1:n-1} + (n-j)c_{l:n-1}c_{j:n-1} \\
&+ (j-l-1)c_{l-1:n-1}c_{j:n-1}]Y_{(l)}Y_{(j)}\,,
\end{aligned} \tag{13.14}$$

with the summation over all (l, j) such that $l < j$, and $(\sum_{l=1}^{n} T_{n-1}^{(l)})^2$ can be easily calculated from (13.14).

13.2.3. *The estimators*

The approach to estimating the k-vector parameter θ^* after reparameterization of θ, for an arbitrary quantile function family $Q(u, \theta)$, is now summarized. We start the construction by selecting k linearly independent functions ϕ_i of θ_i^* and obtain the estimator $\hat{\phi}_i$ of these parametric functions $\phi_i, i = 1, 2, \cdots, k$ as described in Section 13.2.1. The final L-estimator $\breve{\theta}_i^*$s are then obtained in two steps: by first solving the k linear equations $\phi_i(\theta) = \hat{\phi}_i, i = 1, 2, \cdots, k$, and then refining these statistics as in Section 13.2.2. The variances of $\breve{\theta}_i^*$ are calculated using (13.13) and similar expressions can be obtained for the covariances of $(\breve{\theta}_i^*, \breve{\theta}_j^*)$.

13.3. Estimation of Weibull Parameters

The Weibull distribution was introduced in 1939 by W. Weibull for the analysis of breaking strengths of materials but it also appeared earlier in the work of extreme values. The Weibull model is now widely used in modeling lifetime data and has been shown to provide a good fit to several types of data. For a general review of properties and work done in the context of this distribution; [Johnson et al., 1994], (Chapter 21). In this section, we use the approach and results derived in Section 13.2 to derive estimators of the shape and scale parameters of the Weibull distribution.

13.3.1. *Closed form L-estimators*

Now we apply the general approach outlined in Section 13.2, to the estimation of the parameters of the Weibull distribution with quantile function

$$Q_{\mu,\alpha,\sigma}(u) = \mu + \sigma(-\log(1-u))^{\alpha},\ u\varepsilon(0,1), \tag{13.15}$$

and with the *c.d.f* and density function given respectively by

$$F(x) = 1 - \exp(-((x-\mu)/\sigma)^{1/\alpha}), \tag{13.16}$$

and

$$f(x) = \frac{1}{\alpha\sigma}((x-\mu)/\sigma)^{1/\alpha-1}\exp(-((x-\mu)/\sigma)^{1/\alpha}), \tag{13.17}$$

where α, $\sigma > 0$, μ real, $x > \mu$. Specifically, we will use a reparameterized form of this distribution and obtain estimators for $\theta_1^* = \alpha$ and $\theta_2^* = \log(\sigma)$. If $X_1, X_2, \cdots, X_n$ is a random sample from the two parameter Weibull population with $Q_{0,\alpha,\sigma}(u) = Q(u)$ given by (13.15), we start by choosing the pivotal parametric functions:

$$\phi_i = \sum_{i=1}^{2}(1-u)^{i-1}\log Q(u)du = \frac{\log(\sigma)}{i} + \alpha\frac{\Gamma'(1) - \log i}{i}, i = 1, 2. \tag{13.18}$$

The corresponding functions g and h are: $g_i(u) = (1-u)^{i-1}$ and $h_i(x) = \log(x)$, $i = 1, 2$. The parametric functions ϕ_i's, $i = 1, 2$ are estimated consistently by

$$\hat{\phi}_1 = \frac{1}{n}\sum_{i=1}^{n}\log X_{(i)},\ \hat{\phi}_2 = \frac{1}{n}\sum_{i=1}^{n}\left(1 - \frac{i-0.5}{n}\right)\log X_{(i)}, \tag{13.19}$$

and by solving the equations $\hat{\phi}_i = \phi(\theta^*), i = 1, 2$, $\theta^* = (\alpha, \log(\sigma))$, i.e.

$$\frac{1}{n}\sum_{i=1}^{n} \log X_{(i)} = \log(\sigma) - 0.57721\alpha \,, \qquad (13.20)$$

$$\frac{1}{n}\sum_{i=1}^{n}\left(1 - \frac{i-0.5}{n}\right)\log X_{(i)} = 0.5\log(\sigma) - 0.63517\alpha \,, \qquad (13.21)$$

we obtain the pivotal statistics $\hat{\theta}_1^*$ and $\hat{\theta}_2^*$ given as

$$\hat{\theta}_1^* = \hat{\alpha}_n = \frac{1}{0.69313n}\sum_{i=1}^{n}\left(\frac{2(i-0.5)}{n} - 1\right)\log X_{(i)} \qquad (13.22)$$

$$\hat{\theta}_2^* = \widehat{\log(\sigma_n)} = \frac{1}{n}\sum_{i=1}^{n}\left(\frac{1.6655(i-0.5)}{n} + 0.16725\right)\log X_{(i)} \,. \qquad (13.23)$$

These statistics can then each be jackknifed to obtain the estimators.
Note that for the choice of functions g and h for the model (13.15), statistics $\hat{\phi}_i$ as well as the pivots $\hat{\theta}_i^*$ are L-estimators in $Y_i = \log X_{(i)}$. We now use the results from Section 13.2.2 to derive the two-step quantile estimators $\breve{\alpha}_n$ and $\widehat{\log(\sigma_n)}$ and show that these are of a simple form, similar to (13.22) and (13.23) above.
Proposition 3.1. The closed form two-step quantile estimators $\hat{\theta}_1^* = \hat{\alpha}_n$ and $\hat{\theta}_2^* = \widehat{\log(\sigma_n)}$ of the shape and log-scale parameters in Weibull model (13.15) are given by

$$\breve{\alpha}_n = \frac{1}{0.69313n}\sum_{i=1}^{n}\left(\frac{2(i-1)}{n-1} - 1\right)\log X_{(i)} \,, \qquad (13.24)$$

$$\breve{\log(\sigma_n)} + \frac{1}{n}\sum_{i=1}^{n}\left(\frac{1.6655(i-1)}{n-1} + 0.16725\right)\log X_{(i)} \,. \qquad (13.25)$$

Proof. [Proof of the Proposition 3.1] The jackknife estimator $\breve{\alpha}_n$ of the estimator $\hat{\alpha}_n$ is of the form

$$\breve{\alpha}_n = n\hat{\alpha}_n - \frac{n-1}{n}\sum_{i=1}^{n}\hat{\alpha}_{n-1}^{(i)} \,, \qquad (13.26)$$

where $\hat{\alpha}_{n-1}^{(i)}$ is defined ealier. From the proof of Proposition 2.1, we can see that $\sum_{i=1}^{n}\hat{\alpha}_{n-1}^{(i)}$ can be expressed in the form:

$$\sum_{i=1}^{n}\hat{\alpha}_{n-1}^{(i)} = \sum_{i=1}^{n}(i-1)\left(\frac{2(i-1.5)}{n-1}\right)\log X_{(i)} + \sum_{i=1}^{n}(n-i)\left(\frac{2(i-0.5)}{n-1} - 1\right)\log X_{(i)} \,. \qquad (13.27)$$

Using the above, and after several simplifications the jackknife estimators $\breve{\alpha}_n$ reduces to the desired form. The jackknife estimator $\breve{\log(\sigma_n)}$ reduces similarly. □

The asymptotic normality of the estimators given in (13.24) and (13.25) can be easily established by invoking Theorem 1 (ii) in Chapter 19 of Shorack and Wellner[35] and are presented in the following propositions.

Proposition 3.2. If $X_1, X_2, \cdots, X_n$ is a sample from the Weibull model $Q_{0,\alpha,\sigma}$ in (13.15) then estimator $\breve{\alpha}_n$ is asymptotically, as $n \longrightarrow \infty$, normally distributed,

$$\sqrt{n}(\breve{\alpha}_n - \alpha) \xrightarrow{L} N(0, \nu_1^2). \tag{13.28}$$

Proposition 3.3. If $X_1, X_2, \cdots, X_n$ is a sample from the Weibull model $Q_{0,\alpha,\sigma}$ in (13.15) then the asymptotic distribution, as $n \longrightarrow \infty$, of $\breve{\log\sigma_n} = \log(\breve{\sigma}_n)$ is given by the following:

$$\sqrt{n}\left(\log\breve{(}\sigma_n) - \log(\sigma)\right) \xrightarrow{L} N(0, \nu_2^2). \tag{13.29}$$

A sketch of the proof for Propositiona (3.2) and (3.3) is provided below:

Proof. We start by noting that the estimators $\breve{\alpha}_n$ and $\log\breve{(}\sigma_n)$ are of the form $1/n\sum_{i=1}^{n} c_{in} h(X_{(i:n)})$ where $c_{1n}, c_{2n}, \cdots, c_{nn}$ denote the constants and h denotes a known function of the form $h = h_1 - h_2$, with each $h_i \uparrow$ and left continuous. Further, if the assumptions 1 and 2 of Theorem 1 (ii) in [Shorack and Wellner, 1986] (Chapter 19) are satisfied then the asymptotic normality of the estimators and for the Weibull model (13.15) can be easily established.

The constants c_{in} for both estimators are of the form $c_{in} = A\frac{i-1}{n-1} + B$ for some constants A and B. For

$$J_n(t) = c_{in}\,, \quad \frac{i-1}{n} < t \leq \frac{i}{n} \quad 1 \leq i \leq n \tag{13.30}$$

and $J_n(0) = c_{1n}$, we have

$$\lim_{n\longrightarrow\infty} J_n(t) = J(t) = At + B \quad \text{uniformly in} \quad t\varepsilon[0,1] \tag{13.31}$$

and

$$T_n = \int_0^1 \log Q_n(t) J_n(t)\, dt. \tag{13.32}$$

Then from Theorem 1(ii) of Shorack and Wellner[35](Chapter 19) implies that

$$\sqrt{n}(T_n - \mu) \xrightarrow{L} N(0, \nu^2) \tag{13.33}$$

where

$$\mu = \int_0^1 \log Q_{\alpha,\sigma}(t) J(t)\, dt \tag{13.34}$$

and

$$\nu^2 = \int_0^1 \int_0^1 (S \wedge t - st) J(s) J(t) dg(s) dg(t) , \tag{13.35}$$

where $g = \log Q_{\alpha,\sigma}$, provided that the assumptions 1 and $2'$ required for Theorem 1(ii) are satisfied.

Proof of assumption 1: Conditions (16) and (18) on page 662 of [Shorack and Wellner, 1986](Chapter 19) (S&W) are clearly satisfied. For $0 < u < 1$ and $\varepsilon > 0$ we have

$$|\log Q_{\alpha,\sigma}(u)| = \left|\log\left(\sigma\left[\log\frac{1}{1-u}\right]^{\alpha}\right)\right| \leq |\log\sigma| + \alpha\left|\log\log\frac{1}{1-u}\right| \leq M_{\varepsilon} u^{-\varepsilon}(1-u)^{\varepsilon} \tag{13.36}$$

with some $M_{\varepsilon} < \infty$ so that the condition (20) on page 663 or S&W is also satisfied.

Proof of assumption 2$'$: The smoothness condition (29) and the boundedness condition (30)on page 664 of S&W are clearly satisfied for J. Moreover,

$$c_{in} = A\frac{i-1}{n-1} + B = J(\frac{i-1}{n-1}) \text{ with } \frac{i-1}{n} < \frac{i-1}{n-1} < \frac{i}{n} , \tag{13.37}$$

so that (31) on page 664 of S&W also holds true. Consequently, the Theorem 1 (ii) of S&W is applicable. Further, it can be easily seen that

$$\begin{aligned} \mu &= \int_0^1 \log(Q_{\alpha,\sigma}(t)) J(t) dt = A\int_0^1 \log(Q_{\alpha,\sigma}(t))\, t dt + B\int_0^1 \log(Q_{\alpha,\sigma}(t)) dt \\ &= \begin{cases} \alpha & \text{for } T_n = \breve{\alpha}_n \\ \log\sigma & \text{for } T_n = \log(\breve{\sigma}_n. \end{cases} \end{aligned} \tag{13.38}$$

Then, using (13.38) and (13.35) the asymptotic normality of $\breve{\alpha}_n$ and $\log(\breve{\sigma}_n)$ is established and we have propositions (3.2) and (3.3). □

Remark: Based on the propositions 3.2 and 3.3, and the fact that joint distributions of the order statistics are multivariate normal; [Serfling, 1980], (page 80) it is clear that the joint distribution of $T_1 = \breve{\alpha}_n$ and $T_2 = \log(\breve{\sigma}_n)$, given (13.28) and (13.29), would be asymptotically bivariate normal with the mean vector $(\alpha, \log\sigma)'$ and the covariance matrix Σ where the diagonal elements (i.e. variances ν_1^2 and ν_2^2) can be easily obtained from (13.35), and the covariance term $\sigma_{\breve{\alpha}_n, \log(\breve{\sigma}_n)}$, is given by;

$$\begin{aligned} \sigma_{\breve{\alpha}_n, \log(\breve{\sigma}_n)} &= cov(T_1, T_2) = cov\left(\sum_{i=1}^{n} c_{in} Y_{(i)}, \sum_{i=1}^{n} d_{in} Y_{(i)}\right) \\ &= \sum_{i=1}^{n} c_{in} d_{in} var(Y_{(i)}) + 2\sum_{i>j=1}^{n} c_{in} d_{jn} cov\left(Y_{(i)}, Y_{(j)}\right), \end{aligned} \tag{13.39}$$

where $Y_{(i)} = \log X_{(i)}$ and $X_{(i)}$ is the i-th order statistics from Weibull $Weib$ distribution. The asymptotic covariance given in (13.39) can be obtained by substituting the asymptotic variances and covariances for the order statistics found in many standard monographs, e.g. [Serfling, 1980], (page 80) or [Sen and Singer, 1993](Section 4.3). However, utilizing the fact that if $X \sim Weib(1/\alpha, \sigma)$ then $\log X \sim EVD(\log\sigma, \alpha)$, also known as Gompertz distribution, where EVD stands for Extreme-value distribution (a member of the location-scale family of distributions) with location and scale parameters $\log\sigma$ and α, respectively, one can obtain the exact covariance term by obtaining the variances and the covariances of order statistics from EVD. An explicit expression based on Spences function is provided by [Lieblein, 1953] and complete tables for all sample sizes up to 30 have been prepared by [Balakrishnan and Chan, 1992]. For a general discussion one can [Johnson, Kotz and Balakrishnan, 1995], (Chapter 22).

13.4. Comparisons

Because of its practical importance substantial literature on estimation of Weibull parameters exists. A simulation study was undertaken to evaluate and compare the performance of the jackknife estimators discussed in Section 13.3.1 and other competitors available in the literature.

13.4.1. *Competitors*

Some competitors of the jackknife estimators are now outlined. These include the MLEs, a group of estimators based on linear combination of order statistics and estimators based on BLUEs of extreme value distribution. [Zanakis, 1979] conducted a simulation study which identified some preferable estimators of the parameters of the Weibull distribution. These together with the MLE are now briefly reviewed.

(i). **Maximum likelihood.** Major studies of the MLEs of the Weibull parameters include [Cohen, 1965], [Harter and Moore, 1965], [Rockette et al., 1974], [Lemon, 1975], [Prescott and Walden, 1980], [Engelhardt, 1975] and [Mann and Fertig, 1975].
If $X_1, X_2, \cdots, X_n$ is a random sample for the Weibull distribution with probability density (13.17), and we assume that $\mu = 0$, then the maximum likelihood estimators of the parameters $1/\alpha$ and σ satisfy

$$\hat{\alpha}_n^{-1} = \frac{\sum_{i=1}^n X^{1/\hat{\sigma}_n} \log(X_i)}{\sum_{i=1}^n X^{1/\hat{\sigma}_n}} - \frac{1}{n}\sum_{i=1}^n \log(X_i)\,, \tag{13.40}$$

$$\hat{\sigma}_n = \left(\frac{1}{n}\sum_{i=1}^n X_i^{1/\hat{\alpha}_n}\right)^{\hat{\alpha}_n} . \tag{13.41}$$

Solutions of these equations use iterative procedures. Since the likelihood equations can often have multiple local maxima the choice of initial values is critical for identifying the global maxima. To simplify the calculations several approaches to approximations of the MLEs have been suggested e.g. [Engelhardt, 1975]. An excellent account of the large sample distribution of the MLEs for the non-regular case is given by [Smith, 1985].

(ii). **Linear combinations of quantiles.** Many of these estimators due to [Dubey, 1967a] are based on numerical and asymptotic considerations and are based on the linear combination of three percentiles. Specifically, he uses the quantile function $t_k = Q(p_k), k = 1,2,3$, where $0 \le p_1 \le p_2 \le p_3 \le 1$, to estimate the parameters α, σ, and μ, by solving the system of equations given by

$$t_k = \hat{\mu} + \hat{\sigma}(-\log(1-p_k))^{\hat{\alpha}}\ , k = 1,2,3. \tag{13.42}$$

$$\hat{\sigma}_{DB} = \frac{t_1}{(-\log(1-p_1))^{\alpha}}\ ,\ \hat{\alpha}_{DB} = \frac{\log(t_2)-\log(t_1)}{\log(-\log(1-p_2)) - \log(-\log(1-p_1))} . \tag{13.43}$$

[Dubey, 1967a] suggests using the percentiles $p_1 = 0.16731$ and $p_2 = 0.97366$ in (13.43) and claims that the variance of the estimators is then minimized among all estimators of this type. The estimators $\hat{\alpha}_{DB}$ of α is then 66% efficient compared to MLE and it is asymptotically normal. When the threshold parameter is not zero then [Dubey, 1967b] proposes to estimate it using three order statistics $Y_{(1)}, Y_{(2)}$ and $Y_{(3)}$ as

$$\hat{\mu}_{DB} = \frac{y_{(1)}y_{(n)} - y_{(2)}^2}{y_{(1)} + y_{(n)} - 2y_{(2)}}\ , \tag{13.44}$$

and argues that the estimator is almost never impermissible, i.e., especially for large sample sizes.

(iii). **Best linear unbiased estimators.** The BLUEs of the Weibull parameters do not exist in closed form. But the logarithm of the Weibull random variable has an extreme value distribution, and a substantial part of the literature on estimation of Weibull parameters is based upon the BLUEs of the parameters of the extreme value distribution. If has a Weibull distribution with scale parameter σ and shape parameter $1/\alpha$ then $X = \log Y$ has an extreme value distribution with location and scale parameters $\mu = \log\sigma$ and $\sigma = \alpha$. The estimators of these and their modifications are studied, among others, by [Hassanein, 1972], [Bain, 1972], [Mann and Fertig, 1977], and [Engelhardt and Bain, 1977]. Hassaneins estimators of scale and shape parameters based on k order statistics are of the form

$$\hat{\sigma}_{HS} = \exp\left[\sum_{i=1}^{k} V_i \log(y_{(n_i)} - \hat{\mu}_{DB})\right], \quad \hat{\alpha}_{HS} = \sum_{i=1}^{k} W_i \log(y_{(n_i)} - \hat{\mu}_{DB}), \qquad (13.45)$$

where $\hat{\mu}_{DB}$ is as given in (13.44), $n_i = [n\gamma_i]$ and the optimal weights W_i and V_i and spacings γ_i for selected k values were determined by Hassanein (1972) so that the estimators of the extreme value parameters are the best unbiased estimators. Mann and Fertig (1977) and Engelhardt and Bain (1977) propose small-sample unbiased modifications to the earlier estimators. The latter estimators of scale and shape parameters are respectively given by

$$\hat{\sigma}_{EB} = \exp\left[0.5772\hat{\alpha}_{EB} + \frac{1}{n}\sum_{i=1}^{n} \log(y_i - \hat{\mu}_{DB})\right], \qquad (13.46)$$

$$\hat{\alpha}_{EB} = \frac{1}{nk_n}\left[-\sum_{i=1}^{s} \log(y_i - \hat{\mu}_{DB}) + \frac{s}{n-s}\sum_{i=s+1}^{n} n\log(y_i - \hat{\mu}_{DB})\right], \qquad (13.47)$$

where $s = [0.84n]$, the largest integer smaller than or equal to $= 0.84n$, and the unbiasing factor k_n is given by the authors. Simulation comparisons of the above estimators of the Weibull parameters are given, among others, by Zanakis(1979). The results of his simulation combined with new simulation results are used later for comparisons of the above estimators with the new estimator.

13.4.2. *A Monte Carlo experiment and results*

The simulation study mentioned above was conducted in two stages. In the first stage the approach for obtaining jackknife estimates, discussed in Section 13.3.1, was compared to the MLEs. These comparisons were performed by generating 5000 random samples for sample sizes $n = 30, 100$ from $Weib(1/\alpha, \sigma)$, model (3.1), and computing the competing estimates for each sample. For the simulations, the location parameter was set to $\mu = 0$ and the scale parameter was set to $\sigma = 1$, while the shape parameter ranged between $1/\alpha = 0.3$ to 5.0, as appearing in Tables 13.1 and 13.2. The MLEs were computed using FORTRAN routine DNEQNF in IMSL library. The bias and MSE of the estimators for each value of the shape parameter were then calculated from their simulated values. A selection

of the results appears in Tables 13.1-13.2 and Figures 13.1-13.4.
In the second stage of the simulation study we combined simulation results from studies in the literature and our results to compare the small sample behavior of the competing estimators. The new estimators proposed in (13.22) and (13.23) were compared with all other estimators described in previous section, i.e. Dubeys (13.43), Hassaneins (13.45) and Engelhardt and Bains estimators (13.46) and (13.47). An independent simulation experiment, similar to the one conducted by Zanakis[37] and discussed above, was conducted which consisted of generating 5000 samples of various sample sizes $n = 15, 30, 50$, from $Weib(1/\alpha, \sigma)$ distribution and applying all the competing approaches to each sample. The bias and MSE of all the estimators for each value of the shape parameter were then calculated from their simulated values. A selection of results for $n = 30$ appears in Tables 13.3 and 13.4.
In order to assess the merits of all the estimators discussed above as the sample size increases it is necessary to obtain the asymptotic variance expressions and efficiencies and that work is in progress.

Results. Some results gleaned from the simulation experiment above are now summarized:

(i). Figures 13.1 and 13.2 display a comparison of the bias and the MSE for the MLE $\hat{\alpha}_n$ and the proposed new estimator $\breve{\alpha}_n$ of the shape parameter, when the scale is fixed at $\sigma = 1$. Figures 13.3 and 13.4 show the change in the bias and MSE of the MLE $\hat{\sigma}_n$ on the log-scale, and the new estimator $\breve{\sigma}_n$, again as the value of the shape parameter changes with $\sigma = 1$ fixed. From the figures and Tables 13.1 and 13.2, it is seen that the new estimators of the shape and the scale parameters are superior to the MLEs in terms of bias, while their MSE are very close. It is also seen that the performance of the new estimator of the shape parameter is better for smaller values of the shape parameter. Based on the simulation, the MSEs of both the MLE and the new estimate of $\log(\sigma)$ were seen to be almost identical when $\sigma = 1$, regardless of the Weibull shape parameter.

(ii). It is very clear from Tables 13.3 and 13.4, that the new estimator of the shape parameter is superior to all others competitors in terms of bias and MSE. We note that we find the same qualitative result when estimating $1/\alpha$ directly. So the performance does not seem to depend on scale. Similar observations are seen to hold for the estimator of the scale parameter σ when the comparisons are made on the log-scale.

(iii). In order to assess the quality of the jackknife estimators of the variance and covariance of the estimators, these were computed for each of the simulated samples using the pseudovalues as described in Proposition 2.2. The means of the corresponding variances and covariance estimates of $\breve{\alpha}_n$ and $\log\breve{\sigma}_n$ were compared with Monte Carlo estimates of the corresponding variances and covariances. It was seen that the distributions of the pseudovalue estimates of the variance of $\breve{\alpha}_n$ and $\log\breve{\sigma}_n$ are essentially centered around the corresponding simulated value of the variance estimates. Also, we note that as the value of the shape parameter increases, the bias of the jackknife variance estimator increases.

The foregoing observations suggest that the new method yields very accurate and precise estimators of the shape parameter of the Weibull distribution. In view of the asymptotic theory of the linear combination of order statistics, the estimators are consistent, and their asymptotic distributions are readily available. Moreover, they have a simple form. The new estimators are superior to the competing linear estimators based on order statistics, which are generally used in conjunction with extensive tables. Moreover, the estimators proposed above are in closed form and require no tables.

13.5. Conclusions and Miscellaneous Remarks

In this paper we have introduced jackknife based two-step estimators of the two-parameter Weibull population and by empirical comparisons shown that they are superior to most of the estimators in the literature. Furthermore, as compared to the MLEs, we see that the proposed estimators are superior in terms of reduced bias with comparable MSEs. Hence, they are as good or better than others in terms of bias but comparable to the MLEs.

We note that the ideas and approaches used by us have been in past used by other investigators. For example, Reeds (1978) has considered jackknifing MLEs but in case of nonlinear likelihood equations, as in the case of Weibull estimations, it fails to yield closed form estimators. Similarly, Hosking(1990) uses reasoning similar to the first step in our approach but uses L-moments instead of moments. A limitation of this approach is that the exact distribution of the parameter estimators, in general, is difficult to derive. Nelson(1982) considered construction of estimators which, as in our approach, involve guesswork of pivotal quantities but his estimators are not as simple and require the table of weights.

It may be noted that the suggested method would not be appropriate for estimators based on just a few quantiles, because jackknifing them would be similar to jackknifing medians, where the number of distinct pseudovalues is small. The approach seems more suitable therefore for estimators using all the observations. Further, it may also be noted that we have considered only the two parameter

Weibull case and not the three parameter case involving the threshold parameter. However, new issues raised in this context may be approached using strategies as in Smith(1985). Finally, we note that for those who prefer the maximum likelihood method the new estimators can be used as the starting values for obtaining MLEs as they may provide fast convergence even for censored samples, as seen in an example used by Mudholkar et al.(1994).

13.6. Acknowledgments

This research work of Georgia D. Kollia was in part supported by National Cancer Institute grant CA-09168. The research work of Deo Kumar Srivastava was in part supported by the Grant CA21765 and the American Lebanese Syrian Associated Charities. The authors are thankful to anonymous referees for providing many useful suggestions that significantly improved the manuscript.

References

1. Bain, L. J. (1972). Inferences based on censored sampling from the Weibull or extreme-value distribution. *Technometrics*, **14**, 693-702.
2. Balakrishnan, N., and Chan, P. S. (1992). Extended tables of means, variances and covariances of order statistics from the extreme value distribution for sample sizes up to 30. *Report, Department of Mathematics and Statistics, McMaster University, Hamilton, Canada.*
3. Cheng, K. F. (1982). Jackknifing L-estimates. *The Canadian Journal of Statistics*, **10**, 49-58.
4. Chernoff, H., Gastwirth, J. L. and Jones, M. V. (1967). Asymptotic distribution of linear combinations of function of order statistics, with applications to estimation. *The Annals of Mathematical Statistics*, **38**, 52-72.
5. Cohen, A. C. (1965). Maximum likelihood estimation in the Weibull distribution based on complete and on censored samples (Correction: V8, p 570). *Technometrics*, **7**, 579-588.
6. Cox, D. R. and Hinkley, D. V. (1974). *Theoretical Statistics*. Chapman & Hall, New York.
7. Dubey, S. D. (1967a). Some percentile estimators for Weibull parameters. *Technometrics*, **9**, 119-129.
8. Dubey, S. D. (1967b). On some permissible estimators of location parameter for Weibull and certain other distributions. *Technometrics*, **9**, 293-307.
9. Efron, B. (1986). *The Jackknife, the Bootstrap and Other Sampling Plans*. CBMS-NSF Regional Conference Series in Applied Mathematics, Society for Industrial and Applied Mathematics, Philadelphia, PA.
10. Efron, B. and Stein, C. (1981). The jackknife estimate of variance. *The Annals of Statistics*, **9**, 586-596.

11. Engelhardt, M. (1975). On simple estimation of the parameters of the Weibull or extreme-value distribution. *Technometrics*, **17**, 369-374.
12. Engelhardt, M. and Bain, L. J. (1977). Simplified statistical procedure for the Weibull and extreme-value distribution. *Technometrics*, **19**, 323-331.
13. Gardiner, J. C. and Sen, P. K. (1979). Asymptotic normality of a variance estimator of a linear combination of a function of order statistics. *Zeitshrift fr Wahrscheinlichkeitstheorie und Verwandte Gebiete*, **50**, 205-221.
14. Greenwood, J. A., Landwehr, J. M., Matalas, N. C. and Wallis, J. R. (1979). Probability weighted moments: Definition and relation to parameters of several distributions expressible in inverse form. *Water Resources Research*, **15**, 1049-1054.
15. Harter, H. L. and Moore, A. H. (1965). Maximum likelihood estimation of the parameters of gamma and Weibull populations from complete and from censored samples (Correction: V9 p 195; V15 p 341). *Technometrics*, **7**, 639-643.
16. Hassanein, K. M. (1972). Simultaneous estimation of the parameters of the extreme value distribution by sample quantiles. *Technometrics*, **14**, 63-70.
17. Hosking, J. R. M. and Wallis, J. R. (1987). Parameter and quantile estimation for the generalized Pareto distribution. *Technometrics*, **29**, 339-349.
18. Hosking, J. R. M. (1990). L-moments: Analysis and estimation of distributions using linear combination of order statistics. *Journal of the Royal Statistical Society, Series B*, **52**, 105-124.
19. Huang, J. S. (1991). Efficient computation of the performance of bootstrap and jackknife estimators of the variance of L-statistics. *Journal of Statistical Computation and Simulation*, **38**, 45-56.
20. Johnson, N. L., Kotz, S. and Balakrishnan, N. (1994). *C*ontinuous Univariate Distributions, Vol. 1, John Wiley & Sons, New York.
21. Johnson, N. L., Kotz, S. and Balakrishnan, N. (1995). *C*ontinuous Univariate Distributions, Vol. 2, John Wiley & Sons, New York.
22. Lemon, G. H. (1975). Maximum likelihood estimation for the three parameter Weibull distribution based on censored samples. *T*echnometrics, **17**, 247-254.
23. Lieblein, J. (1953). On the exact evaluation of the variances and covariances of order statistics in samples from the extreme-value distribution. *Annals of Mathematical Statistics*, **24**, 282-287.
24. Mann, N. R. and Fertig, K. W. (1975). Simplified efficient point and interval estimation for Weibull parameters. *T*echnometrics, 17, 361-368.
25. Mann, N. R. and Fertig, K. W. (1977). Efficient unbiased quantile estimators for moderate-size complete samples from extreme-value and Weibull distributions; confidence bounds and tolerance and prediction intervals. *T*echnometrics, **1**9, 87-94.
26. Mudholkar, G. S. and Kollia, G. D. (1994). Generalized Weibull family: A structural analysis. *C*ommunications in Statistics Theory and Methods, **2**3, 1149-1171.
27. Nelson, W. (1982). *A*pplied Life Data Analysis. John Wiley & Sons, New York.
28. Parr, W. C. and Schucany, W. R. (1982). Jackknifing L-statistics with smooth weight functions. *J*ournal of the American Statistical Association, **7**7, 629-638.
29. Prescott, P. and Walden, A. T. (1980). Maximum likelihood estimation of the parameters of the generalized extreme-value distribution. *B*iometrika, **6**7, 723-724.
30. Reeds, J. (1978). Jackknifing maximum likelihood estimates. *T*he Annals of Statistics, **6**, 727-739.

31. Rockette, H., Antle, C. and Klimko, L. A. (1974). Maximum likelihood equation with the Weibull model. Journal of the American Statistical Association, **69**, 246-249.
32. Serfling, R. J. (1980). Approximation Theorems of Mathematical Statistics. John Wiley & Sons, New York.
33. Sen, P. K. and Singer, J. M. (1993). Large Sample Methods in Statistics: An Introduction with Applications. Chapman & Hall, New York.
34. Shao, J. and Tu D. (1996). The Jackknife and Bootstrap. Springer-Verlag, New York.
35. Shorack, G. R. and Wellner, J. A. (1986). Empirical Processes with Applications to Statistics. John Wiley & Sons, New York.
36. Smith, R. L. (1985). Maximum likelihood estimation in a class of nonregular cases. Biometrika, **72**, 67-90.
37. Zanakis, S. H. (1979). A simulation study of some simple estimators for the three parameter Weibull distributions. Journal of Statistical Computation and Simulation, **9**, 101-116.

Table 13.1. An Empirical Comparison† of the Maximum Likelihood and the Jackknife Estimators of the Shape Parameter. Sample Size n=100.

Shape	Bias		MSE	
Parameter $1/\alpha$	MLE	JK‡	MLE	JK‡
0.30	0.0046	0.0026	0.0006	0.0008
0.50	0.0066	0.0035	0.0016	0.0022
1.00	0.0133	0.0077	0.0065	0.0084
1.50	0.0191	0.0112	0.0148	0.0188
1.75	0.0220	0.0114	0.0201	0.0254
2.00	0.0292	0.0175	0.0262	0.0333
2.50	0.0330	0.0150	0.0402	0.0509
3.00	0.0398	0.0214	0.0601	0.0761
3.50	0.0443	0.0273	0.0818	0.1037
4.00	0.0551	0.0336	0.1073	0.1378
5.00	0.0764	0.0498	0.1661	0.2114

†Based on a Monte Carlo Experiment with 5000 replications

‡ Jackknife Estimator based on pseudovalues; see Section 13.2.2

Table 13.2. An Empirical Comparison of the Maximum Likelihood and the Jackknife Estimators of the Scale Parameter log(σ), when it equals 0. Sample Size n=100.

Bias		MSE	
MLE	JK‡	MLE	JK‡
-0.0089	0.0039	0.1239	0.1242
-0.0142	-0.0067	0.0439	0.0440
-0.0046	-0.0010	0.0112	0.0111
-0.0024	0.00005	0.0049	0.0049
-0.0013	0.0008	0.0036	0.0036
-0.0016	0.0002	0.0028	0.0028
-0.0016	-0.00003	0.0018	0.0018
-0.0019	-0.0007	0.0012	0.0012
-0.0014	-0.0004	0.0009	0.0009
-0.0008	0.00008	0.0007	0.0007
-0.0012	-0.00042	0.0005	0.0005

†Based on a Monte Carlo Experiment with 5000 replications

‡ Jackknife Estimator based on pseudovalues; see Section 13.2.2

Table 13.3. An Empirical Comparison† of the Exact and Estimated Values of the Shape Parameter using the Jackknife and Three Competing Estimators. Sample Size n=30.

Shape Parameter $1/\alpha$	JK‡	DB◇	HS♠	EB ♣
0.30	0.307	0.310	0.324	0.322
0.50	0.512	0.529	0.507	0.497
1.00	1.026	0.980	0.906	0.889
1.50	1.545	1.416	1.278	1.259
1.75	1.795	1.525	1.400	1.384
2.00	2.058	1.698	1.538	1.506
2.50	2.564	1.956	1.727	1.690
3.00	3.090	2.205	1.982	1.952
3.50	3.590	2.403	2.082	2.035
4.00	4.121	2.515	2.233	2.192
5.00	5.152	2.708	2.345	2.315

†Based on a Monte Carlo Experiment with 5000 replications

‡ Jackknife Estimator based on pseudovalues; see Section 13.2.2

◇ Dubey (13.43), ♠ Hassanein (13.45), ♣ Engelhardt and Bain (13.46) & (13.47)

Table 13.4. An Empirical Comparison† of the MSEs of the Jackknife and Three Competing Estimators of the Shape Parameter. Sample Size n=30.

Shape Parameter $1/\alpha$	JK‡	DB◇	HS♠	EB ♣
0.30	0.003	0.003	0.003	0.003
0.50	0.008	0.009	0.005	0.005
1.00	0.032	0.033	0.030	0.035
1.50	0.071	0.075	0.101	0.107
1.75	0.091	0.149	0.184	0.195
2.00	0.125	0.189	0.287	0.317
2.50	0.190	0.479	0.725	0.782
3.00	0.287	0.946	1.302	1.365
3.50	0.381	1.574	2.352	2.495
4.00	0.488	2.739	3.599	3.745
5.00	0.765	5.989	7.678	7.858

†Based on a Monte Carlo Experiment with 5000 replications

‡ Jackknife Estimator based on pseudovalues; see Section 13.2.2

◇ Dubey (13.43), ♠ Hassanein (13.45), ♣ Engelhardt and Bain (13.46) & (13.47)

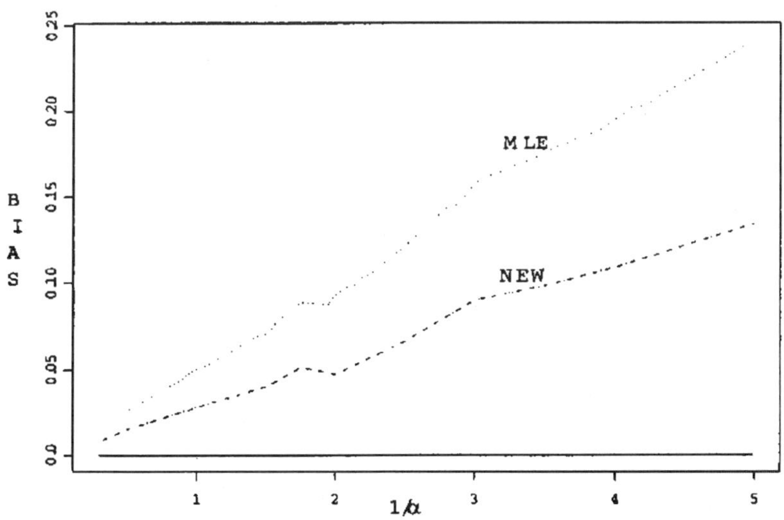

Fig. 13.1. Bias of MLE and New estimator of Shape from 5000W Weibull Samples of n=30 and $\sigma = 1$

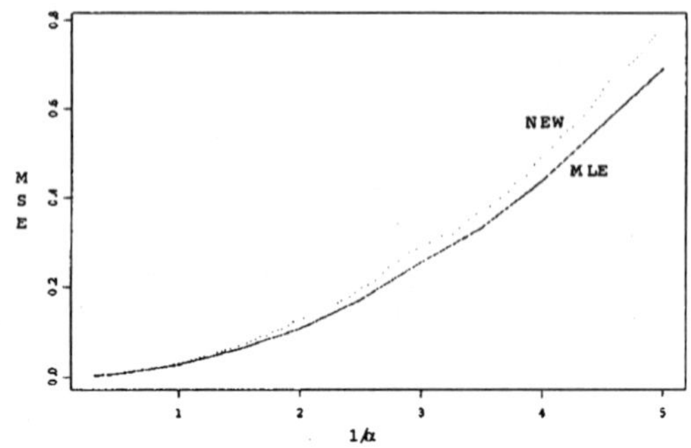

Fig. 13.2. MSE of MLE and New Estimators of Shape from 5000W Weibull Samples of n=30 and $\sigma = 1$

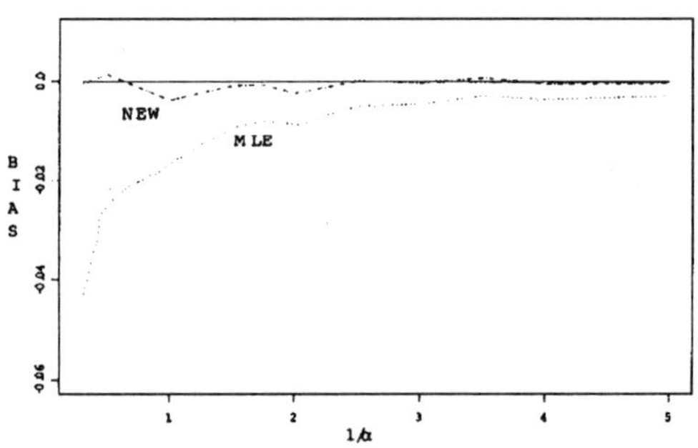

Fig. 13.3. Bias of MLE and New Estimators of the log scale from 5000W Weibull Samples n=30 and $\sigma = 1$

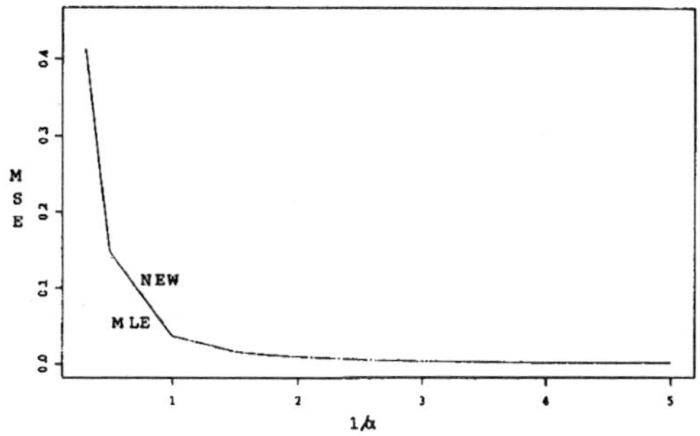

Fig. 13.4. MSE of MLE and New Estimators of the log scale from 5000W Weibull Samples of n=30 and $\sigma = 1$

Chapter 14

On Optimal Estimating Functions in the Presence of Nuisance Parameters

Parimal Mukhopadhyay

Indian Statistical Institute; Kolkata
E-mail: parimal1us@yahoo.com

When there is only one interesting parameter θ_1 and one nuisance parameter θ_2 Godambe and Thompson (1974) showed that the optimal estimating function for θ_1 essentially is a linear function of the θ_1-score, the square of the θ_2-score and the derivative of θ_2-score with respect to θ_2. Mukhopadhyay (2000b) generalized this result to m nuisance parameters. Mukhopadhyay (2000, 2002 a, b) obtained lower bounds to the variance of regular estimating functions in the presence of nuisance parameters. Taking cue from these results we propose a method of finding optimal estimating function for θ_1 by taking the multiple regression equation on θ_1 score and Bhattacharyyas (1946) scores with respect to θ_2. The result is extended to the case of m nuisance parameters.

Contents

14.1. Introduction

Consider a random variable X with distribution function $P(x;\theta)$, dominated by σ-finite measure μ defined on the sample space X (a.e. P) and the probability density function (pdf) $p(x;\theta)$, which but for an unknown parameter vector θ is completely specified for all $x \in X$. Suppose $\theta = (\theta_1;\theta_2)$ where $\theta_1;\theta_2$; are real quantities, $\theta_1 \in \Omega_1;\theta_2 \in \Omega_2;\theta \in \Omega = \Omega_1 X \Omega_2$. We are interested in estimating θ_1 only, θ_2 being a nuisance parameter. Consider an estimating function $g1(x;\theta_1)$ which depends on x and θ_1 only and satisfies certain regularity conditions (given below) and let $G = \{g1(x;\theta_1)\}$, the class of such regular estimating functions. An estimating function $g*1 \in G_1$ is said to be optimal in G_1 if

$$E_\theta[\frac{g*_1}{E_\theta(\partial g*_1/\partial\theta_1)}]^2 \le E_\theta[\frac{g_1}{E_{theta}(\partial g_1/\partial\theta_1)}]^2 \forall g_1 \in G_1 \text{and} \forall\theta \in \Omega$$

The Function

$$\frac{g_1}{E_\theta(\frac{\partial g_1}{\partial\theta_1})}$$

is the standardized estimating function g_1s corresponding to the function g_1.
We assume that
(a) $\Omega_1;\Omega_2$ are open intervals of the real line.
(b) for almost all $x(\mu)$; $(\partial logp/\partial\theta_1), (\frac{1}{p}\frac{\partial^1}{\partial\theta_2^1}), i = 1, \ldots.., K$ exist for all $\theta \in \Omega$ where K is a suitable integer.
(c) $\int pd\mu, \int(\partial logp/\partial\theta_1)pd\mu, \int(\frac{1}{p}\frac{\partial^1}{\partial\theta_2^1}\mu), (i = 1, \ldots..K)$are differentiable under the integral sign $(\theta \in \Omega)$.
(d) Let

$$A = (\partial logp/\partial\theta_1), B_i = \frac{1}{p}\frac{\partial_1 p}{\partial\theta_2^1}, i = 1, ..., k,$$

$$c_{11} = V(A), c_{1i+1} = C(A_iB_i), c_{i+1,i'+1} = C(B_iB_{i'})i, i' = 1, ..., k$$

where C; V denote, respectively, covariance and variance. We assume that the determinant

$$|c_{ij}; i, j = 1, ..., k+1| = |c_k| > 0 \forall\theta, k = 1, ..., K$$

The class of functions G_1 on $X\Omega_1$ is assumed to satisfy the following conditions. For every $g_1 \in G_1$,
(i) $E_\theta(g1) = 0 \forall\theta \in \Omega$.
(ii) for almost all $x(\mu)(\partial g1/\partial\theta_1)$ exist for all $\theta \in \Omega$.
(iii) $\int g1pd\mu$ is once differentiable with respect to θ_1 and differentiable K times with respect to θ_2.
(iv) $[E_\theta(\partial g1/\partial\theta_1)]^2 > 0, \theta \in \Omega$.

The problem of finding optimal estimating functions were earlier considered by Godambe (1960), Kale (1962), Ferreira (1982), Chandrasekar and Kale (1984), Bhapkar and Srinivasan (1994),[Editor's note: also see Kale (2001-2002)] among others. In this paper we propose a method of finding the optimal standardized estimating function g*1s for θ_1 in the presence of nuisance parameter θ_2. When there is only one interesting parameter θ_1 and m nuisance parameters $\theta_{2_i} (i = 1, \ldots., m)$ we modify the method to find the optimal standardized estimating function for 1 in the presence of the nuisance parameters $\theta_{2_i}(i = 1, ..., m)$. The optimal estimating function is shown to attain the lower bound derived in Mukhopadhyay (2000, 2002a, b).

14.2. An Optimal Estimating Function

When there is only one interesting parameter and no nuisance parameter Godambe (1960) showed that the score function A is the optimal estimating function for θ: When there is only one interesting parameter θ_1 and one nuisance parameter θ_2, Godambe and Thompson (1974) obtained the following theorem giving the essential form of the optimal estimating function.

Theorem 2.1: Under some regularity conditions, an optimal estimating function $g*_1 \in G_1$ is given by

$$g*_1 = C_1(\theta_1, \theta_2)\frac{\partial logp}{\partial \theta_1} + C_2(\theta_1, \theta_2)[(\frac{\partial logp}{\partial \theta_2})^2 + (\frac{\partial^2 logp}{\partial \theta_2^2})]$$

where C_1, C_2 are functions of θ_1, θ_2 only and are such that $g*_1$ does not involve θ_2.

Mukhopadhyay(2002b) extended the theorem in two ways.

Assuming that the higher order derivatives, $\frac{\partial log^k p}{\partial \theta_2^k}(k \geq 2)$ exist, under suitable conditions

$$g*_1 = C_1\frac{\partial logp}{\partial \theta_1} + C_2[(\frac{\partial logp}{\partial \theta_2})^2 + (\frac{\partial logp}{\partial \theta_2^2})] + C_3(\theta_1, \theta_2)\psi(x, \theta),$$

where C_1, C_2, C_3 are functions of θ_1, θ_2 and $\psi(x, \theta)$ are functions of $(\partial^k logp = /\partial \theta_2^k); k = 1, 2, ...$ and are such that $g*_1$ does not involve θ_2. Another extension holds in the situation where there is only one interesting parameter θ_1 and r nuisance parameters $\theta_2, ..., \theta_{r+1}$. In this case an optimal estimating function is

$$g*_1 = C_1(\theta_1, ..., \theta_{r+1})\frac{\partial logp}{\partial \theta_1} + \sum_{\partial \theta}^{r+1} C_i(\theta_1, ..., \theta_{r+1})[(\frac{\partial logp}{\partial \theta_1})^2 + \frac{\partial^2 logp}{\partial \theta_i^2}]$$

where $C_i(\theta_1, ...\theta_1)(i = 1, ..., r+1)$ are functions of $\theta_1, ..., \theta_{r+1}$ and are such that $g*_1(x, \theta_1)$ depends on x and θ_1 only and is independent of $\theta_2, ..., \theta_{r+1}$.

It is found in this context that the optimal estimating function should be a linear function of θ_1-score A and high order θ_2 scores. Lindsay and Li (1995) noted that Bhattacharyya's (1946) scores may be used for finding optimal estimating functions.

In finding an optimal standardized estimating function we consider the multiple regression equation of g_1 on the score A and Bhattacharyya's scores $B_1,...,B_k$. The equation is

$$g_1 = v + \alpha A + \beta_1 B_1 + ... + \beta_k B_k \tag{14.1}$$

where $v, \alpha, \beta_1, ...\beta_k$ are suitable constants. Taking expectation of both sides and noting that $E(A) = E(B_i) = 0(i = 1,...,k); v = 0$. Multiplying both sides by $A, B_1,..., B_k$ successively and taking expectation of both sides we have the Normal equations

$$\begin{aligned} E_\theta(g_1 A) &= \alpha c_{11} + \beta_1 c_{12} + ... + \beta_k c_{1k+1} \\ E_\theta(g_1 B) &= \alpha c_{21} + \beta_1 c_{22} + ... + \beta_k c_{2k+1} \quad (14.2) \\ E_\theta(g_1 B_k) &= \alpha c_{k+11} + \beta_1 c_{k+12} + ... + \beta_k c_{k+1k+1} \end{aligned}$$

(14.3)

Now,

$$\int g_1 p d\mu = 0. \tag{14.4}$$

Differentiating both sides with respect to θ_1,

$$E_\theta(g_1 \frac{\partial logp}{\partial \theta_1}) = -E_\theta(\frac{\partial g_1}{\partial \theta_1}).$$

Differentiating both sides of 14.2 with respect to θ_2,

$$\frac{\partial}{\partial \theta_2} \int g_1 p d\mu = \in g_1 \frac{\partial p}{\partial \theta_2} d\mu = E_{theta}(g_1 B_1) = 0. \tag{14.5}$$

Thus, $E_\theta(g_1 B_i) = 0, i = 1,..., K$. Therefore, from 14.2, for a chosen k we have

$$\begin{pmatrix} -E_\theta(\partial g_1/\partial \theta_1) \\ 0 \\ 0 \\ . \\ 0 \end{pmatrix}$$

$$= \begin{pmatrix} c_{11} & c_{12} & \cdots & c_{1k+1} \\ c_{21} & c_{22} & \cdots & c_{2k+1} \\ . & . & \cdots & . \\ . & . & \cdots & . \\ c_{k+11} & c_{k+12} & \cdots & c_{k+1k+1} \end{pmatrix}$$

$$\begin{pmatrix} \alpha \\ \beta_1 \\ . \\ . \\ \beta_k \end{pmatrix} \tag{14.6}$$

Writing $C_k^{-1} = ((c^{ij(k)}))$, we have,

$$\widehat{\alpha} = D_\theta C^{11(k)}, \widehat{\beta}_1 = D_\theta C^{21(k)}, ..., \widehat{\beta}_k = D_\theta c^{k+11(k)} \tag{14.7}$$

where

$$D_\theta = -E_\theta(\frac{\partial g_1}{\partial \theta_1})$$

Hence, we have the standardized estimating function

$$g_{1s(k)} = \frac{g_{1k}}{E_\theta(\partial g_{1k}/\partial\theta_1)} = -[c^{11(k)}A + c^{21(k)}B_1 + ... + c^{k+11(k)}B_k] \tag{14.8}$$

For a suitably chosen k (discussed below) we take $g_{1s(k)}$ given by 14.8 as the optimal standardized estimating function $g*_1$s for θ_1.
Taking $g_1 = g_{1k} = D\theta[c^{11(k)}A + \sum_{i=2}^{k+1} c^{i1}B_{i-1}]$ we have the covariance matrix of $(g_{1k}, A, B_1, ..., B_k)'$ a singular matrix, because g_{1k} is a linear function of $A, B_1, ..., B_k$. Thus,

$$\begin{pmatrix} E_\theta(g_{1k}^2) & -E_\theta(\partial g_{1k}/\partial\theta_1) & 0 & \cdots & 0 \\ -E_\theta(\partial g_{1k}/\partial\theta_1) & c_{11} & c_{12} & \cdots & c_{1k+1} \\ 0 & c_{21} & c_{22} & \cdots & c_{2k+1} \\ . & . & . & \cdots & . \\ 0 & c_{k+11} & c_{k+1k} & . & c_{k+1k+1} \end{pmatrix} = 0 \tag{14.9}$$

Hence,

$$\frac{E_\theta(g_{1k}^2)}{(E_\theta(\partial g_{1s(k)})/\partial\theta_1))^2} = \frac{|C_{11(k)}|}{|C_k|} = c^{(11)k} \tag{14.10}$$

where $|C_{11(k)}|$ is the cofactor of c_{11} in $C_k|$. Thus

$$V(g_{is(k)}) = V(\frac{g_{ik}}{E_\theta(\partial g_{ik}/\partial\theta_1)}) = c^{11(k)} \tag{14.11}$$

where k is suitably chosen.

Mukhopadhyay (2000, 20002a) showed that under regularity conditions (a)-(d) and (i)-(iv), stated in Section 14.1,

$$V(g_{1s}) = \frac{E_\theta(g_1^2)}{(E_{\theta(\partial g_1/\partial\theta_1)})} \geq \frac{|C_{11(m)}|}{|C_m|} = c^{11(m)} = V_m \tag{14.12}$$

where $g_1 \in G_1$ and m is any integer ($\leq K$). Now,

$$\sigma^2_{A.12...m} = V(A - \widehat{A}(1,2,...,m)) = \frac{|C_m|}{|C_{11(m)}|} = \frac{1}{c^{11(m)}}$$

where $\widehat{A}(1,2,...,m)$ is the estimated value of A by taking its multiple regression on $B_1, B_2, ..., B_m$. Hence, as m increases $\sigma^2_{A.12...m}$ decreases, and $c^{11(m)}$ increases. Therefore, V_m is a nondecreasing function of m.

For finding the optimal estimating function $g*_1$s; we calculate Vm for successive values of m and stop at V_k when no further improvement in the lower bound of variance (14.2.11) is possible. Such a k will always exist since V_m is the inverse of the variance of the error in estimating the value of A by its multiple regression on $B_1, B_2, ..., B_m$ and there is a limit to considering the variables $B_1, B_2, ...,$ after which no improvement in regression is possible. We should also check if $g_{1s(k)}$ so obtained is free of nuisance parameter. For this value of k, the standardized estimating function $g_{1sk} = g_{1k} = E_\theta(\partial g_{1k} = \partial\theta_1)$ gives the optimal estimating function $g*_1$s.

Godambe and Thompson (1974), Godambe (1976) have shown the existence and uniqueness of an optimal estimating function in the presence of nuisance parameters. An estimating function is optimal in G1 if the variance of the standardized estimating function is minimum in the class. For the above-mentioned value of k, the standardized estimating function attains this minimum value. Hence, this standardized estimating function is optimum and such a k always exists, if an optimal estimating function exists.

There are many methods of obtaining optimal estimating functions in the presence of nuisance parameters. Some of these procedures require finding a suitable statistic whose marginal distribution is independent of the nuisance parameters or finding a sufficient statistic for the nuisance parameter μ_2, - sufficient for any given value of the interesting parameter θ_1 (see, e.g., Liang and Zeger, 1995).

Sometimes, it may be difficult to obtain such a statistic. The suggested procedure of using multiple regression on θ_1, θ_2-scores is a direct method of obtaining an

optimal estimating function and can be easily worked out.
Remark 1: The suggested method of finding optimal estimating function is basically an extended version of the inequality approach of Chandrasekhar and Kale (1984). It seems natural to call the estimating function with optimal k as MVBSEF(k).
If there exists a MVBSEF(k), then it is optimal in the sense of Godambe and Thompson (1974).

14.3. The Case of More Than One Nuisance Parameter

Suppose now that $\theta = (\theta_1, \theta_2)$ where $\theta_2 = (\theta_{21}, ..., \theta_{2m}), \theta_1, \theta_{2i}(i = 1, ..., m)$ are all real quantities, $\theta_1 \in \Omega_1, \theta_{2i} \in \Omega_2 i(i = 1, ..., m); \theta \in \Omega = \Omega_1 X \prod_{i=1-2i}^{m}$. We are interested in estimating θ_1 only, θ_2 being a vector of m nuisance parameters. We now modify the regularity conditions (a),(b),(c), (d) as follows.
(a) $\Omega_1, \Omega_{2i}(i = 1, ..., m)$ are open intervals of the real line.
(b) for almost all $x(\mu), (\partial logp = \partial\theta_1)), (\partial logp = \partial\theta_{2i})(i = 1, ..., m)$ exist for all $\theta \in \Omega$.
(c) $\int pd\mu, \int(\partial logp = \partial\theta_1)pd\mu, \int(\partial logp = \partial\theta_{2i})d\mu(i = 1, ..., m)$ are differentiable under the integral sign $(\theta \in \Omega)$.
(d) Let $A = (\partial logp = \partial\theta_1), B_i = (\partial logp = \partial\theta_{2i}).(i = 1, ..., m), c_{11} = V(A); c_{1i+1} = C(A, B_i), c_{i+1,i'+1} = C(B_i; B_{i'}), i, i' = 1, ..., m$ where we assume $|C_m| > 0 \forall\theta$.
With these modifications the standardized optimal estimating function is

$$g*_{1s} = -[c^{11(m)}A + c^{21(m)}B_1 + ... + c^{m+1(m)}B_m] \quad (14.13)$$

Here V_m gives the lower bound of the variance of the standardized function of $g_1 \in G_1$.
Remark 1: The method works even if we take high order scores of θ_{2i}(eg. $\frac{1}{p}\frac{\partial^j}{\partial\theta_{2i}^j}, j > 1)(i = 1, ..., m)$ as the regressor variables and calculate the corresponding variance bounds $V_t(t \geq m)$ for finding $g*_1$s. Such problems, including those in the case of several interesting parameters, can be dealt with by considering Lehmann and Hodges inequality on partitioned matrices.

14.4. Examples

Example 4.1 : Let $y = (y_1, ..., y_n)$ be a sample of size n from normal distribution with parameter $\theta = (\theta_2, \theta_1)$,

$$p(y;\theta) = \frac{1}{(2\pi\theta_1)^2} exp[-\frac{1}{2\theta_1}\sum_{i=1}^{n}(y_i - \theta_2)^2]$$

Let $\bar{y} = \sum_{i=1}^{n} y_i/n, s^2 = \sum_{i=1}^{n}(y_i - \bar{y})^2/(n-1)$ we have

$$A = \frac{\partial logp}{\partial\theta_1} = -\frac{n}{2\theta_1} + \frac{1}{2\theta_1^2}[(n-1)s^2 + n(\bar{y} - \theta_2)^2]$$

$$B_1 = \frac{\partial logp}{\partial\theta_2} = \frac{n(\bar{y} - \theta_2)}{\theta_1},$$

$$B_2 = \frac{1}{p}\frac{\partial^2 p}{\partial\theta_2^2} = \frac{n^2(\bar{y} - \theta_2)^2}{\theta_1^2} - \frac{n}{\theta_1},$$

$$B_3 = \frac{n^3(\bar{y} - \theta_2)^3}{\theta_1^3} - \frac{3n^2(\bar{y} - \theta_2)}{\theta_1^2}$$

$$c_{11} = V(A) = \frac{n}{2\theta_1^2}, c_{12} = 0, c_{13} = \frac{n}{\theta_1^2},$$

$$c_{14} = 0, c_{22} = \frac{n}{\theta_1}, c_{23} = 0, c_{24} = 0,$$

$$c_{23} = \frac{2n^2}{\theta_1^2}, c_{34} = 0, c_{44} = \frac{6n^3}{\theta_1^3}$$

Note that for $k = 1, g_{1s}(1)$ is not free of nuisance parameter. For $k = 2$,

$$C_2^{-1} = \begin{pmatrix} \frac{2\theta_1^2}{n-1} & 0 & \frac{-\theta_1^2}{n(n-1)} \\ 0 & \frac{\theta_1}{n} & 0 \\ \frac{-\theta_1^2}{n(n-1)} & 0 & \frac{\theta_1^2}{2n(n-1)} \end{pmatrix}.$$

Hence,

$$g_{1s(2)} = \frac{g_{12}}{E_\theta \frac{\partial g_{12}}{\partial\theta_2}} = -c^{11(2)}A - c^{21(2)}B_1 - c^{31(2)}B_2 \tag{14.14}$$

$$= \theta_1 - s^2 \tag{14.15}$$

Also,

$$V_2 = \frac{|C_{11(2)}|}{|C_2|} = \frac{2\theta_1^2}{n-1}.$$

For $k = 3$

$$c^{11(3)} = \frac{2\theta_1^2}{n-1}, c^{21(3)} = 0, c^{31(3)} = -\frac{\theta_1^2}{n(n-1)}, c^{41(3)} = 0$$

Hence $g_1s(3)$ remains the same as g1s(2) in (14.4.13). Also$V_3 = c^{11(3)} = V_2$ so that no further improvement in the lower bound of variance is possible. Note that $V(g_{1s(3)}) = V_3$. Again, $g_{1s(2)}$ (as well as $g_{1s(3)})$ is free of any nuisance parameter. Thus, the optimal standardized estimating function is $g*_{1s} = g*_{1s(2)} = g_{1s(3)} = \theta_1 - s^2$.
For the same problem Godambe (1976) obtained the optimal estimating function

$$g*_{1G} = \frac{n-1}{2\theta_1^2}(s^2 - \theta_1). \tag{14.16}$$

We note that when standardized, the same function $g*_{1s}$ as in (14.13) is obtained.
Example 4.2: $y = (y_1, ..., y_n)$ is a sample from $N(\theta_1, \theta_2)$ population,

$$p(y;\theta) = \frac{1}{(2\pi\theta_2)^{n/2}} exp[-\frac{1}{2\theta_2}\sum_{i=1}^{n}(y_i - \theta_1)^2]$$

Here

$$A = \frac{n(\bar{y} - \theta_1)}{\theta_2},$$

$$B_1 = -\frac{n}{2\theta_2} + \frac{1}{2\theta_2^2}\sum_{i=1}^{n}(y_i - \theta_1)^2$$

Taking k=1

$$c_{11} = \frac{n}{\theta_2}, c_{12} = 0, c_{22} = \frac{n}{2\theta_2^2}$$

Therefore $g_{1s(1)} = -c^{11}A = theta_1 - \bar{y}$, which is free of any nuisance parameter. Also, $c_{11} = \theta_2/n = V(g*_{1s})$. Here $g*_{1s} = g_{1s(1)}$ is the optimum standardized estimating function.
Example 4.3: Let $(X_i, Y_i)(i = 1, ..., m)$ be 2m independent normal variables with

$$E(X_i) = \theta_1 + \theta_{2i} E(Y_i) = \theta_{2i},$$

$$V(X_i) = V(Y_i) = 1; i = 1, ..., m.$$

Here

$$A = \sum_{i=1}^{m}(x_i - \theta_1 - \theta_{2i}),$$

$$B = x_i + y_i - \theta_1 - \theta_{2i}.$$

Also

$$c_{11} = m; c_{1i+1} = 1,$$

$$c_{i+1,i+1} = 2, ci+1, i'+1 = 0, i \neq i' = 1, ..., m.$$

Thus,

$$|C| = \begin{pmatrix} m\ 1\ 1\ ...\ 1 \\ 1\ 2\ 0\ ...\ 0 \\ .\ .\ .\ ...\ . \\ 1\ 0\ 0\ ...\ 2 \end{pmatrix}$$

Hence

$$c_{11} = \frac{2}{m}, c_{i1} = -\frac{1}{m}, i = 2, ..., m$$

. Therefore

$$\begin{aligned} g*_{1s(1)} &= -\frac{2}{m}\sum_{i=1}^{m}(x_i - \theta_1 - \theta_{2i}) + \frac{1}{m}\sum_{i=1}^{m}(x_i + y_i\theta_1 - 2\theta_{2i}) \\ &= -(\overline{x} - \overline{y} - \theta_1) \end{aligned}$$

Here value of $k = 1$ and $g_{1s(1)} = g*_{1s}$ is free of any nuisance parameter. Also, the lower bound of variance of a standardized estimating function is $c_{11} = \frac{2}{m}$, which is attained by $g*_{1s}$ above.

For the same problem Godambe (1976), using the conditional score function, showed that the optimal standardized estimating function is $g*_{1s}(y;x;\theta) = \overline{x} - \overline{y} - \theta_1$.

Example 4.4 : Let $(X_i, Y_i); (i = 1, ..., m)$ be 2m independent normal variables with

$$E(X_i) = \theta_1, E(Y_i) = \theta_1 + \theta_{2i},$$

$$V(X_i) = V(Y_i) = 1; i = 1, ..., m.$$

Here,

$$A = m(\overline{x} + \overline{y}) - 2m\theta_1 - \sum_{i=1}^{m}\theta_{2i},$$

$$B_i = y_i - \theta_1 - \theta_{2i},$$

$$c_{11} = 2m, c_{1i+1} = 1,$$

$$c_{i+1,i+1} = 1; c_{i+1,i'+1} = 0; i, i' = 1, ..., m(i \neq i').$$

Thus,

$$|C| = \begin{pmatrix} 2 & 1 & 1 & \ldots & 1 \\ 1 & 1 & 0 & \ldots & 0 \\ 1 & 0 & 1 & \ldots & 0 \\ . & . & . & \ldots & . \\ 1 & 0 & 0 & \ldots & 1 \end{pmatrix}$$

Here value of $k = 1$ and $g_{1s(1)}$ is free of any nuisance parameter. The optimal standardized estimating function is $g_{1s(1)} = g*_{1s} = -Ac^{11} - \sum_{i=1}^{m} c^{i+11} B_i = \theta_1 - \bar{x}$. The lower bound of variance is given by $1/m$ which is attained by $g*_{1s}$. The same function $g*_{1s}$ was obtained by Godambe (1976) by using the conditional score function.

References

1. Bhapkar, V.P. and Srinivasan, C. (1994). On Fisher information inequalities in the presence of nuisance parameters. *Ann. Instt. Stat. Math.* 46: 593-604.
2. Bhattacharyya, A. (1946). On some analogues of the amount of information and their use in statistical estimation.*Sankhya A*, 8:1-14.
3. Chandrasekar, B. and Kale, B.K. (1984): Unbiased statistical estimating functions in presence of nuisance parameters. *J. Stat. Pl. Inf.* 9: 45-54. Ferreira, P.E. (1982): Multivariate estimating equations. *Ann. Stat. Math.* 34A: 427-431.
4. Godambe, V.P. (1960). An optimum property of regular maximum likelihood estimation. *Ann. Math. Stat.*, 37: 1208-1211.
5. Godambe, V.P. (1976). Conditional likelihood and unconditional optimum estimating equations. *Biometrika* 63: 277-284.
6. Godambe, V.P. and Thompson, M.E. (1974): Estimating equations in the presence of a nuisance parameter. *Ann. Stat.*, 2, 568-571.
7. Kale, B.K. (1962). An extension of the Cramer-Rao inequality for statistical estimation function. *Skand. Aktur.* 45, 60-89.
8. Kale, B.K. (2001-2002). Estimating functions and equations. *J. Indian Soc. Prob. Statist.*, **6**,1-27.
9. Liang, K.Y. and Zeger, S.L. (1995). Inference based on estimating functions in the presence of nuisance parameters. *Stat. Sciences* 10: 158-173.
10. Lindsay, B.G. and Li, B. (1995): Comments on Inference based on estimating functions in the presence of nuisance parameters by Liang, K-Y and Zager, S.L., *Stat. Sc.* 10: 175.

11. Mukhopadhyay, P. (2000). On some lower bounds for the variance of an estimating function. *Int. J. Math. Stat. Sc.* 9(2).
12. Mukhopadhyay, P. (2002a). On estimating functions in the presence of a nuisance parameter. *Comm. Stat. T & M.*, 31(1): 31-36.
13. Mukhopadhyay, P. (2002b): Some lower bounds on variance of estimating functions. *J. Stat. Res.*, 36(2): 189-197.

Chapter 15

Inference In Exponential Family Regression Models Under Certain Shape Constraints Using Inversion Based Techniques

Moulinath Banerjee

451 West Hall, 1085 South University Ave.; Ann Arbor, MI, 48109–1107. USA
E-mail:moulib@umich.edu

We address the problem of pointwise estimation of a regression function under certain shape constraints, using a number of different statistics that can be viewed as measures of discrepancy from a postulated null hypothesis. Pointwise confidence sets are obtained via the usual inversion technique that exploits the duality between construction of confidence sets for a parameter of interest and testing pointwise hypotheses about that parameter. Monotonicity, unimodality and U–shapes are considered. A major advantage of these proposed methods lies in the fact that the statistics of interest are approximately pivotal for large sample sizes and therefore enable inference to be carried out without the need to estimate difficult nuisance parameters. Multivariate generalizations are briefly discussed.

Contents

15.1. Introduction and Background

Function estimation is a ubiquitous and consequently well–studied problem in nonparametric statistics. In several scientific problems, qualitative background knowledge about the function is available, in which case it is sensible to incorporate such information in the statistical analysis. Shape–restrictions are typical examples of such qualitative knowledge and appear in a large body of applications. In particular, monotonicity is a shape-restriction that shows up very naturally in different areas like reliability, renewal theory, epidemiology and biomedical studies. Closely related to monotonicity constraints are constraints like unimodality or U shapes/bath–tub shapes – functions satisfying such constraints are piecewise monotone. Some of the early work on monotone function estimation goes back to the 1950's. Grenander (1956) derived the MLE of a decreasing density as the slope of the least concave majorant of the empirical distribution function based on i.i.d. observations. The pointwise asymptotic distribution of Grenander's estimator was established by Prakasa Rao (1969). Brunk (1970) studied the problem of estimating a monotone regression function in a signal plus noise model, with additive errors. A key feature of these monotone function problems is the slower pointwise rate of convergence of the MLE ($n^{1/3}$ under the stipulation that the derivative of the monotone function at the point of interest does not vanish), as compared to the faster $\sqrt{n}$ rate in regular parametric models. Moreover, the pointwise limit distribution of the MLE turns out to be a non-Gaussian one, and seems to have first arisen in the work of Chernoff (1964). In this paper, our goal is to introduce and study a variety of statistics for estimating a regression function (at a point) that is either monotone, or unimodal, or U–shaped, in a setting where the conditional distribution of the response, given the covariate, comes from a full rank exponential family. We provide a generic description of such *conditionally parametric models* below. In what follows, we initially assume that the regression function is monotone increasing. Having constructed statistics of interest and studied their limit behavior in this setting, we proceed to demonstrate how our methods extend to decreasing, unimodal and U–shaped regression functions.

15.1.1. *Conditionally parametric response models: least squares and maximum likelihood estimates*

Consider independent and identically distributed observations $\{(Y_i, X_i)\}_{i=1}^n$, where each (Y_i, X_i) is distributed like (Y, X), and (Y, X) is distributed in the following way: The covariate X is assumed to possess a Lebesgue density p_X (with distribution function F_X). The conditional density of Y given that $X = x$ is given by $p(\cdot, \psi(x))$, where $\{p(\cdot, \theta) : \theta \in \Theta\}$ is a one–parameter exponential family of densities (with respect to some dominating measure) parametrized in the natural or canonical form, and ψ is a smooth (continuously differentiable) monotone increasing function that takes values in Θ. Recall that the density $p(\cdot, \theta)$ can be expressed as: $p(y, \theta) = \exp[\theta T(y) - B(\theta)]h(y)$. Also, recall that $E[T(Y)|X = x] =$

$B' \circ \psi(x) \equiv \mu(x)$. Since B is infinitely differentiable on Θ (an open set) and ψ is continuous, $B^{(k)} \circ \psi$ is continuous for every $k > 0$. Moreover, for every θ we have: $B''(\theta) = I(\theta)$ where $I(\theta)$ is the information about θ and is equal to the variance of T in the parametric model $p(y, \theta)$. Therefore $B''(\theta) > 0$, which implies that B' is a strictly increasing function. It follows that B' is invertible (with inverse function, H, say), so that estimating the regression function μ is equivalent to estimating ψ. The function ψ, as shown above, is in one–one correspondence with the monotone regression function μ.

Special cases of this generic formulation abound in the literature.

(a) For example, consider the *monotone regression model*. Here $Y_i = \mu(X_i) + \varepsilon_i$ where $\{(\varepsilon_i, X_i)\}_{i=1}^n$ are i.i.d. random variables, ε_i is independent of X_i, each ε_i has normal distribution with mean 0 and variance σ^2, each X_i has a Lebesgue density $p_X(\cdot)$ and μ is a monotone function. Here, $X \sim p_X(\cdot)$ and $Y \mid X = x \sim N(\mu(x), \sigma^2)$. This conditional density comes from the one–parameter exponential family $N(\eta, \sigma^2)$ (for fixed σ^2 and η varying) and can be readily represented in the canonical form.

(b) Another example is the *binary choice model* under a monotonicity constraint. Here, we have a dichotomous response variable $Y = 1$ or 0 and a continuous covariate X with a Lebesgue density $p_X(\cdot)$ such that $P(Y = 1 \mid X) \equiv G(X)$ is a smooth increasing function of X. Thus, conditional on X, Y has a Bernoulli distribution with parameter $G(X)$. In a biomedical context one could think of Y as representing the indicator of a disease/infection and X the level of exposure to a toxin, or the measured level of a bio-marker that is predictive of the disease/infection. In such cases it is often natural to impose a monotonicity assumption on G.

(c) Finally consider the *Poisson regression model* which is useful for modelling count data: $X \sim p_X(\cdot)$ and $Y \mid X = x \sim \text{Poisson}(\lambda(x))$ where λ is a monotone function. Here, one can think of X as the distance of a region from a hazardous point source (for example, a nuclear processing plant or a mine) and Y the number of cases of disease incidence at distance X (say, cancer occurrences due to radioactive exposure in the case of the nuclear processing plant, or Silicosis in the case of the mine). Given $X = x$, the number of cases of disease incidence Y at distance x from the source is assumed to follow a Poisson distribution with mean $\lambda(x)$ where λ can be expected to be monotonically decreasing in x. Variants of this model have been explored in epidemiological contexts. (Stone (1988), Diggle, Morris and Morton–Jones (1999), Morton–Jones, Diggle and Elliott (1999)).

Let $\hat{\mu}_n$ ($\hat{\psi}_n$) denote the least squares estimate (MLE) of $\mu(\psi)$ and let $\hat{\mu}_n^0$ ($\hat{\psi}_n^0$) denote the constrained least squares estimate (constrained MLE) of $\mu(\psi)$, computed under the null hypothesis that $\mu(x_0) = \eta_0$ (equivalently $\psi(x_0) = \theta_0 \equiv H(\eta_0)$), for some interior point x_0 in the domain of p_X. Each of our proposed statistics is, simply, a measure of discrepancy between the unconstrained and constrained least squares estimates (or MLEs) and can be used to test the null hypothesis above. Confidence sets of a given level for $\mu(x_0)$ will be obtained by inverting these tests, with critical values determined by the quantiles of the corresponding limit distributions under the null hypothesis. Before introducing the statistics of interest, we characterize these estimates below.

Cumulative sum diagram and greatest convex minorant: Consider a set of points in $\mathbb{R}^2$, $\{(x_0,y_0),(x_1,y_1),\ldots,(x_n,y_n)\}$, where $x_0 = y_0 = 0$ and $x_0 < x_1 < \ldots < x_n$. Let $P(x)$ be the left–continuous function such that $P(x_i) = y_i$ and $P(x)$ is constant on (x_{i-1},x_i). We will denote the vector of slopes (left–derivatives) of the greatest convex minorant (henceforth GCM) of $P(x)$ computed at the points $(x_1,x_2,\ldots,x_n)$ by $\operatorname{slogcm}\{(x_i,y_i)\}_{i=0}^n$. The GCM of $P(x)$ is, of course, also the GCM of the function that one obtains by connecting the points $\{(x_i,y_i)\}_{i=0}^n$ successively, by means of straight lines. The slope of the convex minorant plays an important role in the characterization of solutions to least squares problems under monotonicity constraints.

The following proposition characterizes least squares estimates under monotonicity constraints and is adapted from Robertson et. al. (1988).

Proposition: Let $\mathcal{X} = \{x_1 < x_2 < \ldots < x_k\}$ be an ordered set and let w be a positive weight function defined on $\mathcal{X}$. Let g be a real-valued function defined on $\mathcal{X}$ and let $\mathcal{F}$ denote the set of all real–valued increasing functions defined on $\mathcal{X}$. For $\mathcal{G} \subset \mathcal{F}$, denote the least squares projection of g onto $\mathcal{G}$ with respect to the weight function w by $g^\star_{\mathcal{G}}$; i.e: $g^\star_{\mathcal{G}} \equiv \operatorname{argmin}_{f\in\mathcal{G}} \sum_{i=1}^k (f(x_i) - g(x_i))^2\, w(x_i)$. Now, let (i) $\mathcal{F}_{u,\eta}$ denote the set of increasing functions defined on $\mathcal{X}$ that are bounded above by η, (ii) $\mathcal{F}_{l,\eta}$ denote the set of increasing functions defined on $\mathcal{X}$ that are bounded below by η. For $0 \le i \le k$, let $W_i = \sum_{j=1}^i w(x_j)$ and $G_i = \sum_{j=1}^i w(x_j)g(x_j)$. Then:

$$g^\star_{\mathcal{F}_{u,\eta}} \equiv \operatorname{slogcm}\{(W_i,G_i)\}_{i=0}^k \wedge \eta \quad \text{and} \quad g^\star_{\mathcal{F}_{l,\eta}} \equiv \operatorname{slogcm}\{(W_i,G_i)\}_{i=0}^k \vee \eta\,.$$

In the above display, the maximum is interpreted as being taken componentwise and so is the minimum.

Remark: Taking $\eta = \infty$, $\mathcal{F}_{u,\eta}$ becomes $\mathcal{F}$ and it follows that $g^{\star}_{\mathcal{F}} \equiv$ slogcm $\{(W_i, G_i)\}_{i=0}^{k}$.

Least squares estimates of μ: We are now in a position to characterize the least squares estimate of μ. The unconstrained least squares estimate $\hat{\mu}$ is given by $\hat{\mu}_n = \text{argmin}_{\mu \text{ increasing}} \sum_{i=1}^{n} (T(Y_i) - \mu(X_i))^2$. Let $\{X_{(i)}\}_{i=1}^{n}$ denote the ordered values of the X_i's and let $Y_{(i)}$ denote the response value corresponding to $X_{(i)}$. Since μ is increasing, the minimization problem is readily seen to reduce to one of minimizing $\sum_{i=1}^{n} (T(Y_{(i)}) - \mu_i)^2$ over all $\mu_1 \leq \mu_2 \leq \ldots \leq \mu_n$ (where $\mu_i = \mu(X_{(i)})$). Using the above proposition, we find that $\{\hat{\mu}_{ni}\}_{i=1}^{n}$, the minimizer over all $\mu_1 \leq \ldots \leq \mu_n$, is given by $\{\hat{\mu}_{ni}\}_{i=1}^{n} = \text{slogcm}\{G_n(X_{(i)}), V_n(X_{(i)})\}_{i=0}^{n}$, where $G_n(x) = \frac{1}{n}\sum_{i=1}^{n} 1(X_i \leq x)$ and $V_n(x) = \frac{1}{n}\sum_{i=1}^{n} T(Y_i)\, 1(X_i \leq x)$. Using operator notation (for a measure Q on the underlying sample space and a real valued function f defined on the sample space, denote $\int f dQ$ as Qf), $G_n(x) = \mathbb{P}_n 1(-\infty, x]$ and $V_n(x) = \mathbb{P}_n(T(y)\, 1(-\infty, x])$, where $\mathbb{P}_n$ is the empirical measure of the data points that assigns mass $1/n$ to each point (X_i, Y_i). We interpret $X_{(0)}$ as $-\infty$, so that $G_n(0) = V_n(0) = 0$. The (unconstrained) least squares estimate of μ is formally taken to be the piecewise constant right continuous function, such that $\hat{\mu}(X_{(i)}) = \hat{\mu}_{ni}$ for $i = 1, 2, \ldots, n$.

We next consider the problem of determining the constrained least squares estimator, where the constraint is given by $H_0 : \mu(x_0) = \eta_0$. It is easy to see that this amounts to solving two separate optimization problems: (a) Minimize $\sum_{i=1}^{m} (T(Y_{(i)}) - \mu_i)^2$ over all $\mu_1 \leq \mu_2 \leq \ldots \leq \mu_m \leq \eta_0$ and (b) Minimize $\sum_{i=m+1}^{n} (T(Y_{(i)}) - \mu_i)^2$ over all $\eta_0 \leq \mu_{m+1} \leq \mu_{m+2} \leq \ldots \leq \mu_n$; here m is that integer for which $X_{(m)} < x_0 < X_{(m+1)}$. The vector that solves (a) (say $\{\hat{\mu}^0_{ni}\}_{i=1}^{m}$) is given by: $\{\hat{\mu}^0_{ni}\}_{i=1}^{m} = \text{slogcm}\{G_n(X_{(i)}), V_n(X_{(i)})\}_{i=0}^{m} \wedge \eta_0$. On the other hand, the vector that solves (b) (say $\{\hat{\mu}^0_{ni}\}_{i=m+1}^{n}$) is given by:

$$\{\hat{\mu}^0_{ni}\}_{i=m+1}^{n} = \text{slogcm}\{G_n(X_{(i)}) - G_n(X_{(m)}), V_n(X_{(i)}) - V_n(X_{(m)})\}_{i=m}^{n} \vee \eta_0 .$$

The constrained MLE $\hat{\mu}^0_n$ is then taken to be the piecewise constant right–continuous function, such that $\hat{\mu}^0_n(X_{(j)}) = \hat{\mu}^0_{nj}$ for $j = 1, 2, \ldots, n$ and $\hat{\mu}^0_n(x_0) = \eta_0$, and such that $\hat{\mu}^0_n$ has no jumps outside the set $\{X_{(j)}\}_{j=1}^{n} \cup \{x_0\}$.

The characterization of the unconstrained and constrained maximum likelihood estimates (MLEs) of ψ depends heavily on the Kuhn–Tucker theorem. Though this is a very standard theorem in the literature on convex function estimation, we state it briefly, for convenience.

Kuhn-Tucker theorem: Let ϕ be a strictly convex function defined on $\mathbb{R}^n$ and potentially assuming values in the extended real line. Define $R = \phi^{-1}(\mathbb{R})$ and consider the problem of minimizing ϕ on R, subject to a number of inequality and equality constraints that may be written as $g_i(x) \leq 0$ for $i = 1,2,\ldots,k$ and $g_i(x) = 0$ for $i = k+1,\ldots,m$. Here, the g_i's are convex functions. Then $\hat{x} \in R$ uniquely minimizes ϕ subject to the m constraints if and only if there exist non-negative (Lagrange multipliers) $\lambda_1, \lambda_2, \ldots, \lambda_m$ such that (a) $\sum_{i=1}^m \lambda_i g_i(\hat{x}) = 0$ and (b) $\nabla\phi(\hat{x}) + G_{n\times m}^T \lambda_{m\times 1} = 0$, where $G_{m\times n}$ is the total derivative of the function $(g_1, g_2, \ldots, g_m)^T$ at the point $\hat{x}$.

Maximum likelihood estimators of ψ: The likelihood function for ψ, up to a multiplicative factor not depending on ψ, is given by:

$$L_n(\psi, \{Y_i, X_i\}_{i=1}^n) = \prod_{i=1}^n \exp(\psi(X_i)\, T(Y_i) - B(\psi(X_i)))\,,$$

whence the log-likelihood function for ψ is:

$$l_n(\psi, \{Y_i, X_i\}_{i=1}^n) = \sum_{i=1}^n [\psi(X_i)\, T(Y_i) - B(\psi(X_i))] \equiv \sum_{i=1}^n [\psi(X_{(i)})\, T(Y_{(i)}) - B(\psi(X_{(i)}))]\,.$$

Writing $\psi(X_{(i)}) = \psi_i$, it is seen that the problem of computing $\hat{\psi}_n$, the unconstrained MLE reduces to minimizing $\phi(\psi_1, \psi_2, \ldots, \psi_n) \equiv \sum_{i=1}^n [-\psi_i T(Y_{(i)}) + B(\psi_i)]$ over $\psi_1 \leq \psi_2 \leq \ldots \leq \psi_n$. The strict convexity of B implies the strict convexity of ϕ and the Kuhn-Tucker theorem may be invoked with $g_i(\tilde{\psi}) = \psi_i - \psi_{i+1}$ for $i = 1,2,\ldots,n-1$ (here $\tilde{\psi}$ denotes the vector $(\psi_1, \psi_2, \ldots, \psi_n)$). Denoting the minimizer of ϕ by $\hat{\psi}_n = (\hat{\psi}_{n1}, \hat{\psi}_{n2}, \ldots, \hat{\psi}_{nn})$, the conditions (a) and (b) of that theorem translate to: $\lambda_i = \sum_{j=1}^i (T(Y_{(j)}) - B'(\hat{\psi}_{nj})) \geq 0$ for $i = 1,2,\ldots,n-1$, and $\sum_{j=1}^n (T(Y_{(j)}) - B'(\hat{\psi}_{nj})) = 0$.

Setting $\hat{\psi}_{ni} = H(\hat{\mu}_{ni})$ (where H is the inverse function of B'), so that $\hat{\mu}_{ni} = B'(\hat{\psi}_{ni})$ it can be shown that the above conditions are satisifed, whence it follows that the unconstrained MLE $\hat{\psi}_n = H(\hat{\mu}_n)$. For the details of this argument, see Banerjee (2007A).

As in the case of the unconstrained least squares estimator, we can show, by splitting the likelihood maximization problem into two parts, and subsequently invoking the Kuhn-Tucker theorem, that the constrained MLE $\hat{\psi}_n^0$, computed under $H_0 : \psi(x_0) = \theta_0$ (where $\theta_0 = H(\eta_0)$), is given by $\hat{\psi}_n^0 = H(\hat{\mu}_n^0)$.

15.2. Discrepancy Statistics for Testing The Null Hypothesis

We are now in a position to formulate the discrepancy statistics to be used for testing H_0. Inversion of these discrepancy statistics will provide confidence sets for μ (equivalently ψ) at the point x_0.

From the regression point of view, one natural statistic for testing the null hypothesis $H_0 : \mu(x_0) = \eta_0$ is the difference in sum of squares given by:

$$DSS(\eta_0) = \sum_{i=1}^{n} (Y_i - \hat{\mu}_n^0(X_i))^2 - \sum_{i=1}^{n} (Y_i - \hat{\mu}_n(X_i))^2 . \tag{15.1}$$

Large values of $DSS(\eta_0)$ provide evidence against the null hypothesis. To determine what is "large" will require investigation of the large sample distribution of $DSS(\eta_0)$ that we undertake in the next section. Writing the regression model as $T(Y_i) = \mu(X_i) + \tilde{\varepsilon}_i$ (where $\tilde{\varepsilon}_i$ has mean 0), note that $DSS(\eta_0)$ can be interpreted as a pseudo likelihood ratio statistic that is obtained by *pretending* that the $\tilde{\varepsilon}_i$ are (conditionally homoscedastic) normal errors. Another discrepancy statistic can be constructed by computing a global distance between the unconstrained and constrained least squares estimators. More specifically, consider:

$$L_2(\hat{\mu}_n, \hat{\mu}_n^0) = \sum_{i=1}^{n} (\hat{\mu}_n(X_i) - \hat{\mu}_n^0(X_i))^2 = n \int (\hat{\mu}_n - \hat{\mu}_n^0)^2 \, d\, G_n . \tag{15.2}$$

Once again, large values of this quantity provide evidence against the null hypothesis. In similar vein, we can consider:

$$L_2(\hat{\psi}_n, \hat{\psi}_n^0) = \sum_{i=1}^{n} (\hat{\psi}_n(X_i) - \hat{\psi}_n^0(X_i))^2 = n \int (\hat{\psi}_n - \hat{\psi}_n^0)^2 \, d\, G_n . \tag{15.3}$$

While the above provide valid measures of discrepancy, none of these use the actual likelihood function of the data to formulate the notion of discrepancy. The likelihood ratio statistic does precisely that by looking at the difference in the log–likelihood functions evaluated at the unconstrained and constrained MLEs. More precisely, the likelihood ratio statistic for testing H_0 is given by:

$$\begin{aligned} 2 \log \lambda_n(\theta_0) = 2 \Bigg\{ & \sum_{i=1}^{n} [\hat{\psi}_n(X_{(i)})\, T(Y_{(i)}) - B(\hat{\psi}_n(X_{(i)}))] \\ & - \sum_{i=1}^{n} [\hat{\psi}_n^0(X_{(i)})\, T(Y_{(i)}) - B(\hat{\psi}_n^0(X_{(i)}))] \Bigg\} \end{aligned} \tag{15.4}$$

The null hypothesis is rejected for large values of the likelihood ratio statistic. Of course, the maximum likelihood estimate $\hat{\psi}(x_0)$ can be used to make inference about $\psi(x_0)$. As will be shown later $n^{\gamma}(\hat{\psi}_n(x_0) - \psi(x_0))$ converges to a limit distribution, for some positive γ, which depends on the number of derivatives of ψ that vanish at the point x_0.

We finally define versions of the "score statistic" for this class of models. As will be seen later, these have natural connections to the likelihood ratio, least squares and the L_2 statistics introduced above. Consider, the log–likelihood function for the pair (Y,X): $l(Y,\psi(X)) \equiv \log p(Y,\psi(X)) = \psi(X)\,T(Y) - B(\psi(X))$. The log–likelihood function for the data $\{(Y_i,X_i)\}_{i=1}^n$ is given by $l_n(\psi) = n\,\mathbb{P}_n\,l(Y,\psi(X))$, with $\mathbb{P}_n$ denoting the empirical measure of the data vector $\{(Y_i,X_i)\}_{i=1}^n$. Consider a perturbation of ψ in the direction of the monotone function η, defined by the parametric curve $\psi_{\eta,\varepsilon} \equiv (1-\varepsilon)\,\psi(z) + \varepsilon\eta(z)$. Set $l_{n,\varepsilon} = n\,\mathbb{P}_n\,[\psi_{\eta,\varepsilon}(X)\,T(Y) - B(\psi_{\eta,\varepsilon}(X))]$. This can be viewed as the log–likelihood function from a one–dimensional model parametrized by ε, and one can compute a score statistic at $\varepsilon = 0$, by differentiating this parametric log–likelihood at $\varepsilon = 0$. We get:

$$S_{n,\eta,\psi} = \frac{\partial}{\partial\varepsilon}\,l_{n,\varepsilon}\,|_{\varepsilon=0} = n\,\mathbb{P}_n\,[(\eta(X) - \psi(X))\,(T(Y) - B'(\psi(X)))]\,.$$

Our proposed score statistics will be constructed by perturbing $\hat{\psi}_n$ in the direction of $\hat{\psi}_n^0$ and vice versa. Thus, we get:

$$S_{n,\hat{\psi}_n^0,\hat{\psi}_n} \equiv S_{n,1} = n\,\mathbb{P}_n\,\left[(\hat{\psi}_n^0(X) - \hat{\psi}_n(X))\,(T(Y) - B'(\hat{\psi}_n(X)))\right]$$

and

$$S_{n,\hat{\psi}_n,\hat{\psi}_n^0} \equiv S_{n,2} = n\,\mathbb{P}_n\,\left[(\hat{\psi}_n(X) - \hat{\psi}_n^0(X))\,(T(Y) - B'(\hat{\psi}_n^0(X)))\right]\,.$$

It is not difficult to see that $S_{n,1}$ is non-positive with probability one; this is a consequence of the fact that $\hat{\psi}_n$ is the MLE of ψ. The null hypothesis, $\psi(x_0) = \theta_0$, will be rejected for extreme values (both large and small) of the score statistics. Note the contrast with the rejection region using the residual sum of squares, or likelihood ratio or L_2 statistics. With these statistics, it is sensible to reject only for large values, since each of these statistics will tend to increase as the data generating mechanism deviates more and more from the null hypothesis $\psi(x_0) = \theta_0$. In fact, small values are very compatible with the null hypothesis. However, with the score statistics, the same cannot be inferred. The analytical forms of the statistics do not provide any insight regarding the nature of values of the score statistic (large or small) under deviation from the null hypothesis. All that may be inferred is that an atypical value (i.e. either very small or very large) of the score is less consistent with the null. Consequently, both small and large extremes need to be allowed.

In the next section, we study the limit distributions of these statistics under the null hypothesis and show how the results can be used to construct various confidence intervals for ψ (equivalently μ) at a point of interest.

15.2.1. *Relevant stochastic processes and derived functionals*

To study the asymptotic distributions of these competing statistics, we introduce the relevant stochastic processes and certain derived functionals of these. For each $m \geq 1$ and for positive constants c and d, define $X_{c,d,m}(h) = cW(h) + d \mid h \mid^{m+1}$, for $h \in \mathbb{R}$. Here, $W(h)$ is standard two–sided Brownian motion starting from 0. For a real–valued function f defined on $\mathbb{R}$, let $\text{slogcm}(f, I)$ denote the left–hand slope of the GCM (greatest convex minorant) of the restriction of f to the interval I. We abbreviate $\text{slogcm}(f, \mathbb{R})$ to $\text{slogcm}(f)$. Also define:

$$\text{slogcm}^0(f) = (\text{slogcm}\,(f, (-\infty, 0]) \wedge 0)\, 1_{(-\infty,0]} + (\text{slogcm}\,(f, (0, \infty)) \vee 0)\, 1_{(0,\infty)}\,.$$

Set $g_{c,d,m} = \text{slogcm}(X_{c,d,m})$ and $g^0_{c,d,m} = \text{slogcm}^0\,(X_{c,d,m})$. The random function $g_{c,d,m}$ is increasing but piecewise constant with finitely many jumps in any compact interval. Also $g^0_{c,d,m}$, like $g_{c,d,m}$, is a piecewise constant increasing function, with finitely many jumps in any compact interval and differing, almost surely, from $g_{c,d,m}$ on a finite interval containing 0. In fact, with probability 1, $g^0_{c,d,m}$ is identically 0 in some random neighbourhood of 0, whereas $g_{c,d,m}$ is almost surely non-zero in some random neighbourhood of 0. Also, the length of the interval $D_{c,d,m}$ on which $g_{c,d,m}$ and $g^0_{c,d,m}$ differ is $O_p(1)$. The processes $g_{c,d,1}$ and $g^0_{c,d,1}$ in particular (slopes of convex minorants of Brownian motion with quadratic drift) are well-studied in the literature; see, for example, Banerjee and Wellner (2001) and Wellner (2003). The qualitative properties of the convex minorants, as described above, for $m > 1$ are similar to what one encounters in the case $m = 1$ (quadratic drift).

Brownian scaling allows us to relate the processes $(g_{c,d,m}, g^0_{c,d,m})$ to (g_m, g^0_m) (where $g_m \equiv g_{1,1,m}$ and $g^0_m \equiv g^0_{1,1,m}$ are convex minorants of the canonical process $W(t)+ \mid t \mid^{m+1}$), as follows.

Lemma 1 The process $\{(g_{c,d,m}(h), g^0_{c,d,m}(h) : h \in \mathbb{R}\}$ has the same distribution as the process $\{c\,(d/c)^{1/(2m+1)}\,(g_m((d/c)^{2/(2m+1)}\,h), g^0_m((d/c)^{2/(2m+1)}\,h)) : h \in \mathbb{R}\}$ in the space $\mathcal{L} \times \mathcal{L}$. Here $\mathcal{L}$ denotes the space of monotone functions from $\mathbb{R}$ to $\mathbb{R}$ which are bounded on every compact set equipped with the topology of L_2 convergence (with respect to Lebesgue measure) on compact sets.

Define random variables $D_{c,d,m}, T_{c,d,m}, \mathbb{M}_{1,c,d,m}, \mathbb{M}_{2,c,dm}$ as follows:

$$D_{c,d,m} = \int \{(g_{c,d,m}(h))^2 - (g^0_{c,d,m}(h))^2\}\, dh\ ,\ T_{c,d,m} = \int (g_{c,d,m}(h) - g^0_{c,d,m}(h))^2\, dh,$$

$$\mathbb{M}_{1,c,d,m} = \int g^0_{c,d,m}(h)\,(g^0_{c,d,m}(h) - g_{c,d,m}(h))\,dh\ ,$$

$$\mathbb{M}_{2,c,d,m} = \int g_{c,d,m}(h)\,(g_{c,d,m}(h) - g^0_{c,d,m}(h))\,dh$$

and let $\mathbb{D}_m, \mathbb{T}_m, \mathbb{M}_{1,m}, \mathbb{M}_{2,m}$ denote the respective versions with $c = d = 1$. Using Lemma 15.2.1, we can show that the following holds.

Lemma 2 The following scaling relation holds: $c^{-2}(\mathbb{D}_{c,d,m},\ \mathbb{T}_{c,d,m},\ \mathbb{M}_{1,c,d,m},\ \mathbb{M}_{2,c,d,m}) \equiv_d (\mathbb{D}_m,\ \mathbb{T}_m,\ \mathbb{M}_{1,m},\ \mathbb{M}_{2,m})$.

For proofs of these lemmas (which rely on Brownian scaling arguments) see Section 6 of Banerjee (2007A).

15.3. Limit Distributions for The Discrepancy Statistics and Methodological Implications

We will study the limit distribution of the discrepancy statistics at the point x_0 under the assumption that the first $m-1$ derivatives of μ (and equivalently ψ) vanish at the point x_0 but the m'th does not, and is therefore strictly greater than 0 (under our assumption that μ is an increasing function). For $m = 1$, this reduces to the condition that the derivative at x_0 does not vanish. While the assumption of finitely many derivatives vanishing at x_0 is difficult to check from the methodological perspective (unless there happens to be compelling background knowledge that indicates that such is the case) and the case $m = 1$ is the one that can really be used effectively, formulating the results for a general m leads to a unified presentation of results at no additional cost.

We first define localized versions of both the unconstrained and the constrained least squares estimates of μ, and the corresponding MLE's of ψ. Thus, we set:

$$X_n(h) = n^{m/(2m+1)}\,(\hat{\mu}_n(x_0 + h\,n^{-1/(2m+1)}) - \mu(x_0))\,,$$

$$Y_n(h) = n^{m/(2m+1)}\,(\hat{\mu}^0_n(x_0 + h\,n^{-1/(2m+1)}) - \mu(x_0))\,.$$

We also set:

$$\tilde{X}_n(h) = n^{m/(2m+1)}\,(\hat{\psi}_n(x_0 + h\,n^{-1/(2m+1)}) - \psi(x_0))\,,$$

$$\tilde{Y}_n(h) = n^{m/(2m+1)}\,(\hat{\psi}^0_n(x_0 + h\,n^{-1/(2m+1)}) - \psi(x_0))\,.$$

The following facts will be used.

Fact 1. The processes $(X_n(h), Y_n(h) : h \in \mathbb{R})$ converge in distribution to $(g_{a,b,m}(h), g^0_{a,b,m}(h) : h \in \mathbb{R})$ in the space $\mathcal{L} \times \mathcal{L}$, where $a = \sqrt{I(\psi(x_0))/p_X(x_0)}$ and $b = (1/(m+1)!) \mid \mu^{(m)}(x_0) \mid$.

Using the fact that $(\hat{\mu}_n(x), \hat{\mu}^0_n(x)) = (B'(\hat{\psi}_n(x)), B'(\hat{\psi}^0_n(x)))$ and the delta method, it is easily deduced from Fact 1 that $(\tilde{X}_n(h), \tilde{Y}_n(h) : h \in \mathbb{R})$ converge in distribution to $(g_{\tilde{a},\tilde{b},m}(h), g^0_{\tilde{a},\tilde{b},m}(h) : h \in \mathbb{R})$, where $\tilde{a} = \sqrt{1/(I(\psi(x_0))p_X(x_0))}$ and $b = (1/(m+1)!) \mid \psi^{(m)}(x_0) \mid$.

Fact 2. The estimators $\hat{\mu}_n$ and $\hat{\mu}^0_n$ differ on an interval D_n (around x_0) whose length is $O_p(n^{-1/(2m+1)})$.

Fact 3. Let J_n be the set of indices i such that $\hat{\mu}_n(X_{(i)}) \neq \hat{\mu}^0_n(X_{(i)})$. Then J_n can be broken up into ordered blocks of indices $\{B_j\}$ and $\{B^0_j\}$ such that the following holds: (a) For each j, for $i \in B_j$, $\hat{\mu}_n(X_{(i)})$ is constant with the constant value depending on j and given by $n_j^{-1} \sum_{i \in B_j} T(Y_{(i)})$, where n_j is the size of B_j. (b) For each B^0_j, for i in B^0_j, $\hat{\mu}^0_n(X_{(i)})$ is constant with the constant value depending on j; moreover, if the constant value on B^0_j is different from η_0, then it is given by $(n^0_j)^{-1} \sum_{i \in B^0_j} T(Y_{(i)})$, where n^0_j is the size of B^0_j.

We now state our main result.

Theorem 1 Assume that the null hypothesis $\psi(x_0) = \theta_0$ (equivalently $\mu(x_0) = \eta_0$) holds. Then:

(a) The statistic $\mathrm{DSS}(\eta_0)$ defined in (15.1) converges in distribution to $I(\psi(x_0))\,\mathbb{D}_m$, while the likelihood ratio statistic defined in (15.4) converges in distribution to $\mathbb{D}_m$.

(b) The statistic $L_2(\hat{\mu}_n, \hat{\mu}^0_n)$ converges in distribution to $I(\psi(x_0))\,\mathbb{T}_m$ while the statistic $L_2(\hat{\psi}_n, \hat{\psi}^0_n)$ converges in distribution to $(I(\psi(x_0))^{-1}\,\mathbb{T}_m$.

Define weighted versions of the statistics in (b) as follows. Set:

$$L_{2,w}(\hat{\mu}_n, \hat{\mu}^0_n) = \sum_{i=1}^{n} \frac{(\hat{\mu}_n(X_i) - \hat{\mu}^0_n(X_i))^2}{I(\hat{\psi}_n(X_i))},$$

and

$$L_{2,w}(\hat{\psi}_n, \hat{\psi}^0_n) = \sum_{i=1}^{n} (\hat{\psi}_n(X_i) - \hat{\psi}^0_n(X_i))^2 I(\hat{\psi}_n(X_i)).$$

Both $L_{2,w}(\hat{\mu}_n, \hat{\mu}^0_n)$ and $L_2(\hat{\psi}_n, \hat{\psi}^0_n)$ converge to $\mathbb{T}_m$ in distribution.

(c) The statistic $S_{n,1}$ converges in distribution to $\mathbb{M}_{1,m}$ while the statistic $S_{n,2}$ converges in distribution to $\mathbb{M}_{2,m}$. Furthermore, $S_{n,1} + S_{n,2} = I(\psi(x_0))\, L_2(\hat{\psi}_n, \hat{\psi}_n^0) + o_p(1)$ while $S_{n,2} - S_{n,1} = 2 \log \lambda_n + o_p(1)$.

So, both the likelihood ratio and the L_2 statistics can be asymptotically linearly decomposed in terms of the score statistics.

(d) The statistic $n^{m/(2m+1)}(\hat{\mu}_n(x_0) - \eta_0)$ converges in distribution to $(a^{2m}\, b)^{1/(2m+1)}\, g_m(0)$.

Remarks: For a proof outline of Fact 1, see Section 6 of Banerjee (2007A) which uses the "switching relationship" as demonstrated in Examples 3.2.14 and 3.2.15 of Van der Vaart and Wellner (1996). For an alternative approach, that uses continuous mapping arguments for slope-of-greatest-convex-minorant estimators for the case $m = 1$ in a more general model, see the proof of Theorem 2.1 of Banerjee (2007B). Banerjee (2007B) also studies the limit behavior of the likelihood ratio statistic in this general model but does not deal with the other discrepancy measures considered in this paper; indeed, some of the discrepancy measures like the DSS cannot be tackled in the setup of Banerjee (2007B). The natural connection between least squares and maximum likelihood estimates in the current setting (as noted in Section 1.1) is an outcome of the nice structure of exponential family models but is absent in the general setting of Banerjee (2007B). A proof of Fact 2 is given in Banerjee (2007A). Fact 3 is a straightforward consequence of the characterization of isotonic regression estimators as blockwise averages.

Methodological consequences: The above theorem has significant methodological consequences for the estimation of the regression function μ (equivalently, the function ψ). The results in (a), (b) and (c) of Theorem 1 provide a number of pivots through the inversion of which confidence sets for $\mu(x_0)$ can be obtained. Let $d_{m,\beta}, t_{m,\beta}, M_{1,m,\beta}$ and $M_{2,m,\beta}$ denote the β'th quantiles of the distributions of $\mathbb{D}_m, \mathbb{T}_m, \mathbb{M}_{1,m}$ and $\mathbb{M}_{2,m}$ respectively. Consider the null hypothesis $H_\eta : \mu(x_0) = \eta$ (equivalently $\psi(x_0) = H(\eta)$), and denote the constrained least squares estimate of μ under this hypothesis by $\hat{\mu}_n^\eta$ and the corresponding MLE of ψ by $\hat{\psi}_n^\eta$. Let $DSS(\eta)$ denote the residual sum of squares statistic for testing this hypothesis and $2 \log \lambda_n(H(\eta))$ denote the corresponding likelihood ratio statistic. From (a), we obtain two asymptotic level $1 - \alpha$ confidence sets for $\mu(x_0)$ as:

$$\{\eta : I(H(\eta))^{-1}\, \mathrm{DSS}(\eta) \leq d_{m,1-\alpha}\} \text{ and } \{\eta : 2 \log \lambda_n(H(\eta)) \leq d_{m,1-\alpha}\}.$$

Confidence sets for $\mu(x_0)$ based on the first two statistics in Part (b) are given by:

$$\{\eta : I(H(\eta))^{-1}\, L_2(\hat{\mu}_n, \hat{\mu}_n^\eta) \leq t_{m,1-\alpha}\} \text{ and } \{\eta : I(H(\eta))\, L_2(\hat{\psi}_n, \hat{\psi}_n^\eta) \leq t_{m,1-\alpha}\},$$

while, using the weighted versions we get confidence sets:

$$\{\eta : L_{2,w}(\hat{\mu}_n, \hat{\mu}_n^{\eta}) \leq t_{m,1-\alpha}\} \text{ and } \{\eta : L_{2,w}(\hat{\psi}_n, \hat{\psi}_n^{\eta}) \leq t_{m,1-\alpha}\}.$$

Using the results of Part (c), we get the following confidence sets:

$$\{\eta : M_{1,m,\alpha/2} \leq S_{n,\hat{\psi}_n^{\eta},\hat{\psi}_n} \leq M_{1,m,1-\alpha/2}\} \text{ and } \{\eta : M_{2,m,\alpha/2} \leq S_{n,\hat{\psi}_n,\hat{\psi}_n^{\eta}} \leq M_{2,m,1-\alpha/2}\}.$$

The result in (d) can be used to construct a confidence set of the form $[\hat{\mu}(x_0) - n^{-m/(2m+1)}\,(\hat{a}^{2m}\,\hat{b})^{1/(2m+1)}\, q_{m,\alpha/2}, \hat{\mu}(x_0) + n^{-m/(2m+1)}\,(\hat{a}^{2m}\,\hat{b})^{1/(2m+1)}\, q_{m,\alpha/2}]$, where $q_{m,\alpha/2}$ is the $(1-\alpha/2)$'th quantile of the (symmetric) distribution of $g_{1,1,m}(0)$. When $m = 1$, the slope of the greatest convex minorant at 0 is distributed like twice the minimizer of $\{W(h) + h^2 : h \in \mathbb{R}\}$, whose distribution is very well–studied (see, for example, Groeneboom and Wellner (2001)) and is referred to in the literature as Chernoff's distribution. The prime issue with this method lies in the fact that the m'th derivative at the point x_0 needs to be estimated, and this is a difficult affair. However, it is possible to bypass parameter estimation in this case by resorting to resampling techniques. Efron's bootstrap does not work in this situation, but subsampling (see Politis, Romano and Wolf (1999)) does.

15.3.1. *Incorporating further shape constraints*

We now discuss how the above methodology can be extended to incorporate further shape constraints. Our discussion, thus far, has focused on estimating a monotone increasing regression function μ, but works equally well for decreasing functions, and also for unimodal/U–shaped regression functions, provided one stays away from the mode/minimizer of the regression function. We first investigate the case of a decreasing regression function.

Decreasing regression function: The results of Theorem 1 continue to hold for a decreasing regression function (under the assumption that the first $m-1$ derivatives of the decreasing regression function μ vanish at the point x_0 and the m'th does not). In this case, the unconstrained and constrained least squares estimates of μ are no longer characterized as the slopes of greatest convex minorants, but as slopes of least concave majorants. We present the characterizations of the least squares estimates in the decreasing case below. We first introduce some notation. For points, $\{(x_0,y_0),(x_1,y_1),\ldots,(x_k,y_k)\}$ where $x_0 = y_0 = 0$ and $x_0 < x_1 < \ldots < x_k$, consider the right–continuous function $P(x)$ such that $P(x_i) = y_i$ and such that $P(x)$ is constant on (x_{i-1},x_i). We will denote the vector of slopes (left–derivatives) of the LCM (least concave majorant) of $P(x)$ computed at the points $(x_1, x_2, \dots, x_k)$ by slolcm$\{(x_i, y_i)\}_{i=0}^k$. With G_n and V_n as before, it is not difficult to see that $\{\hat{\mu}_n(X_{(i)})\}_{i=1}^n =$ slolcm $\{G_n(X_{(i)}), V_n(X_{(i)})\}_{i=0}^n$. Also, the MLE under $H_0 : \mu(x_0) = \eta_0$ is given by: $\{\hat{\mu}_n^0(X_{(i)})\}_{i=1}^m = \eta_0 \vee \text{slolcm}\,\{G_n(X_{(i)}), V_n(X_{(i)})\}_{i=0}^m$ while $\{\hat{\mu}_n^0(X_{(i)})\}_{i=m+1}^n =$ $\eta_0 \wedge \text{slolcm}\,\{G_n(X_{(i)}) - G_n(X_{(m)}), V_n(X_{(i)}) - V_n(X_{(m)})\}_{i=m}^n$. As with an increasing regression function, the unconstrained and constrained MLE's of ψ are given by $\hat{\psi}_n = H(\hat{\mu}_n)$ and $\hat{\psi}_n^0 = H(\hat{\mu}_n^0)$ respectively.

Unimodal regression functions: Suppose now that the regression function is unimodal. Thus, there exists $M > 0$ such that the regression function is increasing on $[0, M]$ and decreasing to the right of M. The goal is to construct a confidence set for the regression function at a point $x_0 \neq M$ under the assumption that the first $m-1$ derivatives of μ vanish at x_0 and the m'th does not. We consider the more realistic case for which M is unknown.

First compute a consistent estimator, $\hat{M}_n$, of the mode M. With probability tending to 1, $x_0 < \hat{M}_n$ if x_0 is to the left of M and $x_0 > \hat{M}_n$ if x_0 is to the right of M.

Assume first that $x_0 < M \wedge \hat{M}_n$. Let m_n be such that $X_{(m_n)} \leq \hat{M}_n < X_{(m_n+1)}$. Let $\hat{\mu}_n$ denote the unconstrained LSE of μ, using $\hat{M}_n$ as the mode. Then, $\hat{\mu}_n$ is obtained by minimizing $\sum_{i=1}^n (T(Y_{(i)}) - \mu_i)^2$ over all $\mu_1, \mu_2, \dots, \mu_n$ with $\mu_1 \leq \mu_2 \dots \leq \mu_{m_n}$ and $\mu_{m_n+1} \geq \mu_{m_n+2} \geq \dots \geq \mu_n$. It is not difficult to verify that $\{\hat{\mu}_n(X_{(i)})\}_{i=1}^{m_n} = \text{slogcm}\,\{G_n(X_{(i)}), V_n(X_{(i)})\}_{i=0}^{m_n}$ while $\{\hat{\mu}_n(X_{(i)})\}_{i=m_n+1}^n =$ slolcm $\{G_n(X_{(i)}) - G_n(X_{(m_n)}), V_n(X_{(i)}) - V_n(X_{(m_n)})\}_{i=m_n}^n$.
Now, consider testing the (true) null hypothesis that $\mu(x_0) = \eta_0$. Let $m < m_n$ be the number of $X_{(i)}$'s that do not exceed x_0. Denoting, as before, the constrained MLE by $\hat{\mu}_n^0$, it can be checked that $\hat{\mu}_n^0(X_{(j)}) = \hat{\mu}_n(X_{(j)})$ for $j > m_n$, whereas $\{\hat{\mu}_n^0(X_{(i)})\}_{i=1}^m = \eta_0 \wedge \text{slogcm}\,\{G_n(X_{(i)}), V_n(X_{(i)})\}_{i=0}^m$ and

$$\{\hat{\mu}_n^0(X_{(i)})\}_{i=m+1}^{m_n} = \eta_0 \vee \text{slogcm}\,\{G_n(X_{(i)}) - G_n(X_{(m)}), V_n(X_{(i)}) - V_n(X_{(m)})\}_{i=m}^{m_n}.$$

The corresponding unconstrained and constrained MLE's of ψ are obtained by transforming $\hat{\mu}_n$ and $\hat{\mu}_n^0$ by H. The discrepancy statistics from Section 1.2 have the same form as in the monotone function case, with the effective contribution coming from response–covariate pairs for which the covariate lies in a shrinking $(O_p(n^{-1/(2m+1)}))$ neighborhood of the point x_0. The asymptotic distributions of these statistics are identical to those in the monotone function case. For a similar result for the maximum likelihood estimator, in the setting of unimodal density estimation away from the mode, we refer the reader to Theorem 1 of Bickel and Fan

(1996). A rigorous derivation of the asymptotics for the unimodal case involves some embellishments of the arguments in the monotone function scenario and are omitted. Intuitively, it is not difficult to see why the asymptotic behavior remains unaltered. For example, the characterization of the LSE of μ on the interval $[0, M_n]$, with M_n converging to M, is in terms of unconstrained/constrained slopes of convex minorants exactly as in the monotone function case. Furthermore, the behavior at the point x_0, which is bounded away from M_n with probability increasing to 1, is only influenced by the behavior of localized versions of the processes V_n and G_n in a shrinking neighborhood of the point x_0 (where the unconstrained and the constrained MLE's differ) and these behave asymptotically in exactly the same fashion as for the monotone function case. Consequently, the behavior of the LSEs (and consequently, the discrepancy statistics based on these) stays unaffected. An asymptotic confidence interval of level $1-\alpha$ for $\mu(x_0)$ can therefore be constructed in the exact same way as for the monotone function case.

The other situation is when $M \vee \hat{M}_n < x_0$. In this case $\hat{\mu}_n$ has the same form as above. Now, consider testing the (true) null hypothesis that $\mu(x_0) = \eta_0$. Let m be the number of $X_{(i)}$'s such that $\hat{M}_n < X_{(i)} \leq x_0$. Now, $\hat{\mu}_n^0(X_{(j)}) = \hat{\mu}_n(X_{(j)})$ for $1 \leq j \leq m_n$, while

$$\{\hat{\mu}_n^0(X_{(i)})\}_{i=m_n+1}^{m_n+m} = \eta_0 \vee$$

$$\text{slolcm}\left\{G_n(X_{(i)}) - G_n(X_{(m_n)}), V_n(X_{(i)}) - V_n(X_{(m_n)})\right\}_{i=m_n}^{m_n+m},$$

and

$$\{\hat{\mu}_n^0(X_{(i)})\}_{i=m_n+m+1}^{n} = \eta_0 \wedge$$

$$\text{slolcm}\left\{G_n(X_{(i)}) - G_n(X_{(m_n+m)}), V_n(X_{(i)}) - V_n(X_{(m_n+m)})\right\}_{i=m_n+m}^{n}.$$

The discrepancy statistics continue to have the same limit distributions and confidence sets may be constructed in the usual fashion.

U–shaped regression functions: Our methodology extends also to U-shaped regression functions. A U-shaped function is a unimodal function turned upside down (we assume a unique minimum for the function). As in the unimodal case, once a consistent estimator of the point at which the regression function attains its minimum has been obtained, the discrepancy statistics for testing the null hypothesis $\mu(x_0) = \eta_0$ can be constructed in a manner similar to the unimodal case. The alterations of the above formulas that need to be made are quite obvious, given that the regression function is now intially decreasing and then increasing. For the sake of conciseness, we have omitted these formulas. The limit distribution of the competing statistics are identical to those in the unimodal case.

Consistent estimation of the mode: It remains to prescribe a consistent estimate of the mode in the unimodal case. Let $\hat{\mu}^{(k)}$ be the LSE of μ based on $\{(Y_{(j)}, X_{(j)}) : j \neq k\}$, assuming that the mode of the regression function is at $X_{(k)}$ (so the least squares criterion is minimized subject to μ increasing on $[0, X_{(k)}]$ and decreasing to the right of $X_{(k)}$) and let $s_{n,k}$ be the corresponding minimum value of the least squares criterion. Then, a consistent estimate of the mode is given by $X_{(k^\star)}$, where $k^\star = \text{argmin}_{1 \leq k \leq n} s_{n,k}$. Our estimate here is similar to that proposed in Shoung and Zhang (2001). An alternative consistent estimate of the mode could be obtained in the same fashion as above, but replacing minimization of the least squares criterion by the maximization of the likelihood function. An estimator of this type in the setting of a unimodal density is given in Bickel and Fan (1996). An analogous prescription applies to a U–shaped regression function.

15.4. Concluding Discussion

In this paper, we have demonstrated how pointwise inference for a shape constrained regression function may be done in a broad class of regression models. The only structural constraint on the regression models lies in the assumption that the conditional distribution of the response given the covariate belongs to a regular exponential family, with the regression function being monotonic/unimodal/U–shaped in the covariate. The formulation is however shown to be fairly general in the sense that many well known regression models of interest, involving both discrete and continuous responses, can be captured in the framework. For unimodal (or U–shaped) functions, so long as inference is restricted to points away from the mode (or the minimizer), the techniques for estimating a monotone regression function can be conveniently adapted. It should be noted that the monotone regression model $Y = \mu(X) + \varepsilon$ with an additive error ε that is independent of X does not exactly fall in this set-up unless the assumption of Gaussianity (on the error) is made, and even under this assumption the error variance must be estimated for inference on μ to be carried out under the allowed shape constraints. However, the results of this paper continue to hold under fairly general error distributions. Consistent estimation of the error variance is readily accomplished and therefore poses no major challenges.

A natural question here is the extent to which the monotonicity assumption is relevant in making pointwise inference on the regression function, since the monotonicity assumption describes the shape of the function globally. We point out that the monotonicity assumption is *crucial* to the inference strategies that we employ: the discrepancy statistics which are functionals of least squares/ maximum likelihood estimates under the monotonicity constraint turn out to be asymptotically pivotal, thereby enabling us to construct pointwise confidence sets *without the need to estimate nuisance parameters or bandwidth parameters*, a difficulty that smoothing techniques have to contend with. This, in our view, is a major statistical advantage. It must be noted, however, that our methods apply only to globally monotone or piecewise monotone functions (unlike smoothing techniques which apply more generally) and should therefore not be employed if such information is unavailable.

There are several natural directions in which the results of this paper can be extended. Firstly, the proposed estimation strategies do not work at a stationary point of the regression function (like the maximizer of a unimodal function, or the minimizer of a U–shaped function). The isotonic estimates computed in the first section are, in fact, not even consistent at these points. Construction of a confidence set for a unimodal regression function, say, at its modal value will require a penalization based likelihood criterion as in Woodroofe and Sun (1993) or in the more recent work of Pal (2006). Yet another problem is the construction of a likelihood based confidence set for the mode itself, and to our knowledge, a nonparametric solution to this remains unavailable as yet. Secondly, this paper deals with a unidimensional covariate but from the perspective of applications one would like to deal with multivariate generalizations. At the very least, it will be important to incorporate auxiliary covariates into the model whose effect can be modelled parametrically. One way to address this issue is to postulate that the conditional mean of the response Y given covariates (X,W), where X is the primary one–dimensional covariate of interest and W is a vector of supplementary covariates is given by $B'(\beta^T W + \psi(X))$, β being a regression parameter and ψ being constrained by the usual shape restrictions. Models like the semilinear regression model or the logistic regression model can be readily seen to belong to this category. In semiparametric models of this type, the maximum likelihood estimates of ψ do not necessarily admit explicit representations as in the current scenario. In such cases, self–induced characterizations are needed, making the asymptotics more difficult to handle. For some work in this direction, see Banerjee et. al. (2006, 2008). A full discussion of these models is well beyond the scope of this paper and is left as a topic for future research. Finally, an extensive and carefully designed simulation study of the proposed methods will be needed before conclusions can be drawn about the relative optimality of any of these pivots with respect to the others.

15.5. Proof of Theorem 1

For the proof, we refer to $\hat{\mu}_n$ and $\hat{\mu}_n^0$ as $\hat{\mu}$ and $\hat{\mu}^0$ respectively, and to $\hat{\psi}_n$ and $\hat{\psi}_n^0$ as $\hat{\psi}$ and $\hat{\psi}^0$ respectively. We first establish (a). It is easy to see that:

$$\begin{aligned}\mathrm{DSS}(\eta_0) &= \sum_{i\in J_n} [(T(Y_{(i)}) - \mu(x_0)) - (\hat{\mu}^0(X_{(i)}) - \mu(x_0))]^2 - \\ &\quad \sum_{i\in J_n} [(T(Y_{(i)}) - \mu(x_0)) - (\hat{\mu}(X_{(i)}) - \mu(x_0))]^2 \\ &= \sum_{i\in J_n} (\hat{\mu}^0(X_{(i)}) - \mu(x_0))^2 - \sum_{i\in J_n} (\hat{\mu}(X_{(i)}) - \mu(x_0))^2 \\ &\quad -2\sum_{i\in J_n} (T(Y_{(i)}) - \mu(x_0))(\hat{\mu}^0(X_{(i)}) - \mu(x_0)) + \\ &\quad 2\sum_{i\in J_n} (T(Y_{(i)}) - \mu(x_0))(\hat{\mu}(X_{(i)}) - \mu(x_0)) \\ &\equiv I_n - II_n - III_n + IV_n\,.\end{aligned}$$

Consider the term III_n. Let $B_1^0,\ldots,B_L^0$ denote the ordered blocks of indices into which J_n decomposes, such that on each block $\hat{\mu}^0$ assumes the constant value c_j^0 and let B_l^0 denote that single block on which $\hat{\mu}^0$ assumes the constant value η_0. Then, for $j\neq l$, $c_j^0 = m_j^{-1}\sum_{i\in B_j^0} T(Y_{(i)})$, where m_j is the cardinality of B_j^0, by Fact 3. We have: $III_n = 2\sum_{j=1}^L \sum_{i\in B_j^0} (T(Y_{(i)}) - \eta_0)(c_j^0 - \eta_0) = 2\sum_{j=1}^L (c_j^0-\eta_0)\sum_{i\in B_j^0}(T(Y_{(i)})-\eta_0) = 2\sum_{j=1}^L m_j(c_j^0-\eta_0)^2 = 2\sum_{j=1}^L\sum_{i\in B_j^0}(\hat{\mu}^0(X_{(i)})-\eta_0)^2 = 2\sum_{i\in J_n}(\hat{\mu}^0(X_{(i)})-\mu(x_0))^2$. Similarly, it follows that $IV_n = 2\sum_{i\in J_n}(\hat{\mu}(X_{(i)})-\mu(x_0))^2$. Consequently:

$$\begin{aligned}\mathrm{DSS}(\eta_0) &= \sum_{i\in J_n} (\hat{\mu}(X_{(i)}) - \mu(x_0))^2 - \sum_{i\in J_n} (\hat{\mu}^0(X_{(i)}) - \mu(x_0))^2 \\ &= n\mathbb{P}_n[(\hat{\mu}(x)-\mu(x_0))^2 1(x\in D_n)] - n\mathbb{P}_n[(\hat{\mu}^0(x)-\mu(x_0))^2 1(x\in D_n)] \\ &= n^{1/(2m+1)}(\mathbb{P}_n - P)[(X_n^2(h) - Y_n^2(h))1(h\in\tilde{D}_n)] \\ &\qquad + n^{1/(2m+1)} P[(X_n^2(h) - Y_n^2(h))1(h\in\tilde{D}_n)] \quad (15.5)\end{aligned}$$

where $\tilde{D}_n = n^{1/(2m+1)}(x-x_0)$, $h = n^{1/(2m+1)}(x-x_0)$ and X_n and Y_n are as defined at the beginning of Section 2. By Fact 2, $\tilde{D}_n$ is eventually contained in a compact set with arbirtarily high (pre-assigned) probability. Also, the monotone (in h) processes X_n and Y_n are eventually bounded on any compact set with arbitrarily high probability, by Fact 1. Using preservation properties for Donsker classes of functions, it is readily concluded that the class of functions $\{h \mapsto (X_n^2(h) - Y_n^2(h))\, 1(h \in \tilde{D}_n)\}$ is eventually contained in a P–Donsker class of functions with arbitrarily high (pre-assigned) probability. It follows that $n^{1/(2m+1)}\,(\mathbb{P}_n - P)\,[(X_n^2(h) - Y_n^2(h))\, 1(h \in \tilde{D}_n)]$ is $o_p(1)$ and we can write: $\mathrm{DSS}(\eta_0) = n^{1/(2m+1)}\, P\,[(X_n^2(h) - Y_n^2(h))\, 1(h \in \tilde{D}_n)] + o_p(1) = \int_{\tilde{D}_n} (X_n^2(h) - Y_n^2(h))\, p_X(x_0 + h\, n^{-1/(2m+1)})\, dh = p_X(x_0) \int_{\tilde{D}_n} (X_n^2(h) - Y_n^2(h))\, dh + o_p(1) \rightarrow_d p_X(x_0) \int [(g_{a,b,m}(h))^2 - (g_{a,b,m}^0(h))^2]\, dh$, where the last step is a consequence of Fact 1. By Lemma 2, it follows that $\mathrm{DSS}(\eta_0) \rightarrow_d a^2\, p_X(x_0)\, \mathbb{D}_m$; but $a^2\, p_X(x_0) = I(\psi(x_0))$ and this finishes the proof.

Note: In going from the first equality to the second in the above paragraph the factor of $n^{1/(2m+1)}$ *does disappear*. This is due to the fact that if X has density $p_X(x)$, the random variable $R \equiv n^{1/(2m+1)}(X - x_0)$ has density $p_R(r) = p_X(x_0 + r\, n^{-1/(2m+1)})\, n^{-1/(2m+1)}$.

We now establish the limit distribution of the likelihood ratio statistic $2\log\lambda_n(\theta_0)$. This will be done by establishing a simple representation for the likelihood ratio statistic in terms of the slope processes $\tilde{X}_n$ and $\tilde{Y}_n$. We write $2\log\lambda_n(\theta_0) = I_n - II_n$ where $I_n = 2\sum_{i \in J_n} \left(\hat{\psi}(X_{(i)}) T(Y_{(i)}) - \hat{\psi}^0(X_{(i)}) T(Y_{(i)})\right)$ and $II_n = 2\sum_{i \in J_n} (B(\hat{\psi}(X_{(i)})) - B(\hat{\psi}^0(X_{(i)})))$. On Taylor expansion of $B(\hat{\psi}(X_{(i)}))$ and $B(\hat{\psi}^0(X_{(i)}))$ around $\psi(x_0)$, up to the third order, we see that up to an $o_p(1)$ term II_n equals:

$$2 \sum_{i \in J_n} B'(\theta_0) \left\{ (\hat{\psi}(X_{(i)}) - \theta_0) - (\hat{\psi}^0(X_{(i)}) - \theta_0) \right\} +$$

$$\sum_{i \in J_n} B''(\theta_0) \left\{ (\hat{\psi}(X_{(i)}) - \theta_0)^2 - (\hat{\psi}^0(X_{(i)}) - \theta_0)^2 \right\} .$$

Denote the second term on the right side of the above display by Q_n. Combining I_n and II_n we obtain:

$$c2\log\lambda_n(\theta_0) = 2\left[\sum_{i \in J_n} ((\hat{\psi}(X_{(i)}) - \theta_0) - (\hat{\psi}^0(X_{(i)}) - \theta_0))\, (T(Y_{(i)}) - B'(\theta_0)) \right] - Q_n + o_p(1).$$

The first term on the right side of the above display is equal to:

$$\begin{aligned} &2 \sum_{i \in J_n} (\hat{\psi}(X_{(i)}) - \theta_0)\, (B'(\hat{\psi}(X_{(i)})) - B'(\theta_0)) \\ &-2 \sum_{i \in J_n} (\hat{\psi}^0(X_{(i)}) - \theta_0)\, (B'(\hat{\psi}^0(X_{(i)})) - B'(\theta_0)). \end{aligned} \tag{15.6}$$

The above is a direct consequence of Fact 3, and can be established by looking at the blocks of indices on which the unconstrained and constrained MLE's $\hat{\psi}$ and $\hat{\psi}^0$ are constant. Note that these are precisely the same blocks on which the unconstrained and constrained least squares estimates are constant. For example, if B_j is a block of indices contained in J_n on which $\hat{\psi}$ attains the constant value a_0 (and $\hat{\mu}$ attains the constant value $B'(a_0)$), then it is readily checked that $\sum_{i\in B_j}(\hat{\psi}(X_{(i)})-\theta_0)(T(Y_{(i)})-B'(\theta_0)) = \sum_{i\in J_n}(\hat{\psi}(X_{(i)})-\theta_0)(B'(\hat{\psi}(X_{(i)})-B'(\theta_0))$ with the common value being given by $m_j(a_0-\theta_0)(B'(a_0)-B'(\theta_0))$, m_j being the cardinality of B_j. A similar argument applies for the constrained MLE $\hat{\psi}^0$. Next, by Taylor expansion of $B'(\hat{\psi}(X_{(i)}))$ and $B'(\hat{\psi}^0(X_{(i)}))$ around $\psi(x_0)$ (up to the second order), (15.6) can be simplified to:

$$2\left[\sum_{i\in J_n} B''(\theta_0)\left((\hat{\psi}(X_{(i)})-\theta_0)^2-(\hat{\psi}^0(X_{(i)})-\theta_0)^2\right)\right]+o_p(1)\equiv 2Q_n+o_p(1).$$

It follows that $2\log\lambda_n = Q_n+o_p(1)$. Recalling that $B''(\theta_0)=I(\psi(x_0))$, we have:

$$\begin{aligned}2\log\lambda_n &= I(\psi(x_0))\,n\,\mathbb{P}_n\left[\{(\hat{\psi}(x)-\theta_0)^2-(\hat{\psi}^0(x)-\theta_0)^2\}\,1(x\in D_n)\right]+o_p(1)\\ &= I(\psi(x_0))\,n^{1/(2m+1)}\,(\mathbb{P}_n-P)\,[(\tilde{X}_n^2(h)-\tilde{Y}_n^2(h))\,1(h\in\tilde{D}_n)]\\ &\qquad +I(\psi(x_0))\,n^{1/(2m+1)}\,P[(\tilde{X}_n^2(h)-\tilde{Y}_n^2(h))\,1(h\in\tilde{D}_n)]+o_p(1)\end{aligned} \tag{15.7}$$

Apart from the constant $I(\psi(x_0))$ and the $o_p(1)$ term, the above display has the exact same form as the representation for $\mathrm{RSS}(\eta_0)$ in (15.5), but with X_n and Y_n replaced by $\tilde{X}_n$ and $\tilde{Y}_n$ respectively. Following (almost) identical steps to those for $\mathrm{RSS}(\eta_0)$ we conclude that $2\log\lambda_n(\theta_0)\to_d I(\psi(x_0))\,p_X(x_0)\,\tilde{a}^2\,\mathbb{D}_m=\mathbb{D}_m$. This finishes the proof of Part (a).

We next establish Part (b). We only establish the asymptotics for $L_{2,w}(\hat{\mu},\hat{\mu}^0)$. The remaining three statistics can be handled by similar arguments. We have, $L_{2,w}(\hat{\mu},\hat{\mu}^0)=\sum_{i=1}^n(\hat{\mu}(X_i)-\hat{\mu}^0(X_i))^2 I(\hat{\psi}(X_i))^{-1}=A_n+B_n$, where $A_n=\sum_{i=1}^n I(\psi(x_0))^{-1}(\hat{\mu}(X_i)-\hat{\mu}^0(X_i))^2$ and $B_n=\sum_{i=1}^n[(\hat{\mu}(X_i)-\hat{\mu}^0(X_i))^2(I(\hat{\psi}(X_i))-I(\psi(x_0)))]\,[I(\hat{\psi}(X_i))\,I(\psi(x_0))]^{-1}$. Next,

$$\begin{aligned}A_n &= \frac{1}{I(\psi(x_0))}\,n\,\mathbb{P}_n\,[((\hat{\mu}(x)-\mu(x_0))-(\hat{\mu}^0(x)-\mu(x_0)))^2\,1(x\in D_n)]\\ &= \frac{1}{I(\psi(x_0))}\,n^{1/(2m+1)}\,(\mathbb{P}_n-P)\,[(X_n(h)-Y_n(h))^2\,1(h\in\tilde{D}_n)]\\ &\qquad\qquad +\frac{1}{I(\psi(x_0))}\,n^{1/(2m+1)}\,P[(X_n(h)-Y_n(h))^2\,1(h\in\tilde{D}_n)]\\ &\equiv A_{1,n}+A_{2,n}\,.\end{aligned}$$

As in Part (a), $A_{1,n}$ is $o_p(1)$ and the contribution in the limit comes from $A_{2,n}$ alone. As in Part (a), we get:

$$\begin{aligned} A_{2,n} &= \frac{1}{I(\psi(x_0))} \int_{\tilde{D}_n} (X_n(h) - Y_n(h))^2 \, p_X(x_0 + h n^{-1/(2m+1)}) \, dh \\ &= \frac{1}{I(\psi(x_0))} p_X(x_0) \int_{\tilde{D}_n} (X_n(h) - Y_n(h))^2 \, dh + o_p(1) \\ &\to_d \frac{p_X(x_0)}{I(\psi(x_0))} \int \left[(g_{a,b,m}(h) - g^0_{a,b,m}(h))^2 \right] dh \\ &\equiv_d \frac{p_X(x_0)}{I(\psi(x_0))} a^2 \, \mathbb{T}_m = \mathbb{T}_m , \end{aligned}$$

where the convergence in distribution is deduced as in Part (a). The equivalence in distribution is a direct consequence of Lemma 2. It only remains to show that B_n converges in probability to 0. We write:

$$B_n \leq \mathbb{P}_n \left[\frac{n^{1/(2m+1)} (X_n(h) - Y_n(h))^2 \mid I(\hat{\psi}(x_0 + hn^{-1/(2m+1)})) - I(\psi(x_0)) \mid 1(h \in \tilde{D}_n)}{I(\hat{\psi}(x_0 + hn^{-1/(2m+1)})) I(\psi(x_0))} \right] .$$

The expression on the right side of the above display can be eventually bounded, with probability larger than any pre–specified amount, by a constant times

$$\mathbb{P}_n \left[n^{-(m-1)/(2m+1)} (X_n(h) - Y_n(h))^2 \mid \tilde{X}_n(h) \mid 1(h \in [-K, K]) \right],$$

for some large $K > 0$. For $m > 1$ this is $o_p(1)$ by virtue of the facts that the random functions X_n, Y_n and $\tilde{X}_n$ are $O_p(1)$ for $h \in [-K, K]$ and that $n^{-(m-1)/(2m+1)}$ converges to 0. For $m = 1$, this last term is identically 1 and the argument proceeds as follows. We decompose the above display as:

$$\begin{aligned} &(\mathbb{P}_n - P) \left((X_n(h) - Y_n(h))^2 \mid \tilde{X}_n(h) \mid 1(h \in [-K, K]) \right) \\ &\qquad + P \left((X_n(h) - Y_n(h))^2 \mid \tilde{X}_n(h) \mid 1(h \in [-K, K]) \right) . \end{aligned}$$

The first term in the above display goes to 0 in probability yet again using standard arguments from empirical process theory. Also, by direct computation, one finds that the second term in the display is $O_p(n^{-1/(2m+1)})$, and hence $o_p(1)$. Hence B_n does not contribute asymptotically to the limit.

The proof of part (c) uses very similar arguments and is skipped. The details are available in Banerjee (2007A). The proof of Part (d) is a direct consequence of Fact 1; look at the limit distribution of $X_n(0)$ and use the Brownian scaling relations from Lemma 1 to arrive at the result. □

15.6. Acknowledgements

I would like to thank Jayanta Pal for some very helpful discussion. The research for this paper was partly supported by NSF grant DMS-0306235.

References

1. Banerjee, M. and Wellner, J. A (2001). Likelihood ratio tests for monotone functions. *Ann. Statist.* **29**, 1699–1731.
2. Banerjee, M., Biswas, P. and Ghosh, D. (2006). A Semiparametric Binary Regression Model Involving Monotonicity Constraints. *Scandinavian Journal of Statistics* **33,** 4, 673–697.
3. Banerjee, M. (2007A). Competing statistics in exponential family regression models under certain shape constraints. *Technical Report*, **442**, University of Michian, Department of Statistics. Available at http://www.stat.lsa.umich.edu/~moulib/unimodal-expfamily.pdf
4. Banerjee, M. (2007B). Likelihood based inference for monotone response models. *Ann. Statist.*, **35**, 3, 931–956.
5. Banerjee, M., Mukherjee, D. and Mishra, S. (2008). Semiparametric binary regression models under shape constraints with an application to Indian schooling data. To appear in *Journal of Econometrics*.
6. Bickel, P. and Fan, J. (1996). Some problems on the estimation of unimodal densities. *Statist. Sinica* **6**, 23 – 45.
7. Brunk, H.D. (1970). Estimation of isotonic regression. *Nonparametric Techniques in Statistical Inference.*, M.L. Puri, ed.
8. Chernoff, H. (1964). Estimation of the mode. *Ann. Statist. Math* **16**, 31–41.
9. Diggle, P., Morris, S. and Morton–Jones, T. (1999). Case–control isotonic regression for investigation of elevation in risk around a risk source. *Statistics in Medicine* **18**, 1605–1613.
10. Grenander, U. (1956). On the theory of mortality measurements II. *Skand. Akt.* **39**, 125-153.
11. Groeneboom, P. and Wellner J.A. (2001). Computing Chernoff's distribution. *Journal of Computational and Graphical Statistics.* **10**, 388-400.
12. Morton–Jones, T., Diggle, P. and Elliott, P. (1999). Investigation of excess environment risk around putative sources: Stone's test with covariate adjustment. *Statistics in Medicine*, **18**, 189 – 197.
13. Pal, J. (2006). End-point estimation of decreasing densities – asymptotic behavior of penalized likelihood ratio. Available at http://www.samsi.info/jpal/publication.html
14. Politis, D.M., Romano, J.P., and Wolf, M. (1999). *Subsampling*, Springer–Verlag, New York.
15. Prakasa Rao, B.L.S. (1969). Estimation of a unimodal density. *Sankhya. Ser. A*, **31** , 23 - 36.
16. Shoung, J-M. and Zhang, C-H. (2001). Least squares estimators of the mode of a unimodal regression function. *Ann. Statist.*, **29**, 648–665.

17. Stone, R. A. (1988). Investigations of excess environmental risks around putative sources: Statistical problems and a proposed test. *Statistics In Medicine,* **7**, 649–660.
18. Van der Vaart, A. and Wellner, J.A. (1996). *Weak Convergence and Empirical Processes.* Springer, New York.
19. Wellner, J. (2003). Gaussian white noise models: some results for monotone functions. *Crossing Boundaries: Statistical Essays in Honor of Jack Hall,* IMS Lecture Notes-Monograph Series, Vol **43** (2003), 87 – 104. J.E. Kolassa and D. Oakes, editors.
20. Woodroofe, M. and Sun, J. (1993). A penalized likelihood estimate of f(0+) when f is nonincreasing. *Statist. Sinica,* **3**, 501-515.

Chapter 16

Study of Optimal Adaptive Rule in Testing Problem

Subir Kumar Bhandari
Ritabrata Dutta
Rajarshi Guha Niyogi

Indian Statistical Institute,
Kolkata 700 108, India
**E-mail: subir@isical.ac.in*

We have studied the optimal allocation between two treatments, in the context of testing the effectiveness of the treatments with the goal to decrease the two types of error rates along with the number of applications of the less effective treatments. In particular we have seen, that number can be made finite for asymptotically large total sample size, in the context of simple hypothesis testing. Also for other types of problems, the cases are studied analytically with the above goal.

Contents

16.1. Introduction

Adaptive allocation in sequential testing turns out to be challenging problem and has drawn the attention of many statisticians, particularly those working in clinical trials, for the last few decades. For them the goal is to mitigate the ethical problem of randomly assigning an inferior treatment to volunteers in clinical trials. The basic idea is to skew allocation probabilities to reflect the response history of patients, hopefully giving a greater than 50% chance of a patient's receiving the treatment performing better in the trials. Many types of allocation rules has been considered so far in the existing literature.

Historically equal allocation has been used because in some cases this invariant fixed sample rule turns out easier to execute computationally, (see Friedman et al. (1981)) whereas adaptive rules come out better not only from ethical point of view, but also advocates of adaptive design argued that unequal allocation may always result in minimal losses or even gains in power (see Rosenberg et al., (2000)) for trials comparing two treatments.

It has been found allocation proportional to the standard deviation of the parameter estimates is optimal in maximising power for fixed sample size. But the optimal rule among the adaptive rules will be very complicated even in the binary response trials.

In fact, it can be shown by simulation and also by theoretical arguements that for small sample number the two competing densities of the two hypotheses in this context can be mixed with another large point mass with very small probabilities in such a way that exact optimal test for small sample numbers will behave very anomalously from context to context so that it becomes extremely cumbersome to describe them analytically.

Adaptive designs for clinical trials with binary responses have largely focussed on urn models. One of the most widely studied design is the Randomised-Play-The-Winner (RPTW) rule (Wei & Durham, 1978). In the RPTW urn model an urn contains balls representing two treatments (say, A & B). When a patient is ready to be randomised, a ball is drawn and replaced. If the ball is of type A(B), the patient is assigned treatment A(B). A success in treatment generates the addition of another ball of the same type to the urn. A failure on the treatment generates the addition of a ball of the opposite type. Let $N_{A,n}$ be the number of patient assigned to treatment A after n patient is assigned and let $N_{B,n} = n - N_{A,n}$. If $p_A = P(Success|\ treatment\ A)$ and $p_B = P(Success|\ treatment\ B)$, $q_A = 1 - p_A$, $q_B = 1 - p_B$, then the limiting allocation $\frac{N_{A,n}}{N_{B,n}}$ is $\frac{q_B}{q_A}$, a measure of relative risk, as $n \longrightarrow \infty$.

It should be noted that RPTW rule is not constructed on the basis of any formal optimality criterion, but the limiting allocation does have some intuitive appeal, i.e, allocation according to the relative risk of treatments.

Recently, Rosenberg et al. (2001) and Zelen (1969) has considered some procedures which minimises the expected number of failures for some apriori fixed asymptotic variance of some difference function of p_A & p_B.

In any of the references discussed so far, it is found that the ratio of $\frac{N_{A,n}}{N_{B,n}}$ tends to a positive constant. As we have considered in this paper that for simple hypothesis testing between any two populations with sequential adaptive allocation with a fixed N number of patients, we can give a procedure with $\frac{N_{A,n}}{N_{B,n}}$ tending to zero. Moreover $N_{A,n}$ can be made finite for any large values of N.

All the new procedures considered in this paper are permutation invariant procedures and their optimality (in whatever sense considered) is among the set of all permutation invariant procedures. The main idea behind each procedure is that for optimal rules the allocation at some stage should be a function of existing likelihood ratio at that moment. In the procedures at each stage we find what should be the optimal decision for the testing problem at that stage. Then we allocate to the population which is preferred or to those ratios which are in our goal, according to the decision at that stage.

With $p_A < p_B$, for our procedures the finiteness or smaller values of $N_{A,n}$ is only due to the fact that for a sequence of efficient (or good) selection procedures, $E(N_{A,n})$ turns out to be

$$E(N_{A,n}) = \sum_n PICS_n$$

which is effectively smaller ($PICS_n$ denoting probability of incorrect selection at stage n).

16.2. Preliminaries and Main Results: Simple Hypothesis Case

Two populations with probability density functions f_{θ_0}, f_{θ_1} (with respect to some σ-finite measure) respectively ($f_{\theta_0} \neq f_{\theta_1}$) are considered. The total number of units that we can choose from the two populations together is N which is fixed and preassigned. We shall draw samples one by one in such a way that the population chosen at some stage depends on the knowledge of the values of the units chosen earlier. This is known to be the adaptive allocation of the units at each stage.

At each stage n, $N_{0,n}$ and $N_{1,n}$ be the respective total number of units drawn from f_{θ_0} and f_{θ_1}, $N_{0,n} + N_{1,n} = n$. We stop when $n = N$ and take decisions as will be required.

Here we deal with two types of problems. In the first case we consider the competing hypotheses are simple i.e, there are two known fixed densities f_0 and f_1 and we are to test

$$\begin{array}{c} H_0 : (f_{\theta_0}, f_{\theta_1}) = (f_0, f_1) \\ Vs. \\ H_1 : (f_{\theta_0}, f_{\theta_1}) = (f_1, f_0) \end{array} \tag{16.1}$$

In this case let us consider

$$LR_n(0,1) = log[L_n(\underline{x}, H_0)/L_n(\underline{x}, H_1)] \tag{16.2}$$

$$LR_n(0,1/N_{0,n},N_{1,n}) = log[L_n(\underline{x}, H_0|N_{0,n},N_{1,n})/L_n(\underline{x}, H_1|N_{0,n},N_{1,n})] \tag{16.3}$$

where $L_n(\underline{x}, H_i)$ is the likelihood for the observations under H_i and $L_n(\underline{x}, H_i|N_{0,n}, N_{1,n})$ is the conditional likelihood for the observations $\underline{x}$ under H_i given sample numbers $N_{0,n}$ and $N_{1,n}$ at the stage n. We define the following procedures for testing the hypothesis mentioned in 16.1.

16.2.1. *Procedure I*

For the testing problem described above, we consider likelihood ratio $LR_n(0,1)$ as defined in (16.2.2) and at each stage n, we adopt the following allocation rule:
if i) $LR_n(0,1) > 0$ we increase $N_{0,n}$ by 1
if i) $LR_n(0,1) < 0$ we increase $N_{1,n}$ by 1
if i) $LR_n(0,1) = 0$ 1 unit observation is allocated to each population with probability $\frac{1}{2}$ each.
Finally when $n = N$, we accept H_0 with probability 1 if $LR_n > 0$ and with probability $\frac{1}{2}$ if $LR_n = 0$.
Practically as we do not know the distributions of $N_{0,n}$ and $N_{1,n}$ and which are found to be very cumbersome, and also for given $N_{0,n}$ and $N_{1,n}$ the conditional distribution of the unit observations in $\underline{x}$ being independent, we found it preferable to use $LR_n(0,1|N_{0,n},N_{1,n})$ as defined in (16.2.3) in place of $LR_n(0,1)$ as in Procedure I. This is used in the following definition of Procedure II.

16.2.2. *Procedure II*

For the same testing problem as in Procedure I, here we use the similar allocation rule and final rejection/acceptance rule as in Procedure I, excepting that we use $LR_n(0,1|N_{0,n},N_{1,n})$ in place of $LR_n(0,1)$. The rationale behind using the conditional likelihoods in the description of Procedure II is due to the fact that it not only makes computations simple, but also gives consistency and very close to optimal results. Now under the assumption of finiteness of the m.g.f of the likelihood ratio, we state the following theorems about the properties of Procedures I & II as defined above.
Theorem 1. In Procedure II, as $N \to \infty$ $E(N_{1,N}) \to$ a finite number N_{II} (say) under H_0 and $E(N_{0,N}) \to$ the same finite number N_{II} under H_1.

Proof: We have, $N_{1,N} = \sum_{i=1}^{N} I_{LR_i<0}$ and $N_{0,N} = \sum_{i=1}^{N} I_{LR_i>0}$.

$$\begin{aligned}
\text{Under } H_0, E(N_{1,N}) &= \sum_{i=1}^{N} E(I_{LR_i<0}) \\
&= \sum_{i=1}^{N} P(LR_i < 0) \\
&\leq \sum_{i=1}^{N} c_1 \exp(-c_2 i) \text{ [by lemma in appendix,with some } c_1, c_2 \geq 0] \\
&= O(\sum_{i=1}^{N} \exp(-c_3 i)) \text{ [with some } c_3 \geq 0] \\
&= O(1)
\end{aligned}$$

Similarly it can be shown $E(N_{0,N}) = O(1)$, Under H_1. □

Theorem 2. In Procedure I, as $N \to \infty$ $E(N_{1,N}) \to$ a finite number N_I (say) under H_0 and $E(N_{1,N}) \to$ the same finite number N_I under H_1

Proof: proof of this theorem uses similar construction as in theorem 1.

Remark 1: Thus for large N, and under $H_0(H_1)$, on the average all but finitely many values are taken from f_{θ_1} population (f_{θ_0} population). The number $N_I$as in Theorem 2 must be less than N_{II} as in Theorem 1, as $N_{II} = \sum_n PICS_n$ (in Procedure II) uses less information than the corresponding values of $PICS_n$ in Procedure I.

16.2.3. *Simulation studies (tables and results)*

Though we have verified theoretically that for the cases of Procedure I & Procedure II the sample number corresponding to the less effective drug tends to finite, as the ASN tends to ∞ and PCS tends to 1, here we simulate the sample number and PCS for various combinations of (p_0, p_1) and observe the same fact. We have simulated the Bernoulli trials with prob. p_0 and p_1 respectively sequentially by generating random numbers and written programs following Procedure I & II to get the results of correct selection as random binary numbers which are used to estimate PCS in the corresponding case. The corresponding sample numbers are noted as n_1 & n_2. The error in the estimate of PCS is made smaller by repeatation in the program. In the same way the error in averaging $n_1 \vee n_2$ and $n_1 \wedge n_2$ have been minimized. The estimates are displayed in Table 1.

The results for the different pairs of (p_0, p_1) in case of bernoulli trials are displayed to be matched with the results shown. We fixed the sample number first

Table 16.1. Estmates for various (p_0, p_1)

pairs	pcs	n	$E(n_1)$	$E(n_2)$	$E(\frac{n_1 \wedge n_2}{n_1 \vee n_2})$	$E(n_1 \vee n_2)$	$E(n_1 \wedge n_2)$
(.8,.2)	0.97	11	5.31	5.69	0.423027167	7.63	3.27
	0.95	16	8.48	7.52	0.251956182	12.78	3.22
	0.96	21	10.99	10.01	0.175153889	17.87	3.13
	0.98	26	12.69	13.31	0.135867191	22.89	3.11
	0.98	31	14.58	16.42	0.119133574	27.7	3.3
	0.95	36	19.43	16.57	0.091901729	32.97	3.03
	0.98	41	22.2	18.8	0.084943107	37.79	3.21
	0.99	46	23.22	22.78	0.072261072	42.9	3.1
	0.96	51	20.56	30.44	0.064940489	47.89	3.11
	0.98	56	30.58	25.42	0.062215478	52.72	3.28
	0.95	61	30.35	30.65	0.053540587	57.9	3.1
	0.98	66	38.31	27.69	0.055831067	62.51	3.49
	0.99	71	30.9	40.1	0.049209399	67.67	3.33
	0.98	76	36.6	39.4	0.044817157	72.74	3.26
	1	81	34.47	46.53	0.040462428	77.57	3.43
	0.98	86	43.76	42.24	0.037895245	82.86	3.14
	0.95	91	48.06	42.94	0.035385141	87.89	3.11
	0.99	96	52.46	43.54	0.035598706	92.7	3.3
	0.99	101	48.55	52.45	0.034306196	97.65	3.35
	0.98	106	54.09	51.09	0.030026237	102.91	3.09

and simulated for each (p_0, p_1) pair, shown in the tables, 1000 times. Each time we have taken H_0 and H_1 to be true with probability 0.5. It is to be noted that the samples assigned to the medicine with lower values of p remains finite with the increase of total sample number. We listed that value in the column of $E(n_1 \wedge n_2)$. The values of PCS, $E(n_1)$, $E(n_2)$ and related values are the estimates obtained with 1000 iterations (as mentioned earlier), each value being very close to the actual mathematical values with small standard errors (as verified). However if the estimates were obtained with averaging over a considerable number, say 10 individual estimates then the precision would have been increased further.

At the beginning of each of the 1000 loops, we started with small, fixed constant equal values of n_1 and n_2 so that the small constant values do not affect the asymptotic properties, or we could have also assume that $LR_0 \equiv 0$.

We have taken only scattered values of (p_0, p_1) pair and verified the results proved in Theorem 1. Our simulation result only support Theorem 1 & also shows the bahaviour of $E(n_1 \vee n_2)$, $E(n_1 \wedge n_2)$ w.r.t PCS & ASN. This shows that convergence of $E(n_1 \wedge n_2)$ to a finite value occurs moderately fast in each case of the (p_0, p_1) values. However verification of the same for Procedure I (i.e for Theorem 2) is extremely complicated and is not practicable due to difficulties in computations of the likelihood functions at each stage. These difficulties have been overcome using conditional likelihood for the verification of Theorem 1.

pairs	**pcs**	n	$E(n_1)$	$E(n_2)$	$E(\frac{n_1 \wedge n_2}{n_1 \vee n_2})$	$E(n_1 \vee n_2)$	$E(n_1 \wedge n_2)$
(.7,.3)	0.87	11	5.65	5.35	0.500682128	7.33	3.67
	0.87	16	8.32	7.68	0.307189542	12.24	3.76
	0.93	21	9.5	11.5	0.297096973	16.19	4.81
	0.92	26	13.12	12.88	0.224105461	21.24	4.76
	0.97	31	16.79	14.21	0.167608286	26.55	4.45
	0.98	36	17.96	18.04	0.135646688	31.7	4.3
	0.97	41	18.67	22.33	0.137624861	36.04	4.96
	0.93	46	23.32	22.68	0.124144673	40.92	5.08
	0.98	51	24.25	26.75	0.106050748	46.11	4.89
	0.97	56	26.46	29.54	0.087801088	51.48	4.52
	0.96	61	31.61	29.39	0.075079309	56.74	4.26
	0.99	66	28.19	37.81	0.068825911	61.75	4.25
	0.94	71	36.65	34.35	0.064946753	66.67	4.33
	0.95	76	39.15	36.85	0.055115924	72.03	3.97
	0.99	81	37.04	43.96	0.063969526	76.13	4.87
	0.95	86	40.64	45.36	0.05083089	81.84	4.16
	0.98	91	41.88	49.82	0.055072464	86.25	4.75
	0.95	96	46.17	49.83	0.053208996	91.15	4.85
	0.95	101	56.32	44.68	0.045548654	96.6	4.4
	0.98	106	50.06	55.04	0.044952681	101.44	4.56

Remark 2: The hypothesis to be tested in the usual problems of clinical trials is being considered in the above context as a special case. It has been checked theoretically through the aforesaid theorems that the no. of applications of the less effective treatments can be made finite, by the procedures we have suggested, in an asymptotic case. In fact by extensive simulation studies in this particular context strongly reinforce our result as the tables shows above. In a nutshell the studies reveal the interesting fact that at bigger size of the sequential sample, following the procedures suggested here, PCS is going upwards and almost reaches 1 as well as attaining finiteness of the no. of applications of the less effective treatment and the expected ratio between n_1, n_2 tending to zero, uniformly over all the values of (p_0, p_1). It is important to note that this has not been attained earlier.
Remark 3: It is important to note that sample no. corresponding to f_{θ_0} or f_{θ_1} whichever we want to make smaller (finite) should be incorporated accordingly in the description of Procedure I and II and Theorems 1 and 2 will automatically be enforced to that context.

pairs	pcs	n	$E(n_1)$	$E(n_2)$	$E(\frac{n_1 \wedge n_2}{n_1 \vee n_2})$	$E(n_1 \vee n_2)$	$E(n_1 \wedge n_2)$
(.6,.4)	0.78	11	5.32	5.68	0.632047478	6.64	4.26
	0.74	16	7.63	8.47	0.516587678	10.63	5.37
	0.72	21	9.56	11.44	0.612903226	13.02	7.98
	0.82	26	12.88	13.12	0.413043478	18.4	7.6
	0.9	31	15.79	15.21	0.301427372	23.82	7.18
	0.9	36	16.03	19.97	0.292639138	27.85	8.15
	0.91	41	19.84	21.16	0.273291925	32.2	8.8
	0.88	46	22.36	23.64	0.275651692	36.06	9.94
	0.88	51	24.25	26.75	0.293103448	39.81	11.19
	0.92	56	30.22	25.78	0.234567901	45.36	10.64
	0.96	61	35.62	31.48	0.232323232	49.5	11.5
	0.89	66	33.59	32.41	0.221995927	54.01	11.99
	0.86	71	35.63	35.37	0.222451791	58.08	12.92
	0.93	76	42.32	33.68	0.183431953	64.22	11.78
	0.94	81	32.41	48.59	0.160292222	69.81	11.19
	0.9	86	45.93	40.07	0.200279135	71.65	14.35
	0.93	91	45.07	45.93	0.18289354	76.93	14.07
	0.95	96	52.61	43.39	0.137845206	84.37	11.63
	0.9	101	57.05	43.95	0.145124717	88.2	12.8
	0.86	106	58.36	47.64	0.150173611	92.16	13.84
	0.91	111	57.05	53.95	0.127475876	98.45	12.55
	0.95	116	68.91	47.09	0.087669948	106.65	9.35
	0.97	121	68.45	52.55	0.10948102	109.06	11.94
	0.96	126	66.02	59.98	0.110523533	113.46	12.54
	0.94	131	59.55	71.45	0.104459995	118.61	12.39
	0.94	136	73.43	62.57	0.103090275	123.29	12.71
	0.97	141	77.77	66.23	0.081039638	130.43	10.57

16.3. Appendix

Theorem 3. Under the existance of the moment generating function of log likelihood ratio, $P(LR_n < 0) \leq c_1 \exp(-c_2 n)$, for some constants $c_1, c_2 \geq 0$.

Proof: $LR_n = \sum_{i=0}^{N_{0,n}} Y_{01i} + \sum_{i=0}^{N_{1,n}} Y_{10i}$, where Y_{01i} are i.i.d loglikelihood vectors for unit sample and Y_{10i} are similar. Let, $\mu_1 = E_{H_0}(Y_{01i}) > 0$, $\mu_2 = E_{H_0}(Y_{10i}) > 0\ \forall i$ and $\mu = \mu_1 \wedge \mu_2$

pairs	pcs	n	$E(n_1)$	$E(n_2)$	$E(\frac{n_1 \wedge n_2}{n_1 \vee n_2})$	$E(n_1 \vee n_2)$	$E(n_1 \wedge n_2)$
(.55,.45)	0.71	11	5.9	5.1	0.729559748	6.36	4.64
	0.64	16	8.25	7.75	0.756311745	9.11	6.89
	0.67	21	11.16	9.84	0.692183723	12.41	8.59
	0.67	26	13.58	12.42	0.610904585	16.14	9.86
	0.73	31	14.2	16.8	0.483253589	20.9	10.1
	0.74	36	19.21	16.79	0.540436457	23.37	12.63
	0.61	41	20.46	20.54	0.616081987	25.37	15.63
	0.77	46	22.06	23.94	0.509186352	30.48	15.52
	0.71	51	25.21	25.79	0.387755102	36.75	14.25
	0.74	56	26.57	29.43	0.459473547	38.36	17.64
	0.87	61	34.49	26.51	0.394285714	43.75	17.25
	0.73	71	36.7	34.3	0.442307692	51.98	19.02
	0.79	76	35.59	40.41	0.365909965	54.37	21.63
	0.82	86	44.1	41.9	0.397829685	64.9	21.1
	0.86	91	45.56	45.44	0.401869159	68.7	22.3
	0.8	96	49.31	46.69	0.325115562	71.43	24.57
	0.84	106	56.61	49.39	0.323251418	76.71	29.39
	0.87	111	53.17	57.83	0.343973121	84.32	26.68
	0.8	116	56.66	59.34	0.2783192	84.9	31.1
	0.82	121	59.16	61.84	0.38363138	93.3	27.7
	0.88	126	65.86	60.14	0.32111402	99.38	26.62
	0.87	131	67.04	63.96	0.36631331	100.48	30.52
	0.9	136	74.19	66.81	0.296474874	108.78	27.22
	0.78	146	76.66	69.34	0.267860737	108.66	37.34
	0.76	151	75.12	75.88	0.303742038	109.98	41.02
	0.81	156	84.09	71.91	0.36765889	125.61	31.39
	0.88	161	91.08	69.92	0.239233609	129.97	31.03
	0.81	166	82.45	83.55	0.343640714	130.99	35.01
	0.83	176	84.43	91.57	0.372976905	137.51	38.49
	0.89	181	95.74	85.26	0.251905947	145.88	35.12
	0.9	186	97.03	88.97	0.238747403	153.65	32.35
	0.84	191	107.8	83.2	0.267272311	153.26	37.74
	0.82	196	93.35	102.65	0.328258506	151.01	44.99
	0.9	201	119.02	81.98	0.278048072	168.46	32.54
	0.89	206	93.89	112.11	0.240745818	165.43	41.57

$$
\begin{aligned}
P(LR_n < 0) &\le \frac{E(\exp(\frac{t}{n}[\sum_{i=0}^{N_{0,n}}(\mu_1 - Y_{01i}) + \sum_{i=0}^{N_{1,n}}(\mu_2 - Y_{10i})]))}{\exp(\mu t)} \text{ [where } t > 0]\\
&\le \frac{(1+k_1\frac{t^2}{n^2})^n}{\exp(\mu t)} \text{[with some } k_1 > 0 \text{ and for all sufficiently small value of } \tfrac{t}{n}]\\
&\le \exp(k_1\frac{t^2}{n} - \mu t)\\
&\le c_1\exp(-c_2 n) \text{[with suitable choice of particular small values of } \tfrac{t}{n}]
\end{aligned}
$$

16.4. Acknowledgement

The authors thank with gratitude annonymous referee of the paper for suggestions and help to revise the paper.

References

1. An-lin, Chenga, Anand N. Vidyashankar, (2006) *Journal of Statistical Planning and Inference* **136**, 1875.
2. Antognini, A.B. (2003) 'Optimal' Randomized Designs for Sequential Experiments with Two Treatments, *Proceedings of the Seventh Young Statisticians Meeting*, Andrej Mrvar (Editor), Metodolo ski zvezki, 21, Ljubljana: FDV.
3. Bai, Z.D. Feifang Hu and William F. Rosenberger, (2002) *The Annals of Statistics* **30**, 122.
4. Bandyopadhyay, U. Atanu, Biswas, (1999) *Journal of Statistical Planning and Inference* **79**, 45.
5. Bandyopadhyay, U. Atanu, Biswas, (2000) *Journal of Statistical Planning and Inference* **83**, 441.
6. Bandyopadhyay, U. Atanu, Biswas, (2004) *Journal of Statistical Planning and Inference* **123**, 207.
7. Berry D.A. & B. Fristedt, Bandit Problems, *Chapman and Hall* (1985).
8. Brannath, W., P. Bauer, M. Posch, *Journal of Statistical Planning and Inference* **136**, 1956 (2006).
9. Chang, M. (2005) Adaptive Design for Clinical Trials, *Invited paper for International Symposium on Applied Stochastic Models and Data Analysis*, France.
10. Coad, D.S. Z. Govindarajulu, (2000) *Journal of Statistical Planning and Inference* **91**, 53.
11. Friedman, L.M., C.D. Furgberg, and D.L. DeMets, (1981) Fundamentals of Clinical Trials. *Boston: Wright PSG.*
12. Haines, L.M., I. Perevozskaya, William F. Rosenberger, (2003) Bayesian Optimal Designs for Phase I Clinical Trials, *Biometrics* **59**, 591.

13. Hardwick, J., R. Oehmke, Quentin F. Quentin Stout, (2006) *Journal of Statistical Planning and Inference* **136**.
14. Hu, F. and William F. Rosenberger, (2003) *Journal of the American Statistical Association* **98**, 463.
15. Louis, T.A. (1977) *Biometrics* **33**, 627.
16. *Mathwork* 2006 [http://www.mathworks.com].
17. Perevozskaya, I., William F. Rosenberger; Linda M. Haines, (2003) *Canad. J. Statist.*, 225.
18. Rosenberger, W.F. (1999) *Controlled Clinical Trials* **20**, 328.
19. Perevozskaya, I., William F. Rosenberger; Linda M. Haines, (2003) *Canad. J. Statist.*, 225.
20. Rosenberger, W. F., Feifang. Hu, (2004) *Clinical Trials* **1**, 141.
21. Vladimir Mats, A.; William F. Rosenberger; Nancy, Flournoy, (1998) Restricted optimality for phase I clinical trials. *New developments and applications in experimental design*, Seattle, WA, 50 (1997), IMS Lecture Notes Monogr. Ser., 34, *Inst. Math. Statist.*, Hayward, CA.
22. Wei, L.J., S. Durham, (1978) *Journal of The American Statistical Association* **73**.
23. William F., Rosenberger; N., Stallard; A., Ivanova; Cherice N. Harper; Michelle L., (2001) *Biometrics* **57**, 909.
24. Zelen, M. (1969) *Journal of the American Statistical Association* **64**, 131.

Chapter 17

The G-IG Analogies and Robust Tests for Inverse Gaussian Scale Parameters

Govind S. Mudholkar[1], Hongyue Wang[1], and Rajeswari Natarajan[2]

[1] *Departments of Statistics and Biostatistics*
University of Rochester, Rochester, NY, 14627, USA
E-mail:mudholkarg5@aol.com
E-mail:cookie@bst.rochester.edu
[2] *Genentech Inc. One DNA Way, South San Francisco CA 94080, USA*
E-mail:raj_nat2003@yahoo.com

The two parameter inverse Gaussian distribution $IG(\mu, \lambda)$ with expectation μ and scale parameter λ is often more appropriate and convenient for modeling and analysis of nonnegative right skewed data as compared to the better known and now ubiquitous Gaussian distribution. Its convenience stems from its analytic simplicity and the striking similarities of its methodologies with those employed with the normal theory models. These similarities, known as the G-IG analogies, include the IG-analogues of the classical one and two sample t-tests, the ANOVA F-test and the basic results involving the estimation and testing of normal variances. The purpose of this article is two-fold: first to review the spectrum of currently known G-IG analogies, and the second to present and examine two robust tests for homogeneity of IG scale parameters, namely, an analogue of Box-Andersen (1955) and a jackknife based test.

Contents

17.1. Introduction

The Inverse Gaussian distribution known as the first passage time distribution of Brownian motion with positive drift, derived by Schrödinger and Smoluchowsky, is now widely used to model positively skewed data. Wald derived it in the context of the distribution of average sample number (ASN) in a sequential probability ratio test. Tweedie observed that the log Laplace transform of the distribution is the inverse of that of Gaussian distribution and named it the inverse Gaussian distribution. Furthermore, Tweedie investigated the basic properties and methodologies associated with the IG distribution and noted their striking similarities to those associated with the Gaussian theory. These G-IG analogies were first highlighted in Folks and Chhikara and tabulated in Iyengar and Patwardhan. Further developments of these appear, e.g., in Mudholkar and Tian, Mudholkar and Natarajan, Natarajan and Mudholkar, Natarajan and Mudholkar and Wang. In addition to the existing procedures and analogies, all these works give compelling reasons for advocating more extensive use of the Inverse Gaussian distribution for analyzing right-skewed data.

Inferences on location and scale parameters serve as guiding paradigms for a broad array of statistical decisions. Although problems involving location parameters are the most basic and common in statistics, the scale parameters can be of intrinsic importance in their own right and problems involving them have played a pivotal role in the development of statistics. The scale parameter of the normal distribution played a central role in the development of such basic ideas as the two-sample t-tests, bias in estimates, and likelihood ratio tests, sufficiency, completeness, and critical regions with Neyman structure. In the second half of the twentieth century Box's analysis of the optimal test for the hypothesis of equality of variances inaugurated the modern study and theory of robust inference.

In the ANORE, the analysis of reciprocals, in the IG-methodology, as in its classical counterpart ANOVA, equality of the scale parameters is a basic assumption and its verification is obviously important. Interestingly, inference procedures concerning the inverse Gaussian scale parameters are also strikingly similar to those concerning the normal variances. Box showed that normal theory inferences about variances are very sensitive to the assumption of normality. Miller studied various approaches with jackknife technique showing promise towards developing a robust solution. A general review of the jackknife methods appears in Miller and Efron. A consideration of the IG-analogue of these ideas is a part of this report.

A current overview of the G-IG analogies appears in Section 17.2. In Section 17.3, analogues of tests due to Bartlett, Cochran and Hartley for testing homogeneity of λ are shown to be non-robust in ways similar to the normal case. Construction of robust methods using the Box-Anderson approach and using analysis of variance on jackknife pseudovalues, similar to Miller's approach is also discussed in Section 17.3. Section 17.4 presents a Monte Carlo study that evaluated the validity and power properties of these tests. Robustness of the procedures is examined for variety of alternatives including lognormal and the recently introduced contaminated Inverse Gaussian distribution, see Mudholkar and Natarajan.

17.2. G-IG Analogies

The main appeal of the inverse Gaussian distribution lies in its analytical simplicity and the striking similarities of its methodologies with those employed with the normal theory models. These G-IG analogies were first compiled by Iyengar and Patwardhan in a table consisting of 14 basic facts including some by Tweedie. However, many more analogies have since been discovered. Some of them were included in Mudholkar and Natarajan; see also Natarajan. We believe that only surface has been scratched on the issue and the facts deserve to be more widely known. In this section we present a brief overview of these analogies and summarize them in Table 2.

1. Nomenclature. Although the IG distribution is originally due to Schrödinger and Smoluchowsky, its name is due to Tweedie who based it on the inverse relationship between the cumulant generating functions of the Gaussian and inverse Gaussian distribution. The probability density function of an IG(μ, λ) random variable X is given by:

$$f_X(x;\mu,\lambda) = \sqrt{\frac{\lambda}{2\pi x^3}} \exp\left\{-\frac{\lambda}{2\mu^2 x}(x-\mu)^2\right\}, \quad x > 0, \mu > 0, \lambda > 0. \qquad (17.1)$$

Noteworthy alternative derivations of the distribution are due to Wald(1947), Halphen (published as Dugue), Huff and Whitemore and Seshadri. Some versions of the distribution are also known as Halphen's laws, see Seshadri's for an accounting of Halphen's laws and the war-time circumstances surrounding their publication under the authorship Dugue.

2. The Basics. Many items of Table 2 may be attributed to Tweedie. He investigated some basic characteristics of the IG distribution, established some important statistical properties and pinpointed certain analogies between the two families of distributions. If $X \sim$ N(μ, σ^2), then $E(X) = \mu$, Var$(X) = \sigma^2$. Analogously, if $X \sim$ IG(μ, λ), then $E(X) = \mu$. However, IG(μ, λ) is not a location-scale family and Var$(X) = \mu^3/\lambda$ is not invariant of the expectation μ, and therefore is not a convenient dispersion measure of the family. However, the parameter λ, with $\lambda^{-1} = E(X^{-1}) - \{E(X)\}^{-1}$, acts as an appropriate dispersion parameter and in many ways, as discussed below and listed in the table, plays the role that σ^2 plays in the normal family. The role of λ, as a dispersion parameter, may be understood by appealing to the AM-GM-HM inequality. λ^{-1} exceeds or equals 0, with equality if and only if X is degenerate.

3. Descriptive Measures and Statistics. Let $X_i, i = 1,\ldots,n$ denote a random sample from either $N(\mu,\sigma^2)$ or $IG(\mu,\lambda)$ population. Then in the normal case the basic statistics $\bar{X} = \sum_{i=1}^{n} X_i/n$ and $S^2 = \sum_{i=1}^{n}(X_i - \bar{X})^2/(n-1)$ are complete and sufficient. Parallely, in the IG case $\bar{X} = \sum_{i=1}^{n} X_i/n$ and $V = \sum_{i=1}^{n}(1/X_i - 1/\bar{X})/n$, which are basic, are also complete and sufficient.

Furthermore in the normal case, the coefficients b_1 of skewness and b_2 of kurtosis play useful role in the basic exploratory data analysis. In this context, Mudholkar and Natarajan defined the concept of IG-symmetry, which depends on a certain moment relationships and proposed the coefficients of IG-skewness δ_1 and IG-kurtosis δ_2, where

$$\delta_1 = \frac{\mu_2'/\mu^2 - \mu\nu}{(\mu\nu - 1)\sqrt{\mu_2'/\mu^2 - 1}}, \tag{17.2}$$

$$\delta_2 = \frac{\eta^2\mu^2}{(\nu\mu - 1)^2} + 1, \tag{17.3}$$

where $\nu = E(1/X)$, $\mu = E(X)$, $\mu_2' = E(X^2)$, $\eta^2 = \mathrm{Var}(1/X) + 2(1-\mu\nu)/\mu^2 + \mathrm{Var}(X)/\mu^4$. These are analogous to the classical coefficients of skewness $\sqrt{\beta_1}$ and kurtosis β_2 (see items 15, 19, and 20), in the sense of that, their sample counterparts agree exactly with sample coefficients b_1 and b_2 in terms of the asymptotic distributions; see items 22 and 23. We note that Mudholkar and Natarajan(2002) also present a (δ_1, δ_2)-chart, as the IG analog of classical (β_1, β_2)-chart (see item 21).

4. Maximum Likelihood Estimators. The MLE's of μ and σ^2 based on a random sample $X_1,\ldots,X_n$ from a $N(\mu,\sigma^2)$ population are respectively, $\hat{\mu} = \sum_{i=1}^{n} X_i/n$ and $\hat{\sigma}^2 = \sum_{i=1}^{n}(X_i - \bar{X})^2/n$. For a sample from $IG(\mu,\lambda)$ population, the MLE's of μ and λ^{-1} are $\hat{\mu} = \sum_{i=1}^{n} X_i/n$ and $V = \sum_{i=1}^{n}(1/X_i - 1/\bar{X})/n$, respectively. The two sets of MLE's are analogous in many ways. In the IG case, the sample mean, for a fixed sample size n, retains the distribution of the parent population with the same μ but with λ replaced by $n\lambda$, that is, $\bar{X} \sim IG(\mu, n\lambda)$. The distribution of V, after being suitably scaled, is a chi-square with $df = n-1$, the same as does the MLE of σ^2 in the normal case; see item 2. Furthermore, $\bar{X}$ and V are independent just as $\bar{X}$ and S^2 are. Tweedie(1941) used this fact to develop significance tests for IG means in a manner analogous to the independence of $\bar{X}$ and S^2 in development of student's t test.

5. Characterizations. The two well known characterizations of the normal distribution have close analogues in the IG context. First, it is well known that $\bar{X}$ and S^2 are independent if and only if the sample is from a normal population. Khatri(1962) showed that the sample statistics $\bar{X}$ and V are independent if and only if the population is Inverse Gaussian; see items 3 and 9. Another important characterization of the two families involves entropy. The entropy $H(f)$ of a random variable X with the probability density function $f(\cdot)$ is defined by:

$$H(f) = E[-\log f(x)] = -\int f(x)\log f(x)dx. \tag{17.4}$$

Shannon, generally recognized as the father of the Information theory, characterized the normal distribution by showing that, among all the distributions with fixed variance σ^2, the normal distribution has the maximum possible entropy. Mudholkar and Tian gave an IG-analogue of this characterization by showing that among all nonnegative, absolutely continuous distributions with a fixed value of λ, where $\lambda^{-1} = E(X^{-1}) - E(X)$, the Root-reciprocal Inverse Gaussian distribution has the maximum entropy; see item 29.

6. Significance Tests. The earliest and most striking analogies between the Gaussian and Inverse Gaussian families relate to the significance tests for the parameters of the two families. The IG-analogues of the classical one-sample and two-sample t-tests for the means and the k-sample ANOVA F-test for the the homogeneity of the means and also the basic variance ratio test for homogeneity of variances, originally due to Tweedie, were highlighted by Folks and Chhikara(1975) in a paper presented to the Royal Statistical Society. It evoked a surprise in the form of a rhetorical question by a discussant; see the discussion part of the paper. Specifically, Tweedie showed that the significance tests of IG means and the homogeneity of k-sample IG means use t-distribution and F distribution as the null distributions respectively; see items 6, 7 and 8. Suppose $X_i, i = 1, \ldots, n$ is a random sample from IG(μ, λ). The test statistic for $H_0 : \mu = \mu_0$ is $T = \sqrt{n}(\bar{X} - \mu_0)/\mu_0\sqrt{\bar{X}V}$, where $V = \sum_{i=1}^{n}(X_i^{-1} - 1/\bar{X})$. Given significance level α, the acceptance region for H_0, in this case, is $T \leq c = \sqrt{n/(n-1)}\,t_{1-\alpha/2}$, where $t_{1-\alpha/2}$ is the $(100(1-\alpha/2))$th percentile of the t distribution with $n-1$ degrees of freedom. Similarly, for k-sample case, the test statistic for $H_0 : \mu_1 = \ldots = \mu_k$ is

$$T = \frac{\sum_{i=1}^{k}\sum_{j=1}^{n_i}\left(x_{i\cdot}{}^{-1} - x_{\cdot\cdot}{}^{-1}\right)/(k-1)}{\sum_{i=1}^{k}\sum_{j=1}^{n_i}\left(x_{ij}{}^{-1} - x_{\cdot}{}^{-1}\right)/(\sum n_i - k)} \tag{17.5}$$

which has an F distributions with degrees of freedom $k-1$ and $\sum_{i=1}^{k} n_i - k$. Subsequently, Chhikara(1989) established the UMP unbiased character of the one sample and two sample t-test for the IG means. Furthermore, it is interesting to note that the IG-analogue of the ANOVA decomposition of sums of squares in the IG case involves the reciprocals. Hence it is named as Analysis of Reciprocals, abbreviated ANORE.

7. Goodness-of-fit Test Lin and Mudholkar(1980) developed the Z-test of the composite hypothesis of normality using the independence of the mean and variance of normal samples. It is based on the Fisher transformation of the correlation coefficient between sample means and the cube roots of the corresponding sample variances, obtained by deleting one observation at a time. Mudholkar, Natarajan and Chaubey(2001) developed an analogous goodness-of-fit test for the IG models. The asymptotic null distributions of the test statistics for the two goodness-of-fit tests are exactly the same, namely $N(0,3/n)$; see items 12, 13, and 14.

Vasicek was the first to use a characterization result explicitly to develop a goodness-of-fit test. He used a non-parametric estimator of the entropy and Shannon's characterization result to construct a test for the composite hypothesis of normality. An analogous entropy-based goodness-of-fit test is developed in Mudholkar and Tian; see item 30. It is interesting to note that the maximum possible values of entropies in the two characterization results are remarkably analogous and the null distributions of the two statistics have interesting similarity.

There is another interesting fact associated with the two entropy tests. Vasicek found the null distribution of the entropy test of normality analytically intractable as it remains so to date. Hence, he presented a Monte Carlo tabulation of the critical values of the test. The asymptotic distribution of the entropy test for the IG hypothesis is also intractable. Hence, Mudholkar and Tian also gave a Monte Carlo tabulation of the IG entropy test. They noted that the critical values for the normal and IG test were very close, but not exactly the same. A closer look at Vasicek's work showed that he had used the average of 12 pseudo-uniforms to generate normal deviates for his Monte Carlo experience. Mudholkar and Tian redid the simulation for the Vasicek's test of normality using the state-of-art normal generator and used various statistical procedures to show that the null distribution of the two entropy tests are exactly the same.

Mudholkar, Marchetti and Lin constructed a new test statistic denoted by Z_3 using the characteristic independence of the mean and the third central moment of normal samples. They find that the Z_3 and the Lin and Mudholkar test statistic can be shown to be asymptotically normal and mutually independent in even small samples, $n \geq 5$, and may be combined effectively to test for normality against restricted or unrestricted skewness and kurtosis alternatives. In the same spirit, another class of goodness-of-fit test for the IG models is presented by Natarajan and Mudholkar. They gave a systematic development of tests based on the IG analogs d_1 and d_2 of b_1 and b_2. The development closely parallels the skewness and kurtosis based tests of normality due to Bowman and Shenton and others usefully presented by D'Agostino and Stephens in their monograph on the goodness-of-fit tests.

8. Bayesian Inference In the Bayesian framework for parametric statistical analysis, the parameter is considered random about which the available priori information is formalized in terms of the prior distribution. The prior is thus to be elicited and modeled using the prior belief. Yet the analytical notions such as conjugate prior, Jeffrey's prior, noninformative and maximum entropy priors have played important theoretical roles. Thus for the normal case, with the variance known, the natural conjugate prior for the mean is normal. For the IG family, when λ is known, the natural conjugate for the standard parametrization IG(μ, λ) does not exist (Palmer, 1973). However, if reparameterized by $\delta = 1/\mu$, the natural conjugate prior for δ is also the normal family; see item 11.

9. Order Restricted Inference. The order restricted inference in normal theory began with the work of Chernoff. A systematic investigation of the problem following a major advance by Bartholomew which led to the monograph by Barlow et al. Many of the developments in these work rely on the likelihood methods. The implementation involves null distribution, which varies with the order constraint from sample size to sample size. Yet the problem and the solution have great practical significance since tests for homogeneity of means with order restriction are essentially generalizations of the concept of one-tail or two-tail tests.

Mudholkar and McDermott, McDermott and Mudholkar, and Mudholkar, McDermott and Aumont gave simple approaches to conducting order restricted inference in the classical normal theory framework; see items 32 and 33. These approaches involve classical methods for combining independent tests and use of simple easily available null distributions such as student's t and F distribution. Mudholkar and Tian and Tian and Mudholkar gave simple solutions to the IG-analogs of the problem involving order-restricted IG means, whereas Natarajan, Mudholkar and McDermott gave the solutions of homogeneity of order restricted IG dispersion parameter λ's.

10. Extreme Value Distributions. Extreme value distributions are theoretically attractive because of their historical background and mathematical elegance. They have also applied importance in view of their role in modeling entities such as breaking strength materials in engineering and floods in hydrology.

Freimer et al offered a simple approach based on expansions of the quantile function of a probability distribution to obtain the associated distributions of the extremes of a random sample from it. They illustrated the method using the generalized Tukey's Lamda and the Weibull families. The method was further used by Mudholkar and Kollia and Mudholkar and Hutson to derive the extreme value distributions associated with the generalized Weibull family and the exponentiated Weibull family respectively. The method permits the derivation of not only the extreme value distributions, but also to obtain the asymptotic distribution of the extreme spacings, which have the tail length connotations. The basic idea is originally due to Shuster who considered only convergence in probability of the extreme spacings to refine the extreme value distribution classification as ES-long,

ES-medium and ES-short. Freimer et al, on the other hand, considered convergence in law of the extreme spacings and showed that if there exist constants a_n and b_n such that $a_n X_{n:n} + b_n$, where $X_{n:n}$ denotes the maximum of the sample, has an asymptotic distribution, then the tail of the distribution as measured by $X_{n:n} - X_{n-1:n}$ is $O_p(1/a_n)$.

Freimer et al showed that the extreme tail of a normal sample satisfies $X_n - X_{n-1} = O_p\left((\log n)^{-1/2}\right)$. More recently, Mudholkar and Tian (2002) have showed that if a random sample is from RRIG population, then analogously, the extreme value distribution and extreme spacings are the same with the normal case. Furthermore, the tail length of the RRIG random variable is also of the same high probability of magnitude as in the normal case; see items 34 and 35.

11. Unimodality and Related Ideas. Items 36-40 in Table 2 summarizes the most recent theoretical results related to unimodality and R-symmetry; see Mudholkar and Wang(2007).

Corresponding to the classical symmetry, a new concept of R-symmetry for non-negative right-skewed families is defined. A random variable $X \geq 0$ with p.d.f. $f(\cdot)$ is R-symmetric about the R-center θ if X/θ is reciprocally symmetric about 1, or equivalently if $f_X(\theta/x) = f_X(\theta \cdot x)$. Several examples of R-symmetric distributions are given, which include some popular distributions such as lognormal and RRIG distributions. R-symmetry is a more physically transparent, conceptually clear notion than IG-symmetry and can be used to explain the IG-symmetry. It is shown that If X R-symmetric about θ, then $Y = 1/X^2$ is IG-symmetric.

It is well known that any symmetric unimodal distribution is a mixture of symmetric uniform distributions. Similarly, it can be shown that all R-symmetric unimodal distributions form a convex cone generated by R-symmetric uniform distributions. Wintner proved that the convolution of symmetric unimodal distributions is symmetric unimodal. Analogously, it is shown that the product convolution of R-symmetric unimodal distributions is R-symmetric unimodal.

According to the concept of peakedness defined by Birnbaum, the normal populations with fixed common mean μ and standard deviation σ have a peakedness ordering with respect to σ. Mudholkar and Wang(2007) gives a parallel concept of R-peakedness and show that RRIG random variables with parameter 1 and λ are R-peaked ordered with respect to the dispersion parameter λ.

It is well known that the power functions of the UMP unbiased Z and student's t tests for the hypotheses involving normal means are monotone increasing functions of $\mu - \mu_0$, where μ and μ_0 are the population means under alternative and null hypothesis respectively. Utilizing the properties of R-symmetric distributions, it can be shown that the power functions of the UMP unbiased Z and t tests for the means of IG populations are monotone increasing functions of the non-centrality parameters.

17.3. Inferences Regarding Inverse Gaussian Scale Parameters

It is well known that the F test for equality of two normal variances, as well as the multi-sample tests of homogeneity of several normal variances due to Bartlett, Cochran and Hartley, depend heavily on the normality assumption and are strongly non-robust in the sense that they are even asymptotically inaccurate if the normality assumption fails. The tests due to Box-Andersen and those based on Jackknife method are robust alternatives for the problem. In this section we present the IG-analogues of these results in the context of the tests involving IG dispersion parameter λ's. Specifically, these tests are shown to be nonrobust, and robust procedures based on Box-Andersen approach and Jackknife methodology are developed.

Let $X_{ij}, j = 1,2,\ldots,n_i$ be random samples of size n_i from inverse Gaussian populations $IG(\mu_i, \lambda_i), i = 1,2,\ldots,k$. Let $\Sigma_i n_i = N$. We consider the problem of testing the hypothesis $H_0 : \lambda_1 = \lambda_2 = \ldots = \lambda_k$, or equivalently, $H_0 : {\lambda_1}^{-1} = {\lambda_2}^{-1} = \ldots = {\lambda_k}^{-1}$ against the alternative that the λ_i are not all equal. The maximum likelihood estimates of the λ_{i-1} are given by $V_i = \Sigma_{j=1}^{n}(1/X_{ij} - 1/\bar{X}_i)/(n_i - 1), i = 1,2,\ldots,k$. Also, $(n_i - 1)\lambda_i V_i \sim \chi^2_{n_i-1}$. Let $\omega_i = n_i - 1$ and let $V = \Sigma_i \omega_i V_i/\omega$, where $\omega = \Sigma_i \omega_i$. The IG analogues for testing homogeneity of Inverse Gaussian scale parameters corresponding to the three classical statistics for testing homogeneity of normal variances are due to Bartlett, Cochran and Hartley, respectively; the null percentiles of these statistics may be found in the Biometrika Tables for Statisticians. These are respectively:

$$M_{IG} = \omega \log V - \Sigma_{i=1}^{k} \omega_i \log V_i, \tag{17.6}$$

$$G_{IG} = \frac{\max_{1 \le i \le k} V_i}{\Sigma_{i=1}^{k} V_i}, \tag{17.7}$$

$$F_{IG} = \frac{\max_{1 \le i \le k} V_i}{\min_{1 \le i \le k} V_i}. \tag{17.8}$$

These can be unified as test statistic $T_{r,s}$:

$$T_{r,s} = \frac{\{\Sigma_{i=1}^{k} \frac{\omega_i}{\omega}(V_i)^r\}^{1/r}}{\{\Sigma_{i=1}^{k} \frac{\omega_i}{\omega}(V_i)^s\}^{1/s}}, \tag{17.9}$$

where (17.3.6)-(17.3.8) correspond to $\log T_{1,0}$, $T_{\infty,-\infty}$, $T_{\infty,1}$, respectively.

Following an argument similar to that of Bartlett (1937), it can be shown that for moderate size samples from Inverse Gaussian populations M_{IG} is approximately $(1 + A)\chi^2_{k-1}$, where A is given by $A = \Sigma_i(1/\omega_i - 1/\omega)/3(k-1)$. Thus, as in

Box (1953), each of the modified statistics given in equations (17.3.6)-(17.3.8) is nonrobust. To see this, note that the asymptotic distribution of V_i from a single population with the first two positive and first two negative moments finite, see Mudholkar and Natarajan(2002), is given by

$$\sqrt{n}(V_i - [\nu_i - (1/\mu_i)]) \rightarrow_d N(0, \eta_i^2), \tag{17.10}$$

where

$$\nu_i = E(X_{ij}^{-1}), \mu_i = E(X_{ij}), \ \ \tau_i^2 = Var(X_{ij}^{-1}), \ \ \sigma_i^2 = Var(X_{ij}),$$

$$\eta_i^2 = \tau_i^2 + 2(1 - \mu_i \nu_i)/\mu_i^2 + \sigma_i^2/\mu_i^4.$$

Consequently, the logarithmic transformation yields

$$\sqrt{n}(\log V_i - \log[\nu_i - (1/\mu_i)]) \rightarrow_d N\left(0, \frac{\eta_i^2 \mu_i^2}{(\nu_i \mu_i - 1)^2}\right), \tag{17.11}$$

or $N(0, \delta_{2i} - 1)$, which reduces to $N(0,2)$ under the IG assumption.
If the coefficient of IG-kurtosis δ_{2i}, the IG analogue of the classical coefficient of kurtosis β_{2i}, as introduced in Mudholkar and Natarajan(2002) can be assumed to be equal to δ_2 across the k populations, then M_{IG} can be shown to be asymptotically $(1 + 0.5(\delta_2 - 3))\chi_{k-1}^2$ for any parent population having finite first two positive and negative moments. Thus, the Type I error control of M_{IG} is not even asymptotically controlled if δ_2 is unequal to three. In particular, in large samples the asymptotic mean of M_{IG} would be $(1+0.5(\delta_2 - 3))(k-1)$ instead of $k-1$ and asymptotic standard deviation $(1+0.5(\delta_2 - 3))\sqrt{2(k-1)}$ instead of $\sqrt{2(k-1)}$. Hence, as in the normal case the asymptotic sampling distribution of V_i may be used in constructing the analogue of the robust Box-Andersen test. The analogue of the well-known jackknife test of Miller for homogeneity of the variances of normal populations will also be described.

Analogue of the two-sample Box-Andersen test. The Box-Andersen test for equality of σ^2 adjusts the degrees of freedom so that the asymptotic variance of the F test statistic under normal theory matches the asymptotic variance of F under sampling distribution with general kurtosis. Thus, for developing the analogue of the Box-Andersen test for testing equality of the scale parameters of two Inverse Gaussian populations, we adjust the degrees of freedom of the F-statistic based on the asymptotic distribution of V (see Mudholkar and Natarajan). Toward this end we use the fact that, as $n \rightarrow \infty$:

$$\sqrt{\frac{n}{0.5(\delta_2 - 1)}}(V - (1/\lambda)) \rightarrow_d N\left(0, \frac{2}{\lambda^2}\right), \tag{17.12}$$

where $1/\lambda = (\nu\mu - 1)/\mu$. Also,

$$F_{IG} = \frac{V_2}{V_1} \sim F[\delta(n_2 - 1), \delta(n_1 - 1)], \tag{17.13}$$

under H_0, where $\delta = [0.5(\delta_2 - 1)]^{-1}$. As in the normal case, the sample *IG*-kurtosis

$$d_2 = \frac{m_2'/\bar{X}^2 + m'_{-2}\bar{X}^2 - 3\bar{X}^2\bar{Y}^2 + 2\bar{X}\bar{Y} - 1}{(\bar{X}\bar{Y} - 1)^2} + 3. \tag{17.14}$$

based on corresponding sample moments, i.e., $m_{-2} = \Sigma_{i=1}^n (1/X_i)^2/n$, $\bar{Y} = \Sigma_{i=1}^n (1/X_i)/n$ and $m_2' = \Sigma_{i=1}^n X_i^2/n$ can be used in place of δ_2 above when conducting the test. For two samples, d_2 can be estimated by weighted sample IG-kurtosis for each sample.

Multi-sample Jackknife test. We consider the relevant parameter to be $\log \lambda_i^{-1}$ since, as in the case of normal scale parameters, the logarithmic transformation is the "the pivotal quantity" as stated in Cressie stabilized the asymptotic variance. Then, the 'delete-one' jackknife yields the following pseudovalues:

$$W_{ij} = n_i \log V_i - (n_i - 1) \log V_{i(-j)}), \tag{17.15}$$

where $V_{i(-j)}$ is the sample estimate of λ^{-1} from group i with observation j deleted, $i = 1, 2, \ldots, k$, $j = 1, 2, \ldots, n_i$. As can be seen from the jackknife theory (e.g. see Thorburn [57, 58]), the asymptotic distribution of $\bar{W}_i$, the mean of the pseudo values in group i, is asymptotically normal with mean $\log \lambda_i^{-1}$ and constant variance $\delta_2 - 1, i = 1, 2, \ldots, k$, $j = 1, 2, \ldots, n_i$, based on the approximate independence of the pseudovalues. Hence, to conduct a robust test, one may perform analysis of variance on these pseudovalues.

Remark 3.1. The analogue of Levene's test in the IG case would be to consider $U_i = 1/X_{1i} - 1/\bar{X}_1$ and $W_i = 1/X_{2i} - 1/\bar{X}_2$ as approximately i.i.d. observations instead of the original $(X_{1i}, X_{2i}), i = 1, 2, \ldots, n_j$, $j = 1, 2$. We did consider the analogue of Levene's test but it performed poorly in our empirical study in terms of Type I error control. This may be because the correlations among the U_i and W_i are not small enough to be ignored, unlike in the original test.

17.4. A Monte Carlo Study

In this section we present a small Monte Carlo experiment performed in order to understand the operating characteristics of the robust tests proposed in section 3 in the context of two populations. In the robustness literature involving normal variances, symmetric populations have been used as the alternatives to the normal

Table 17.1. Monte Carlo Power* Function of Competing Tests for $H_0 : \lambda_1 = \lambda_2$ vs. $H_1 : \lambda_1 \neq \lambda_2$

$n = 25$								
	$\alpha = 0.05$				$\alpha = 0.01$			
$\Delta = \lambda_2/\lambda_1$	1	3	5	10	1	3	5	10
Inverse Gaussian Distribution								
F-test	0.057	0.755	0.970	0.999	0.013	0.522	0.897	0.998
Box-Anderson	0.051	0.678	0.935	0.997	0.008	0.361	0.737	0.960
Jackknife	0.054	0.704	0.950	0.998	0.015	0.476	0.847	0.991
Contaminated IG Distribution: $IG(1,1) + 20\%IG(1,10)$								
F-test	0.089	0.730	0.953	0.998	0.024	0.519	0.870	0.998
Box-Anderson	0.053	0.583	0.874	0.987	0.010	0.281	0.615	0.987
Jackknife	0.059	0.618	0.899	0.994	0.017	0.389	0.755	0.994
Lognormal Distribution								
F-test	0.126	0.712	0.934	0.997	0.042	0.524	0.844	0.989
Box-Anderson	0.047	0.585	0.893	0.995	0.075	0.293	0.687	0.963
Jackknife	0.068	0.572	0.838	0.968	0.019	0.366	0.679	0.920
$n = 10$								
Inverse Gaussian Distribution								
F-test	0.050	0.335	0.624	0.902	0.010	0.129	0.347	0.730
Box-Anderson	0.062	0.299	0.530	0.797	0.012	0.097	0.228	0.487
Jackknife	0.066	0.326	0.575	0.843	0.022	0.167	0.365	0.673
Contaminated IG Distribution: $IG(1,1) + 20\%IG(1,10)$								
F-test	0.082	0.355	0.611	0.875	0.021	0.161	0.366	0.707
Box-Anderson	0.061	0.243	0.436	0.698	0.011	0.074	0.167	0.375
Jackknife	0.071	0.282	0.492	0.758	0.027	0.143	0.298	0.567
Lognormal Distribution								
F-test	0.102	0.373	0.601	0.858	0.029	0.183	0.378	0.693
Box-Anderson	0.055	0.252	0.496	0.826	0.010	0.074	0.205	0.535
Jackknife	0.079	0.294	0.492	0.744	0.027	0.152	0.303	0.563

* Based on 50,000 replications for each entry, $SE \leq 0.002$

populations. In an analogous manner, in this experiment we use IG-symmetric populations such as the inverse Gaussian, lognormal, and contaminated Inverse Gaussian, a special case of the IG scale-mixtures introduced in Mudholkar and Natarajan. The density functions of these populations appear similar to those of the Inverse Gaussian densities.

Fifty thousand pairs of samples each of size n for several values of n were generated from the types of above populations. The Inverse Gaussian samples were generated using the algorithm by Michael et al and the Bernoulli random variables were used to generate the mixtures. The IMSL C subroutine library was used to obtain the lognormal samples. For the case of two populations, the tests due to Bartlett, Cochran and Hartley coincide with the classical F-test. Hence, the IG analogue of the F-test, the jackknife test and the Box-Anderson test, as presented in section 3, were applied to each of the samples described above. A selection of the percentages of samples in 50,000 trials for which the tests rejected the null hypothesis $H_0 : \lambda_1 = \lambda_2$ where $\lambda^{-1} = E(1/X) - 1/E(X)$, for various distributions and combinations of $\Delta = \lambda_2/\lambda_1$ and α, the level of significance, is presented in Table 1.

We considered $\Delta = 1, 3, 5, 10$, where $\Delta = 1$ corresponds to the null hypothesis for which the percentages should be close to α. For $\Delta > 1$, the percentages are Monte Carlo estimates of the powers of the tests at particular alternatives. The following are some of the observations gleaned from the results of the simulation study:

IG analogue of F-test: As in the normal case, the IG analogue of F-test is non-robust in the sense that its type I error is highly sensitive to departures from the *IG* assumptions. It may be noted that the IG populations and the lognormal population are hard to distinguish from each other. When the population is lognormal, the type I error of the IG analogue of the classical F-test with 5% and 1% level of significance are 12.6% and 4.2% respectively for sample size 25. Similar discrepancies is also noted with sample size 10.

The jackknife test: Although the test seems to offer better type I error control than the analogue of F-test, its type I error is larger than the level of significance and may be termed as anti-conservative. We note that the theoretical justification of the jackknife approach is in terms of asymptotics and this is reflected in the fact that the type I error control of the jackknife test improve with the increasing sample size.

Analogue of Box-Andersen Test: This test appears to offer the best type I error control in general and reasonable power properties among the tests compared in our empirical study.

In conclusion, the analogue of Bartlett, Cochran and Hartley tests are hypersensitive to the inverse Gaussian assumption in a manner similar to their normal counterparts. Interestingly, the analogue of Box-Anderson's test provides a reasonably robust alternative to Bartlett, Cochran and Hartley's analogues for testing homogeneity of IG scale parameters. The jackknife test albeit anti-conservative seems to have adequate Type I error control for samples of sizes at least as large as 25 and also has satisfactory power. It is expected to offer a reasonable tool for developing tests for homogeneity of scale parameters of several inverse Gaussian populations.

References

1. Bartholomew. D.J. (1959) A test of homogeneity for ordered alternatives, *Biometrika*, **46**, 36-48.
2. Bartholomew. D.J.(1959) A test of homogeneity for ordered alternatives II, *Biometrika*, **46**, 328-335.
3. Barlow, R.E., D. J. Bartholomew, J. M. Bremner and H. D. Brunk. (1972) *Statistical Inference Under Order Restrictions* Wiley, London.
4. Bartlett, M.S. (1937). Properties of sufficiency and statistical tests, *Proceedings in Royal Statistical Society*, A, **160**, 268-282.
5. Birnbaum. Z.W. (1948). On random variables with comparable peakedness, *Annals of Mathematical Statistics*, **19**,76-81.
6. (1966). *Biometrika Tables for Statisticians* , Volume I. Edited by Pearson and Hartley, Third Edition, Cambridge University Press, London.
7. Bowman, K.O. and L. R. Shenton. (1975). Omnibus test contours for departures from normality based on $\sqrt{b_1}$ and b_2, *Biometrika*, **62**, 243-250.
8. Box, G. E. P.(1953). Non normality and tests on variances, *Biometrika*, **40**, 318-335.
9. Box. G. E. P. and S. L. Andersen. (1955). Permutation theory in the derivation of robust criteria and the study of departures from assumption, *Journal of the Royal Statistical Soceity, Series B*, **17**, 1-26.
10. Chhikara, R.S. (1975). Optimum tests for comparison of two inverse gaussian distribution means, *Australian Journal of Statistics*, **17 (2)**, 77-38.
11. Chhikara R.S. and J. L. Folks. (1989). The inverse Gaussian distribution, theory, methodology and applications, (Marcel Dekker Inc., New York).
12. Chernoff, H. (1954). Testing homogeneity against ordered alternative, *Annals of Mathematical Statistics*, **34**, 945-956.
13. Cochran, W.G. (1941). The distribution of the largest of a set of estimated variances as a fraction of their total, ıThe Annals of Eugenics, **11**, 47-52.
14. Cressie. N. (1981). Transformations and the Jackknife, *Journal of the Royal Statistical Society, Series B*, **43**, 177-182.
15. D'Agostino, R.B. and M. A. stephens. (1986). *Goodness-of-Fit Techniques* Dekker, New York.
16. Dugué, D. (1941). Sur un nouveau type de courbe de fré, *Comptes Rendus de l'Academie des Sciences*,Paris, Tome 213, 634-635.
17. Efron, B. (1982). The jackknife, the bootstrap and other resampling plans, *CBMS Regional Conference Series in Applied Mathematics*, **38**.
18. Freimer. M., G. Kollia, G. S. Mudholkar and C. T. Lin. (1989). Extremes, extreme spacings and outliers in the Tukey and Weibull families, *Communications in Statistics: Theory and Methods*, **18**, 4261-4274.
19. Folks, J.L. and R. S. Chhikara. (1978) The inverse Gaussian distribution and its statistical applications – A Review, *Journal of Royal Statistical Society*, B, **40**, 263-289.
20. Hartley, H.O. (1939). Testing homogeneity of a set of variances, *Biometrika*, 31, 249-255 .
21. Huff, B.W. (1975). The inverse Gaussian distribution and Root's barrier construction, *Sankhyā*, 37A, 345-353.
22. Iyengar, S. and G. Patwardhan. (1988.) Recent developments in the inverse Gaussian

distribution, *Handbooks of Statistics*, 7, (eds. P.R. Krishnaiah and C. R. Rao), 479-490, Elsevier, North-Holland, Amsterdam.

23. Khatri, C.G. (1962). A characterization of the inverse Gaussian distribution, *Annals of Mathematical statistics*, 800-803.
24. Levene. H. (1960). Robust tests for equality of variances, *Contributions to probability and Statistics*, 278-292, (Stanford University Press, Stanford, CA).
25. Lin, C.C. and G. S. Mudholkar. (1980) A simple test for normality against asymmetric alternatives, *Biometrika*, **67**, 455-461.
26. McDermott. M.P. and G. S. Mudholkar. (1993). A simple approach to testing homogeneity of order-constrained means, *Journal of the American Statistical Association*, **88**, 1371-1379.
27. Michael. J.R., W. R. Schucany and R. W. Haas. (1976). Generating random variables using transformation with multiple roots, *American Statistician*, **30**, 88-90.
28. Miller, R.G. (1968). Jackknifing variances, *The Annals of Mathematical Statistics*, 39, 567-582.
29. Miller,R.G. (1974). The Jacknife - A Review, *Biometrika*, **61**, 1-15.
30. Mudholkar. G.S. and M. P. McDermott, (1989). A class of tests for equality of ordered means, *Biometrika*, **76**, 161-168.
31. Mudholkar G.S., M. P. McDermott and J. Aumont. (1993). Testing homogeneity of ordered variances, *Metrika*, **40**, 271-281.
32. Mudholkar. G.S. and G. Kollia. (1994). Generalized Weibull family: A structural analysis, *Communications in Statistics: Theory and Methods*, **23**, 1149-1171.
33. Mudholkar. G.S. and A. Hutson. (2000). The epsilon-skew-normal distribution for analyzing near-normal data, *J. Statist. Plann. Inference*, **83**, 291-309.
34. Mudholkar, G.S., R. Natarajan and Y. P. Chaubey. (2001). A goodness-of-fit test for the inverse Gaussian distribution using its independence characterization, *Sankhyā*, Series B, **63**, 362-374.
35. Mudholkar, G.S. and L. Tian. (2002). An entropy characterization of the inverse Gaussian distribution and related goodness-of-fit test, *Annals of the Institute of Statistical Mathmatics*, 102(2), 211-221.
36. Mudholkar, G.S. and L. Tian. (2002). On the null distribution of entropy tests for the Gaussian and inverse Gaussian models, *Communications in Statistics, Theory and Methods*, **30**, 1507-1520.
37. Mudholkar, G.S. and R. Natarajan. (2002). The Inverse Gaussian models: Analogues of symmetry, skewness and kurtosis, *Annals of the Institute of Statistical Mathematics*, **54**,138-154.
38. Mudholkar, G.S., C. E. Marchetti, and C. T. Lin.(2002). Independence characterizations and testing normality against restricted skewness-kurtosis alternatives, *Journal of Statistical Planning and Inference*, **104**, 485-501.
39. Mudholkar, G.S. and H. Wang. (2007). IG-symmetry and R-symmetry: Interrelations and Applications to the Inverse Gaussian Theory, *J. Statist. Plann. Inference*. (in press)
40. Natarajan, R. and G. S. Mudholkar. (2004). Moment Based Goodness-of-fit Tests for Inverse Gaussian Distribution, *Technometrics*, **46**, 339-347.
41. Natarajan, R. (2005). Inverse Gaussian and Gaussian analogies, *Encyclopedia of Statistical Sciences*2nd ed.
42. Natarajan. R., G. S. Mudholkar and M. P. McDermott. (2005). Order restricted infer-

ence for the scale-like inverse Gaussian parameters, *Journal of Statistical Planning and Inference*, **127**, no. 1-2, 229-236.

43. Schrödinger, E. (1915) Zur theorie der fall-und steigversuche an teilchenn mit Brownsche bewegung, *Physikalische Zeitschrift*, **16**, 289-295 .
44. Seshadri, V. (1997). "Halphen's laws", *Encyclopedia of Statistical Science*. 1, 302-306.
45. Shannon, C.E. (1949). *The Mathematical Theory of Communication*. Wiley, New York, 55.
46. Shuster. J.J. (1968). On the inverse Gaussian distribution function, *Journal of the American Statistical Association*. 63, 1514-1516.
47. Smoluchowski,M.V. (1915). Notiz über die Berechning der Brownschen Molkularbewegung bei des Ehrenhaft-millikanchen Versuchsanordnung, *Physikalische Zeitschrift*, 16, 318-321.
48. Thorburn. D. (1976). Some asymptotic properties of jackknife statistics, *Biometrika*, 63 , 305-313.
49. Thorburn. D. (1977). On the asymptotic normality of the Jackknife, *Scandinavian Journal of Statistics*, 4, 113-118.
50. Tian. L. and G. S. Mudholkar.(2003). The likelihood ratio tests for homogeneity of inverse Gaussian means under simple order and simple tree order, *Communications in Statistics: Theory and Methods*, 32 , 791-805.
51. Tweedie, M.C.K. (1941). A mathematical investigation of some electrophoretic measurements on colloids, *Unpublished M.Sc. Thesis*, University of Reading, England.
52. Tweedie, M.C.K. (1945). Inverse statistical variates, *Nature*, 155, 453.
53. Tweedie, M.C.K. (1946). The regression of the sample variance on the sample mean, *Journal of London Mathematical Society*, 21, 22-28.
54. Tweedie, M.C.K. (1947). Functions of a statistical variate with given means, with special reference to Laplacian distributions, *Proceedings of the Cambridge Philosophical Society*, 43, 41-49.
55. Tweedie, M.C.K. (1956). Statistical properties of inverse Gaussian distributions, *Virginia Journal of Science*, 7, 160-165 .
56. Tweedie, M.C.K. (1957). Statistical properties of inverse Gaussian distributions-I, *Annals of Mathematical Statistics*, 28, 362-377 .
57. Tweedie, M.C.K. (1957). Statistical properties of inverse Gaussian distributions-II, *Annals of Mathematical Statistics*, 28, 696-705.
58. Vasicek, O. (1976) A test for normality based on sample entropy, *Journal of Royal Statistical Society*, B, 38, 54-59.
59. Wald, A. (1947). *Sequential Analysis* Wiley, New York.
60. Whitmore, G. A. and V. Seshadri. (1978). A heuristic derivation of the inverse Gaussian distribution, *The American Statistician*, 41, 280-281.
61. Wintner. A. (1938). *Asymptotic Distributions and Infinite Convolutions*, Ann Arbor, MI: Edwards Broth.

Table 17.2: Some Well Known Analogies between the Normal and the Inverse Gaussian Distributions

Item	Normal Framework	Inverse Gaussian Framework
0.	$X_i, i=1,\ldots,n$ i.i.d. $N(\mu,\sigma^2)$	$X_i, i=1,\ldots,n$ i.i.d. $\mathrm{IG}(\mu,\lambda)$
1.	Ind. $X_i \sim N(\mu_i,\sigma_i^2), \sum X_i \sim N(\sum\mu_i, \sum\sigma_i^2)$	Ind. $X_i \sim \mathrm{IG}(\mu_i,\lambda_i)$, $\sum X_i \sim \mathrm{IG}(\sum\mu_i, \psi(\sum\mu_i)^2)$ if $\lambda_i/\mu_i^2 = \psi$, $\forall i$
2.	MLE: $\hat{\mu} = \bar{X}$, $\widehat{\sigma^2} = S^2 = n^{-1}\sum(X_i - \bar{X})$	MLE: $\hat{\mu} = \bar{X}$, $\widehat{\lambda}^{-1} = V = n^{-1}\sum(X_i^{-1} - \bar{X}^{-1})$
	$\bar{X} \sim N(\mu, \sigma^2/n)$, $nS^2/\sigma^2 \sim \chi^2_{n-1}$	$\bar{X} \sim \mathrm{IG}(\mu, n\lambda)$, $n\hat{\lambda}^{-1}/\lambda^{-1} = n\lambda V \sim \chi^2_{n-1}$
3.	$\bar{X}$ and S^2 are independent	$\bar{X}$ and V are independent
4.	$(\bar{X}, S^2)$ complete, sufficient for (μ,σ^2)	$(\bar{X}, V)$ complete, sufficient for (μ,λ)
5.	$(X-\mu)^2/\sigma^2 \sim \chi^2_1$	$\lambda(X-\mu)^2/(\mu^2 X) \sim \chi^2_1$
6.	UMPU t-test exists for $H_0: \mu = \mu_0$	UMPU t-test exists for $H_0: \mu = \mu_0$
7	UMPU two sample t-test exists for $H_0: \mu_1 = \mu_2$	UMPU two sample t-test exists for $H_0: \mu_1 = \mu_2$
	against one-side and two-side alternatives	against one-side and two-side alternatives
8.	ANOVA F-test for homogeneity of k means	ANORE F-test for homogeneity of k means
9.	$\bar{X}$ and S^2 independent iff Normal	$\bar{X}$ and V independent iff IG
10.	Saddlepoint approximation for p.d.f.	Saddlepoint approximation for p.d.f.
	of $\bar{X}$ is exact up to scaling	of $\bar{X}$ is exact up to scaling
11.	Bayesian context: Conjugate families for	Bayesian context: Conjugate families for
	μ, σ^{-2} and (μ, σ^{-2}) jointly, are	μ^{-1}, λ and (μ^{-1}, λ) jointly, are
	normal, Gamma, and bivariate	truncated normal, Gamma, and bivariate
	normal-Gamma, respectively	truncated normal-Gamma, respectively

Table 17.2 (Continued)

Item	Normal Framework	Inverse Gaussian Framework
12.	Independence-based goodness-of-fit test	Independence-based goodness-of-fit test
	$Z(G) = \tanh^{-1}(r(G)),\ r(G) = \text{Corr}(\bar{X}_{-i}, U_i)$	$Z(\text{IG}) = \tanh^{-1}(r(\text{IG})),\ r(\text{IG}) = \text{Corr}(\bar{X}_{-i}, V_{-i})$
	$U_{-i} = \left[\sum_{j\neq i} - (\sum_{j\neq i} X_j)^2/(n-1)\right]^{1/3}$	$V_{-i} = \sum_{j\neq i}(1/X_j - 1/\bar{X}_{-i})/(n-1)$
13.	$E(r(G)) = -\sqrt{\beta_1}/\sqrt{\beta_2 - 1}$	$E(r(\text{IG})) = -\sqrt{\delta_1}/\sqrt{\delta_2 - 1}$
14.	Under H_0, $\sqrt{n}\,r(G) \to_D N(0,3)$, $\sqrt{n}Z(G) \to_D N(0,3)$	Under H_0, $\sqrt{n}\,r(\text{IG}) \to_D N(0,3)$, $\sqrt{n}Z(\text{IG}) \to_D N(0,3)$
15.	Symmetry about $\mu = 0 \Rightarrow$ all odd order moments=0	IG-symmetry about $\mu :\Leftrightarrow E(X/\mu)^{r+1} = E(X/\mu)^{-r}$
16.	$Z(G)$-test suitable for detecting skew alternatives	$Z(IG)$-test suitable for detecting IG-skew alternatives
17.	Contaminated normal distribution models	Contaminated IG distribution models
18.	Scale mixtures of normals are symmetric about μ	Scale mixtures of IG are IG-symmetric about μ
19.	Coefficient of skewness $\sqrt{\beta_1}$ in normal theory	Coefficient of IG-skewness δ_1 in IG theory
20.	Coefficient of kurtosis $\beta_2 \geq 1$ with equality for	Coefficient of IG-kurtosis $\delta_2 \geq 1$ with equality for
	symmetric two-point distributions	IG-symmetric two-point distributions
21.	Pearson's (β_1, β_2)-chart	(δ_1, δ_2)-chart
22.	Sample versions $(\sqrt{b_1}, b_2)$ in modeling	Sample version (d_1, d_2) in modelling
23.	Asymptotic null distributions under normality,	Asymptotic null distributions for IG population,
	$\sqrt{nb_1} \to_d N(0,6)$, $\sqrt{n}(b_2 - 3) \to_d N(0,24)$	$\sqrt{n}d_1 \to_d N(0,6)$, $\sqrt{n}(d_2 - 3) \to_d N(0,24)$
24.	$\sqrt{b_1}$ and b_2 asymptotically independent	d_1 and d_2 asymptotically independent
	under normality	under IG model

Table 17.2 (Continued)

Item	Normal Framework	Inverse Gaussian Framework
25.	Inferences regarding σ^2 are nonrobust	Inferences regarding λ are nonrobust
26.	Jackknife-based methods for σ^2	Jackknife-based methods for λ
27.	Role of β_2 in Box-Anderson-type test for σ^2's	Role of δ_2 in Box-Anderson-type test for λ's
28.	Box-Cox transformations in normal-theory	Box-Cox transformations in IG-theory
29.	Maximum entropy characterization of normality	Maximum entropy characterization of IG
30.	Entropy test $K_{mn}(G)$: asymptotic null distribution	Entropy test $K_{mn}(IG)$: asymptotic null distribution
	$\sqrt{n}(K_{mn}-\sqrt{2\pi e}) \to_D N(0,\pi e)$ (scale Parameter known)	$\sqrt{n}(K_{mn}-\sqrt{2\pi e}) \to_d N(0,\pi e)$ (scale Parameter known)
31.	Kullback-Leibler g-o-f tests	Analogous Kullback-Leibler g-o-f tests
32.	Test for ordered means (LRT): null distributions	Test for ordered means (LRT): null distributions
33.	Tests for ordered means using combination method	Tests for ordered means using combination method
34.	Extreme value distribution of $N(0,1)$ population	Extreme value distribution of $RRIG(1,1)$ population
	$\sqrt{2\log n}\,X_{n:n} - 2\log n \to_d -\log Y$	$\sqrt{2\log n}\,X_{n:n} - 2\log n \to_d -\log Y$
35.	Extreme spacing of $N(0,1)$: $S_{n:n} = O_P\left((\log n)^{1/2}\right)$	Extreme spacing of $RRIG(1,1)$: $S_{n:n} = O_P\left((\log n)^{1/2}\right)$
36.	Symmetry about center	R-symmetry about R-center
37.	Symmetric unimodal distribution (e.g. normal)	R-symmetric unimodal distribution (e.g. RRIG)
38.	Convolutions of symmetric unimodal	Product convolutions of R-symmetric unimodal
	distributions are symmetric unimodal	distributions are R-symmetric unimodal
39.	Power function of t test for normal mean monotone	Power function of t test for IG mean monotone
40.	Peakedness	R-peakedness

Chapter 18

Clusterwise Regression Using Dirichlet Mixtures

Changku Kang[1] and Subhashis Ghosal[2]

[1] *Economics Statistics Department, The Bank of Korea; 110, 3-Ga, Namdaemun-Ro, Jung Gu, Seoul, Korea. E-mail: koncap@gmail.com*

[2] *Department of Statistics, North Carolina State University; 2501 Founders Drive, Raleigh, North Carolina 27695, U.S.A. E-mail: sghosal@stat.ncsu.edu*

The article describes a method of estimating nonparametric regression function through Bayesian clustering. The basic working assumption in the underlying method is that the population is a union of several hidden subpopulations in each of which a different linear regression is in force and the overall nonlinear regression function arises as a result of superposition of these linear regression functions. A Bayesian clustering technique based on Dirichlet mixture process is used to identify clusters which correspond to samples from these hidden subpopulations. The clusters are formed automatically within a Markov chain Monte-Carlo scheme arising from a Dirichlet mixture process prior for the density of the regressor variable. The number of components in the mixing distribution is thus treated as unknown allowing considerable flexibility in modeling. Within each cluster, we estimate model parameters by the standard least square method or some of its variations. Automatic model averaging takes care of the uncertainty in classifying a new observation to the obtained clusters. As opposed to most commonly used nonparametric regression estimates which break up the sample locally, our method splits the sample into a number of subgroups not depending on the dimension of the regressor variable. Thus our method avoids the curse of dimensionality problem. Through extensive simulations, we compare the performance of our proposed method with that of commonly used nonparametric regression techniques. We conclude that when the model assumption holds and the subpopulation are not highly overlapping, our method has smaller estimation error particularly if the dimension is relatively large.

Contents

18.1. Introduction

Consider the problem of estimating a regression function $m(x) = \mathrm{E}(Y|X = x)$ based on sampled data $(X_1, Y_1), \ldots, (X_n, Y_n)$, where X is d-dimensional. Typically $m(x)$ is estimated under the homoscedastic signal plus noise model

$$Y_i = m(X_i) + \varepsilon_i, \quad \varepsilon_1, \ldots, \varepsilon_n \text{ i.i.d. with } E(\varepsilon) = 0, \quad \mathrm{var}(\varepsilon) = \sigma^2, \tag{18.1}$$

assuming that $m(x)$ has a parametric form $\psi(x;\theta)$ or $m(x)$ is estimated nonparametrically based only on smoothness assumptions. Numerous methods such as those based on the nearest neighbors, kernel, local linearity, orthogonal series or spline smoothing for estimating the regression function exist in the literature; see any standard text such as Ref.6. These methods are extremely powerful in lower dimension. In higher dimension, because observations are relatively sparsely distributed over the space, one typically needs to assume additional structure in the regression function. Generalized additive models (GAM) [Hastie and Tibshirani,1990], classification and regression tress (CART) [Breiman et al.,1984] and multivariate adaptive regression splines (MARS) [Friedman,1991] are among the popular estimation methods in higher dimension.

Bayesian methods for the estimation of a regression function have been developed by putting priors directly on the regression function. A popular method of assigning a prior on functions is by means of a Gaussian process. Ref.21 considered an integrated Wiener process prior for $m(x)$. The resulting Bayes estimate, in the sense of a noninformative limit, is a smoothing spline. Priors based on series expansions with normally distributed coefficients, which are also Gaussian priors, have been investigated well for various choices of basis functions. An indirect method of inducing a prior on the regression function from that on the joint density $f(x,y)$ of (x,y) was considered by Ref. 16, who used a Dirichlet mixture of normal prior for the joint density. The idea is to use the relation $m(x) = \int y f(x,y)dy / \int f(x,y)dy$. This method requires estimating density functions in a higher dimensional space and the growth of the regression function is restricted by that of a linear function, since multivariate normal regressions are linear. In addition, the method suffers heavily from curse of dimensionality in higher dimension.

Sometimes, however, X is a sampled from a population which can be best described as a mixture of some hidden subpopulations. In this case, it is plausible that the regression function has different parameter values when the samples come from different groups. This situation will typically arise if there is one unobservable categorical label variable which influences both X and Y, but conditional on that missing label variable, the relationship between Y and X is linear up to an additive error. However, as we do not observe the group labels, the overall regression function of Y on X is a weighted average of the individual regression functions, where the weights are given by the probabilities of the observation being classified into different groups given its value. In such a situation, an easy calculation (see the next section) shows that the model (18.1) fails. The mixture distribution arises naturally in biology, economics and other sciences. For example, in a marketing survey, suppose that consumers rate the quality of a product. Different consumers may give different weights to various factors depending on their background and mentality. In other words, the population actually consists of different subpopulations where different regression functions are in effect, but subpopulation membership is abstract and is not observable. Had we observed the labels, regression analysis would have been straightforward. In the absence of the labels, we use the auxiliary measurements to impute the lost labels and use a simple regression analysis within each hypothetical group. The number of groups and group membership labels are nuisance parameters in the analysis, which should be integrated out with respect to their posterior distribution.

In this paper, we propose a method to estimate an unknown regression function by viewing it as a mixture of an unknown number of some simple regression functions associated with each subpopulation. These corresponding subgroups in the sample are estimated by identifying clusters in the data cloud. The clusters are automatically produced by a Dirichlet mixture model, a popular Bayesian nonparametric method. Standard regression methods are used in fitting a regression function within each cluster. Each iteration of the Markov chain Monte-Carlo (MCMC) sampling based on the Dirichlet mixture process produces a partitioning of the data cloud and thus an estimate of the regression function. The final regression estimate is obtained by averaging out the estimate obtained in each iteration thus integrating out the nuisance label parameters with respect to their posterior distribution obtained from the Dirichlet mixture process prior. Thus our method may be viewed as an ensemble method like bagging [Breiman, 1996] and random forest [Breiman, 2001], but here the ensemble is produced by MCMC iterations rather than by bootstrap resamples.

The similarity and difference of our method with other nonparametric methods are interesting to note. Regression estimates based on free-knot splines, kernels or nearest neighbors partition the space and fit some standard regression function in each part and combine these individual estimates. We also combine individual estimates based on different components of the sample. However, unlike other standard methods, we partition the observed samples rather than the space where they lie. Under the assumed condition that the samples come from a fixed (although unknown) number of groups, our method requires partitioning the sample in a number of groups that essentially remains fixed independent of the dimension of the space. In contrast, many standard nonparametric methods in higher dimension require splitting the space in a relatively high number of subregions fueling the demand for a larger sample size, thus leading to the curse of dimensionality problem. The simulations conducted in Section 18.3 confirm that our method performs better than standard nonparametric methods, particularly in higher dimension, provided that the mixture model representation is appropriate. In addition, we could get conservative pointwise credible bands very easily which are usually difficult with other methods. Another operational advantage of our method is that there is no need to choose any smoothing parameter.

There have been attempts to utilize clusters to partition data and estimate regression. The term clusterwise regression was first used by Ref. 18. Ref 4 proposed a conditional mixture maximum likelihood methodology to identify clusters and partition the sample before applying linear regression in each piece. The method estimates the parameters of the mixture distribution by maximum likelihood using the EM algorithm. The number of components k in the mixture is chosen by the Akaike information criterion. Their method is closest in spirit to ours. However, they do not take into account the uncertainty in the number of groups and each group membership. In the smoothly mixing regression method approach of Ref. 10, one estimates the probability of an observation with value x belonging to different subpopulations by a multinomial probit model. The main reason for using this model is that it produces simple and tractable posterior distributions within the Gibbs sampling algorithm. However, there is no room for a data driven choice of k and the method suffers from the curse of dimensionality problem in higher dimension.

We shall compare our method with standard regression estimates. More specifically, by means of extensive simulation studies, we shall compare the performance of our method with the kernel and spline methods in one dimension. In higher dimension, we shall compare with the estimates obtained from GAM and MARS.

The paper is organized as follows. Section 18.2 provides detailed description of our method, including the MCMC algorithms and the construction of confidence bands. Simulation results are presented and discussed in Section 18.3.

18.2. Method Description

Suppose that we have k regression models

$$Y_i = m_j(X_i) + \varepsilon_i, \quad \varepsilon_i \text{ i.i.d. with } \mathrm{E}(\varepsilon_i) = 0,\ \mathrm{var}(\varepsilon_i) = \tau_j^2, \quad i = 1, \ldots, n, \quad (18.2)$$

corresponding to the jth subgroup in the populations, $j = 1, \ldots, k$. Let J be an unobserved random variable indicating the subgroup label. The prior distribution of J is given by $\pi_j = P(J = j)$ and thus the posterior distribution given $X = x$ is given by $\pi_j(x) = P(J = j|X = x)$. Then, after eliminating the unobserved J, the regression of Y on X is given by

$$m(x) = \sum_{j=1}^{k} \mathrm{P}(J = j|X = x)\mathrm{E}(Y|X = x, J = j) = \sum_{j=1}^{k} \pi_j(x) m_j(x). \quad (18.3)$$

Also observe that the conditional variance is given by

$$\begin{aligned} \mathrm{var}(Y|X = x) &= \mathrm{E}_J(\mathrm{var}(Y|X = x, J)) + \mathrm{var}_J(\mathrm{E}(Y|X = x, J)) \\ &= \sum_{j=1}^{k} \tau_j^2 \pi_j(x) + \sum_{j=1}^{k} m_j^2(x)\pi_j(x) - \Big(\sum_{j=1}^{k} m_j(x)\pi_j(x)\Big)^2, \end{aligned}$$

which is not constant in x even if all τ_j's are equal. Hence the commonly assumed nonparametric regression model (18.1) fails to take care of the simple situation of missing labels.

The regression function $m(x)$ given by (18.2) may be estimated by its posterior expectation given the observed data. This can be evaluated in two stages — in the first stage we can obtain the posterior expectation given the hidden subpopulation labels and the observed data, while in the second stage we average the resulting quantity with respect to the conditional distribution of the labels given the data. The second stage averaging may be taken conditional on the X values only. For a discussion on how the Y observations may also be used to detect the hidden labels, see Ref. 20. Thus we need to model the joint distribution of the X and the missing subpopulation labels, which is done below. The first stage analysis reduces to k independent parametric regression problems. Therefore the posterior expectation of $m_j(x)$, $j = 1, \ldots, k$, may essentially be replaced by the corresponding least square estimates. Alternatively, one may view m_js as hyperparameters and the resulting substitution corresponds to an empirical Bayes approach. While a fully Bayesian analysis at this stage is quite possible, the minor difference between the posterior expectations and the least square estimates, particularly when the sample size is not very small, makes the additional computing requirement unappealing. Henceforth we shall consider this hybrid approach of least square estimation in the first stage coupled with a posterior analysis of missing subpopulation labels.

The joint distribution of X and the missing label J will be specified somewhat indirectly, through another set of latent variables. The technique is based on one of most popular methods in Bayesian nonparametrics, namely that of Dirichlet mixture (DM). Let $X_1,\ldots,X_n$ be i.i.d. from a density

$$f(x;G,\Sigma) = \int \phi(x;\theta,\Sigma)dG(\theta), \tag{18.4}$$

where $\phi(\cdot;\theta,\Sigma)$ is the density function of the d-dimensional normal density with mean θ and dispersion matrix Σ. Let G follow $\mathrm{DP}(M,G_0)$, the Dirichlet process with precision parameter M and center measure G_0 in the sense of Ref. 8. The model can be equivalently written in terms of latent variables $\theta_1,\ldots,\theta_n$ as

$$X_i|\theta_i \overset{\text{ind}}{\sim} N(\cdot;\theta_i,\Sigma), \quad \theta_i|G \overset{\text{iid}}{\sim} G, \quad G \sim \mathrm{DP}(M,G_0). \tag{18.5}$$

Because Dirichlet samples are discrete distributions, the Dirichlet mixture prior has the ability to automatically produce clusters. More precisely, we may identify X_i and X_j as the member of a common cluster if $\theta_i = \theta_j$. Note that the last event occurs occasionally because the posterior distribution of $(\theta_1,\ldots\theta_n)$ given $(X_1,\ldots,X_n)$ is given by the Polya urn scheme described by

$$\begin{gathered}
\theta_i|(\theta_{-i},\Sigma,X_1,\ldots,X_n) \propto q_{i0}dG_i(\theta_i) + \textstyle\sum_{j\neq i} q_{ij}\delta_{\theta_j}(\theta_i), \\
dG_i(\theta) \propto \phi(X_i;\theta,\Sigma)dG_0(\theta), \\
q_{i0} \propto M \textstyle\int \phi(X_i;\theta,\Sigma)dG_0(\theta), \quad q_{ij} \propto \phi(X_i;\theta_j,\Sigma), \quad q_{i0} + \textstyle\sum_{j\neq i} q_{ij} = 1;
\end{gathered}$$

here and throughout, the subscript "$-i$" stands for all but i. This way a probability distribution on partitions of $\{1,2,\ldots,n\}$ according to subpopulation membership is obtained, which in effect describes a joint distribution of X observations and the corresponding labels. In particular, the distribution of the number of hidden subpopulations is also described by the process.

Taking advantage of the many ties in θ_i's, a more efficient algorithm can be constructed based on a reparameterization $(k,\varphi_1,\ldots,\varphi_k,s_1,\ldots,s_n)$, where k is the number of distinct θ_i values, $\varphi_1,\ldots,\varphi_k$ are the set of distinct θ_i's and $s_i = j$ if and only if $\theta_i = \varphi_j$. In fact, to identify the clusters, we need not even know the values of $\varphi_1,\ldots,\varphi_k$; it suffices to know the configuration vector $\mathbf{s} = (s_1,\ldots,s_n)$. This will allow substantial reduction of computational complexity at the MCMC step. The clusters are given by $I_j = \{i : s_i = j\}$, $j = 1,\ldots,k$, and $H = \{I_1,\ldots,I_k\}$ is a partition of $I = \{1,\ldots,n\}$. Integrating out $\varphi_1,\ldots,\varphi_k$ in the above step, we obtain the following algorithm to generate the clusters through a Gibbs sampler when M,

G_0 and Σ are given:

$$\mathrm{P}(s_i = j|\mathbf{s}_{-i}, X_1, \ldots, X_n) \propto \begin{cases} n_{-i,j} \int \phi(X_i;\theta,\Sigma) dH_{-i,j}(\theta), & j = 1, \ldots, k_{-i}, \\ M \int \phi(X_i;\theta,\Sigma) dG_0(\theta), & j = k_{-i}+1, \end{cases} \tag{18.6}$$

where $H_{-i,j}$ is the posterior distribution based on the prior G_0 and all observations X_l for which $l \in I_j$ and $l \neq i$, that is,

$$dH_{-i,j}(\theta) \propto \prod_{l \in I_j, l \neq i} \phi(X_l;\theta,\Sigma) dG_0(\theta),$$

$n_{-i,j} = \#I_j \setminus \{i\}$, $j = 1, \ldots, k_{-i}$, k_{-i} is the number of distinct elements in $\{\theta_j : j \neq i\}$ and $\mathbf{s}_{-i} = (s_1, \ldots, s_{i-1}, s_{i+1}, \ldots, s_n)$. Letting G_0 have density $\phi(\cdot;\xi,\Phi)$, it follows that

$$\mathrm{P}(s_i = j|s_{-i}, X_1, \ldots, X_n) \propto M\phi(X_i;\xi,\Sigma+\Phi), \quad j = k_{-i}+1.$$

For $j = 1, \ldots, k_{-i}$, the corresponding expression can be simplified using

$$\begin{aligned} &\int \phi(X_i;\theta,\Sigma) dH_{-i,j}(\theta) \\ &= \frac{\int \phi(X_i;\theta,\Sigma) \prod_{l \in I_j, l \neq i} \phi(X_l;\theta,\Sigma)\phi(\theta;\xi,\Phi) d\theta}{\int \prod_{l \in I_j, l \neq i} \phi(X_l;\theta,\Sigma)\phi(\theta;\xi,\Phi) d\theta} \\ &= \frac{n_{-i,j}}{n_{-i,j}+1} \phi(X_i;\bar{X}_{+i,j}, \frac{n_{-i,j}}{n_{-i,j}+1}\Sigma) \times \frac{\phi(\xi;\bar{X}_{+i,j}, \frac{1}{n_{-i,j}+1}\Sigma+\Phi)}{\phi(\xi;\bar{X}_{-i,j}, \frac{1}{n_{-i,j}}\Sigma+\Phi)}, \end{aligned} \tag{18.7}$$

where

$$\bar{X}_{-i,j} = \frac{1}{n_{-i,j}} \sum_{l \in I_j, l \neq i} X_l \quad \text{and} \quad \bar{X}_{+i,j} = \frac{1}{n_{-i,j}+1} \left(\sum_{l \in I_j, l \neq i} X_l + X_i \right). \tag{18.8}$$

The above expressions are controlled by the hyper-parameters M, ξ, Φ and Σ. In a Bayesian setting, it is natural to put priors on them too and update during the MCMC stage. However, as we have integrated out $\varphi_1, \ldots, \varphi_k$, the complicated form of the marginal likelihood of ξ, Φ and Σ based on only $(X_1, \ldots, X_n)$ and $(s_1, \ldots, s_n)$ rules out existence of natural conjugate priors. One may update the hyperparameters using the Metropolis-Hastings algorithm, but it slows down the process. To avoid that, we take an empirical Bayes approach and estimate these parameters within an MCMC step by taking advantage of the knowledge of the partition. As ξ is the marginal expectation of X, we estimate it by $\hat{\xi} = \bar{X}$. Note that Σ is the dispersion matrix within group, and hence can be estimated by

$$\hat{\Sigma} = n^{-1} \sum_{j=1}^{k} \sum_{l \in I_j} (X_l - \bar{X}_j)(X_l - \bar{X}_j)^T, \qquad \bar{X}_j = n_j^{-1} \sum_{l \in I_j} X_l. \tag{18.9}$$

The matrix Φ stands for the between variation and hence can be estimated by

$$\hat{\Phi} = n^{-1} \sum_{j=1}^{k} n_j (\bar{X}_j - \bar{X})(\bar{X}_j - \bar{X})^T. \tag{18.10}$$

Clearly both $\hat{\Sigma}$ and $\hat{\Phi}$ are positive definite provided that $n_j > 1$ for all j. Note that, as we are dealing with a relatively few parameters, for large sample size, the difference between the empirical and full Bayesian estimates are likely to be small. The precision parameter M can be updated by Gibbs sampling using a clever data augmentation technique [Escobar and West, 1995]. However, to reduce computational time, we shall work with a fixed moderate value of M such as 1 or 2.
It may be noted that each step of MCMC changes the membership of at most one observation. Since sample means and dispersions can be calculated recursively when an observation is added to or deleted from a group, it is clear that one can avoid calculation from the scratch every time in 18.8.
Also it follows that we could form location-scale mixtures in 18.5 by allowing Σ to vary in each group as Σ_j, $j = 1, \ldots, k$. The same methodology will work if Σ_j's are estimated separately, for instance, by $\hat{\Sigma}_j = n_j^{-1} \sum_{l \in I_j} (X_l - \bar{X}_j)(X_l - \bar{X}_j)^T$.
Once the clusters have been identified, we can use a standard regression method, such as the method of least squares for linear regression, within each cluster giving estimates $\hat{m}_j(x)$ of regression functions $m_j(x)$, $j = 1, \ldots, k$. Polynomial regression may be used if we suspect the lack of linearity in the model. Variants of the least square estimate such as the ridge regression estimate [Hoerl and Kennard,1970] or the LASSO [Tibshirani,1996] may be used in place of ordinary least squares to lower the variation in the estimate if there is a lot of multicolinearity in the data. The latter is particularly very useful in higher dimension where typically a variable selection is desirable before estimation. Further, as the clusters produced by the Dirichlet mixture will invariably differ from the true clusters, the use of a robust alternative to the least square estimate is desirable to prevent misclassified observations altering the estimates significantly. These robust estimates require some tuning and are generally harder to compute compared to the least squares estimate. For the sake of simplicity, we do not pursue these modifications in the present paper.
To complete the estimation of $m(x)$, it remains to estimate $\pi_j(x)$, $j = 1, \ldots, k$. In order to do that, we consider a Bayesian model based on the given observations and the clusters. The prior probability of a new observation coming from the jth subgroup is given by the empirical fraction of observations falling in the group, that is n_j/n, $j = 1, \ldots, k$. Under the empirical model, the jth subpopulation is assumed to be normal with center at $\bar{X}_j$ and dispersion matrix $\hat{\Sigma}$. If the value of the new observation is x, then the likelihood of the jth subgroup is $\phi(x; \bar{X}_j, \hat{\Sigma})$.

Applying the Bayes theorem, the posterior probability of x coming from the jth subpopulation, based on the empirical model, is proportional to $n_j\phi(x;\bar{X}_j,\hat{\Sigma})$. To estimate the regression function at x given the clusters, one may do a "model averaging" with respect to the empirical posterior probabilities to yield the estimate

$$\hat{m}(x) = \frac{\sum_{j=1}^{k} n_j\phi(x;\bar{X}_j,\hat{\Sigma})\hat{m}_j(x)}{\sum_{j=1}^{k} n_j\phi(x;\bar{X}_j,\hat{\Sigma})}. \tag{18.11}$$

Alternatively, one may do a "model selection" to select the most plausible submodel and then use the estimate within that submodel to yield the estimate

$$\hat{m}(x) = \hat{m}_{\hat{j}}(x), \quad \hat{j} = \arg\max\{j : n_j\phi(x;\bar{X}_j,\hat{\Sigma})\}. \tag{18.12}$$

Both of the above estiamtes 18.11 and 18.12 are based on a given value of k and the given break-up of the clusters. Of course, all of these are unknown. In order to remove the dependence on a particular cluster formation, we form an ensemble of estimates given by 18.11 or 18.12 — a fresh estimate in each MCMC iteration of the DM process. The final estimate may thus be formed by a simple averaging of these $\hat{m}(x)$. We call these ensemble based estimates corresponding to 18.11 and 18.12 as DM-AVE (Dirichlet mixture — average) and DM-ML (Dirichlet mixture — most likely) respectively.
Piecing all the previous discussions together, the complete algorithm to calculate our estimate may be described as follows:

- Choose a sufficiently fine grid of x-values.
- Start with an initial configuration $(k, s_1, \ldots, s_n)$.
- Let $\mathbf{s}_{-i}$, k_{-i} and $n_{-i,j}$ be as defined above. Given $\mathbf{s}_{-i}$ and $X_1, \ldots, X_n$, sample s_i from $\{1, \ldots, k_{-i}, k_{-i}+1\}$ with probabilities

$$\begin{cases} c\frac{n_{-i,j}^2}{n_{-i,j}+1}\phi(X_i;\bar{X}_{+i,j},\frac{n_{-i,j}}{n_{-i,j}+1}\hat{\Sigma}) \times \frac{\phi(\bar{X};\bar{X}_{+i,j},\frac{1}{n_{-i,j}+1}\hat{\Sigma}+\hat{\Phi})}{\phi(\bar{X};\bar{X}_{-i,j},\frac{1}{n_{-i,j}}\hat{\Sigma}+\hat{\Phi})}, & j = 1, \ldots, k_{-i}, \\ cM\phi(X_i;\bar{X},\hat{\Sigma}+\hat{\Phi}), & j = k_{-i}+1, \end{cases}$$

  where $\bar{X}_j$, $\bar{X}_{+i,j}$, $\bar{X}_{-i,j}$, $\hat{\Sigma}$ and $\hat{\Phi}$ are defined in equations 18.8, 18.9 and 18.10, and c is the reciprocal of the following expression:

$$M\phi(X_i;\bar{X},\hat{\Sigma}+\hat{\Phi}) + \sum_{j\neq i}\frac{n_{-i,j}^2}{n_{-i,j}+1}\phi(X_i;\bar{X}_{+i,j},\frac{n_{-i,j}}{n_{-i,j}+1}\hat{\Sigma}) \times \frac{\phi(\bar{X};\bar{X}_{+i,j},\frac{1}{n_{-i,j}+1}\hat{\Sigma}+\hat{\Phi})}{\phi(\bar{X};\bar{X}_{-i,j},\frac{1}{n_{-i,j}}\hat{\Sigma}+\hat{\Phi})}.$$

- Repeat the above procedure for $i = 1, \ldots, n$ to complete an MCMC cycle.

- Compute DM-AVE and DM-ML as defined by 18.11 and 18.12 for each cycle.
- Repeat the cycle sufficiently large number of times to let the chain mix well.
- Compute the averages of the estimates DM-AVE or DM-ML for each cycle after a burn-in for each point on a chosen grid.
- Obtain the estimated function by joining the average values over the grid points.

In density estimation applications, one usually works with the initial configuration $k = n$, that is, all clusters are singleton. Here one cannot use that because $\hat{\Sigma}$ will then be the zero matrix, and thus the formula will break down. As an alternative, one may split the domain into small cubes and form the initial clusters by looking at the sample values falling into these regions, where singleton clusters will have to be merged with some neighboring one.
Note that we substitute ξ, Σ, Φ, and $m_1(\cdot), \ldots, m_k(\cdot)$, as well as the group mean values $\theta_1, \ldots, \theta_k$ within an MCMC iteration by the corresponding empirical estimates in the analysis, thus adopting an empirical Bayes approach. Implementation of a fully hierarchical Bayesian approach is possible provided that we do full updating of the Dirichlet mixture process. However, the fully Bayesian approach takes much longer time to run and does not seem to lead to more accurate estimates.
It may be observed that we used the conjugate normal center measure G_0 for the Dirichlet process. In principle, it is possible to use any center measure, but the equation 18.7 and the formula preceding that will not have any closed-form expression. One can carry out the fully Bayesian method of updating each $\varphi_1, \ldots, \varphi_k$ by using specially devised algorithms for the non-conjugate case like the "no gaps" algorithm [MacEachern and Müller, 1998]. However, the simplification resulting from integrating out the latent variables $\varphi_1, \ldots, \varphi_k$ and working only with the configuration vector $(s_1, \ldots, s_n)$ will not be possible, unless one calculates integrals numerically in equation 18.6. Numerical integration at each iteration of the Gibbs sampling scheme will be a daunting task and will result in extremely slow and inefficient computing.
Our method can be expanded further to construct a pointwise band for the regression function. Let x be a fixed point in the space of the regressor. A confidence interval for $m(x) = \mathrm{E}(Y|X = x)$ is easily obtained from the method of least squares given the subgrouping of the samples and the group to which the new observation $X = x$ belongs. However, since both the group to which x belongs and the break-up of clusters are unknown, we can integrate them out according

to their posterior distributions, leading to a "posterior expected pointwise confidence band" as described below. Treat $X = x$ as a new observation and suppose that J stands for the label of the cluster where x belongs. Given the clusters, let $[L_j, U_j]$ be $100(1-\alpha_0)\%$ confidence limits for a $m_j(x)$, $j = 1, \ldots, k$. This confidence interval may be thought of as a posterior credible interval for $m_j(x)$. Especially, in large sample sizes the two notions will tend to agree. Let L be the $\gamma/2$-quantile of $L_1, ..., L_k$ and U be the $1-\gamma/2$-quantile of $U_1, \ldots, U_k$, respectively, where $\gamma = \alpha - \alpha_0$ and the associated probabilities are $\pi_1(x), \ldots, \pi_k(x)$. Let $m^*(x) = m_j(x)$ if $J = j$, $j = 1, \ldots, k$. Then

$$\mathrm{P}(m^*(x) \leq L) = \sum_{j:L_j \leq L} \mathrm{P}(m_j(x) \leq L)\pi_j(x) + \sum_{j:L_j > L} P(m_j(x) \leq L)\pi_j(x) \leq \frac{\gamma}{2} + \frac{\alpha_0}{2}.$$

Similarly $\mathrm{P}(m^*(x) \geq U) \leq (\gamma + \alpha_0)/2$, and so $\mathrm{P}(L \leq m^*(x) \leq U) \geq 1 - \alpha$, $\alpha = \gamma + \alpha_0$. Thus by the convexity of $[L, U]$, $m(x)$, being an average of $m^*(x)$, has the same credible bounds given the clusters. Finally, the bounds L and U may be averaged out over MCMC iterations leading the final posterior expected pointwise confidence band. The obtained band may not have frequentist validity though, and it more closely resembles a posterior credible band. Asymptotically, Dirichlet mixtures estimate densities consistently [cf. Ghosal et al., 1999], which implies that the mixing distribution is estimated consistently in the weak topology. In view of a result of Ref. 5, consistency essentially implies that the number of clusters in the pattern obtained by the posterior is asymptotically at least as much as the true number of groups. In general, the obtained number can be considerably bigger, which is also confirmed by the simulation study. However, the superfluous breaking up of groups causes only a minor loss of precision because of smaller usable sample size. On the other hand, an inappropriate merging of two groups introduces serious bias in the estimates.

18.3. Simulation Study

In this section we do several simulations to evaluate the performance of our method. First, data are generated from a mixture of normal distributions in the univariate and the multivariate case, and in different groups different regression functions are given with several parameters. We proceed to estimate the fitted regression function using our proposed method and also some other methods such as those based on kernel and spline smoothing in the univariate case and MARS and GAM in the multivariate case.

We compare our method with other methods by the empirical L_1-error and L_2-error between estimates and the true regression function, defined respectively by

$\mathrm{EL}_1(\hat{f}) = n^{-1}\sum_{i=1}^{n}|\hat{f}(x_i) - f(x_i)|$ and $\mathrm{EL}_2(\hat{f}) = \{n^{-1}\sum_{i=1}^{n}(\hat{f}(x_i) - f(x_i))^2\}^{1/2}$. Since the success of our method is intimately linked with the ability to discover the correct patterns in the data as well as the number of clusters, in the simulation study, we shall monitor the distribution of the number of groups found by the posterior and the similarity of the obtained groupings with the actual ones.

In order to monitor how close is the partition induced by the Dirichlet mixture process to the true partition, we shall use Rand's measure [Rand,1971] of similarity of clusterings. A more refined measure is provided by Ref. 14. Given n points, $\theta_1,\ldots,\theta_n$ and two clustering vectors, $\mathbf{s} = \{s_1,\ldots,s_{k_1}\}$ and $\mathbf{s}' = \{s'_1,\ldots,s'_{k_2}\}$, Rand's measure d is defined by

$$d(\mathbf{s},\mathbf{s}') = \frac{\sum_{i=1}^{n}\sum_{j=i+1}^{n}\gamma_{ij}}{\binom{n}{2}}, \tag{18.13}$$

where

$$\gamma_{ij} = \begin{cases} 1, \text{ if there exist } k \text{ and } k' \text{ such that both } \theta_i \text{ and } \theta_j \text{ are in both } s_k \text{ and } s'_{k'}, \\ 1, \text{ if there exist } k \text{ and } k' \text{ such that } \theta_i \text{ is in both } s_k \text{ and } s'_{k'} \text{ while } \theta_j \\ \quad \text{is in neither } s_k \text{ or } s'_{k'}, \\ 0, \text{ otherwise.} \end{cases} \tag{18.14}$$

In practice, we use a simple computational formula for d:

$$d(\mathbf{s},\mathbf{s}') = \Big[\binom{n}{2} - \frac{1}{2}\Big\{\sum_{i=1}^{n}\Big(\sum_{j=i+1}^{n} n_{i,j}\Big)^2 + \sum_{j=1}^{n}\Big(\sum_{i=j+1}^{n} n_{i,j}\Big)^2\Big\} + \sum_{i=1}^{n}\sum_{j=i+1}^{n} n_{i,j}^2\Big] \Big/ \binom{n}{2}, \tag{18.15}$$

where $n_{i,j}$ is the number of points simultaneously classified in the ith cluster of $\mathbf{s}$ and the jth cluster of $\mathbf{s}'$. Note that d ranges from 0 to 1. When $d = 0$, the two clusterings have no similarities, and when $d = 1$ the clusterings are identical. Rand's measure will be computed in each MCMC step. A boxplot showing its distribution will also be displayed in one instance.

Simulation studies are conducted under several different combinations of the true model and different sample sizes. Throughout this section and in the next, we shall work with the choice $M = 1$ for the precision measure of associated Dirichlet process. To see how accurate the clustering is, we also monitor Rand's measure and the number of clusters formed by the Dirichlet mixture process. All of simulation work is done by the package R (version 2.0.1).

18.3.1. *One dimension*

To give a simple example we consider a univariate predictor X having the following distribution.

$$X \sim 0.3N(-0.9,\sigma^2)+0.2N(-0.3,\sigma^2)+0.4N(0.4,\sigma^2)+0.1N(1.0,\sigma^2),$$

where $\sigma^2 = 0.01, 0.02, 0.03, 0.04$ are considered.
First generate 100 samples of X from above mixture of normals and generate Y using four linear functions with different coefficients are given within each group.

$$f_1(x) = 2.3+1.1x, \quad f_2(x) = 1.5-0.6x, \quad f_3(x) = 0.8+0.9x, \quad f_4(x) = 1.7-0.2x,$$

with an additive noise term having mean 0 and variance $\tau^2 = 0.03$.
In the MCMC sampling step, we generate 5000 samples and ignore the first 1000 as a burn-in period. For appropriate selection of burn-in period, the trace plot of the number of distinct clusters was used and 1000 steps were found to be sufficient. This simulation work is repeated 100 times. Each simulation took about 5 minutes. It is found that the performances of our two methods DM-AVE and DM-ML are almost identical in every occasion.

For kernel estimation, we use a normal kernel with the help of the R function "ksmooth" and choose bandwidth by the default mechanism in that program. We also monitor the empirical L_2-error and L_1-error. It shows that our method has good performance for all the cases we investigated. Larger the value of σ^2 is, more overlapped the groups are, and smaller is the value of Rand's measure. This is expected because if some of clusters are severely overlapped, it is hard to separate them.
Comparing the errors by means of Wilcoxon rank test in the cases $\sigma^2 = 0.01$ and 0.02, we conclude that the median error of DM-AVE method is significantly smaller than that of other two methods at the 5% level of significance. For $\sigma^2 = 0.03$, the difference between DM-AVE and spline is significant but the difference with the kernel estimate is not significant. For $\sigma^2 = 0.04$, none of the differences are significant.
For a typical case, we monitor the number of clusters is given which in Figure 18.2. The graph of the confidence band for one such instance is shown in Figure 18.3.

Now let us consider the case that the true regression function is not linear. Let X be generated by the following distribution

$$X \sim 0.3N(-1.9,\sigma^2)+0.3N(-0.3,\sigma^2)+0.4N(1.4,\sigma^2),$$

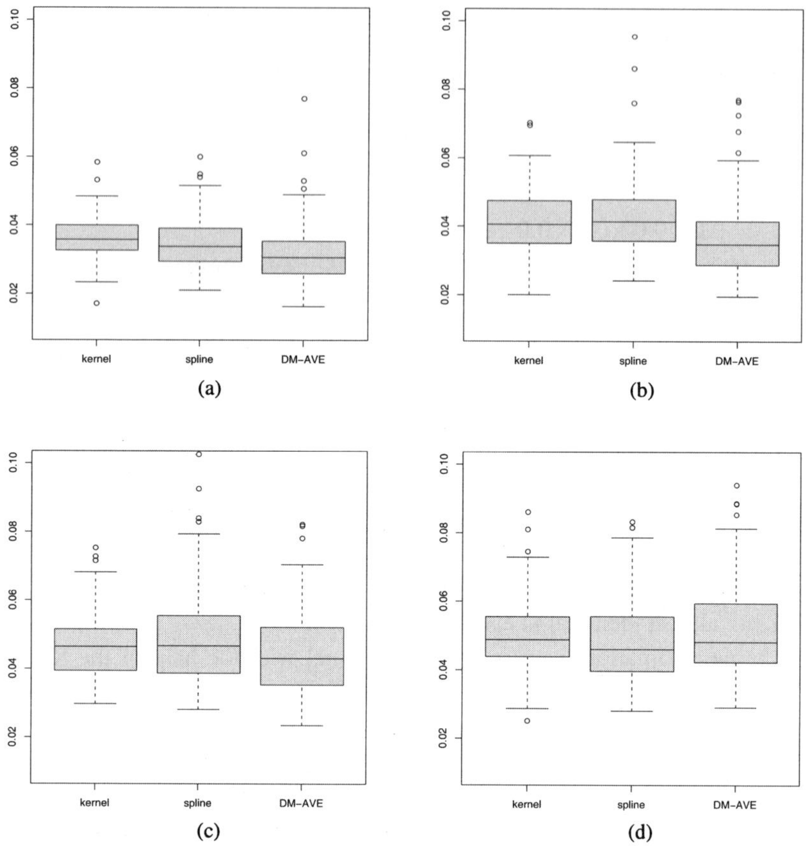

Fig. 18.1. (One dimension) : The boxplots of L_2-error. Rand's measures on average are 0.94, 0.87, 0.83 and 0.81, respectively.

where $\sigma^2 = 0.01, 0.02, 0.03, 0.04$ are considered. The regression functions in each cluster are given by

$$f_1(x) = 0.3 + 0.3\sin(2\pi x), \quad f_2(x) = 1.2 - 0.6\sin(2\pi x), \quad f_3(x) = 0.5 + 0.9\sin(2\pi x).$$

Note here that we have three clusters in this case. Again we generate 5000 MCMC samples after 1000 burn-in and repeated the experiment 100 times. In this case, we use the cubic regression as well as the linear regression. Figure 18.4 tells us that the cubic regression performs better than the linear regression method and other methods. Formal rank tests shows that cubic DM-AVE method is significantly better than the kernel method. The spline method here performs slightly better than cubic DM-AVE although the difference is not significant. It may be noted

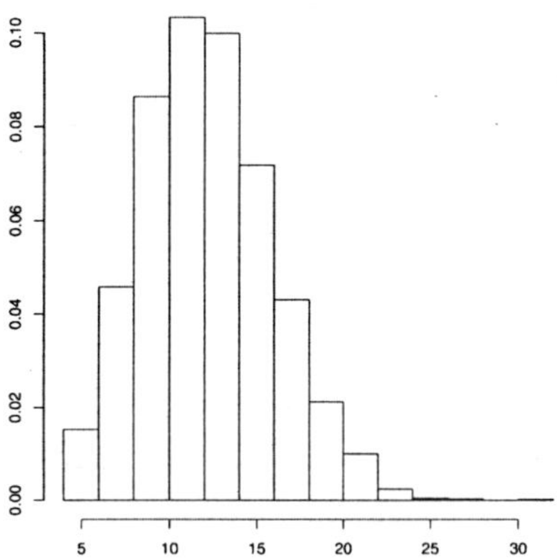

(a) for $\sigma^2 = 0.01$

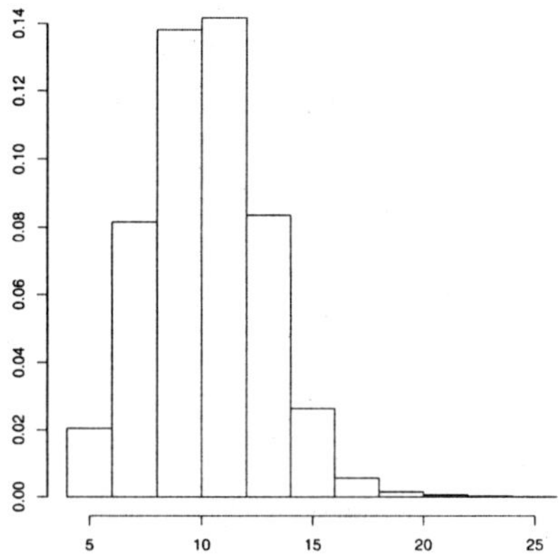

(b) for $\sigma^2 = 0.04$

Fig. 18.2. (One dimension) : Histograms of the distribution of k

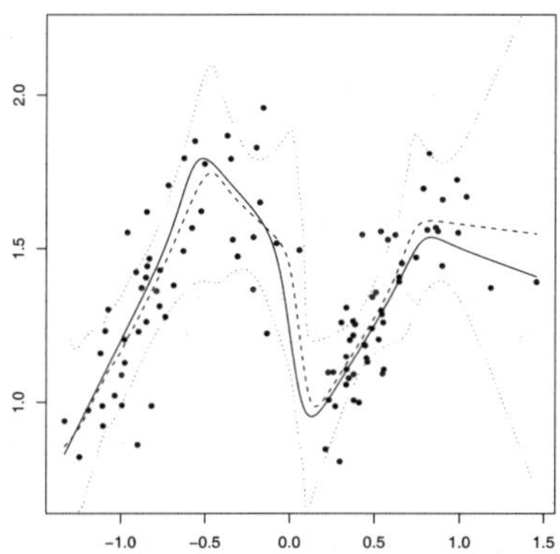

Fig. 18.3. (One dimension) : The plots of data and fitted regression function with $\sigma^2 = 0.03$ and $\tau^2 = 0.03$. The solid line is the true mean function and dotted line is fitted line using DM-AVE.

that the cubic is also not the correct functional form of the regression functions in the groups. Thus the DM-AVE method performs reasonably well even under misspecified models provided that a flexible function like a cubic is used.

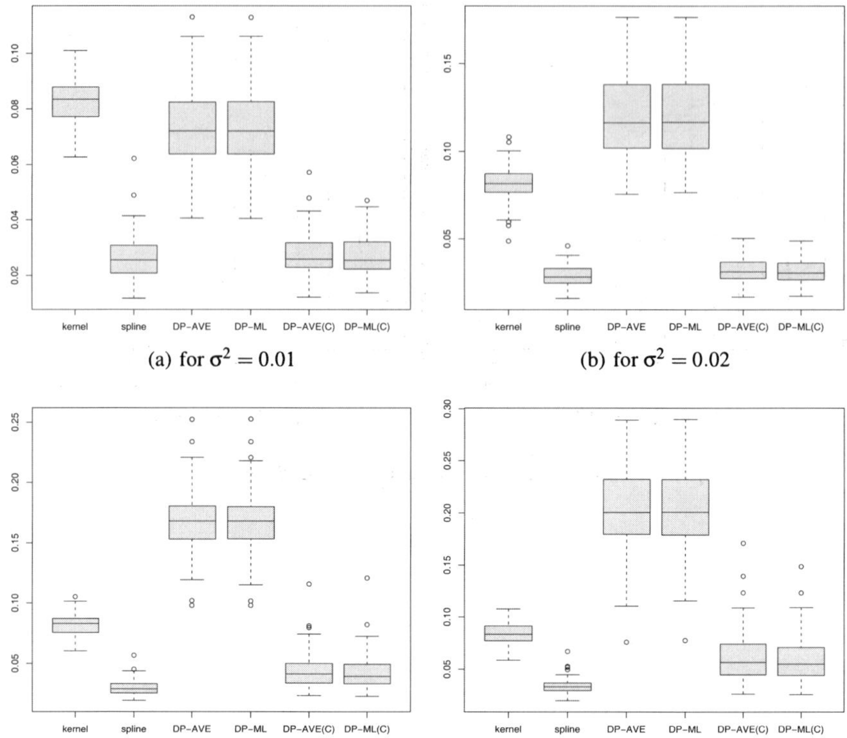

(a) for $\sigma^2 = 0.01$ (b) for $\sigma^2 = 0.02$

(c) for $\sigma^2 = 0.03$ (d) for $\sigma^2 = 0.04$

Fig. 18.4. (One dimension) : The case where the true model is not linear. The plots of L_1-error are shown. Average Rand's measures are 0.97, 0.96, 0.95 and 0.94, respectively.

18.3.2. *Two dimension*

We consider $X = (X_1, X_2)^T$ distributed as a mixture of bivariate normal

$$X \sim 0.3N_2\left(\begin{pmatrix}1\\1\end{pmatrix},\Sigma\right) + 0.2N_2\left(\begin{pmatrix}2\\5\end{pmatrix},\Sigma\right) + 0.4N_2\left(\begin{pmatrix}4\\0.5\end{pmatrix},\Sigma\right) + 0.1N_2\left(\begin{pmatrix}5\\3\end{pmatrix},\Sigma\right)$$

where the regression functions in the subpopulations are given by

$$f_1(X) = 2.3 + 1.1X_1 - 2X_2, \quad f_2(X) = 1.5 - 0.6X_1 + 1.2X_2,$$
$$f_3(X) = 0.8 + 0.9X_1 - 1.1X_2, \quad f_4(X) = 1.7 - 0.2X_1 + 1.2X_2,$$

and the dispersion matrix Σ for the error given the groups label is taken of the form $\sigma^2 I_2$. We consider five different choices 0.2, 0.4, 0.6, 0.8 and 1.0 of σ^2 in the simulations. The error variance, τ^2, is taken to be 0.1. We generate $n = 100$

samples from the distribution and obtain the estimate of the regression function and the related quantities based on the sample. We generate 5000 MCMC samples ignoring the first 1000 as burn-in. The whole experiment is repeated 100 times. For one such replication, Figure 18.5 shows the data plot and the boxplot of Rand's measure obtained over different MCMC iterations.

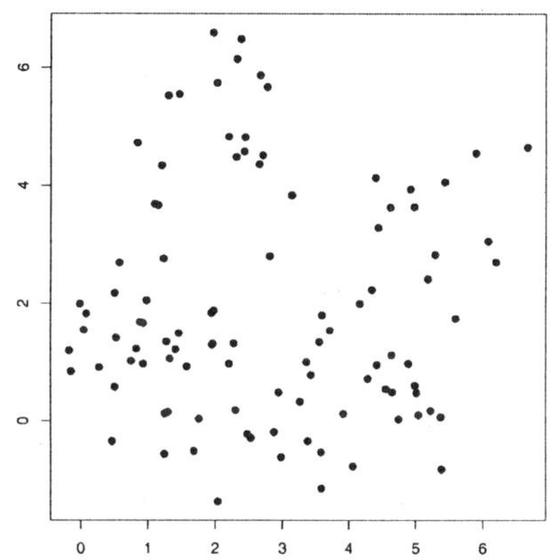

(a) The plot of X

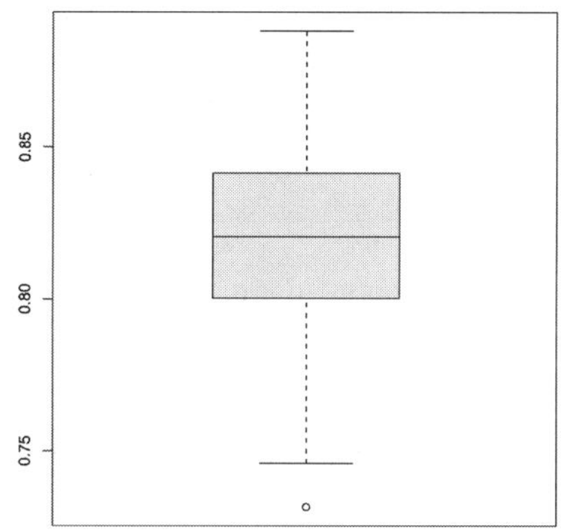

(b) Boxplot of Rand's measure

Fig. 18.5. (Two dimension): Data plot and boxplot of Rand's measure with $\sigma^2 = 0.8$.

In this case, the true number of clusters is four and those clusters are overlapped with each other. As the value of σ^2 determines how much the clusters are overlapped, we consider moderate values of σ^2.

We compare the L_2-error of DM-AVE and DM-ML with those of GAM and MARS. To allow certain flexibility, we set the degree of interaction to two, which means that it allows interaction between two variables.

Wilcoxon rank test comparing the medians of L_2-errors shows that DM-ML and DM-AVE are comparable, while those of GAM and MARS are higher at 5% level of significance except for $\sigma^2 = 1.0$, where the advantage over MARS is not statistically significant.

18.3.3. *Higher dimension*

We consider the case when the predictor X has 10 variables and distributed as a mixture of four multivariate normal:

$$X \sim \sum_{j=1}^{4} \omega_j N_{10}\left(\mu_j, \Sigma\right).$$

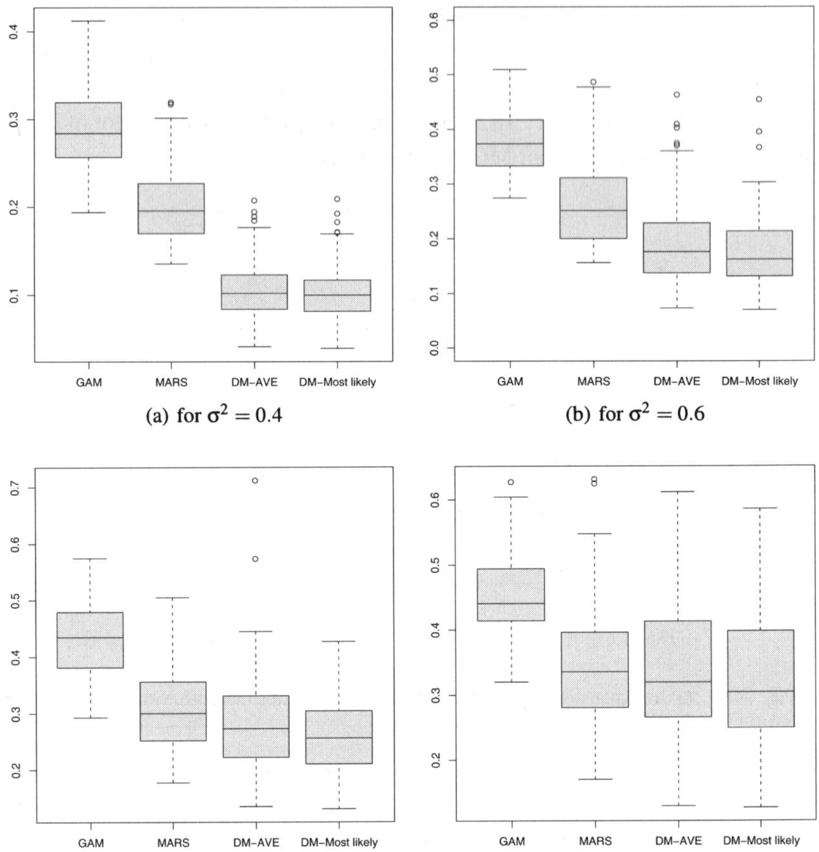

(a) for $\sigma^2 = 0.4$ (b) for $\sigma^2 = 0.6$

(c) for $\sigma^2 = 0.8$ (d) for $\sigma^2 = 1.0$

Fig. 18.6. (Two dimension) : The boxplots of the L_2-error.

We let $\Sigma = \sigma^2 I_{10}$, $\sigma^2 = 0.2, 0.3, 0.4, 0.5$ and $\tau^2 = 0.1$. The four mean vectors are given by

$$\mu_1 = (1,1,1,1,2,4,2,4,4,0.5)^T, \quad \mu_2 = (4,0.5,5,3,5,3,1,1,1,1)^T,$$
$$\mu_3 = (2,5,2,5,4,0.5,4,0.5,5,3)^T, \quad \mu_4 = (5,3,1,1,1,1,2,5,2,5)^T.$$

Let the weights be $\omega = (0.3, 0.2, 0.4, 0.1)$ and the within subpopulation regression functions are given by

$$f_j(x) = \alpha_j + \beta_j^T x, \quad j = 1, \ldots, 4,$$

where $\alpha_1 = 2.3$, $\alpha_2 = 1.5$, $\alpha_3 = 0.8$, $\alpha_4 = 1.7$ and β_j's are the vectors of length

10 given by

$$\beta_1 = (1.1, -2, 1.1, -2, -0.6, 1.2, -0.6, 1.2, 0.9, -1.1)^T,$$
$$\beta_2 = (0.9, -1.1, -0.2, 1.2, -0.2, 1.2, 1.1, -2, 1.1, -2)^T,$$
$$\beta_3 = (-0.6, 1.2, -0.6, 1.2, 0.9, -1.1, 0.9, -1.1, -0.2, 1.2)^T,$$
$$\beta_4 = (-0.2, 1.2, 1.1, -2, 1.1, -2, -0.6, 1.2, -0.6, 1.2)^T.$$

We generated 200 sample data from the resulting population and obtain estimates of regression. The whole simulation is repeated 100 times. We consider four different σ^2 values 0.2, 0.3, 0.4 and 0.5. Our method DM-AVE is compared with GAM and MARS. In the MCMC sampling scheme, we generate 5000 samples after ignoring first 1000 as burn-in. Figure 18.7 shows that DM-AVE is better than GAM and MARS for all four choices of σ^2. The differences are significant at 5% significance level according to the Wilcoxon rank test.

18.4. Conclusions

We argued that in many natural applications, the population may be viewed as an aggregate of some hidden subpopulations, in each of which a possibly different but simple regression regime works. The number of hidden subpopulations is also considered as unknown. The overall regression function may then be estimated by identifying the missing subpopulation labels and estimating regression function parametrically in each group of data corresponding to same subpopulation labels. The approach automatically leads to estimators not affected by the curse of dimensionality problem. Moreover, assuming that the data clusters are identified fairly accurately, the method seems to enjoy nearly parametric rate of convergence. However, the uncertainty in identifying the clusters should be taken into account. In order to find the data clusters, we follow the Bayesian approach and consider the Dirichlet process mixture prior on the distribution of the regressor variable, which has the ability to automatically identify clusters in MCMC iterations. By averaging over possible estimates corresponding to different clustering in these MCMC iterations, we obtain a very stable and accurate estimator of the regression function. We compare the performance of our estimator with some popular nonparametric estimators and find that when the model assumptions hold, our estimator has significantly smaller estimation error. The effect is particularly pronounced in higher dimension. There it outperforms the GAM and MARS estimators which are not even prone to curse of dimensionality problem. The strength of our method seems to come from the assistance of the model and the near parametric rate of precision compared to other nonparametric methods.

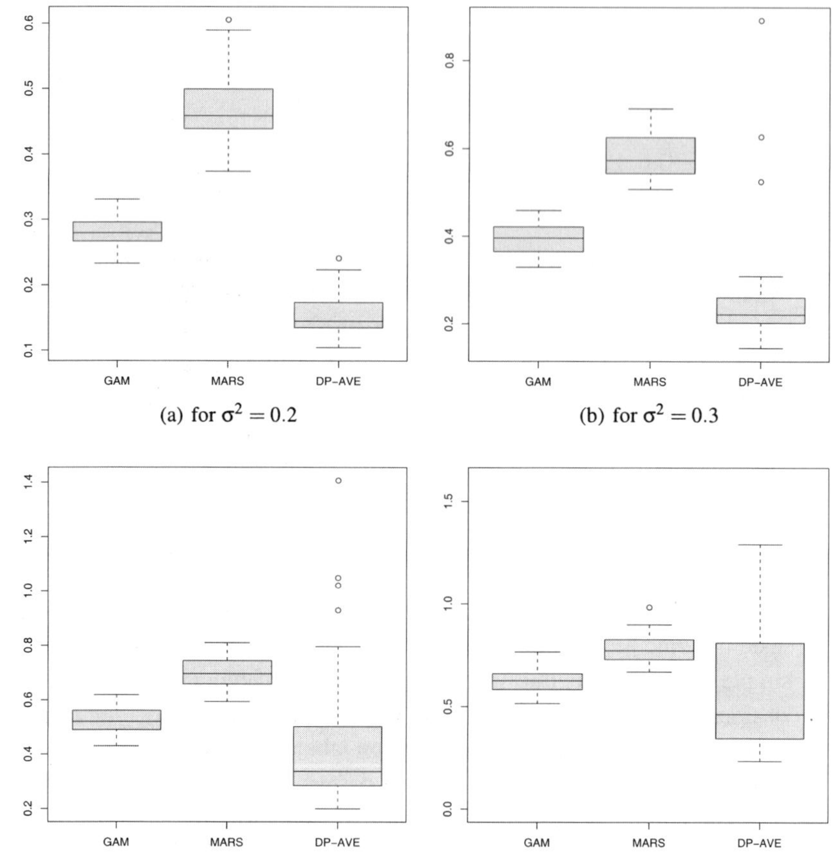

(a) for $\sigma^2 = 0.2$

(b) for $\sigma^2 = 0.3$

(c) for $\sigma^2 = 0.4$

(d) for $\sigma^2 = 0.5$

Fig. 18.7. (Ten dimension): Boxplots of L_1-error with respect to σ^2 values. Average Rand's measures are 0.84, 0.77, 0.76 and 0.76, respectively.

18.5. Acknowledgments

Research of both authors are partially supported by NSF grant number DMS-0349111 awarded to the second author.

References

1. Breiman, L. J. H. Friedman, R. Olshen and C. J. Stone, (1984). *Classification and Regression Trees*, Belmont, CA: Wadsworth.

2. Breiman, L. (1996). Bagging predictors. *Machine Learning* **24**, 123–140
3. Breiman, L. (2001). Random forests. *Machine Learning* **45**, 5–32 .
4. DeSarbo W.S.and W. L. Cron, (2001). A maximum likelihood methodology for clusterwise linear regression. *J. Classification* **5**, 249–282, (2001).
5. Donoho, D.L. (1988). One-sided inference about functionals of a density. *Ann. Statist.* **16**, 1390–1420.
6. Efromovich,S. (1999). *Nonparametric Curve Estimation: Methods, Theory and Applications.* Springer, New York.
7. Escobar, M.D. (1995)*J. Amer. Statist. Assoc.* **90**, 557–588.
8. Ferguson,T.S. (1973). A Bayesian analysis of some nonparametric problems. *Ann. Statist.* **1**, 209–230.
9. Friedman, J.H. (1991). Multivariate adaptive regression splines (with discussion). *Ann. Statist.* **19**, 1–141.
10. Geweke, J. and M. Keane, (2005). Smoothly Mixing Regressions, *J. Econometrics*, **138**, 252-291.
11. Ghosal, S., J. K. Ghosh and R. V. Ramamoorthi, (1999). Posterior Consistency of Dirichlet Mixtures in Density estimation. *Ann. Statist.* **27**, 143–158 .
12. Hastie, T.J. and R. I. Tibshirani, (1990). *Generalized Additive Models.* (Chapman and Hall).
13. Hoerl, A. and R. Kennard, (1970). Ridge regression: Biased estimation for nonorthogonal problems. *Technometrics* **12**, 55–67 .
14. Hubert L. and P. Arabie, (1985). Comparing partitions. *J. Comput. Graph. Statist.* **7**, 233–228.
15. MacEachern S. and P. Müller, (1998). Estimating mixture of a Dirichlet process models, *J. Comput. Graph. Statist.* **7** 223–228.
16. Müller, P. A. Erlkanli and M. West, (1996). Bayesian curve fitting using multivariate normal mixtures. *Biometrika* **83**, 67–79 .
17. Rand, W.M. (1971). Objective criteria for the evaluation of clustering methods. *J. Amer. Statist. Assoc.* **66**, 846–850 .
18. Spath, H. (1979). Algorithm 39: Clusterwise Linear Regression. *Computing* **22**, 367–373 .
19. Tibshirani, R.J. (1996). Regression shrinkage and selection via the Lasso. *J. Roy. Statist. Soc. Ser. B* **58**, 267–288.
20. Van Aelst, S. X. Wang, R. H. Zamar and R. Zhu, (2006). Linear grouping using orthogonal regression. *Computational Statistics and Data Analysis* **50**, 1287–1312.
21. Wahba, G. (1978). Improper priors, spline smoothing and the problem of guarding against model errors in regression. *J. Roy. Statist. Soc., Ser. B* **40**, 364–372.

Chapter 19

Bayesian Analysis of Rank Data Using SIR

Arnab Kumar Laha[1] and Sourabh Dongaonkar[2]

[1] *Indian Institute of Management, Ahmedabad*

[2] *Indian Institute of Technology, Kanpur*

E-mail:arnab@iimahd.ernet.in

Rank data occurs quite regularly in the context of market research studies, opinion polls, sports, etc. In this paper, we discuss the problem of estimation of the true rank when the rank given by the respondents are subject to error. Various error structures are discussed like, transposition errors, cyclic errors etc. and methods to analyze data with these error structures are derived. Analysis of data with a completely general error structure is also discussed. It is seen that in all the cases considered, the posterior marginal distribution of the true rank can be conveniently obtained using Sampling Importance Resampling (SIR) technique. A real life rank data set is analysed to illustrate the methodology.

Contents

19.1. Introduction

In various real life situations data are in the form of ranks. Rank data occurs frequently in market research studies, opinion polls, sports, educational testing, and psychometric studies, etc. where the respondents are often asked to rank a set of items in ascending or descending order. Some concrete examples of situations where rank data arises are (a) in a market research study where the respondents are asked to rank a set of attributes of a car in descending order of their importance on purchase decision (b) in an interview where several experts are asked to rank a set of candidates according to their suitability for a particular job (c) in an opinion poll where the respondents are asked to rank a set of candidates according to their preference for being the president of a nation and (d) in a gymnastic competition where judges are asked to rank a set of gymnasts in ascending order based on their performance in an event. A discussion on some aspects of modeling and analysis of rank data can be found in Marden(1995). In all of the above situations it is important for the decision maker to draw conclusions regarding the "true rank" of each item based on the available data.

Let the m items to be ranked be denoted as $\{1,2,\ldots,m\}$. Any permutation of these m items is a bijective function from $\{1,2,\ldots,m\}$ to itself and can be represented as

$$\sigma = \begin{pmatrix} 1 & 2 & \cdots & m \\ \sigma(1) & \sigma(2) & \cdots & \sigma(m) \end{pmatrix} \tag{19.1}$$

where $\sigma(i) \in \{1,2,...,m\}$ for all $i = 1,2,...,m$ and $\sigma(i) \neq \sigma(j)$ if $i \neq j$. The interpretation of the above notation is that item 1 is ranked $\sigma(1)$, item 2 is ranked $\sigma(2)$ and so on. We will denote the set of all permutations of $\{1,2,...,m\}$ as Λ_m. For a comprehensive discussion on permutations from an algebraic standpoint see Fraleigh (1982).

In this paper, we assume that the ranks given by the respondents are permutations of the true rank of the items. Let π denote the true rank of the items which is unknown. We suppose that the rank given by the ith respondent is $\sigma_i = \tau \circ \pi$ where $\tau_i \in \Lambda_n$ and $\circ$ denotes composition of functions. As in Abstract Algebra, we say that σ_i is the product of τ_i and π. The permutation τ_i can be thought of as "error" analogous to the case of the linear model $y_i = \mu + e_i$ where it is assumed that the observation y_i is mean μ plus some error e_i. The Sampling Importance Resampling (SIR) method is used to derive the posterior distribution of the true rank. In the SIR methodology a prior (joint) distribution is specified for the parameter(s). In this paper our parameter of interest is the true rank π. Samples are then drawn from this prior distribution and the likelihood is calculated for each such sample. The prior is then resampled using the likelihoods as weights. The resample constitutes a sample from the (joint) posterior distribution of the parameter(s). The approximate posterior marginal distribution of each of the unknown parameters can then be easily obtained. For an elegant discussion of SIR methodology the reader may look into Smith and Gelfand (1992).

In section 19.2, we consider situations where ranks given by the respondents possibly differ from the true rank by at most a transposition. In section 19.3, we consider a generalization of the case discussed in section 19.2 and suppose that the ranks given by the respondents possibly differ from the true rank by two or more disjoint transpositions. In section 19.4, we consider the case when ranks given by the respondents possibly differ from the true rank by one or more disjoint cycles. Since transpositions are cycles of order two this case is a generalization of the case discussed in section 19.3. In section 19.5, we discuss the completely general case where we allow the ranks given by the respondents to be any permutation of the true rank. In section 19.6, we illustrate the above methodology using some real life examples. In section 19.7, we make some concluding remarks.

19.2. One Transposition Error

In this section we discuss a simple situation where the rank given by a respondent possibly differs from the true rank by at most one transposition. A permutation τ is called a transposition if there exist i, j such that $\tau(i) = j, \tau(j) = i$ and $\tau(k) = k$ for all $k \neq i, j$. As an example suppose that we are interested to rank the four countries USA, India, Brazil and Tanzania in terms of their influence on world affairs. We may ask a group of experts on world affairs to rank the four countries based on the given criterion. In this case the ranks of USA and Tanzania can be decided upon unanimously but there may be a difference of opinion among the experts on the ranks of India and Brazil. Thus all the experts will give the same rank to USA and Tanzania but the ranks given to India and Brazil by some may be a transposition of the ranks given by the others.

Let there be n respondents and suppose that the rank given by i^{th} respondent $\sigma_i = \tau_i \circ \pi$ where τ_i is either the identity permutation or is a transposition. We first determine the possible values of π. Since $\tau_i \circ \tau_i = I$ (the identity permutation) we have $\pi = \tau_i \circ \sigma_i$. Let $T = \{\alpha : \alpha = I$ or α is a transposition$\}$ and define $S_i = \{\alpha_j \circ \sigma_i : \alpha_j \in T\}$ for $i = 1, 2, ...n$. Note that $\pi \in S_i$ for $i = 1, 2, ..., n$ and hence $\pi \in \bigcap_{i=1}^{n} S_i$. Thus all elements in $S = \bigcap_{i=1}^{n} S_i$ are possible values of π. If $S = \phi$ (empty set) then we can conclude that this model is not appropriate for the given data set. If $S = \{\pi\}$ then π is uniquely determined and no further analysis is required. In what follows we will assume that S contains at least two elements.

Suppose $S = \{\pi_1, ..., \pi_s\}$ and let us write $T = \{\alpha_0 = I, \alpha_1, ..., \alpha_t\}$. For each, $\pi_k \in S$ we find transpositions $\alpha_i^k \in T$ such that $\alpha_i^k \circ \sigma_i = \pi_k$ for $i = 1, 2, ..., n$. Now among the α_i^ks let α_r occur m_{rk} times, for $r = 0, 1, ..., t$. We assume that τ_is are independent identically distributed(i.i.d) T-valued random variables with $P(\tau_i =$

$\alpha_j) = p_j, j = 0, 1, ..., t, \sum_{j=0}^{t} p_j = 1$. Then the log-likelihood is

$$l(\pi_k, p_o, p_1, ..., p_t | \sigma_i, i = 1, 2, ..., n) \sum_{i=0}^{t} m_{ik} ln p_i. \quad (19.2)$$

We can now choose a joint prior distribution $(\pi, p_0, p_1, ..., p_t)$ and apply the SIR methodology to obtain samples from the joint posterior distribution of $(\pi, p_0, p_1, ..., p_t)$. An approximate posterior marginal distribution of π can then be obtained.

19.3. Disjoint Transposition Errors

We now consider a generalization of the situation discussed in the previous section. Instead of assuming that the ranks given by a respondent may differ from the true rank by only a transposition as in the previous section we now allow the possibility that it may differ by several disjoint transpositions. Two transpositions are said to be disjoint if they act on different pairs of elements i.e. if τ_1 and τ_2 are two disjoint transpositions with $\tau_1(i) = j, \tau_1(j) = i$ and $\tau_2(k) = t, \tau_2(t) = k$ then $\{i, j\} \cap \{k, t\} = \phi$. A collection of transpositions is said to be disjoint if any two of them are disjoint. Note that disjoint transpositions commute i.e. if τ_1 and τ_2 are two disjoint transpositions then $\tau_1 \circ \tau_2 = \tau_2 \circ \tau_1$ and hence we have $(\tau_1 \circ \tau_2) \circ (\tau_2 \circ \tau_1) = l$. Let, $T_d = \{\alpha : \alpha = l$ or α is a transposition or α is a product of (two or more) disjoint transpositions$\}$
It can be easily seen that if the permutation $\alpha \in T_d$ then $\alpha \circ \alpha = l$. Let the rank given by the ith respondent be $\sigma_i = \tau_i \circ \pi$ where $\tau_i \in T_d$. We can now proceed as in section 2 to first determine the possible values of π. Define $S_i = \{\alpha_j \circ \sigma_i : \alpha_j \in T_d\}$ for $i = 1, 2, ..., n$. Then all elements in $S = \cap_{i=1}^{n} S_i$ are possible values of π. If $S = \phi$ then we can conclude that this model is not appropriate for the given data set. If $S = \{\pi\}$ then π is uniquely determined and no further analysis is required. If S contains more than one element let $S = \{\pi_1, ..., \pi_u\}$ and write $T_d = \{\alpha_0 = l, \alpha_1, ..., \alpha_q\}$. For each $\pi_k \in S$, we find $\alpha_i^k \in T_d$ such that $\alpha_i^k \circ \sigma_i = \pi_k$ for $i = 1, 2, ..., n$. Now among the α_i^k's let α_r occur m_{rk} times, for $r = 0, 1, ..., q$. We assume that τ_i's are i.i.d T_d- valued random variables with $P(\tau_i = \alpha_j) = p_j, j = 0, 1, ..., q, \sum_{j=0}^{q} p_j = 1$. Then the log-likelihood is

$$l(\pi_k, p_0, p_1, ..., p_q | \sigma_i, i = 1, 2, ..., n) = \sum_{i=0}^{q} m_{ik} ln p_i \quad (19.3)$$

We can now use the SIR methodology as in section 19.2 to obtain an approximate posterior marginal distribution of π.

19.4. Cyclic Errors

In this section we discuss the case when the errors are cycles of length less than or equal to $g(2 \leq g)$ or product of disjoint cycles of length less than or equal to g. If $g = 2$ we get the situation discussed in section 3 since cycles of length 2 are transpositions. We assume that $\sigma_i = \tau_i \circ \pi$ where τ_i is a cycle of length less than or equal to g or is a product of disjoint cycles of length less than or equal to g. Let, $T_c = \{\alpha_1^{V_1} \circ \alpha_2^{V_2} \circ ... \circ \alpha_w^{V_w} | \alpha_1, ..., \alpha_w$ are disjoint cycles of length less than or equal to g, $0 \leq v_i \leq g, i = 1, ..., w, w \geq 1\}$

We can now proceed as in the earlier sections to determine the possible values of π. Define $S_i = \{\alpha_j \circ \sigma_i : \alpha_j \in T_c\}$ for $i = 1, 2, ..., n$. It is easy to see that (a) if τ_i is a cycle of length less than or equal to g then at least one of $\tau_i^V \circ \sigma_i, v = 1, ..., g-1$ equals π (b) disjoint cycles commute and (c) the inverses of two disjoint cycles are also disjoint. Hence, we can find $\alpha_i \in T_c$ such that $\alpha_i \circ \sigma_i = \pi$ for $i = 1, 2, ..., n$. Thus, $\pi \in S_i$ for $i = 1, 2, ..., n$ and hence in $S = \cap_{i=1}^n S_i$. As in earlier cases, here also, all elements of S are the possible values of π. If $S = \phi$ then we can conclude that this model is not appropriate for the given data set. The above procedure can then be repeated with a higher value of g. If $S = \{\pi\}$ then π is uniquely determined and no further analysis is required.

If S contains more than one element let $S = \{\pi_1, ..., \pi_z\}$ and write $T_c = \{\alpha_0, = l, \alpha_1, ..., \alpha_c\}$. For each $\pi_k \in S$, we find $\alpha_i^k \in T_c$ such that $\alpha_i^k \circ \sigma_i = \pi_k$ for $i = 1, 2, ..., n$. Now among the α_i^k's let α_r occur m_{rk} times, for $r = 0, 1, ..., c$. We assume that τ_i's are i.i.d T_c -valued random variables with $P(\tau_i = \alpha_j) = p_j, j = 0, 1, ..., c, \sum_{j=0}^c p_j = 1$. Then the log-likelihood is

$$l(\pi_k, p_0, p_1, ..., p_c | \sigma_i, i = 1, 2, ..., n) = \sum_{i=0}^{c} m_{ik} ln p_i \tag{19.4}$$

We can now use the SIR methodology as in section 19.2 above to obtain an approximate marginal posterior distribution of π.

19.5. The General Case

In this section we discuss the completely general case where we do not impose any restriction on the nature of the error. Thus,$\sigma_i = \tau_i \circ \pi$ where τ_i can be any permutation in Λ_m. Since $\{\alpha_j^{-1} \circ \sigma_i | \alpha_j \in \Lambda_m\} = \Lambda_m$ for all i, we have all permutations in Λ_m as possible values of π. We now proceed in the same spirit as in the earlier sections: let $\Lambda_m = \{\alpha_0 = l, \alpha_1, ..., \alpha_f\} (f = m! - 1)$ and for each $\pi_k \in \Lambda_m$ we find $\alpha_i^k \in \Lambda_m$ such that $\alpha_i^k \circ \sigma_i = \pi_k, i = 1, 2, ..., n$. Now among the

Table 19.1. Rankings given by the 25 doctoral students

Sl. No.	Rank	Sl. No.	Rank	Sl. No.	Rank	Sl. No.	Rank	Sl. No.	Rank
1	1,2,5,3,4	6	1,2,5,3,4	11	1,2,5,3,4	16	1,2,5,3,4	21	1,2,3,5,4
2	2,1,5,3,4	7	1,2,5,3,4	12	2,5,1,4,3	17	1,2,5,3,4	22	1,2,3,5,4
3	1,2,3,5,4	8	1,2,3,5,4	13	1,2,3,5,4	18	1,2,3,5,4	23	1,2,3,5,4
4	1,2,5,3,4	9	1,2,5,4,3	14	1,2,3,5,4	19	1,2,5,3,4	24	1,2,3,5,4
5	1,2,5,3,4	10	1,5,2,3,4	15	1,2,3,4,5	20	1,2,3,5,4	25	2,5,1,3,4

α_i^k's let α_r occur m_{rk} times, for $r = 0, 1, 2, ..., f$. We assume that τ_i's are i.i.d Λ_m -valued random variables with $P(\tau_i = \alpha_j) = p_j, j = 0, 1, ..., f, \sum_{j=0}^{f} p_j = 1$. Then the log-likelihood is

$$l(\pi_k, p_0, p_1, ..., p_f | \sigma_i, i = 1, 2, ..., n) = \sum_{i=0}^{f} m_{ik} ln p_i \tag{19.5}$$

We can now use the SIR technique to obtain the approximate posterior marginal distribution of π.

Remark: Since every permutation can be represented as a product of disjoint cycles we can get the general case from the cyclic error case discussed in section 19.4 by taking $g = n$.

19.6. Example

In this section we illustrate the techniques discussed in the preceding sections using a real-life example. We asked 25 doctoral students of management to rank, in decreasing order of the effect, the following qualities and accomplishments that they thought may have had on the decision of the interview panel for their selection to the doctoral programme of this premier management institute in India-

1. Academic Accomplishments
2. Communication Skills
3. Extra Curricular Activities (sports, cultural etc.)
4. Physical Appearance
5. Body Language

The responses of the students are given in Table 19.1. We are interested to find the true rank of the qualities and accomplishments as perceived by the students who were selected in the doctoral programme. This information can be a valuable input to the design of the interview process of this institute.

Since the data contains several responses which are transpositions of one another we first attempt to fit the disjoint transposition errors model discussed in

Table 19.2. The posterior probabilities of the ranks in S.

$\pi =$	π_1	π_2	π_3	π_4
Posterior Probability	0.8196	0.1216	0.0384	0.0204

section 19.3. We find that the model is not appropriate for analyzing this data set as the set of all possible values of the true rank S turns out to be the empty set. We then attempt to fit the disjoint cyclic errors model discussed in section 4. We restrict ourselves to cycles of length at most 3 (i.e. g = 3). In this case we find that the set S is non-empty. In fact, $S = \{\pi_1, \pi_2, \pi_3, \pi_4\}$ where $\pi_1 = 2,1,5,3,4; \pi_2 = 2,1,5,4,3; \pi_3 = 1,2,5,4,3; \pi_4 = 1,2,5,4,3$. We assume that the prior on π is independent of the prior on $(p_0, p_1, ..., p_c)$ (where $c+1$ is the total number of permutations of desired type i.e. those which are product of disjoint cycles of length at most 3; in this case c=65) and hence the joint prior on $(\pi, p_0, p_1, ..., p_c)$ is the product of the above two priors. We put an uniform prior on the possible values of π and a Dirichlet $(1,1,...,1)$ prior on $(p_0, p_1, ...p_c)$. Note that prior expectation of all the p_i's are equal since $E(p_i) = \frac{1}{c+1}$, for all $i = 0, 1, ..., c$. The samples from the prior joint distribution of $(\pi, p_0, p_1, ..., p_c)$ can be easily drawn using standard methods. 10,000 samples from the joint prior distribution of $(\pi, p_0, p_1, ..., p_c)$ are generated and the likelihood is computed for each of these. 10,000 samples from the posterior joint distribution of $(\pi, p_0, p_1, ..., p_c)$ are then generated by resampling the prior with probabilities proportional to the likelihood. The approximate posterior marginal distribution of π is given in Table 19.2. Since the posterior probability $P(\pi = \pi_1) = 0.8196$ one may think of π_1 as the true rank under this model.

An alternative approach can be to apply the model discussed in section 19.5 which makes no assumptions regarding the nature of errors. Analysis based on this model is computationally much more demanding as the number of parameters is much larger than in the earlier ones. We generated 100,000 samples from the joint prior distribution of $(\pi, p_0, p_1, ..., p_f)$ (in this case f=119) and the likelihood is computed for each of these. Then 100,000 samples from the posterior joint distribution of $(\pi, p_0, p_1, ..., p_f)$ are generated by resampling the prior with probabilities proportional to the likelihood. The result obtained is displayed in the histogram given in figure 19.1. We see that though some ranks have higher posterior probability than others the magnitude of the difference is not that large. This is reflected in the fact that 90% highest posterior density (HPD) credible set consists of 109 ranks out of 120. Thus not much useful information can be gained from this approximate marginal distribution. This possibly indicates the need for a much larger number of prior and posterior samples since the number of unknown parameters is quite large.

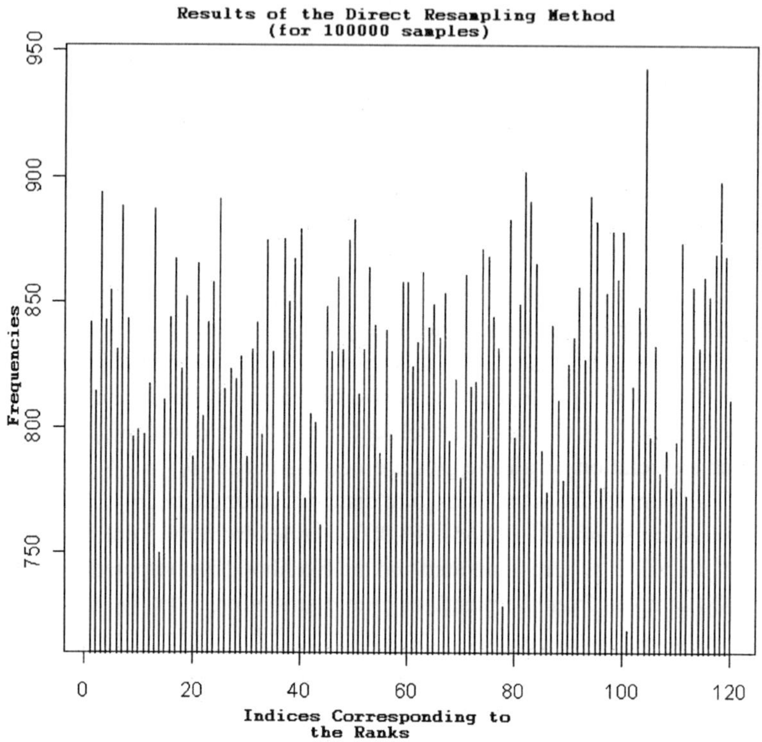

Fig. 19.1. Histogram of the ranks in the general case (based on 100000 posterior samples)

19.7. Concluding Remarks

In this paper we have discussed a Bayesian analysis of rank data under various kinds of error structures using SIR. In section 19.5 above, we have presented a completely general formulation which can be adopted if there is no reason to assume a simple structure for errors such as those discussed in sections 19.2-19.4. While the formulation given in section 19.5 is completely general, a disadvantage of it is that, it is extremely computationally intensive. So, whenever possible one should attempt to provide a simple structure for errors for a given data set. This will reduce the computation load substantially. In this context it may be helpful to observe that error structures different from those discussed in sections 19.2-19.4 can be easily handled by using similar ideas.

19.8. Acknowledgement

The second author wishes to thank the Indian Institute of Management, Ahmedabad for all the help and support provided to him during his visit.

References

1. Fraleigh, J. B. (1982). *A First Course in Abstract Algebra*, 3rd Edition, Addison Wesley, USA.
2. Marden, J. I. (1995). *Analyzing and Modelling Rank Data*, Chapman and Hall, London.
3. Smith, A.F.M and Gelfand, A.E. (1992). Bayesian statistics without tears: A sampling-resampling perspective, *The American Statistician*, **46**, 2, 84-88.

Chapter 20

Bayesian Tests of Equality of Stratified Proportions for a Multiple Response Categorical Variable

Balgobin Nandram

Department of Mathematical Sciences, Worcester Polytechnic Institute, 100 Institute Road, Worcester, MA 01609-2280

E-mail: balnan@wpi.edu

In many sample surveys, there are items that require individuals in different strata to make at least one of a number of choices. Within each strata, we assume that each individual has made all his/her choices, and the number of individuals with none of these choices is not reported. The analyses of such survey data are complex, and the categorical table with mutually exclusive categories can be sparse. We use a simple Bayesian product multinomial-Dirichlet model to fit the count data both within and across strata. Using the Bayes factor we show how to test that the proportions of individuals with each choice are the same over the strata. Because the Bayesian test is sensitive to prior specification, we also obtain a simpler test based on Bayesian estimation, a procedure that is barely sensitive to the prior specifications (i.e., reasonable departures from a noniformative prior). Using data from the Kansas Farm Survey and a simulation study, we have compared our two Bayesian tests with other non-Bayesian tests.

Contents

20.1. Introduction

In many surveys, multiple categorical responses or measurements are made on members of different populations, sub-groups or strata. This often arises in surveys where individuals can mark all answers that apply when responding to a multiple-choice question. Typically, it is of interest to estimate the proportions of individuals responding in different categories and to determine whether the distributions of responses differ among strata. As pointed out by Loughin and Scherer(1998), the standard chi-squared test is inadequate for this purpose, and they developed an adjusted chi-squared statistic for which a bootstrap procedure is needed to obtain the p-value of the test. However, they did not study the number of individuals who might have responded 'no' to all categories. We develop a Bayesian method to test the hypothesis of equality of proportions across the strata using the Bayes factor and an estimation procedure.

The study in Loughin and Scherer(1998) was motivated by the survey conducted by the Department of Animal Sciences at Kansas State University. Livestock farmers in Kansas were asked a variety of questions regarding their usual practices. One matter of interest to the researchers was the availability of veterinary information to farmers of all education backgrounds. For this reason, farmers were classified according to their education level (high school or less, vocational school, 2-year college, 4-year college, or other), and they were asked, "What are your primary sources of veterinary information?" Boxes could be marked for (A) professional consultant, (B) veterinarian, (C) state or local extension services, (D) magazines, and (E) feed companies and representatives. The question of interest, then, was whether the distribution of information sources differs among education levels. There were 262 farmers who provided responses to education levels and sources of veterinary information. However, the numbers of farmers who said 'no' on all sources are unknown. Also, the number of nonrespondents is unknown to us and we ignore them in this paper.

We consider a categorical variable with c levels, and we assume that an individual can respond in at least none of these levels. For example, for a categorical variable with two levels, A and B, the response of an individual can be none, denoted by ϕ, only A, only B, or both A and B, denoted by AB. A response sequence consists of zeros and ones with exactly one of them representing which of ϕ, A, B, AB an individual chooses. Thus, with c categories there are $L = 2^c$ response sequences an individual can have. Suppose also that there are r strata; in this simple example assume $r = 2$. Thus, there are four possible response sequences, and they are $(0,0)$, $(1,0)$, $(0,1)$, $(1,1)$ (e.g., (0,0) represents 'no' on both categories and $(1,1)$ 'yes' on both categories), and this is a partition of the space of all possible sequences of responses. In Stratum 1 let $n_{11}, n_{12}, n_{13}, n_{14}$ denote the number of individuals with the $(0,0), (1,0), (0,1), (1,1)$ sequences respectively. A 2×4 table with the $n_{j\ell}$ is called the n-table. In general, the n-table has r rows for the strata and L columns for the counts of the response sequences, and this table is usually sparse making the routine use of the standard Pearson chi-squared statistic for the test of equality of the strata proportions questionable. Note that in our formulation the n-table contains the missing "information" in the sequence with 'no' in all categories. Let C_k denote the set of sequences that have a one for the kth category. Then, $m_{jk} = \sum_{\ell \in C_k} n_{j\ell}, j = 1, \ldots, r,\ k = 1, \ldots, c$, form the m-table. So that while the m-table is a $r \times c$ table, the n-table is a $r \times L$ table. It is always easy to obtain the m-table from the n-table. An illustration of the two tables is given in Table 20.1. NOTE: n_{11} is the number of sequences with 'no' in all categories, and m_{11}

Table 20.1. Illustration of the n-table and the m-table for two categories and two strata

	n-table				m-table	
Stratum	ϕ	A	B	AB	A	B
1	n_{11}	n_{12}	n_{13}	n_{14}	m_{11}	m_{12}
2	n_{21}	n_{22}	n_{23}	n_{24}	m_{21}	m_{22}

is the number of individuals in category A (e.g., $m_{11} = n_{12} + n_{14}$).

One typically analyzes the m-table, a table in which the counts are the total number of individuals in each category. For example, for two variables the counts are $m_{11} = n_{12} + n_{14}$, $m_{12} = n_{13} + n_{14}$ in the first stratum. Thus, in the m-table an individual can belong to two categories, so that the standard Pearson chi-squared test cannot be applied to these data. Note that the n-table does not contain the information in the sequence with 'no' in all categories (i.e., missing data). Since it is the m-table that is usually analyzed, the uncertainty associated with the sequence of 'no' in all categories is usually ignored.

Thus, analysis of data from either the n-table or the m-table is problematic. It is apparent that one needs to use the n-table to make inference in the m-table. That is, we need to express the cell probabilities in the n-table in terms of the cell probabilities in the m-table; the parameters in the m-table being a much reduced set for reasonable values of c, the number of categories. Let $\pi_{j\ell},\ j = 1, \ldots, r,\ \ell = 1, \ldots, L$ denote the cell probabilities in the n-table and p_{jk} denote the cell probabilities in the m-table (i.e., the p_{jk} are the probabilities of selecting the kth category). Then, $p_{jk} = \sum_{\ell \in C_k} \pi_{j\ell},\ j = 1, \ldots, r,\ k = 1, \ldots, c$. The $\pi_{j\ell}$ can be related to the p_{jk} in

a similar manner (see (20.1) below). More importantly, equality of the stratum proportions in the m-table is equivalent to equality of the stratum proportions in the n-table. Thus, in theory one can use either the n-table or m-table to make inference of both sets of parameters.
Loughin and Scherer(1998) developed a large sample weighted chi-squared test and a small sample bootstrap test for the hypothesis that the probability of selecting any given veterinary information source is identical among the five education levels. As noted, ordinary statistical inference based on the standard chi-squared statistic is inappropriate because the 453 entries in Table 20.2 are not independent. See Loughin and Scherer(1998) for the n-table; we note that it should not be used for inference because it is extremely sparse with 89 cells having zero counts. To test the null hypothesis of multiple marginal independence with Table 20.2, Loughin and Scherer(1998) constructed a modified Pearson statistic that compares the cell counts in Table 20.2 with their proper expected values under the null hypothesis. They showed that its null asymptotic distribution is that of a linear combination of chi-squared random variables each with one degree of freedom. Unfortunately, the coefficients of the linear combination in their statistic depend on the unknown cell probabilities for the complete table, making their procedure impractical. Thus, they use the bootstrap to obtain the p-value of the test. Also, Agresti and Liu(1999) pointed out a second difficulty with the use of the modified Pearson chi-squared statistic in that it is not invariant to switching the 'yes' and 'no' labels for all items.

Table 20.2. Tabulation of veterinary information sources by education level (EDL) for farmers in Kansas

	Information source					
EDL	A	B	C	D	E	Total
HS	19	38	29	47	40	173
VS	2	6	8	8	4	28
2YC	1	13	10	17	14	55
4YC	19	29	40	53	29	170
Other	3	4	8	6	6	27
Total	44	90	95	131	93	453

NOTE: Data are taken from Loughin and Scherer[7]; see their Table 2. The number of farmers, who did not select none of the sources in each education level, is unknown. Note that the expanded table, given in Loughin and Scherer[7], is collapsed into each of the five sources; a farmer could be in at least one of the sources.

Agresti and Liu(1999) used a marginal logit approach to help overcome these problems with the modified chi-squared statistic. They view the response as a cross-classification of c binary responses. For example, in the Kansas Farm Survey, variable A indicates that a farmer said 'yes' to source A, variable B indicates that a farmer said 'yes' to source B, variable C indicates that a farmer said 'yes' to source C, and so forth. The complete table is the $r \times L$ contingency table (the n-table) showing the counts of possible response sequences at each level of the independent variable (e.g., education level), and the m-table is the marginal table showing the counts at each source (i.e., binary data). As pointed out by Agresti and Liu(1999), proper analyses use the n-table rather than the m-table. Let p_{jk}, $j = 1,\dots,r$, $k = 1,\dots,c$, denote the proportion of 'yes' responses in the m-table, Agresti and Liu(2001) describe the marginal logit model approach, in which $(p_{jk}, 1 - p_{jk})$, $k = 1,\dots,c$, are the c marginal distributions for the L cross-classification of responses at the jth level of the independent variable. For example, they use the model $\log(p_{jk}/(1-p_{jk})) = \beta_k, j = 1,\dots,k = 1,\dots,c$, as a baseline model (multiple marginal independence model in their terminology), assuming that for each level of the independent variable the cell counts in the n-table follow independent multinomial distributions over the strata. Then, they use goodness-of-fit tests to test this baseline model against various alternatives. For example, one alternative is the saturated model $\log(p_{jk}/(1-p_{jk})) = \beta_{jk}$, $j = 1,\dots,r$, $k = 1,\dots,c$, and another alternative is the unsaturated model $\log(p_{jk}/(1-p_{jk})) = \alpha_j + \gamma_k$, $j = 1,\dots,r$, $k = 1,\dots,c$, with $\alpha_r = 0$. As is evident this procedure uses overlapping data (i.e., the logits of the first and second category use common data so that simultaneous inference is needed to figure out the p-value of the test of marginal independence). [See Agresti and Liu(2001)] for further development of this procedure to include random effects and a discussion of the generalized estimating equations approach.

We can improve the statistical methodology by removing the difficulties associated with the use of the modified chi-squared statistic, the bootstrap procedure, and asymptotic approximations associated with the marginal logit model approach (i.e., maximum likelihood estimation procedure and the asymptotic normality associated with the goodness-of-fit tests). This can be done using a Bayesian approach. In addition, none of the above-mentioned authors has considered the missing cell problem (i.e., number of individuals answering 'no' to all categories); see Nandram and Zelterman(2007) for a Bayesian methodology to estimate a missing cell in a multinomial distribution. Thus, using a Bayesian methodology we show how to accommodate the missing data, and use the Bayes factor to test for equality

of the cell proportions across strata.
However, it is well-known that the Bayes factor is sensitive to the specification of prior distribution. One alternative to the Bayes factor is the intrinsic Bayes factor (Berger and Pericchi(1996)). Unfortunately there are difficulties with the intrinsic Bayes factor as well; specifically in our application it is difficult to define a minimal training sample. The Bayes factor still has useful features such as it is calibrated (Kass and Raftery(1995)), and if we can show that in a particular application it is not so sensitive, it is not unreasonable; see Nandram and Choi(2006). In addition, we have constructed an alternative test based on Bayesian estimation, rather than Bayesian hypothesis testing, an idea originally discussed in Nandram and Choi(2007). Thus, we have two Bayesian tests.
The rest of the paper discusses a novel Bayesian methodology to the modeling of a categorical variable for which individuals can select any number of categories, and these individuals are stratified as in the Kansas Farm Survey. In Section 20.2 we present a key assumption needed to simplify our method, and we describe the methodology in which we show how to test for marginal independence using the Bayes factor and Bayesian estimation. Technical details for the test based on the Bayes factor are discussed in Nandram, Toto and Katzoff(2009). In Section 20.3 we describe an innovative analysis of the data from the Kansas Farm Survey.

20.2. Bayesian Methodology

In Section 20.1 we describe the assumptions used to develop the likelihood function. In Section 20.2 we discuss pertinent posterior distributions which are needed to calculate the Bayes factor, and to construct the test based on Bayesian estimation. In Section 20.3 we discuss hypothesis testing of equality of the cell proportions over the strata. We will discuss two models, one in which the p_{jk} are unrestricted (called the *unrestricted* model) and the other in which the $p_{jk} = p_k,\ k = 1,\dots,c,\ j = 1,\dots,r$ (called the *restricted* model). In Section 20.4 we discuss the test based on Bayesian estimation.

20.2.1. *Construction of the likelihood function*

In a sample of size n_j from the jth strata, $j = 1,\dots,r$, let $n_{j1},\dots,n_{jL}$, denote the number of farmers with these responses. We also let $\sum_{\ell=1}^{L} n_{j\ell} = n_j,\ j = 1,\dots,r$, and note that the n_{j1} are unobserved.
Let E_{jk} denote the event that within the jth stratum a farmer chooses the kth category, $j = 1,\dots,r,\ k = 1,\dots,c$. A key assumption in our methodology is that the

E_{jk} are independent, and

$$P(E_{jk}) = p_{jk},\ j = 1,\ldots,r,\ k = 1,\ldots,c.$$

For example, the probability that an individual in the jth stratum chooses all categories is $\prod_{k=1}^{c} p_{jk}$, and letting C denote a set of categories in the jth stratum, the probability that an individual chooses the categories in C but not those outside C is $\prod_{k\in C} p_{jk} \prod_{k\notin C}(1-p_{jk})$.

Now, recall that there are $L = 2^c$ response sequences $I_{j\ell}$ in the jth stratum with each sequence having c elements corresponding to the categories; note that these sequences are the same in each stratum and they are fixed known vectors with each element being 0 or 1. Thus, letting $\pi_{j\ell}$ be the probability that a farmer within the jth stratum has the ℓth response sequence, by independence of the E_{jk} we have

$$\pi_{j\ell} = \prod_{k=1}^{c} p_{jk}^{I_{jk\ell}}(1-p_{jk})^{1-I_{jk\ell}},\ \ell = 1,\ldots,L,\ j = 1,\ldots,r. \tag{20.1}$$

Note again that $p_{jk} = \sum_{\ell\in C_k} \pi_{j\ell},\ j = 1,\ldots,r,\ k = 1,\ldots,c.$

Now, letting $n_j' = (n_{j1},\ldots,n_{jL})$ and $\pi_j' = (\pi_{j1},\ldots,\pi_{jL})$, like Agresti and Liu (1999) within the Bayesian framework, we make the relatively mild assumption that

$$n_j \mid \pi_j, N_j \overset{ind}{\sim} \text{Multinomial}(N_j,\pi_j),\ j = 1,\ldots,r,\ N_j = \sum_{\ell=1}^{L} n_{j\ell}.$$

Note that the N_j are unknown quantities. Then, letting $p_j' = (p_{j1},\ldots,p_{jc})$ and using (20.1), we have

$$p(n_j \mid p_j, N_j) = \frac{n_j!}{\prod_{\ell=1}^{L} n_{j\ell}!}\prod_{k=1}^{c}\left\{p_{jk}^{\sum_{\ell=1}^{L} n_{j\ell}I_{jk\ell}}(1-p_{jk})^{\sum_{\ell=1}^{L} n_{j\ell}(1-I_{jk\ell})}\right\},\ j = 1,\ldots,r.$$

Also, because $I_{jk1} = 0,\ j = 1,\ldots,r,\ k = 1,\ldots,c$, we have

$$p(n_{j(1)}, n_{j1} \mid p_j, N_j) =$$

$$\frac{(n_{j1}+\sum_{\ell=2}^{L} n_{j\ell})!}{n_{j1}!\prod_{\ell=2}^{L} n_{j\ell}!}\prod_{k=1}^{c}\left\{p_{jk}^{\sum_{\ell=2}^{L} n_{j\ell}I_{jk\ell}}(1-p_{jk})^{n_{j1}+\sum_{\ell=2}^{L} n_{j\ell}(1-I_{jk\ell})}\right\},$$

where, for convenience, we let $n_j = (n_{j(1)}, n_{j1})$ with $n_{j(1)}' = (n_{j2},\ldots,n_{jL}),\ j = 1,\ldots,r$.

Finally, assuming independence over strata, letting $n_1' = (n_{11},\ldots,n_{r1})$, $n_{(1)}' = (n_{1(1)}',\ldots,n_{r(1)}')$, $p' = (p_1',\ldots,p_r')$, we have

$$p(n_{(1)}, n_1 \mid p, N) =$$

$$\prod_{j=1}^{r}\left[\frac{(n_{j1}+\sum_{\ell=2}^{L}n_{j\ell})!}{n_{j1}!\prod_{\ell=2}^{L}n_{j\ell}!}\prod_{k=1}^{c}\left\{p_{jk}^{\sum_{\ell=2}^{L}n_{j\ell}I_{jk\ell}}(1-p_{jk})^{n_{j1}+\sum_{\ell=2}^{L}n_{j\ell}(1-I_{jk\ell})}\right\}\right]. \quad (20.2)$$

To obtain a full Bayesian analysis, we need a joint prior distribution on p and N, and *a priori* we will assume that p and N are independent. We will discuss the choices of priors for p and N in Sections 20.2 and 20.3. We note that *a posteriori* sometimes it is more convenient to work with the n_{j1} rather than the $N_j = n_{j1} + \sum_{k=2}^{c} n_{jk}$, $j = 1,\dots,r$, and vice versa.

20.2.2. *Pertinent posterior distributions*

For the unrestricted model, a priori with independence for all parameters, we take

$$p_{jk} \overset{ind}{\sim} \text{Uniform}(0,\ 1),\ j=1,\dots,r,\ k=1,\dots,c$$

$$p(N_j)=1,\ N_j=1,2,\dots;\ j=1,\dots,r. \quad (20.3)$$

Thus, using Bayes' theorem in (20.2) and (20.3), the joint posterior density of $p, n_1 \mid n_{(1)}$ is

$$\pi(p,n_1 \mid n_{(1)}) \propto$$

$$\prod_{j=1}^{r}\left[\frac{(n_{j1}+\sum_{\ell=2}^{L}n_{j\ell})!}{n_{j1}!\prod_{\ell=2}^{L}n_{j\ell}!}\prod_{k=1}^{c}\left\{p_{jk}^{\sum_{\ell=2}^{L}n_{j\ell}I_{jk\ell}}(1-p_{jk})^{n_{j1}+\sum_{\ell=2}^{L}n_{j\ell}(1-I_{jk\ell})}\right\}\right]. \quad (20.4)$$

To fit (20.4) to the data, we will use a sampling-based method to obtain samples from the joint posterior density. First, note that the posterior conditional density of p is simply

$$p_{jk} \mid n_{j1}, n_{j(1)} \overset{ind}{\sim} \text{Beta}\left\{\sum_{\ell=2}^{L} n_{j\ell}I_{jk\ell}+1,\ n_{j1}+\sum_{\ell=2}^{L} n_{j\ell}(1-I_{jk\ell})+1\right\}, \quad (20.5)$$

$j=1,\dots,r,\ k=1,\dots,c$. Thus, once samples of n_1 are obtained from the posterior density of $\pi(n_1 \mid n_{(1)})$, samples are easily obtained from the posterior density of p using the composition method.

How do we obtain samples from $\pi(n_1 \mid n_{(1)})$? By construction $\pi(n_1 \mid n_{(1)}) = \prod_{j=1}^{r}\pi(n_{j1} \mid n_{j(1)})$, so we describe how to obtain samples from $\pi(n_{j1} \mid n_{j(1)})$, $j = 1,\dots,r$, without using Markov chain Monte Carlo methods. Thus, integrating out p from (20.4), we have

$$p(n_{j1} \mid n_{j(1)}) \propto \frac{(n_{j1}+\sum_{\ell=2}^{L}n_{j\ell})!}{n_{j1}!}\prod_{k=1}^{c}\left\{\frac{\{n_{j1}+\sum_{\ell=2}^{L}n_{j\ell}(1-I_{jk\ell})\}!}{(n_{j1}+\sum_{\ell=2}^{L}n_{j\ell}+1)!}\right\},\ j=1,\dots,r. \quad (20.6)$$

Nandram, Toto and Katzoff(2009) showed in their Appendix A that the posterior density in (20.6) is proper (i.e., the joint posterior density of $(p_j,\ n_{j1})$ is a well-defined probability measure). Also, it is easy to show that $\lim_{n_{j1}\to\infty} p(n_{j1} \mid n_{j(1)}) = 0$. This latter statement shows that there exists a point which has negligible probability to its right; this is important for the testing procedure using the Bayes factor.

The function in (20.6) is easy to compute for all values of the n_{j1}, and therefore, it is easy to draw a sample from it. In our example, we have taken the largest value of n_{j1} to be $1,000$; so that we compute this function at 1,001 points, $n_{j1} = 0,1,\ldots,1,000$, to obtain the probability mass function. In practice, this probability mass function has its support in a much narrower range of values near to zero. Thus, we obtain an estimate of the posterior probability mass functions of the n_{j1}.

For the restricted model (i.e., $p_{jk} = p_k,\ j = 1,\ldots,r,\ k = 1,\ldots,c$), the situation is slightly more complex. A composition method can be used to draw p, but it is now computationally more efficient to draw the n_{j1} using a Gibbs sampler (Gelfand and Smith 1990). The posterior conditional density of p is even simpler, and it is

$$p_k \mid n_1, n_{(1)} \overset{ind}{\sim} \text{Beta}\left\{ \sum_{j=1}^{r} \{ \sum_{\ell=2}^{L} n_{j\ell} I_{jk\ell} \} + 1,\ \sum_{j=1}^{r} \{ n_{j1} + \sum_{\ell=2}^{L} n_{j\ell}(1 - I_{jk\ell}) + 1 \} \right\}, \tag{20.7}$$

$k = 1,\ldots,c$. Thus, once samples of n_1 are obtained from the posterior density of $\pi(n_1 \mid n_{(1)})$, samples are easily obtained from the posterior density of p again using the composition method. Now, the n_{j1} are correlated because their joint posterior density is

$$p(n_1 \mid n_{(1)}) \propto \left\{ \prod_{j=1}^{r} \frac{(n_{j1} + \sum_{\ell=2}^{L} n_{j\ell})!}{n_{j1}!} \right\} \left\{ \prod_{k=1}^{c} \frac{[\sum_{j=1}^{r} \{ n_{j1} + \sum_{\ell=2}^{L} n_{j\ell}(1 - I_{jk\ell}) \}]!}{[\sum_{j=1}^{r} \{ n_{j1} + \sum_{\ell=2}^{L} n_{j\ell} \} + 1]!} \right\}. \tag{20.8}$$

Nandram, Toto and Katzoff(2009) showed in their Appendix B that the joint posterior density, $p(n_1 \mid n_{(1)})$ in (20.8), is proper provided that $\sum_{j=1}^{r} \sum_{\ell=2}^{L} n_{j\ell} > r - c$. This condition can be easily satisfied.

It is easy to draw $p(n_{j1} \mid n_{j(1)}),\ j = 1,\ldots,r$. In fact, it now easy to run a Gibbs sampler. We have used 500 iterates to "burn in" the sampler, and took everyone thereafter (negligible autocorrelations).

For both cases, $p_{jk} = p_k,\ j = 1,\ldots,r,\ k = 1,\ldots,c$, and in the unrestricted model, we have used a conservative sample of 10,000 iterates for inference; this runs in less than half a minute.

20.2.3. *Test based on the Bayes factor*

First, we note that the test of $p_{jk} = p_k,\ j = 1,\ldots,r,\ k = 1,\ldots,c$ is equivalent to the test of $\pi_{j\ell} = \pi_\ell,\ j = 1,\ldots,r,\ \ell = 1,\ldots,L$; Nandram, Toto, and Katzoff(2009) in their Appendix C proved this equivalence. Thus, we can simply test the hypothesis $p_{jk} = p_k,\ j = 1,\ldots,r,\ k = 1,\ldots,c$.
To test the hypothesis, $p_{jk} = p_k,\ j = 1,\ldots,r,\ k = 1,\ldots,c$,we first use the Bayes factor which is the ratio of the marginal likelihoods under the two models. Let M_1 represent the unrestricted model and M_0 the restricted model. Then, the marginal likelihoods are $p_{M_1}(n_{(1)})$ and $p_{M_0}(n_{(1)})$, and the Bayes factor is $BF = p_{M_0}(n_{(1)})/p_{M_1}(n_{(1)})$. Kass and Raftery(1995) gave a comprehensive description of Bayes factors and their interpretation.
Because hypothesis-testing, which uses the Bayes factor, requires a proper prior on n_1 we will take n_{j1} to be independent with

$$p(n_{j1}) = \frac{1}{a_j^{(0)}+1},\ n_{j1} = 0,\ldots,a_j^{(0)}, \tag{20.9}$$

where $a_j^{(0)},\ j = 1,\ldots,r$, are to be specified. Also, because the Bayes factor may be sensitive to the priors $p_{jk} \overset{iid}{\sim} \text{Uniform}(0,1)$ in the unrestricted model and $p_k \overset{iid}{\sim} \text{Uniform}(0,1)$ in the restricted model, we use more flexible priors

$$p_{jk} \overset{ind}{\sim} \text{Beta}(\alpha_k,\beta_k),\ \ p_k \overset{ind}{\sim} \text{Beta}(\alpha_k,\beta_k). \tag{20.10}$$

We specify the $a_j^{(0)}$ in the estimation procedure. After we fit the posterior distribution $(p, n_1 \mid n_{(1)})$ in (20.4), we have the posterior density of $n_1 \mid n_{(1)}$. Thus, we have the range of the n_{j1} at our disposal. We take $a_j^{(0)} = \text{maximum}\{n_{j1} : P(n_{j1}) \geq .001,\ n_{j1} = 0,1,\ldots\},\ j = 1,\ldots,r$. In addition, the $(\alpha_k,\ \beta_k)$ are to be specified; one choice is to take $\alpha_k = \beta_k = 1$, and we will study sensitivity to this choice later. Thus, by independence of (20.9) and (20.10), the joint probability density of $n_{(1)}, p, n_1$ is

$$p_{M_1}(n_{(1)}, p, n_1) = \prod_{j=1}^{r}\left[\frac{1}{a_j^{(0)}+1}\frac{(n_{j1}+\sum_{\ell=2}^{L} n_{j\ell})!}{n_{j1}!\prod_{\ell=2}^{L} n_{j\ell}!}\right.$$

$$\left.\times\prod_{k=1}^{c}\frac{p_{jk}^{\sum_{\ell=2}^{L} n_{j\ell}I_{jk\ell}+\alpha_k-1}(1-p_{jk})^{n_{j1}+\sum_{\ell=2}^{L} n_{j\ell}(1-I_{jk\ell})+\beta_k-1}}{B(\alpha_k,\beta_k)}\right], \tag{20.11}$$

and under the hypothesis, $p_{jk} = p_k,\ j = 1,\ldots,r,\ k = 1,\ldots,c$,the joint probability density of $n_{(1)}, p, n_1$ is

$$p_{M_0}(n_{(1)}, p, n_1) = \left\{ \prod_{j=1}^{r} \frac{1}{a_j^{(0)}+1} \frac{(n_{j1} + \sum_{\ell=2}^{L} n_{j\ell})!}{n_{j1}! \prod_{\ell=2}^{L} n_{j\ell}!} \right\}$$

$$\times \prod_{k=1}^{c} \frac{p_k^{\sum_{j=1}^{r}\{\sum_{\ell=2}^{L} n_{j\ell} I_{jk\ell}\}+\alpha_k-1} (1-p_k)^{\sum_{j=1}^{r}\{n_{j1}+\sum_{\ell=2}^{L} n_{j\ell}(1-I_{jk\ell})\}+\beta_k-1}}{B(\alpha_k,\beta_k)}. \quad (20.12)$$

Note that unlike (20.11), in (20.12) there is a single vector $p' = (p_1,\ldots,p_c)$. To obtain the marginal likelihoods from (20.11) and (20.12), we must integrate out p and n_1. The integration of p is straightforward, but that of n_1 is not so simple. We have worked with the logarithm of the Bayes factor.

Let $\mathcal{N} = \{(n_{11},\ldots,n_{r1}),\ 0 \le n_{j1} \le a_j^{(0)},\ j = 1,\ldots,r\}$. Then, under the unrestricted model the marginal likelihood is

$$p_{M_1}(n_{(1)}) = \sum_{n_1 \in \mathcal{N}} \left[\prod_{j=1}^{r} \left\{ \frac{1}{a_j^{(0)}+1} \frac{(n_{j1} + \sum_{\ell=2}^{L} n_{j\ell})!}{n_{j1}! \prod_{\ell=2}^{L} n_{j\ell}!} \right.\right.$$

$$\left.\left. \times \prod_{k=1}^{c} \left\{ \frac{(\sum_{\ell=2}^{L} n_{j\ell} I_{jk\ell} + \alpha_k - 1)! \{n_{j1} + \sum_{\ell=2}^{L} n_{j\ell}(1-I_{jk\ell}) + \beta_k - 1\}!}{(n_{j1} + \sum_{\ell=2}^{L} n_{j\ell} + \alpha_k + \beta_k - 1)! B(\alpha_k,\beta_k)} \right\} \right\} \right], \quad (20.13)$$

and under the restricted model the marginal likelihood is

$$p_{M_0}(n_{(1)}) = \sum_{n_1 \in \mathcal{N}} \left[\prod_{j=1}^{r} \left\{ \frac{1}{a_j^{(0)}+1} \frac{(n_{j1} + \sum_{\ell=2}^{L} n_{j\ell})!}{n_{j1}! \prod_{\ell=2}^{L} n_{j\ell}!} \right\} \times \right.$$

$$\left. \prod_{k=1}^{c} \frac{\{\sum_{j=1}^{r} \sum_{\ell=2}^{L} n_{j\ell} I_{jk\ell} + \alpha_k - 1\}! [\sum_{j=1}^{r} \{n_{j1} + \sum_{\ell=2}^{L} n_{j\ell}(1-I_{jk\ell})\} + \beta_k - 1]!}{[\sum_{j=1}^{r} \{n_{j1} + \sum_{\ell=2}^{L} n_{j\ell}\} + \alpha_k + \beta_k - 1]! B(\alpha_k,\beta_k)} \right]. \quad (20.14)$$

It is not efficient to compute the Bayes factor $BF = p_{M_0}(n_{(1)})/p_{M_1}(n_{(1)})$ by computing these two marginal likelihoods separately if this computation is to be done repeatedly. Also, the cardinality of the set $\mathcal{N}$ is very large so that it is very time-consuming to compute the marginal likelihoods directly, even though one can do so. We describe how to compute the Bayes factor in Appendix A.

20.2.4. *Test based on Bayesian estimation*

This method uses Bayesian estimation to construct a 95% credible hyper-rectangle in rc-space. Inference is obtained for p_{jk} under the unrestricted model with posterior density given in (20.4). We obtain a procedure similar to the one in Nandram and Choi(2007) for a test of independence in a two-way categorical table.
Under the hypothesis $p_1 = \ldots = p_r$, we have $\frac{1}{r}\sum_{j=1}^{r} p_j = p_1 = \ldots = p_r$. Then, we consider the parameters $\rho_{jk} = p_{jk}/\frac{1}{r}\sum_{j=1}^{r} p_{jk} - 1,\ j = 1,\ldots,r,\ k = 1,\ldots,c$. Thus, under the hypothesis $p_1 = \ldots = p_r$, we have $\rho_{jk} = 0,\ j = 1,\ldots,r,\ k = 1,\ldots,c$. Note that the parameters ρ_{jk} inherit their prior and posterior densities from the p_{jk}.
We need a 95% simultaneous credible region for $\rho_{jk},\ j = 1,\ldots,r,\ k = 1,\ldots,c$. We can do so using the method of Besag et al.(1995). As we discussed already, it is easy to draw a random sample $p_{jk}^{(h)},\ j = 1,\ldots,r,\ k = 1,\ldots,c,\ h = 1,\ldots,M$, where M is large, and we take it to be $M = 10,000$. This provides the needed sample $\rho_{jk}^{(h)},\ j = 1,\ldots,r,\ k = 1,\ldots,c,\ h = 1,\ldots,M$. The method of Besag et al.(1995) provides a 95% credible hyper-rectangle in a manner similar to the one-dimensional problem in which the [.025M] and [.975M] ordered values of $\rho_{jk}^{(h)},\ h = 1,\ldots,M$ are used as the end points of a 95% credible interval.
We can form a test that $\rho_{jk} = 0,\ j = 1,\ldots,r,\ k = 1,\ldots,c$, easily. This is done by simply checking that the hyper-rectangle contains $\rho = 0$. If each of the components of ρ contains 0, then the hyper-rectangle does. That is, if all components of ρ contain 0, there is equality of the cell proportions across strata, and if at least one component of ρ does not contain 0, the null hypothesis is rejected.
One can quantify the strength of the evidence against the null hypothesis by counting the number γ of the rc credible intervals that do not contain zero; so that $\gamma = 0, 1, \ldots, rc$. If $\gamma = rc$, there is very strong evidence against the null hypothesis; if $\gamma = 0$, there is no evidence against the null hypothesis. That is, the degree of evidence against the null hypothesis depends on the magnitude of γ.

20.3. Data Analysis

Using our methods in Section 20.2 we now analyze the data in Table 1. We also make comparisons with other analyses of these data. In Section 20.3.1 we discuss testing equality of the proportions over education levels. In Section 20.3.2 we perform a simulation study to further compare the tests based on the Bayes factor, the bootstrap statistic and the hyper-rectangles.

20.3.1. *Testing equality of the proportions over education levels*

We consider the test of equality of the sample proportions across education levels. We have considered two scenarios, one using the original data and the other using collapsed data. Vocational school and other are collapsed to form four levels of education (a 4×5 categorical table); we do so because the sparseness of the original table is mostly removed (cannot do much with the expanded table). This will give a slightly better assessment of the tests.
Our method gives a log-Bayes factor of 12.21 with a numerical standard error of .0268 for the original data; for the collapsed data the log-Bayes factor is 8.22 with a numerical standard error of .0265. Thus, in either case there is very strong evidence for the restricted model (i.e., equality of cell proportions); see Kass and Raftery(1995).
For the original data set, our method based on Bayesian estimation, shows that only one of the twenty-five 95% simultaneous credible intervals does not contain zero; it is ρ_{45} with its 95% credible interval being $(-.703,\ -.013)$. Thus, there is mild evidence against the null hypothesis that the strata proportions are equal. For the collapsed data set, our method based on Bayesian estimation, shows that two of the twenty 95% simultaneous credible intervals do not contain zero; they are ρ_{32}, with its 95% credible interval being $(-.690,\ -.021)$, and ρ_{45}, with its 95% credible interval being $(-.688,\ -.048)$. Thus, the evidence against the null hypothesis that the strata proportions are equal is slightly stronger for the collapsed data. In any case, this is very weak evidence against the null hypothesis of equality of proportions across education levels. Having near twenty-five credible intervals for the original data not containing zero, or twenty for the collapsed data, will be very strong evidence against the null hypothesis.
Our bootstrap method differs from that of Louglin and Scherer(1998) because we use a product-multinomial model. Referring to the expanded table, we obtain maximum likelihood estimates by fitting $n_j \mid \pi \overset{iid}{\sim} \text{Multinomial}(n, \pi),\ j = 1, \ldots, r$; the n_{j1} are estimated from our Bayesian procedure. Then, we drew B samples from $n_j \overset{iid}{\sim} \text{Multinomial}(n, \hat{\pi}),\ j = 1, \ldots, r$, where $\hat{\pi}_\ell = \sum_{j=1}^{r} n_{j\ell}/n,\ \ell = 1, \ldots, L$ and we computed the m-table for each bootstrap sample. Finally, we computed the test statistic of Louglin and Scherer(1998). Our bootstrap method gives a p-value of .052 for the original data, not significant at the 5% significance level; it gives a p-value of .015 for the collapsed data, now significant at the 5% significance level. The bootstrap method of Louglin and Scherer(1998) gives a p-value of .047 with a numerical standard error of .003 for the original data. At the 5% significance level, in our case the null hypothesis is marginally not significant, and in the case of Louglin and Scherer(1998) it is marginally significant.

Further, we have considered looking at the reduced table that consists of only the first part of the expanded table (i.e., only A, only B, only C, only D and only E), both original and collapsed; of the 262 individuals there are 162 ($\approx 64.1\%$) of the data in this reduced table. We also consider looking at the m-table as though there are no 'double counts'. Momentarily, let n_{jk}, $j = 1,\ldots,r$, $k = 1,\ldots,c$, denote the cell counts of either table. Also, with standard notation we let $n_{j\cdot} = \sum_{k=1}^{c} n_{jk}$, $n_{\cdot k} = \sum_{j=1}^{r} n_{jk}$ and $n_{\cdot\cdot} = \sum_{j=1}^{r}\sum_{k=1}^{c} n_{jk}$. To test for equality of cell proportions, we test $n_j \mid p \overset{ind}{\sim} \text{Multinomial}(n_{j\cdot}, p)$, $j = 1,\ldots,r$, $p \sim \text{Dirichlet}(1,\ldots,1)$ versus $n_j \mid p_j \overset{ind}{\sim} \text{Multinomial}(n_{j\cdot}, p_j)$, $j = 1,\ldots,r$, $p_j \overset{iid}{\sim} \text{Dirichlet}(1,\ldots,1)$. Then, it is easy to show that the Bayes factor is

$$BF = \frac{\prod_{j=1}^{r}(n_{j\cdot}+c-1)!\prod_{k=1}^{c} n_{\cdot k}!}{\{(c-1)!\}^{r-1}(n_{\cdot\cdot}+c-1)!\prod_{j=1}^{r}\prod_{k=1}^{c} n_{jk}!}.$$

In Table 2 for the reduced table, the chi-squared test shows no evidence against equality of the cell proportions whereas the log-Bayes factor shows strong evidence for the equality of the cell proportions, contrary to the evidence from the collapsed data. For the m-table, the log-Bayes factor shows very strong evidence for the equality of cell proportions, but for the collapsed data the evidence is very weak.

Nandram, Toto and Katzoff[10] presented a sensitivity study of the log-Bayes factor to assess the choice of the prior density of the $p_{jk} \overset{ind}{\sim} \text{Beta}(\alpha_k, \beta_k)$ in the unrestricted model and $p_k \overset{ind}{\sim} \text{Beta}(\alpha_k, \beta_k)$ in the restricted model. For this purpose, they have taken $\alpha_k = \beta_k = \alpha$, $k = 1,\ldots,c$, and studied the log-Bayes factor as a function of α. Here $\alpha = 1$ corresponds to the prior distribution, which is usually used in the binomial-beta model, and $\alpha = .50$ to Jeffreys' prior. Nandram, Toto and Katzoff(2009) in their Figure 1 plotted the log-Bayes factors for both the original and the collapsed data as a function of alpha; the one for the original data is slightly higher. The two curves decrease and then increase, and the standard error at each point is relatively small. For the original data the minimum is at .78 with a Bayes factor of 8.02 and a numerical standard error of .027. For the collapsed data the minimum is at .78 (same point) with a Bayes factor of 4.70 and a numerical standard error of .027. The log-Bayes factor is slightly more sensitive for the original data. However, this sensitivity does not matter because there is very strong evidence for the equality of the strata proportions. That is, the minimum of the log-Bayes factor is larger than 5 (very strong evidence) for the original data and almost so (strong evidence) for the collapsed data for the equality of cell proportions in either case; see Kass and Raftery(1995).

Finally, we compare our method with the model-based approach of Agresti and

Liu(1999). To represent *multiple marginal independence*, they use the baseline model $\log(p_{jk}/(1-p_{jk})) = \beta_k,\ j=1,\ldots,r,\ k=1,\ldots,c$. One alternative they tested is the saturated model $\log(p_{jk}/(1-p_{jk})) = \beta_{jk},\ j=1,\ldots,r,\ k=1,\ldots,c$, which reduces to the baseline model under the null hypothesis $\beta_{1k} = \ldots = \beta_{rk},\ k = 1,\ldots,c$. They reported a likelihood-ratio statistic of $G^2 = 30.9$ and a chi-squared statistic of $\chi^2 = 26.7$ with 20 degrees of freedom each, and respective p-values of .06 and .14. They concluded that there is "weak evidence against the hypothesis." Another alternative is the unsaturated model $\log(p_{jk}/(1-p_{jk})) = \alpha_j + \gamma_k,\ j = 1,\ldots,r,\ k=1,\ldots,c$, with $\alpha_r = 0$ which reduces to the baseline model under the null hypothesis $\alpha_j = 0,\ j=1,\ldots,r$. They reported a likelihood-ratio statistic of $G^2 = 8.1$ (no chi-squared statistic was reported) with 4 degrees of freedom, and p-value of .09. They concluded that there is "weak evidence of improvement over the multiple marginal independence model." See Agresti and Liu(1999) for other models with similar p-values. Clearly, these tests are not significant at the 5% significance level. Thus, the results of Agresti and Liu(19999) are in concordance with those from the Bayes factor and the hyper-rectangles.

Unfortunately, the main issue with the method of Agresti and Liu(1999)is that it is not clear how to tell what is the specific correct alternative model; there will always be uncertainty in picking such a model. The method of Louglin and Scherer(1998) is attractive, but their bootstrap test statistic is defective (see Agresti and Liu(1999) for detailed discussion). Though sensitive to prior specification, the Bayes factor is still a possible alternative, and the test based on Bayesian estimation may be preferred.

20.3.2. *A simulation study*

We have performed a simulation study to compare the test based on the Bayes factor, the bootstrap statistic, and the 95% credible hyper-rectangle.

We have generated the data in the following manner. First, we have estimated the cell probabilities using maximum likelihood estimators in the expanded table (i.e., n-table with five levels of education and five choices) using the data from the Kansas Farm Survey. We have done so for each education level and for all educational levels combined into a single multinomial distribution. Let $\pi^{(a)}$ denote the vector of cell probabilities corresponding to the combined multinomial distribution. We have drawn $\pi_j \overset{iid}{\sim} \text{Dirichlet}(n_j + .001),\ j = 1,\ldots,5$, and we have computed

$$\Delta = \frac{1}{5}\sum_{j=1}^{5}\left\{\sum_{\ell=1}^{L}(\pi_{j\ell} - \pi_\ell^{(a)})^2\right\}^{1/2}.$$

We have repeated the process 1000 times, to get 1000 values of Δ which we have used to identify choices of π_j. We have selected three cases associated with values of Δ (a) $\Delta = 0.00$ corresponding to no difference between the π_j (i.e., take $\pi_{j\ell} = \pi_\ell^{(a)}$, $j = 1,\ldots,5$); (b) $\Delta = 0.13$ (i.e., this corresponds to the case in which the $\pi_{j\ell}$ are maximum likelihood estimates based on the raw data; and (c) $\Delta = 0.24$, the median of the 1000 Δ values; this corresponds to reasonable large differences among the π_j. Then, we draw $n_j \overset{ind}{\sim} \text{Multinomial}(n_j, \pi_j)$, $j = 1,\ldots,5$, and the $m_{jk}, j = 1,\ldots,5$, $k = 1,\ldots,5$, are easily obtained. Note that the n_j are the column totals (i.e., sum over rows) corresponding to the observed data from the Kansas farm survey.

We have drawn 1000 data sets at each of (a), (b) and (c). For each data set we have computed the logarithm of the Bayes factor (LBF), the p-value of the bootstrap statistic, and the number γ of the 95% simultaneous credible intervals of the ρ_{jk} not containing zero. For cases (a) and (c) the results are clear, but for case (b) the results are not so clear.

When $\Delta = 0.00$, there are no differences among the education levels. Of the 1000 data sets, the LBFs are larger than 5 for 999 of them, thereby showing very strong evidence for H_0. This is in concordance with the bootstrap statistic which shows very weak evidence against H_0 for 911 of these data sets. The results from the Bayes factor and the bootstrap statistic are also in concordance with the value of γ; 914 of the 1000 hyper-rectangles contain zero.

When $\Delta = 0.24$, there are significant differences between the education levels. Each of the 1000 data sets gives a p-value smaller than .01 (i.e., there is very strong evidence against H_0). Of the 1000 data sets, the LBFs are smaller than -5 for 952 of them, thereby showing very strong evidence against H_0. Here only 7 of the hyper-rectangles contain zero, and 993 hyper-rectangles show different degrees of departure from the null hypothesis. However, there are no hyper-rectangles with at least ten of the 95% simultaneous credible intervals not containing zero. Again, there is a reasonable degree of concordance among the three procedures.

However, when $\Delta = 0.13$ as in case (b), there are substantial differences among the test based on the Bayes factor, the bootstrap statistic, and the hyper-rectangles. First, of the 1000 data sets there are 823 showing very strong evidence against H_0 under the bootstrap statistic, but the evidence under the Bayes factor varies from very strong evidence for H_0 to very strong evidence against H_0. For example, of the 823 data sets that show very strong evidence for H_0, under the bootstrap statistic there are 184 data sets showing very strong evidence against H_0 and 206 data sets showing very strong evidence for H_0. Here 210 of the hyper-rectangles contain zero, and 790 hyper-rectangles show different degrees of departure from the null hypothesis. In this case, the procedure based on the hyper-rectangles and the bootstrap statistic are more in concordance than the test based on the Bayes factor.

Under extreme situations as in (a) and (c), the results from the bootstrap statistic, the Bayes factor and the hyper-rectangles are expected to be in concordance. On the other hand, in marginal situations as in case (b), the bootstrap statistic and the hyper-rectangles are more in concordance than the test based on the Bayes factor. Tests based on the Bayes factor are defective because they are well-known to be sensitive to the prior specifications and the difficulty in computing the Bayes factor may not worth the effort. In short, Bayesians should avoid using them! Alternative Bayesian tests such as the one based on the hyper-rectangles are much-needed.

20.4. Acknowledgement

The author thanks Criselda Toto and Myron Katzoff for their assistance.

Appendix A: Computation of the Bayes Factor

First, letting $a_{jk} = \sum_{\ell=2}^{L} n_{j\ell} I_{jk\ell}$ and $b_{jk} = n_{j1} + \sum_{\ell=2}^{L} n_{j\ell}(1 - I_{jk\ell})$, it is interesting to note that

$$BF = E\left\{ \frac{W_0(n_1)}{W_1(n_1)} \mid n_{(1)} \right\}, \tag{20.15}$$

where

$$W_0(n_1) = \prod_{k=1}^{c} \left\{ \frac{(\sum_{j=1}^{r} a_{jk} + \alpha_k - 1)!(\sum_{j=1}^{r} b_{jk} + \beta_k - 1)!}{(\sum_{j=1}^{r} a_{jk} + b_{jk} + \alpha_k + \beta_k - 1)!} \right\}$$

and

$$W_1(n_1) = \prod_{j=1}^{r} \prod_{k=1}^{c} \left\{ \frac{(a_{jk} + \alpha_k - 1)!(b_{jk} + \beta_k - 1)!}{(a_{jk} + b_{jk} + \alpha_k + \beta_k - 1)!} \right\}.$$

Note that in (20.15) expectation is taken over the unrestricted model with posterior density

$$p(n_{j1} \mid n_{j(1)}) \propto$$

$$\frac{(n_{j1} + \sum_{\ell=2}^{L} n_{j\ell})!}{n_{j1}!} \prod_{k=1}^{c} \left\{ \frac{(a_{jk} + \alpha_k - 1)!(a_{jk} + \beta_k - 1)!}{(a_{jk} + b_{jk} + \alpha_k + \beta_k - 1)!} \right\}, \tag{20.16}$$

$n_{j1} = 0, \ldots, a_j^{(0)}$, $j = 1, \ldots, r$. Also, note that (20.16) is the same as (20.6) with $\alpha_k = \beta_k = 1$, and (20.16) is convenient importance function because the posterior densities of the n_{j1} are independent. We have available estimates of the posterior densities of the n_{j1} in (20.16).

We have two choices to calculate the Bayes factor. First, we can compute BF directly (without using a sampling-based method). Because of the large cardinality of $\mathcal{N}$, this is a time-consuming process, and we would need to do this calculation repeatedly in a simulation study. Thus, in the second method, it is much more efficient to use Monte Carlo integration to compute BF. We can easily draw a large sample $n_{j1}^{(h)}$, $h = 1,\ldots,M$, from $p(n_{j1} \mid n_{j(1)})$, $j = 1,\ldots,r,$, $h = 1,\ldots,M$. Then, a simulation consistent estimator of BF is

$$\widehat{BF} = \frac{1}{M}\sum_{h=1}^{M} e^{W(n_1^{(h)})},$$

where $W(n_1^{(h)}) = \log\{W_0(n_1^{(h)})\} - \log\{W_1(n_1^{(h)})\}$, $h = 1,\ldots,M$ (i.e., $\widehat{BF}$ approaches BF almost surely as M approaches infinity). We start with $M = 10000$, and we increase M until $\widehat{BF}$ stabilizes; stabilization occurs around 30,000. We note that the answers obtained from the two methods (the exact one and the one using Monte Carlo integration) are roughly the same. Also note that by construction of the $a_j^{(0)}$, BF is almost non-sensitive to reasonable departure from the selected $a_j^{(0)}$.

Finally, note that we have worked with the logarithm of the Bayes factor, and using a first-order Taylor's series expansion, an approximate numerical standard error of $\log(\widehat{BF})$ is $S_W/\sqrt{M}$ where $S_W^2 = \sum_{h=1}^{M}\{W(n_1^{(h)}) - \bar{W}\}^2/(M-1)$ with $\bar{W} = \sum_{h=1}^{M} W(n_1^{(h)})/M$.

References

1. Agresti,A. and I. Liu,(1999) Modeling a categorical variable allowing arbitrarily many category choices, *Biometrics*, **55**, 936.
2. Agresti, A. and I. Liu, (2001) Strategies for modeling a categorical variable allowing multiple category choices, *Sociological Methods and Research*, **29**, 403.
3. Berger, J.O. and L. R. Pericchi,(1996) The intrinsic Bayes factor for model selection and prediction, *Journal of the American Statistical Association*, **91**, 122.
4. Besag,J., P. Green, D. Higdon, and K. Mengersen,(1995) Bayesian computation and stochastic systems (with discussion), *Statistical Science*, **10**, 3 .
5. Gelfand, A.E. and A. F. M. Smith, (1990) Sampling-based approaches to calculating marginal densities, *Journal of the American Statistical Association*, **85**, 398..
6. Kass, R. and A. Raftery,(1995) Bayes factors, *Journal of the American Statistical Association*, **90**, 773.
7. Loughin, T.M. and P. N. Scherer, (1998) Testing for association in contingency tables with multiple column responses, *Biometrics*, **54**, 630.

8. Nandram,B. and J. W. Choi, (2007) Alternative tests of independence in two-way categorical tables, *Journal of Data Science*, 5, 217.
9. Nandram, B. and J. W. Choi,(2006) A Bayesian analysis of a two-way categorical table incorporating intra-class correlation, *Journal of Statistical Computation and Simulation*, **76**, 233.
10. Nandram,B. M. C. S. Toto, and M. Katzoff,(2009) Bayesian inference for a stratified categorical variable allowing all possible category choices, *Journal of Statistical Computation and Simulation*, **79**, 161-179.
11. Nandram, B. and D. Zelterman, (2007) Computational Bayesian inference for estimating the size of a finite population, *Computational Statistics and Data Analysis*, **51** 2945..

Chapter 21

Respondent-Generated Intervals in Sample Surveys: A Summary

S. James Press[1] and Judith M. Tanur[2]

[1] *University of California, Riverside*
E-mail: jpress@ucr.edu
[2] *State University of New York, Stony Brook*
E-mail: jtanur@notes.cc.sunysb.edu

Contents

21.1. Introduction

This paper summarizes research on Respondent-Generated Intervals (RGI) as a new methodology for recall in factual sample surveys. The theoretical development involves Bayesian hierarchical modeling, while the motivation and applications involve cognitive aspects of surveys. We are assuming:
(1) in a factual survey with a recall-type question, each respondent (R) has a distinct recall distribution for the fact being requested;
(2) R knew the true value at some time in the past, or knew enough to construct an accurate answer, but because of imperfect recall, he/she is not now certain of the true value;
(3) R does not deliberately deceive.
Each individual's degree of uncertainty about an unknown parameter could be assessed by assessing many points on that individual's subjective probability distribution, using a sequence of elicitation questions. Then we could connect the points by a smooth curve representing the underlying distribution. Such an approach would be impractical in sample survey because of the time constraints

during a survey interview. But we can compromise by using three points: basic answers and bounds. That is, we can ask for a basic answer to an important question, such as: " How many more or fewer times did you visit the doctor last year compared with the previous year?" (We have phrased the question in this way in order to obtain both positive and negative values of the variable, so that the assumption of normality is more plausible.) The basic answer is called the "usage quantity". But we could also ask: What is the smallest you are almost certain that difference could be, and what is the largest you are almost certain that difference could be? We call these answers the bounds.

Thus R's themselves generate intervals in which their true beliefs lie. Their true beliefs are not forced into pre-assigned intervals, as is often done in other survey protocols. The RGI protocol uses these bounds to improve estimation.

21.2. Modeling

Suppose R_i gives a usage quantity, y_i and bounds $(a_i, b_i), i = 1, ..., n$, as his/her answers.

The random quantities y_i, a_i, b_i are jointly distributed.

Assume: $(y_i|\theta_i, \sigma_i^2) \sim N(\theta_i, \sigma_i^2)$.

Assume the means of the usage quantities, are themselves exchangeable and normally distributed about some unknown population mean of fundamental interest, θ_0, so that

$$(\theta_i|\theta_0, \tau^2) \sim N(\theta_0, \tau^2).$$

Thus, respondent i has a recall distribution whose true mean value is θ_i (e.g., each R is attempting to recall his/her particular change in number of visits to the doctor last year).

We desire to estimate θ_0. Assume $(\sigma_1^2, ..., \sigma_n^2, \tau^2)$ are known; they will be assigned later.

Denote the column vector of usage quantities by $\mathbf{y} = (y_i)$ and the column vector of means by $\theta = (\theta_i)$.

Let $\sigma^2 = (\sigma_i^2)$ denote the column vector of data variances. The joint density of the y_i's is then given in summary form by:

$$p(\mathbf{y}|\theta, \sigma^2) \propto \exp\{(-\frac{1}{2})\sum_1^n(\frac{y_i - \theta_i}{\sigma_i})^2\}.$$

The joint density of the θ_i's is given by

$$p(\theta|\theta_0,\tau^2) \propto \exp\{(-\frac{1}{2})\sum_1^n(\frac{\theta_i-\theta_0}{\tau})^2\}.$$

So the joint density of $(\mathbf{y},\theta)$ is given by

$$p(\mathbf{y},\theta|\theta_0,\tau^2,\sigma^2) = p(\mathbf{y}|\theta,\sigma^2)p(\theta|\theta_0,\tau^2).$$

The Final Result

Square the terms in the exponent of $p(\mathbf{y},\theta|\theta_0,\tau^2,\sigma^2)$ combine them, and integrate out over all θ_i. Then, adopt a vague prior distribution for θ_0 ($p(\theta_0) \propto$ constant). Using Bayes' theorem, the posterior distribution for θ_0 becomes.

$$(\theta_0|\mathbf{y},\tau^2,\sigma^2) \sim N(\widetilde{\theta},\omega^2)$$

where

$$\widetilde{\theta} = \sum_1^n \lambda_i y_i$$

$$\lambda_i \equiv \frac{(\frac{1}{\sigma_i^2+\tau^2})}{\Sigma_1^n(\frac{1}{\sigma_i^2+\tau^2})}, \sum_1^n \lambda_i = 1, \quad 0 \le \lambda_i \le 1.$$

and

$$\omega^2 = \frac{1}{\Sigma_1^n(\frac{1}{\sigma_i^2+\tau^2})}.$$

Thus the mean $\widetilde{\theta}$ of the conditional posterior density of the population mean, θ_0, is a convex combination of the respondents' usage quantities; i.e., it is a weighted average.

Interpret $(\sigma_i^2+\tau^2)^{-1}$ as the precision attributable to R_i's response, and interpret $\Sigma_1^n(\sigma_i^2+\tau^2)^{-1}$ as the total precision attributable to all respondents. Then, λ_i is interpretable as the proportion of total precision attributable to respondent i.

Thus, the greater R_i's precision proportion, the greater the weight that is automatically assigned to respondent's usage response.

Assessing the Variances

We must still assess the variances $(\sigma_1^2,\sigma_2^2,...,\sigma_n^2,\tau^2)$. Suppose that:

$$a_i|a_{i0}.\psi_{ai}^2 \sim N(a_{i0},\psi_{ai}^2) \text{ and } b_i|b_{i0},\psi_{bi}^2 \sim N(b_{i0},\psi_{bi}^2).$$

Next, using the structure of the normal distribution, assume the offered bounds for all subjects in the sample are approximately 2 standard deviations on either side of their respective means. (This is equivalent to assuming that when we ask R's to

give an interval they are almost certain covers truth they interpret almost certain as 95% sure.)
Accordingly, take approximately,

$$4\sigma_i = b_i - a_i, \quad i = 1,...,n$$

as our assessments for the σ_i's. Since

$$\lambda_i = \frac{(\frac{1}{\sigma_i^2+\tau^2})}{\Sigma_1^n \frac{1}{\sigma_i^2+\tau^2}}$$

and

$$4\sigma_i = b_i - a_i,$$

Rs who give shorter intervals have heavier weights. Hence we hope accurate Rs give short intervals, and we developed questioning that encourages that tendency (see Chu, Press and Tanur, 2004).
Then, define:

$$a* = \frac{1}{N}\Sigma_1^N a_{i0};\ b* = \frac{1}{N}\Sigma_N^1 b_{i0};$$

$$\overline{a} = \frac{1}{n}\Sigma_1^n a_i; \qquad \overline{b} = \frac{1}{n}\Sigma_1^n b_i;$$

where $a*, b*$ are averages of the true (unobserved) values of these bounds over the entire population; $\overline{a}, \overline{b}$ are the averages of the observed values of the bounds over the sample.
Assume approximately:

$$\psi_{a1}^2 = \psi_{a2}^2 = ... = \psi_a^2 \qquad \psi_{b1}^2 = \psi_{b2}^2 = ... = \psi_b^2.$$

Then

$$\overline{a} \sim N(a*, \frac{\psi_a^2}{n}); \qquad \overline{b} \sim N(b*, \frac{\psi_b^2}{n})$$

Next note that the true population mean value for respondent i must be between its bounds:

$$a* \leq \theta_0 \leq b*$$

Credibility Intervals
For 95% credibility on $a*$ with respect to a vague prior we have (approximating 1.96 by 2, here and throughout, for convenience):

$$\overline{a} - 2\frac{\psi_a}{\sqrt{n}} \leq \theta_0 | \text{data} \leq \overline{b} + 2\frac{\psi_b}{\sqrt{n}}$$

Table 21.1. Comparing Sample and RGI Posterior Means for Estimating Population Means in Campus Experiments Using Normal Priors and Range Estimator-Stony Brook

	Truth	X-bar	Conf. Int. (95%)	Post Mean	**Cred. Int. (95%)**
CREDI TS	67.53	63.13	**(56.12, 70.14)**	**63.69**	**(55.54, 71.84)**
GPA	2.91	2.99	**(2.89, 3.09)**	**2.97**	**(2.85, 3.09)**
SAT-M	570.80	593.72	(572.40,615.00)	**591.97**	**(553.15,630.79)**
SAT-V	503.20	526.00	(503.80, 548.20)	**519.01**	**(478.52,559.50)**
TICKETS	0.53	**0.92**	(.56, 1.28)	0.95	**(.32, 1.58)**
FINES	1.52	2.25	**(0, 5.41)**	**1.00**	**(.03, 1.96)**

From the normality and 95% credibility,

$$4\tau = (\bar{b} - \bar{a}) + \frac{2}{\sqrt{n}}(\psi_a + \psi_b)$$

But ψ_a and ψ_b are unknown. Estimate them by their sample quantities:

$$S_a^2 \equiv \overline{\Psi}_a^2 \equiv \frac{1}{n}\sum_n^1 (a_i - \bar{a})^2; \qquad S_b^2 \equiv \overline{\Psi}_b^2 \equiv \frac{1}{n}\sum_n^1 (b_i - \bar{b})^2$$

Then, the assessment procedure for τ becomes:

$$4\tau = (\bar{b} - \bar{a}) + \frac{2}{\sqrt{n}}(S_a - S_b)$$

21.3. Empirical Work

We ran parallel record-check surveys on our campuses, asking students questions about their lives on campus. If R's consented, answers were verified through appropriate campus offices. On both campuses we asked (and can construct normal priors for) the student's GPA (Grade Point Average), SAT (Scholastic Aptitude Test) math and verbal scores (SATM, SATV), and the number of oncampus parking tickets R had been given (TICKETS). At SUSB we were also able to use the number of credits R had earned (CREDITS), and the number of library fines assessed (FINES).

Campus Experiment Results

We calculated the posterior mean, using a proper prior and the range of the bounds to estimate the population variance; it was closer to truth than the sample mean for 8 of the 10 items (see Tables 21.1 and 21.2). The Bayesian credibility interval covered truth for all 10 items; the traditional confidence interval covered truth only for 6 of the 10. In the tables, boldface point estimates denote winners; boldface interval estimates denote intervals that cover truth.

Nonnormality

What if the data suggest that a normal distributional assumption isn't met? If the

Table 21.2. Comparing Sample and RGI Posterior Means for Estimating Population Means in Campus Experiments Using Normal Priors and Range Estimator-Riverside

	Truth	x-bar	Conf. Int.	Post Mean (95%)	Cred. Ind. (95%)
GPA	3.05	3.10	**(3.00,3.20)**	**3.04**	**(2.88,3.21)**
SAT-m	574.08	572.60	**(549.50, 595.60)**	**574.05**	**(537.54, 610.56)**
SAT-v	485.40	503.00	**(481.20,524.80)**	**500.38**	**(463.74, 537.02)**
TICKETS	0.21	**0.51**	(.27, .75)	0.63	**(.09, 1.16)**

distribution of the data is gamma, we've used the Wilson-Hillferty transformation on the data first (see McCullagh and Nelder, 1983, p. 153). It is:

$$Z_i = 3\left[(\frac{y_i}{\theta_i})^{1/3} - 1\right]$$

Table 21.2 shows some results from another classroom record check survey, before and after using this transformation. The various conditions reflect the different modes of questioning designed to encourage accurate respondents to give short intervals and less accurate ones to give longer ones (see Chu, Press and Tanur, 2004). Here two different Bayesian estimators are used. The extended average estimator uses the assessment of τ as described above,

$$4\tau = (\overline{b} - \overline{a}) + \frac{2}{\sqrt{n}}(S_a - S_b)$$

The extended range estimator substitutes for $(\overline{b}, \overline{a})$ by using the range of the sample bounds $(b_0 - a_0)$, that is, the distance between the highest upper bound and the lowest lower bound offered by respondents.

Before the use of the gamma transformation we see that the sample mean was the closest to truth in 5 of the 9 comparisons. After transformation one of the Bayesian estimators was closest to truth in 7 of the 9 comparisons.

Nonresponse

Although respondents to the RGI protocol are asked to give both a usage quantity and bounds, some respondents ignore these instructions and while they don't give a usage quantity, they do give bounds. When this happens, we can use the bounds information for imputation. In our original classroom record check surveys, we found that for some questions a large proportion of the respondents who left out usage quantities did supply us with bounds. Figure 21.1 shows these proportions.

We can use the midpoint of bounds these respondents supplied as point estimates for those respondents. We are then faced with the question of how good the imputation is. Unfortunately, verification data were not collected for respondents who did not give full responses, but fortunately for our purposes, the true amounts for the data on fees the students paid (variables HFEE/SB,

Table 21.3. Effect of Gamma Transformation

	Using Normal Assumption					Using Gamma Transformation				
Condition	Midterm					Midterm				
	n	Av.Truth	x-bar	ext.av.	ext.range	n	Av.Truth	x-bar	ext.av.	ext.range
UnAnch.	80	77.04	**78.00**	78.78	78.10	80	77.04	78.00	78.18	**77.41**
Nar/Wide	46	81.87	81.74	83.83	**81.89**	46	81.87	**81.74**	80.52	80.77
Wide/Wide	25	78.88	**80.20**	80.94	80.38	25	78.88	80.20	**79.59**	79.65
	Registration Fee					Registration Fee				
UnAnch.	78	$239	$1,495	**$1,302**	$1,468	75	$239	$1,555	**$959**	$1,461
Nar/Wide	51	$239	$882	**$856**	$874	50	$239	$900	**$594**	$848
Wide/Wide	44	$239	$1,029	**$815**	$967	42	$239	$1,078	**$681**	$924
			Movies			Movies				
UnAnch.	61	5.75	**4.64**	3.90	4.54	55	5.80	5.15	4.32	**5.35**
Nar/Wide	51	4.61	**4.53**	4.02	4.42	50	4.70	4.62	4.43	**4.76**
Wide/Wide	38	5.82	**5.18**	4.13	5.00	35	6.11	**5.63**	4.13	5.50

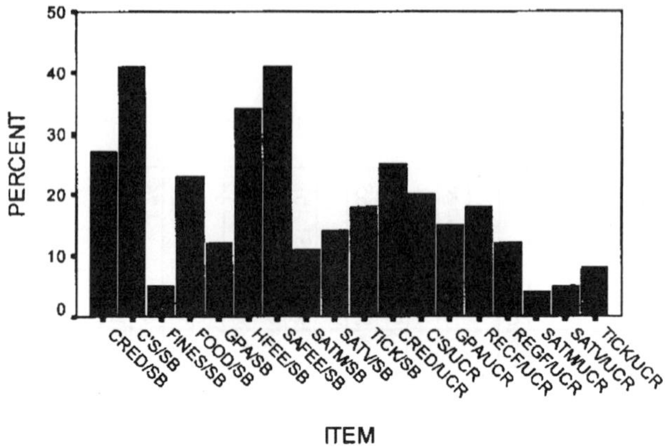

Fig. 21.1. Percentage of Cases with Missing Usage Quantiles that have Respondent-Generated Intervals

SAFEE/SB, RECF/UCR, and REGF/UCR in Figure 21.1) are uniform across student-respondents. Hence we can use the fee data to evaluate the accuracy of using the midpoint of the bounds for imputation. Table 21.4 shows the results. We see that the average midpoint is quite close to truth for two of the four variables.

21.4. Conclusions

RGI has advantages over conventional methods of fielding/analyzing surveys.
1) It often supplies estimates that are more accurate than conventional sample means.
2) R's can answer to sensitive questions which they might not otherwise answer.
3) R's generally feel that providing merely bounds to a question that has a numerical answer is less revealing to the interviewer than is answering a question that requires a specific point estimate for an answer.
4) Hence RGI can be useful in reducing item nonresponse.

Table 21.4. Accuracy of midpoint of bounds

Item	N	Truth	Average Midpoint
SUSB Health Fee	74	\$70.00	\$123.98
SUSB Activity Fee	102	\$78.75	\$87.14
UCR Rec Fee	26	\$59.00	\$121.71
UCR Reg Fee	24	\$1,371.25	\$1,343.65

References

1. Liping, Chu, S.James Press, and Judith M. Tanur (2004) Confidence Elicitation and Anchoring in the Respondent-Generated Intervals (RGI) Protocol. *Journal of Modern Applied Statistical Methods*,**3**,417-431.
2. McCullagh,P. and A.J. Nelder (1983) *Generalized Linear Models.* London: Chapman and Hall.
3. Wilson,E.B. & Hilferty, M. M. (1931). The distribution of chi-square. *Proceedings of the National Academy of Sciences of the United States of America*, **17**, 684-688.

Chapter 22

Quality Index and Mahalanobis D^2 Statistic

Ratan Dasgupta

Indian Statistical Institute, Kolkata

E-mail:ratandasgupta@gmail.com

In conformity with Bhattacharyya affinity or Hellinger affinity between two probability distributions, definition of Mahalanobis distance is extended to the case when the dispersion matrices of two populations are not necessarily equal. This includes the case where some of the variables are discrete or multinomial, and the mean vector is asymptotically normal after suitable standardization. We show that the Mahalanobis distance is unperturbed by inclusion of highly correlated additional variables. A 'quality index' is proposed to compare individuals or objects having multiple characteristics. As for example, a group of students at entrance level is ranked on the basis of a real data set on several tests.

Contents

22.1. Introduction

Let $x = (x_1, x_2, ..., x_p)'$ be a p dimensional random variable with mean $\mu = (\mu_1, \mu_2, ..., \mu_p)'$ and a positive definite dispersion matrix $\sum = ((\sigma_{ij}))\text{pxp}$, where $\sigma_{ij} = \text{cov}(x_i, x_j)$. The Mahalanobis distance squared between x and μ is defined as,

$$\begin{aligned}\Delta^2(x,\mu) &= (x-\mu)'\,\Sigma^{-1}\,(x-\mu)\\ &= (x-\mu)'(x-\mu) = ||x-\mu||^2, \text{ if } \Sigma = I.\end{aligned}$$

This distance differs from the usual Euclidian distance $||x-\mu||$; here the relative dispersions and the correlations of the elements of x are taken into account.
For two populations with means $\mu^{(1)}$ and $\mu^{(2)}$, and *a common dispersion matrix* Σ, the Mahalanobis distance squared between the two means or, two populations is defined as,

$$\Delta^2(\mu^{(1)},\mu^{(2)}) = (\mu^{(1)}-\mu^{(2)})'\Sigma^{-1}(\mu^{(1)}-\mu^{(2)}) \tag{22.1}$$

Let C be a matrix such that $\Sigma = CC'$ and let $\nu^{(i)} = C^{-1}\mu^{(i)},\ i=1,2$. Then,

$$\Delta^2 = (\nu^{(1)}-\nu^{(2)})'(\nu^{(1)}-\nu^{(2)}) = \sum_{i=1}^{p}(\nu_i^{(1)}-\nu_i^{(2)})^2 \tag{22.2}$$

which is the Euclidian distance squared, in terms of transformed co-ordinates.

Let $x_1^{(1)},\cdots,x_{n_1}^{(1)}$ be a sample of size n_1, from a population with mean vector and dispersion matrix as $(\mu^{(1)},\Sigma)$ and $x_1^{(2)},\cdots,x_{n_2}^{(2)}$ be a sample of size n_2, from $(\mu^{(2)},\Sigma)$. Then, estimate of $\mu^{(1)}$ is $\bar{x}^{(1)} = \sum_{i=1}^{n_1} x_i^{(1)}/n_1$, of $\mu^{(2)}$ is $\bar{x}^{(2)} = \sum_{i=1}^{n_2} x_i^{(2)}/n_2$ and an unbiased estimate S of the common dispersion matrix Σ is given by,

$$\begin{aligned}(n_1+n_2-2)S = &\sum_{i=1}^{n_1}(x_i^{(1)}-\bar{x}^{(1)})(x_i^{(1)}-\bar{x}^{(1)})' \\ &+\sum_{i=1}^{n_2}(x_i^{(2)}-\bar{x}^{(2)})(x_i^{(2)}-\bar{x}^{(2)})'\end{aligned} \tag{22.3}$$

$$\text{i.e.,}\quad n\,S = (n_1-1)S_1+(n_2-1)S_2,\ \ n=n_1+n_2-2 \tag{22.4}$$

An estimate of Δ^2 in 22.1 is provided by sample Mahalanobis distance squared,

$$D^2 = (\bar{x}^{(1)}-\bar{x}^{(2)})'S^{-1}(\bar{x}^{(1)}-\bar{x}^{(2)}) \tag{22.5}$$

Under the assumption of multivariate normal distribution, the bias term goes to zero, when sample sizes n_1 and n_2 are large, i.e.
$ED^2 = \{\Delta^2+\frac{1}{p}(n_1^{-1}+n_2^{-1})\}\frac{n}{n-p+1} \simeq \Delta^2,\ n=n_1+n_2-2.$
Mahalanobis distance 22.1, 22.1 and 22.5 arise, for example, in Hotelling's $T^2 = n(\bar{x}-\mu_0)'S^{-1}(\bar{x}-\mu_0)$, for testing $H_0:\mu=\mu_0$ versus $H_1:\mu\neq\mu_0$. This is an extension of univariate t statistic squared viz. $n(\bar{x}-\mu_0)^2/s^2$, when $p=1$.
The D^2 statistic has application in discriminant analysis, cluster analysis and in robust estimation by multivariate trimming of outliers.
Here, we derive a quality index based on a number of relevant quality characteristics, while assessing individuals or products. To this end, extension of D^2

statistic is made, using the concept of Bhattacharyya affinity, for the case when the dispersion matrices are possibly unequal in two populations. This includes the case where some of the variables are discrete or multinomial, and the mean vector is asymptotically normal after suitable standardization by an application of multivariate central limit theorem. The proposed technique may be used in grading computer software, ranking of individuals or industrial products and in general, for objects with multiple characteristics.

In section 22.2, we prove a result that D^2 statistic is robust with respect to inclusion of highly correlated variables. The error caused by deleting the correlated coordinate, while calculating the distance, is estimated in Proposition 22.2.1 and Remark 2. In sections 22.3 and 22.4, the distance is generalized and a quality index is proposed. In section 22.5, we use the technique to a real data set, for ranking a group of entrance students, based on their marks in three different tests.

22.2. D^2 Statistic with Highly Correlated Variables

Mahalanobis distance takes into account the covariance structure of the co-ordinate random variables $x_i, i = 1, \cdots, p$. If the correlation coefficient between co-ordinates is near to 1, essentially there is a single co-ordinate. The effect of such strong dependence on the sample version of Mahalanobis distance is of interest.

Without loss of generality, let there be two co-ordinates and the covariance structure be standardized, i.e., $\Sigma = \begin{pmatrix} 1 & \rho \\ \rho & 1 \end{pmatrix}$.

Then, $\Sigma^{-1} = (1-\rho^2)^{-1} \begin{pmatrix} 1 & -\rho \\ -\rho & 1 \end{pmatrix}$.

We wish to study the behavior of D^2 statistic as $|\rho| \to 1$.

Consider the distance between $(\bar{x}_1, \bar{y}_1) = \hat{\mu}^{(1)}$ and $(\bar{x}_2, \bar{y}_2) = \hat{\mu}^{(2)}$; i.e., the D^2 statistic based on a sample of size n each; from two populations $(\mu^{(1)}, \Sigma)$ and $(\mu^{(2)}, \Sigma)$, where $\mu^{(1)} = (\mu_x^{(1)}, \mu_y^{(1)})$, $\mu^{(2)} = (\mu_x^{(2)}, \mu_y^{(2)})$. An estimate of common dispersion matrix Σ is S given by 22.3. Now $S^{-1} \to \Sigma^{-1}$ with probability 1, as $n \to \infty$ by strong law of large numbers. Thus,

$$\begin{aligned}
&\lim_{n \to \infty} D^2(\hat{\mu}^{(1)}, \hat{\mu}^{(2)}) \\
&= \lim_{n \to \infty} (\bar{x}_1 - \bar{x}_2, \bar{y}_1 - \bar{y}_2) S^{-1} (\bar{x}_1 - \bar{x}_2, \bar{y}_1 - \bar{y}_2)' \\
&\qquad = (1-\rho^2)^{-1} \lim_{n \to \infty} (d_1, d_2) \begin{pmatrix} 1 & -\rho \\ -\rho & 1 \end{pmatrix} (d_1, d_2)'
\end{aligned}$$

$$= (1-\rho^2)^{-1} \lim_{n\to\infty} (d_1^2 - 2\rho d_1 d_2 + d_2^2)$$

$$= \lim_{n\to\infty} \{d_1^2 + \frac{(d_2-\rho d_1)^2}{(1-\rho^2)}\} \tag{22.6}$$

where, $d_1 = \bar{x}_1 - \bar{x}_2$, $d_2 = \bar{y}_1 - \bar{y}_2$.

Let the regression of y on x be linear, i.e.,
$Y = \mu_y + \rho\frac{\sigma_y}{\sigma_x}(x-\mu_x) = \mu_y + \rho(x-\mu_x)$, since $\sigma_x = \sigma_y = 1$.
Thus, $y = Y + e$ where, $e = e_i$ is the error of prediction in the ith sample.
So, $(y-\mu_y) = \rho(x-\mu_x) + e$, $Var(e) = (1-\rho^2), E(e) = 0$.
i.e., $(\bar{y}-\mu_y) = \rho(\bar{x}-\mu_x) + \bar{e}$, $\bar{e} = \sum_i e_i/n \to 0$, with probability 1; as $n \to \infty$.
i.e., $(\bar{y}^{(1)} - \mu_y^{(1)}) = \rho(\bar{x}^{(1)} - \mu_x^{(1)}) + \bar{e}^{(1)}$, for the first sample. Similarly,
$(\bar{y}^{(2)} - \mu_y^{(2)}) = \rho(\bar{x}^{(2)} - \mu^{(2)}) + \bar{e}^{(2)}$, for the second sample.

Thus, $d_2 - \mu_y^{(1)} + \mu_y^{(2)} = \rho(d_1 - \mu_x^{(1)} + \mu_x^{(2)}) + \bar{e}^{(1)} - \bar{e}^{(2)}$, and
$d_2 - \rho d_1 = (\mu_y^{(1)} - \rho\mu_x^{(1)}) - (\mu_y^{(2)} - \rho\mu_x^{(2)}) + O_p(n^{-1/2}) = O_p(n^{-1/2})$; i.e.,

$$\begin{aligned} D^2(\hat{\mu}^{(1)}, \hat{\mu}^{(2)}) &= (\bar{x}_1 - \bar{x}_2)^2 + O_p(n^{-1}(1-\rho^2)^{-1}) \\ &= (\mu_x^{(1)} - \mu_x^{(2)})^2 + O_p(n^{-1}(1-\rho^2)^{-1}) \\ &= \Delta^2(\mu_x^{(1)}, \mu_x^{(2)}) + O_p(n^{-1}(1-\rho^2)^{-1}) \end{aligned} \tag{22.7}$$

The r.h.s of 22.7 is based on a single variable x. So, D^2 statistic is unperturbed when variables with high correlations are added in the set of variables. For limiting D^2 approximation, the error in deleting a correlated coordinate in Δ^2 is $O_p(n^{-1}(1-\rho^2)^{-1})$, under the assumption of linear regression. Thus, D is robust with respect to inclusion of highly correlated variables. The result is stated below.

22.2.1. *Proposition*

The D^2 statistic is unperturbed when an additional variable with high correlation is included to the set of variables. Under the assumption of linear regression, the error term in the limiting D^2 approximation by Δ^2, deleting the correlated co-ordinate from Δ^2, is $O_p(n^{-1}(1-\rho^2)^{-1})$.

Remark 1. The condition of linear regression is satisfied for normal distribution. When $|\rho| = 1$, then both Σ and S are singular and the Mahalanobis distance is not defined. If the absolute value of correlation between any two variables in the set of variables is high; $|\rho|$ is near 1, any one of the two correlated variables may be dropped. The growth of n has to be proportional to $(1-\rho^2)^{-1}$, $|\rho| \to 1$, so that error of approximation is small.

Remark 2. The above order of approximation may be obtained in the 'almost sure' form as well. By Marcinkiewicz-Zygmund strong law of large numbers (MZSLLN); see e.g., Chow and Teicher (1978), one may obtain $\bar{X}_n - \mu = o(n^{-\gamma/(1+\gamma)})$ a.s., where $X, X_1, X_2, \cdots$ are iid random variables with $EX = \mu$ and $E|X|^{1+\gamma} < \infty$, $0 < \gamma < 1$. Thus, assuming the existence of fourth moment of coordinate random variables, each element of the matrix S converges to corresponding population value at the rate $o(n^{-\gamma^*})$ a.s., $0 < \gamma^* < 1$, with an application of MZSLLN to x_i^2, y_i^2, $x_i y_i$, etc. The matrix S is then 'almost surely' within $o(n^{-\gamma^*})$ neighborhood of the parameter Σ. Next, use the relationship, $(A+UV')^{-1} = A^{-1} - \frac{(A^{-1}U)(V'A^{-1})}{1+V'A^{-1}U}$, where A is nonsingular and U and V are two column vectors. The error rate in approximation mentioned in Proposition 22.2.1, is then $o(n^{-\gamma^*}(1-\rho^2)^{-1})$ a.s., $0 < \gamma^* < 1$.

22.3. A Quality Index Based on D^2 Statistic

Consider a multivariate population with mean μ and dispersion matrix Σ. The Mahalanobis distance of the point μ from origin 0, (considered as the base point or ground level) is $\mu' \Sigma^{-1} \mu = \Delta_0^2$, $\mu' = (\mu_1, ..., \mu_p)$. Let μ_i, the i th co-ordinate of μ, denote the mean of i th quality characteristic x_i of an industrial product; higher the value of characteristic, better is the product.

One may interpret the distance Δ_0 as the (unknown) quality index of the product, with respect to the base level 0.

To estimate Δ_0 on the basis of repeated quality measurements $x = (x_1, \cdots, x_p)$ by a group of assessors, one may proceed as follows.

Let there be k assessors and let n_i independent observations be made by the i th assessor; $i = 1, 2, \cdots, k$. The model we shall assume is $x_{ij} = \mu + \alpha_i + e_{ij}$, $j = 1, 2, \cdots, n_i$, $i = 1, 2, \cdots, k$, where x_{ij} is the j th (vector valued) observation by i th assessor. For i- th assessor, assume that e_{ij} has mean vector 0 and dispersion matrix $\sum$. The component α_i may be interpreted as random (bias) effect due to i th assessor. Let α_i be uncorrelated with $e_{ij}, j = 1, 2, \cdots, n_i, i = 1, 2, \cdots, k$. Also, let e_{ij} be uncorrelated with $e_{i'k}$, $i \neq i', \forall j, k$.

Now, suppose that $E\ \alpha_i = 0$, i.e., mean bias over all the assessors is zero. An estimate of μ is $\hat{\mu} = n^{-1}\Sigma_{i,j} x_{ij}$; $n = \Sigma_{i=1}^{k} n_i$. An unbiased estimator of Σ, based

on the observations taken by the ith assessor is S_i, where $(n_i - 1)S_i = \Sigma_{j=1}^{n_i}(x_{ij} - \bar{x}_i)(x_{ij} - \bar{x}_i)'$, and $\bar{x}_i = n_i^{-1}\Sigma_{j=1}^{n_i} x_{ij}$. The pooled estimate of Σ is S, where $(n-k)S = \Sigma_{i=1}^{k}(n_i - 1)S_i$.
i.e.,

$$S = (n-k)^{-1}\Sigma_{i=1}^{k}(n_i - 1)S_i = (n-k)^{-1}\Sigma_{i,j}(x_{ij} - \bar{x}_i)(x_{ij} - \bar{x}_i)' \qquad (22.8)$$

Here, $(\bar{x}_i - \hat{\mu})$ is an estimate of α_i, the bias term of i th assessor. If there is a single observation x_i taken by the i th assessor, the bias term α_i has to be dropped from the model; i.e., $x_i = \mu + e_i$, $i = 1, ..., k$. Let $D_0^2 = \hat{\mu}'\, S^{-1}\, \hat{\mu}$, where $\hat{\mu} = \Sigma_{i=1}^{k} x_i/k$, $(k-1)S = \Sigma_{i=1}^{k}(x_i - \bar{x})(x_i - \bar{x})'$.
One may estimate the unknown quality index $\Delta_0 = (\mu'\, \Sigma^{-1}\mu)^{1/2}$; *by* $D_0 = (\hat{\mu}'\, S^{-1}\hat{\mu})^{1/2}$, *the sample Mahalanobis distance statistic.*
An unsatisfactory feature in using the above index is that a criterion of the type $(\theta - t)'J_t(\theta - t)$, may decrease as $|\theta - t|$ becomes large, where J_θ is the Fisher information matrix. See, e.g. Hauck and Donner (1977). However, in the context of quality index, this problem may be circumvented by requiring that, the finally computed index has to be non-decreasing in all coordinates; and such corrections are incorporated after D_0 is calculated with real data.

Let the measurement of the characteristics of two products have unknown mean and dispersion as $(\mu^{(1)}, \Sigma)$ and $(\mu^{(2)}, \Sigma)$. The *difference in quality index square* is $\Delta^2 = (\mu^{(1)} - \mu^{(2)})'\Sigma^{-1}(\mu^{(1)} - \mu^{(2)})$.

The statistic $D^2 = (\bar{x}^{(1)} - \bar{x}^{(2)})'S^{-1}(\bar{x}^{(1)} - \bar{x}^{(2)})$ of (22.6), may be considered as an estimate of Δ^2.
In subjective assessment of products by assessors, outliers may be present in the collected measurements. To eliminate outliers, one may adopt multivariate trimming.
Next we consider the situation when the dispersion matrices of two populations are not equal.

22.4. Bhattacharyya Affinity and D^2 Statistic

Let F_1, F_2 be two distribution function with densities $f_1(x)$ and $f_2(x)$ with respect to some measure ν. Bhattacharyya affinity between these two populations is defined as,

$$d = d_{f_1,f_2} = \int \{f_1(x)f_2(x)\}^{1/2} d\nu \qquad (22.9)$$

Bhattacharyya (1946) considered $\theta^2 = (\cos^{-1} d)^2$ as a measure of dissimilarity of f_1 and f_2 in the context of multinomial population. This measure has application in genetics and computer science. Consideration of the function d goes back to Hellinger(1909), and sometimes this is also refereed as Hellinger affinity. The index 22.9 for two probability distributions F_1, F_2 is independent of the reference measure ν.

Kolmogorov distance δ and Hellinger distance d_H in terms of d can be expressed as, $\delta = 1 - d$, and

$$d_H(F_1,F_2) = [\frac{1}{2}\int \{f_1^{1/2}(x) - f_2^{1/2}(x)\}^2 d\nu]^{1/2}$$
$$= (1-d)^{1/2} \qquad (22.10)$$

The function $d_H(F_1,F_2) = (1-d)^{1/2}$ satisfies all the properties of a distance, see e.g., Kraft (1955).
In particular, when $f_1(x)$ and $f_2(x)$ are normal densities: $\phi_1 = N_p(\mu^{(1)},\Sigma_1)$ and $\phi_2 = N_p(\mu^{(2)},\Sigma_2)$ then,

$$-\log d = \tfrac{1}{8}(\mu^{(1)} - \mu^{(2)})'\Sigma^{-1}(\mu^{(1)} - \mu^{(2)}) + \tfrac{1}{2}\log\frac{|\Sigma|}{(|\Sigma_1||\Sigma_2|)^{1/2}}, \quad \Sigma = (\Sigma_1 + \Sigma_2)/2$$

$$= \frac{1}{8}\Delta^2, \quad \text{when} \quad \Sigma_1 = \Sigma_2 = \Sigma \qquad (22.11)$$

In view of 22.11, one may extend the definition of Δ^2 for *two arbitrary densities* f_1 and f_2 as follows:

$$\Delta^2_{f_1,f_2} = -8\log d_{f_1,f_2} \qquad (22.12)$$

The measures $\delta, d_H, \Delta^2_{f_1,f_2}$ are monotone functions of d. Thus, order of ranking is same under these measures.
The extended definition of Δ^2 to the case $\Sigma_1 \neq \Sigma_2$, for *two multivariate normal densities* $\phi_1 = N_p(\mu^{(1)},\Sigma_1)$ and $\phi_2 = N_p(\mu^{(2)},\Sigma_2)$, by 22.11 and 22.12 reduces to,

$$\Delta^2_{\phi_1,\phi_2} = (\mu^{(1)} - \mu^{(2)})'\Sigma^{-1}(\mu^{(1)} - \mu^{(2)}) + 4\log\frac{|\Sigma|}{(|\Sigma_1||\Sigma_2|)^{1/2}} \qquad (22.13)$$

$$= \Delta^2_{\phi_1,\phi_2;I} + \Delta^2_{\phi_1,\phi_2;II},$$
say. The above can be estimated from sample by,

$$D^2_{\phi_1,\phi_2} = (\bar{x}^{(1)} - \bar{x}^{(2)})'\bar{S}^{-1}(\bar{x}^{(1)} - \bar{x}^{(2)}) + 4\log\frac{|\bar{S}|}{(|S_1||S_2|)^{1/2}} \tag{22.14}$$

$$\text{where } \bar{S} = \hat{\Sigma} = (\hat{\Sigma}_1 + \hat{\Sigma}_2)/2, \\ = (S_1 + S_2)/2.$$

The measure D_{ϕ_1,ϕ_2} can be used in the cases where, it is known apriori that $\Sigma_1 \neq \Sigma_2$.
Observe that $D^2_{\phi_1,\phi_2} = \Delta^2_{\hat{\phi}_1,\hat{\phi}_2}$ where $\hat{\phi}_1 = N_p(\bar{x}^{(1)}, S_1)$ and $\hat{\phi}_2 = N_p(\bar{x}^{(2)}, S_2)$.

Taking conditional expectation with respect to nested sub σ-fields, Kraft(1955) showed that the function d of the two sequence of restricted measures over the sequence of increasing sub σ-fields, are non decreasing. By the same result, the components $\Delta^2_{\phi_1,\phi_2;I}$ and $\Delta^2_{\phi_1,\phi_2;II}$ of 22.13 are non-negative and non decreasing in p, the number of co-ordinates in x; see e.g., Rao and Varadarajan (1963).
By an application of SLLN, $D^2_{\phi_1,\phi_2}$ converges to the population distance squared $\Delta^2_{\phi_1,\phi_2}$. The results are summarized below.

Result 1. *The sample distance squared statistic $D^2_{\phi_1,\phi_2}$ based on iid samples of sizes m and n from $\phi_1 = N_p(\mu^{(1)}, \Sigma_1)$ and $\phi_2 = N_p(\mu^{(2)}, \Sigma_2)$ respectively converges a.s. to the corresponding population distance squared. The two components of distances 22.13 and 22.14 are non decreasing in p. i.e., $\Delta^2_{\phi_1,\phi_2;I} \uparrow p$, $\Delta^2_{\phi_1,\phi_2;II} \uparrow p$, $D^2_{\phi_1,\phi_2;I} = \Delta^2_{\hat{\phi}_1,\hat{\phi}_2;I} \uparrow p$, $D^2_{\phi_1,\phi_2;II} = \Delta^2_{\hat{\phi}_1,\hat{\phi}_2;II} \uparrow p$ and $D^2_{\phi_1,\phi_2} = D^2_{\phi_1,\phi_2;I} + D^2_{\phi_1,\phi_2;II} \to \Delta^2_{\phi_1,\phi_2;I} + \Delta^2_{\phi_1,\phi_2;II} = \Delta^2_{\phi_1,\phi_2}$ a.s., as $m \to \infty$ and $n \to \infty$.*

Remark 3. The assumption of identical distribution may be relaxed. The parameters (μ, Σ) may be different for independent observations in a manner that SLLN still holds for the elements in S, and as a consequence D^2 converges to Δ^2. This may occur in industrial productions, where process parameters may vary a little over time.

Next, we prove a result stating that the extended Mahalanobis distance defined via 22.9 and 22.12 *between the means of sample observations* from two populations having non-normal densities f_1 and f_2 converges to $\Delta^2_{\phi_1,\phi_2}$.

Let $X, X_1, X_2, \ldots$ be independent and identically distributed p variate random variable from a density f with $EX_f = \mu$ and $\mathcal{D}_f(X) = \Sigma$, where Σ is nonsingular. By multivariate central limit theorem the distribution F_n of normalized sum $S_n = n^{-1/2}(X_1 + X_2 + ... + X_n - n\mu)$ converge to the normal distribution $N_p(0,\Sigma)$. Assume that the distribution of X has an absolutely continuous component with respect to Lesbesgue measure. Then the distribution F_n has an absolutely continuous component and this component becomes dominant as the sample size $n \to \infty$, and the corresponding density $\bar{f}$ converges to the normal density ϕ in L_1 metric; i.e.,

$$\int |\bar{f} - \phi| \, d\lambda \to 0 \tag{22.15}$$

as $n \to \infty$. Mamatov and Halikov(1964) proved this result in R^p extending the local limit theorem by Prokhorov(1952) in R. Bloznelis(2002) showed that the result is not necessarily true in infinite dimension.
Note that $\bar{f} \to \phi$ almost everywhere in the sample space. The result 22.15 ensures the normal approximation $\Phi(B)$ of the probabilities $P(S_n \in B)$ uniformly over the class of Borel sets, i.e.,

$$\sup\{|P(S_n \in B) - \Phi(B)|\} \to 0 \tag{22.16}$$

For bounded f, it is easy to see that, as $n \to \infty$

$$\int |\bar{f}^{\,1/2} - \phi^{\,1/2}| \, d\lambda \to 0 \tag{22.17}$$

where $\bar{f}$ is the density of sample mean.
Now consider iid samples of sizes m and n from two populations with bounded densities f_1 and f_2 respectively. Let $\bar{f}_1$ and $\bar{f}_2$ be the densities of the means of sample observations from f_1 and f_2. Write

$$\int |\bar{f}_1^{\,1/2}\bar{f}_2^{\,1/2} - \phi_1^{1/2}\phi_2^{1/2}| \, d\lambda = \int |\bar{f}_1^{\,1/2}(\bar{f}_2^{\,1/2} - \phi_2^{1/2}) + \phi_2^{1/2}(\bar{f}_1^{\,1/2} - \phi_1^{1/2})| \, d\lambda \tag{22.18}$$

From 22.17 with f as f_1 and f_2 ; ϕ as ϕ_1 and ϕ_2, it now follows that 22.18 tends to zero as the sample sizes $m \to \infty$, and $n \to \infty$. Hence the following proposition.
Proposition 2. *The extended Mahalanobis distance squared $\Delta^2_{\bar{f}_1,\bar{f}_2}$ defined via 22.12 between the two mean of iid samples from bounded densities f_1 and f_2, with sizes m and n respectively, converges almost surely to $\Delta^2_{\phi_1,\phi_2}$ of 22.13, i.e.,*

$\Delta^2_{\bar{f}_1,\bar{f}_2} \to \Delta^2_{\phi_1,\phi_2}$, *a.s., as* $m \wedge n \to \infty$; *provided* f_1 *and* f_2 *have absolutely continuous components. Thus,* $D^2_{\phi_1,\phi_2}$ *may be considered as an estimate of* $\Delta^2_{\bar{f}_1,\bar{f}_2}$.

Remark 4. The condition of bounded density in R^p, may be relaxed for a general estimate T_n, including the sample mean. Then, one need to assume conditions like (B) and (C) of Le Cam (1990), which ensure that distance between the distribution of T_n and Φ tend to zero in a uniform manner. A rate of convergence may also be obtained from (22.18) and (22.19). When the density f is *bounded*, Korolev and Shevtsova (2005) proved the Berry-Esseen theorem $||F_n - \Phi|| \leq \frac{1}{6\sqrt{2\pi}} \frac{E|X|^3}{\sigma^3\sqrt{n}} + O(n^{-3/4}(\log\, n)^{1/4})$ for iid random variables, achieving optimal constant in the bound.

Remark 5. Quality Index: The definition of Δ^2 given in 22.13 may be extended to asymptotically normal distributions, e.g., when some of the variables are discrete or multinomial and the mean vector is asymptotically normal by central limit theorem. One may then define the *difference in quality* index between two objects or production processes having measurement-parameters $(\mu^{(1)}, \Sigma_1)$ and $(\mu^{(2)}, \Sigma_2)$ as Δ; to be estimated from sample by D, given in 22.14.

The proposed quality index remains invariant under the transformations: $x \to Bx$, where B is a nonsingular matrix. If the quality characteristics are measured on the positive side of the axis then product's quality can be compared by the respective distances from 0. When quality characteristics are measured as categorical variable, with ordering like 'fair' 'good', 'excellent' etc., data analysis may be carried out by assigning numbers 0, 1, 2, ... for such variables. The minimum quality index, the origin, refers to a 'product at the base level' in the group; remaining products are compared with respect to it, by the distance measure.

22.5. An Example

For admission in a reputed academic institution, aspiring students undergo a multistage screening. The first test is objective type with multiple choices which is followed by a short answer type written test. The maximum one can score in tests 1 and 2 are 116 and 100 respectively.

Only those students, who score above 80 in test 1, are scrutinized for second test answer-scripts of written examination. From the assessed students at the second stage, those who did well in both the examinations are further screened by an interview of about 30 minutes duration per candidate; for final selection. The maximum one can score in the interview is 20. To interview a particular candidate, a subset of about seven members from the total board-members is assigned. The interview-marks are given by the assigned members on a 21 point scale $\{0, 1, 2, \ldots, 20\}$ and the median-score is considered as the consolidated score of interview for that candidate.

Let (x_1, x_2, x_3) denote the scores in the three tests. The problem is to rank the students on the basis of the set of three scores per candidate. Ranking of students may be done by the sum total of the three test scores, but that does not take into account the correlation structure. The variables are correlated in general and one may use Mahalanobis distance as an alternative criterion for ranking the students. A student who is more 'Mahalanobis-distant' from the minimum level of quality, viz. the origin $(0,0,0)$, is considered to be a better student.

The estimated (3x3) dispersion matrix based on 68 observations is given below.

$$\hat{\Sigma} = \begin{pmatrix} 51.0070237 & 0.9749781 & 1.089333 \\ 0.9749781 & 63.7565847 & 15.605026 \\ 1.0893327 & 15.6050263 & 10.937390 \end{pmatrix}.$$

The sample correlations are, $r_{12} = .0171$, $r_{13} = .0461$, $r_{23} = .5909$.

The three eigenvalues and corresponding eigenvectors are as follows.

$e_1 = 68.110625$, $e_2 = 50.933056$, $e_3 = 6.657319$,
$y_1 = (0.07163967,\ 0.96188469,\ 0.26390453)'$,
$y_2 = (0.997269891,\ -0.073824582,\ -0.001642073)'$,
$y_3 = (0.01790316,\ 0.26330168,\ -0.96454741)'$.

The score x_3, has major effect only in y_3, which explains 5.3% of the variation.

The sample partial correlations of two random variables eliminating the linear effect of the remaining one, are as follows. $r_{12.3} = -.0126$, $r_{13.2} = .0446$, $r_{23.1} = .5908$. The values are not much perturbed by conditioning, except probably the first one where the value (near zero) changes sign.

Test for $H_o : \rho_{12.3} = 0$ vs. $H_1 : \rho_{12.3} < 0$ is provided by $t = r_{12.3}\sqrt{n-3}/\sqrt{1-r_{12.3}^2} \sim t_{n-3}$, a t distribution with $(n-3)$ degress of freedom under H_o. Calculated $t = -0.102$, with p-value of significance $> .45$. We may accept $H_o : \rho_{12.3} = 0$. The same conclusion holds while testing $H_o : \rho_{12} = 0$; vs. $H_o : \rho_{12} > 0$; $t = r\sqrt{n-2}/\sqrt{1-r^2} \sim t_{n-2}$, under H_o. Calculated $t = 0.139$, p-value of significance is $p = .445$. Thus one may conclude that the above change of sign for r_{12} and $r_{12.3}$ is statistically insignificant.

The (2x2) dispersion matrix of the first two scores, based on 429 observations is

$$\hat{\Sigma}_{2X2} = \begin{pmatrix} 64.18853 & 38.56986 \\ 38.56986 & 149.95984 \end{pmatrix},$$

with $r_{12} = .3931$. This is higher than $r_{12} = .0171$ with $n = 68$. However, the latter value is based on further extreme tail of the distribution of marks.
The two eigenvalues and corresponding eigenvectors are,
$e_1 = 164.75272$, $e_2 = 49.39564$,
$y_1 = (0.3581000,\ 0.9336832)'$,
$y_2 = (0.9336832,\ -0.3581000)'$.

Table 22.1. Mahalanobis distance from origin of 68 individuals

Individual	x_1	x_2	x_3	M.D
3	108	70	18.0	17.35519
1	103	68	19.5	16.63087
63	113	39	8.0	16.53114
10	108	53	17.0	16.43427
62	112	39	8.0	16.39737
4	104	61	18.0	16.33837
29	108	49	15.0	16.23101
28	109	40	15.0	16.03475
31	108	44	15.0	16.03445
5	97	69	18.0	15.97606
7	104	53	17.0	15.92561
53	108	41	10.5	15.89675
15	108	32	16.0	15.76241
17	101	55	16.0	15.62961
27	104	46	15.0	15.58889
32	104	45	15.0	15.54977
18	104	41	16.0	15.44210
46	104	43	12.0	15.44173
58	104	42	10.0	15.41858
38	104	41	14.0	15.38291
14	94	64	16.0	15.30179
45	104	39	12.0	15.28452
54	104	38	10.0	15.24941
8	96	56	17.0	15.07052
34	100	44	14.0	14.97306
35	100	42	14.0	14.89511
55	100	41	10.0	14.84677
2	93	56	18.0	14.72283
40	95	51	13.5	14.66060
66	99	37	6.0	14.64906
19	95	46	15.0	14.43008
61	96	42	8.0	14.42569
9	96	40	17.0	14.42140

continued to the next page.

Individual	x_1	x_2	x_3	M.D
41	96	43	13.0	14.40061
30	92	53	15.0	14.39337
13	93	50	16.0	14.37861
68	96	39	6.0	14.35502
67	91	49	6.0	14.32861
59	97	38	10.0	14.32060
50	97	38	11.0	14.31658
12	93	47	16.0	14.24363
11	94	42	17.0	14.22468
56	96	37	10.0	14.14712
57	96	37	10.0	14.14712
33	95	38	14.5	14.11832
26	90	52	15.0	14.09083
25	91	49	15.0	14.06095
37	93	43	14.0	14.02684
43	93	43	13.0	14.01243
36	94	37	14.0	13.93870
60	94	38	9.0	13.93848
52	94	36	10.5	13.84152
22	91	43	15.0	13.79494
47	92	41	12.0	13.78858
24	92	39	15.0	13.77938
64	92	39	8.0	13.74863
6	88	49	17.0	13.73774
44	89	46	12.0	13.64390
42	89	46	13.0	13.64151
39	88	48	14.0	13.62242
21	90	41	15.0	13.59071
23	88	44	15.0	13.45368
16	88	42	16.0	13.41207
49	87	46	11.0	13.40295
51	88	43	11.0	13.36945
48	86	46	11.0	13.27652
65	86	41	8.0	13.07614
20	84	45	15.0	12.99333

In Table 22.1, we present the three scores per individual and the Mahalanobis distance (M.D) of the individuals from origin. The entries are sorted in decreasing order of M.D. The individual number in the first column refers to the rank of the individual with respect to the third variable, the interview score.
In Figures 22.1 and 22.2, we provide a graphical representation of the pattern in which the Mahalanobis distances gradually decrease from the highest value. These may help to locate a cut off point for selecting candidates through the entrance tests. In Figure 22.1, there seems to be a break after the M.D. score 14.72283. There are 28 individuals scoring greater than or equal to that value.
Figure 22.2 refers to M.D. with first two test scores, viz. objective type and short answer type test scores. There is a break after the M.D. score 11.84701. There are 174 individuals scoring greater than or equal to that value. The third quartile of M.D. scores is 12.5033, with 107 individuals scoring greater than or equal to the value. One may decide about the number of candidates to be called for interview regarding final selection, from the quartiles.

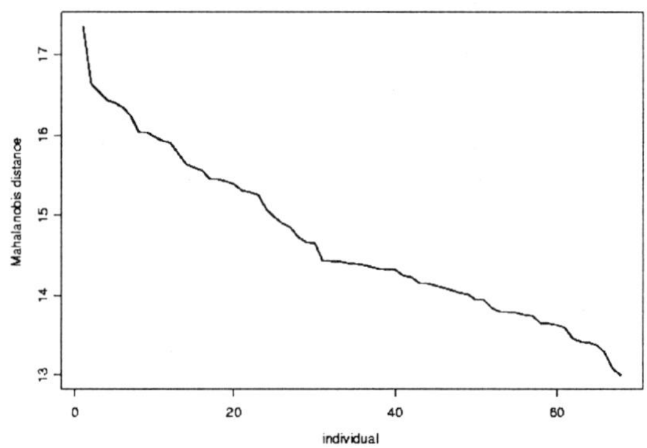

Fig. 22.1. MD in descending order for 68 individuals

As already mentioned, major effect from the interview score x_3 is towards the third principal component y_3, which explains only 5.3% of total variation. To this end, interview performance may be recorded in an enlarged set, rather than grading within the present set with 21 elements only, viz. $\{0,1,2,...,20\}$. This may enable finer distinction between the candidates, thus increasing the variability of scores. The same interview-board should preferably interview all the candidates to reduce subjective variation. Finally, for each candidate, consolidation of interview

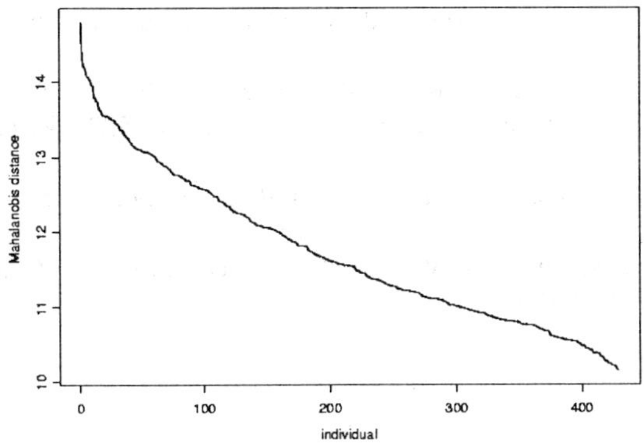

Fig. 22.2. MD in descending order for 429 individuals

score from different board members should be done in robust manner preserving efficiency as well; e.g., one may use trimmed mean or linear combination of order statistics of the set of interview scores for each candidate, with smooth weight function, allotting less weight towards the extremes.

22.6. Acknowledgements

The author thanks Professor J.K. Ghosh for interesting discussions. Thanks are also due to Professor Debasis Sengupta for his help in SAS programming.

References

1. Bhattacharyya, A. (1946). On a measure of divergence between two multinomial populations, *Sankhyā*, A. 7., 401-406.
2. Chow, Y. S. and Teicher, H. (1978). *Probability Theory: Independence, Interchanability, Martingales*. Springer, New York.
3. Hauck, W. W. and Donner, A. (1977). Walds test as applied to hypotheses in logit analysis. *J. Amer. Statistical Assoc.* **72**, 851-853. Corrigendum (1980) 75, 482.
4. Hellinger, E. D. (1909). Neue Begrundung der Theorie quadratischen Formen von unendlichen vielen Veranderlichen. *Journal fur Reine und Angewandte Mathematik*, **136**. 210-271.
5. Korolev, Yu. V. and Shevtsova, I. G. (2005). On the accuracy of normal approximation.*I. Theor. Probab. Appl.* **50**, 298-310.
6. Kraft, C. (1955). Some conditions for consistency and uniform consistency of statistical procedures. *Univ. California Publ. Statist.* **2**. 125-142.

7. Le Cam, L. (1990). On the standard asymptotic confidence ellipsoids of Wald. *ISI Review*, Vol 58. **2**. 129-152.
8. Mamatov, M. and Halikov, M. K. (1964). Global limit theorems for distribution functions in the higher dimensional case. *Izv. Akad. Nauk UzSSR, Ser. Fiz. Mat. Nauk* **1** , 13-21.
9. Prokhorov, Yu. V. (1952). *A local theorem for densities*. Dokl. Akad. Nauk SSSR (N.S) 83 , 797-800.
10. Rao, C. R. and Varadarajan, V. S. (1963). Discrimination of Gaussian processes, *Sankhyā*, A. **25**. 303-330.

Chapter 23

AQL Based Multiattribute Sampling Scheme

Anup Majumdar

SQC & OR Unit, Indian Statistical Institute, Kolkata 700108
E-mail: anupmajumdar@email.com

Contents

23.1. Introduction

23.1.1. *The multiattribute acceptance sampling*

Irrespective of the type of product, evaluation of conformity to specified requirements of its quality characteristics is an integral part of quality assurance. Although they form a set of necessary verification activities almost at all stages of production, these activities, known as inspection do not add value to the product on their own and are to be kept at their minimum. The sampling inspection where a portion of a collection of product units is inspected on a set of characteristics with a view to making decision about acceptance or otherwise becomes relevant in this context. The number of elements of the set of characteristics to be verified for this purpose is unlikely to be just one in most of the practical situations. We may call this as multiattribute inspection as employed for verification of materials procured from outside and further at all stages of production, through semi finished and finished or assembly stages to final despatch to the customers. At all such stages consecutive collections of products called lots, are submitted for acceptance or alternative disposition.

23.2. Scope of The Present Discussion

Many authors have considered construction of multiattribute sampling inspection plans from different considerations. The economic design of multiattribute sampling schemes taking account of Bayesian principles based on appropriate prior distribution was considered by Schmidt and Bennett (1972), and further by Case, Schmidt and Bennett (1975), Ailor, Schimdt and Bennet (1975), Majumdar (1990, 1997), Moskowitz, Plante and Tang (1986), Moskowitz, Plante and Tang and Ravindran (1984) and Tang, Plante and Moskowitz (1986). Gadre, M.P. and Rattihalli, R.N. (2004) have considered multiattrbute plans of given strength. However there has not been any attempt to construct Acceptable Quality Level (AQL) based plans for ready use of the industry taking into account the the features specific to the multiattribute situations. In this discussions we try to construct such plans.

23.3. Inspection Scenarios

To understand the special features of the multiattribute situation we first note that in a multiattribute situation the quality characteristics are decidedly unequal in their effect on fitness for use. A relatively few are serious i.e. of critical importance; many are minor. Clearly, the more important the characteristic, the greater the attention it should receive in such matters as: extent of quality planning, precision of processes, tooling and instrumentation, sizes of samples, strictness of conformance etc. To this end many companies utilize formal systems of seriousness classification. Juran and Gryna(1996) tabulate one such classification scheme in food industry.

The general practice of the industry has been to assign different AQL values and employ effectively a parallel system of sampling inspections. For example, Hansen (1957) reported adaptation of MIL-STD-105A to the acceptance inspection of the M38A1 truck commonly called 'the jeep' manufactured by Wiley's Motors Inc, in Toledo, Ohio. Two hundred and four characteristics were classified in four classes, namely: special defects (one hundred per cent inspection) compris-

Table 23.1. Checklist for verification of presentation defects in finished and packed garments

Sl.No.	Defect type
1.	Omission of tags
2.	Torn / crushed poly bag
3.	Main label center out
4.	Puckering in the folding of the garment
5.	Bending of the garment
6.	Scratch marks on the box
7.	Deformation of the box
8.	Twisted collar
9.	Button down up down
10.	Bubbling
11.	Incomplete button stitching
12.	Visible fusing marks
13.	Wrong size tag
14.	Wrong bar code
15.	Wrong hang tag

ing of 11 attributes, major defects (AQL 15 per one hundred vehicle) comprising of 14 attributes, minor defects (AQL 150 defects per 100 vehicle) comprising of 69 attributes, incidental defects (AQL 400 defects per 100 vehicles) comprising of 110 attributes.

We cite another example where, a sample of finished garment is verified for 15 characteristics (all attribute type) before delivery as listed in Table 23.1.

We would of course keep in mind that, in all situations, attributes need not be classified in mutually exclusive category as has been done for the "jeep" inspection. In some situations, defect occurrence may be mutually exclusive by nature. For example the defect 'Omission of tags' and the defect 'Wrong size tag' of garment verification can not occur simulteneously i.e. their occurances are mutually exclusive. On the other hand, defect of different types may occur as jointly independently. For example, a functional defect and a surface defect for a metal closure might occur as independent of each other. We also neeed to consider the situation where inspection scenario may change from identification of a defective unit to the counting of defects in a unit.

23.4. The type B Operation Characteristic (OC) Function in a Multiattribute Situation

We suppose that there are r attribute characteristics inspectable separately for any product unit and if we take a sample of size n from a lot of size N, all the r characteristics are observable in any order. The sampling scheme we are interested in should be applicable when lots are submitted consecutively in a more or less continuous manner. For a single attribute, Dodge and Romig (1959) defined product of quality p as the the conceptually infinite output of units from a stable process for which the probability of producing a defective unit is p, which is also called the long term fraction defective in the process or the process average. The concept of type B OC curve of a sampling plan as formally defined by Dodge and Romig (1959) is the curve showing probability of accepting a lot as a function of product quality, where product quality is expressed as fraction defective p. So lots are outcomes of a production process, which constitutes as described a Bernoullian sequence with parameter p i.e. in a lot of size N, the number of defectives in the lot, X follows binomial distribution with parameter N, p and if a sample of size n is drawn at random without replacement from the lot, then the number of defectives observable in such a sample, x follows binomial distribution with parameters n, p. Thus,

$$Pr(x|p) = b(x,n,p) = \binom{n}{x} p^x (1-p)^{n-x}.$$

In this case for a single sampling plan with acceptance criterion accept if $x \leq c$; reject otherwise the type B OC function is given as,

$$B(c,n,p) = \sum_{x=0}^{c} b(x,n,p).$$

In case of $r(\geq 2)$ attributes, $\mathbf{p} = (p_1, p_2, ..., p_r)$ denotes the process average vector, where p_i is the long term fraction defective in the process in respect of the i th attribute, $i = 1,2,...,r$. For a lot of size N, the quality is measured by the vector $\mathbf{X} = (X_1, X_2, ..., X_r)$, where X_i = number of defectives in the lot in respect of the i th attribute, $i = 1,2,...,r$. A random sample of size n is drawn from a lot and the verification of non-conformity with respect to (w.r.t.) r specified attributes can be carried out in the sample in any order. The number of defectives vector for the sample is denoted as $\mathbf{x} = (x_1, x_2, ..., x_r)$.

Two situations are considered viz., (i) Defective units w.r.t. the different attributes occur independently i.e. $X_1, X_2, ..., X_r$ are jointly independently distributed and $x_1, x_2, ..., x_r$ are consequently jointly independently distributed. (ii) Defective units w.r.t. the different attributes occur mutually exclusively i.e. the occurrence of one kind of defect precludes the occurrence of another kind of defect.

a) When defect occurences are jointly independent

When the defect occurrences of different types are jointly independent the probability of any $(X_1, X_2, ..., X_r)$, X_i being the number of defectives in a lot of type $i, i = 1, 2, ..., r$ is

$$Pr(X_1, X_2, ..., X_r) = \prod_{i=1}^{r} Pr(X_i). \tag{23.1}$$

The probability of obtaining x_i defectives corresponding to a characteristic i from a sample of size n drawn from a lot of size N is

$$Pr(x_i \mid X_i) = \binom{n}{x_i}\binom{N-n}{X_i - x_i} \Big/ \binom{N}{X_i}. \tag{23.2}$$

If now X_i is assumed to follow binomial with parameters N and p_i and they are independent as given by (23.1), the unconditional probability of obtaining $(x_1, x_2, ..., x_r)$ defectives in sample of size n is

$$\sum_{X_1}\sum_{X_2} \cdots \sum_{X_r} \prod_{i=1}^{r} \left[\binom{n}{x_i}\binom{N-n}{X_i - x_i} \Big/ \binom{N}{X_i}\right] \left[\binom{N}{X_i} p_i^{X_i}(1-p_i)^{N-X_i}\right]$$

$$= \prod_{i=1}^{r} \left[\binom{n}{x_i} p_i^{x_i}(1-p_i)^{n-x_i}\right].$$

If b(., ., .) denotes an individual term of the binomial distribution, then

$$Pr(x_1, x_2 ..., x_r \mid p_1, p_2 ... p_r) = \prod_{i=1}^{r} b(x_i, n_i, p_i), x_i = 0, 1, 2, ..., n;$$
$$0 < p_i < 1; i = 1, 2, ..., r. \tag{23.3}$$

b) When defect occurrences are mutually exclusive

When defect occurrences are mutually exclusive, the expression for the probability of observing $(x_1, x_2, ..., x_r)$ defective in a sample of size n from a lot of size N containing $(X_1, X_2, ..., X_r)$ defectives will be multivariate hypergeometric as

$$Pr(x_1, x_2, ..., x_r \mid X_1, X_2, ..., X_r)$$
$$= \binom{X_1}{x_1}\binom{X_2}{x_2} \cdots \binom{X_r}{x_r}\binom{N - X_1 - X_2 ... - X_r}{n - x_1 - x_2 ... - x_r} \Big/ \binom{N}{n} \tag{23.4}$$

At any process average vector $(p_1, p_2, ..., p_r)$ the joint probability distribution of $(X_1, X_2, ..., X_r)$ can be assumed to be multinomial $(N, p_1, p_2, ..., p_r)$ such that

$$Pr(X_1, X_2, ..., X_r \mid p_1, p_2, ..., p_r)$$
$$= \binom{N}{X_1}\binom{N - X_{(1)}}{X_2} ... \binom{N - X_{(r-1)}}{X_r} p_1^{X_1} ... p_r^{X_r}(1 - p_{(r)})^{(N - X_{(r)})}$$

where

$$X_{(i)} = X_1 + X_2 + ... + X_i \quad \textit{and} \quad p_{(i)} = p_1 + p_2 + ... + p_i; \quad i = 1, 2, ..., r. \quad (23.5)$$

From (23.4) and (23.5) it follows that the unconditional probability is

$$Pr(x_1, x_2, ..., x_r \mid p_1, p_2, ..., p_r)$$
$$= \binom{n}{x_1}\binom{n - x_{(1)}}{x_2} ... \binom{n - x_{(r-1)}}{x_r} p_1^{x_1} ... p_r^{x_r}(1 - p_{(r)})^{(n - x_{(r)})},$$

where

$$x_{(i)} = x_1 + x_2 + ... + x_i, i = 1, 2, ..., r; \quad 0 \le x_{(1)} \le x_{(2)} ... \le x_{(r)} \le n;$$
$$0 < p_i < 1, \quad i = 1, 2, ..., r; p_{(r)} < 1. \quad (23.6)$$

Thus the average probability follows multinomial distribution with parameter $(n, p_1, p_2, ..., p_r)$.

23.4.1. *Poisson conditions*

Hald (1981) has used the phrase 'Poisson conditions' when Poisson probability can be used as an approximation to the binomial in the expressions of type B OC function. We use the phrase to cover the situations as described below.

(i) Poisson as approximation to binomial and multinomial

If $p_i \to 0$, $n \to \infty$ and $np_i \to m_i$ then the binomial probability $b(x_i, n, p_i)$ tends to Poisson probability $g(x_i, np_i)$ where

$$b(x_i, n, p_i) = \binom{n}{x_i} p_i^{x_i}(1 - p_i)^{n - x_i}$$

and

$$g(x_i, np_i) = e^{-np_i}(np_i)^{x_i}/(x_i)!.$$

Under these conditions for $i = 1,2,...,r$, the expression given as (23.3) can be modified as

$$Pr(x_1,x_2,...,x_r \mid p_1,p_2,...,p_r) = \prod_{i=1}^{r} g(x_i,np_i); \quad x_i \geq 0, i = 1,2,...,r. \qquad (23.7)$$

If we also make an additional assumption that $\sum_{i=1}^{r} p_i \to 0$ then the equation (23.6) can also be modified as (23.7).

(ii) Poisson as an exact distribution and occurrences of defect types independent

In case we count the number of defects per item for each characteristic, we may construct a model for which the expected number of defects for a item for i th the characteristic equals to p_i in the long run. We assume the number of defects for r distinct characteristics in a item are independently distributed. The output of such a process is called a product of quality $(p_1,p_2,...,p_r)$, the parameter vector representing the mean occurrence rates (of defects) per observational unit. The total number of defects for any characteristic in a lot of size N from such a process will vary at random according to a Poisson law with parameter Np_i for the i th characteristic under usual circumstances.
Similarly, the distribution of number of defects on attribute i, in a random sample of size n drawn from a typical lot will be be a Poisson variable with parameter np_i. Independence of the different characteristics will be naturally maintained in the sample, so that the joint probability of occurrence can be expressed as (23.7). Here $p_i, i = 1,2,...,r$ denotes the average number of defects per item in respect of characteristics i instead of proportion defectives, since in this situation we are dealing with defects rather than defectives.

23.4.2. *Expressions of type B OC function under Poisson conditions*

Suppose now we employ a sampling plan given hereunder.
We draw a random sample of size n from a lot of size N; observe the number of defectives/defects for each of r attribute characteristics, $x = (x_1,x_2,...,x_r)$. giving the vector of defectives/defects w.r.t. the r attribute types.
Let A be the set of $(x_1,x_2,...,x_r)$ combinations for which we decide to accept the lot i.e. if $(x_1,x_2,...,x_r) \in A$, accept the lot and reject it otherwise.
From the discussions in the above preceeding sections it follows that irrespective

of whether the defectives (defects) occurrences are independent or mutually exclusive, the probability of acceptance under Poisson conditions at a process average $(p_1, p_2, ..., p_r)$ is given by

$$\mathbf{P}(\mathbf{p}) = \sum_{x \in A} \prod_{i=1}^{r} g(x_i, m_i) \tag{23.8}$$

23.5. Multiattribute Sampling Schemes Based on AQL

At the outset it is to be made clear that there is no published plan specifically for a multiattribute situation based on Acceptable Quality Level (AQL). We, therefore, look at the plans available for single attribute and examine the consequences of using them in a multiattribute situation.

23.5.1. *The basic features of MIL-STD-105D and its derivatives*

The published set of plans most widely used is the US Military standard 105D (1963), which has been adopted as the international standard ISO 2859 (1974). We denote this standard as MIL-STD-105D. The standard is based on what is known as Acceptable Quality Level (AQL) and considered to be most suitable for the consumers.

The Acceptable Quality Level (AQL) is defined as the maximum percent defective (or the number of defects per hundred units) that for purpose of acceptance sampling can be considered satisfactory as a process average.

Thus the lots produced at process average of AQL or better level should have a high probability of getting accepted. A producer can always increase his acceptance probability by improving his process average.

This system of selecting an appropriate sample size is not based on an explicit mathematical model. For any lot size, the table gives the corresponding sample size. The relationship is based on what is considered reasonable in practice. There is no theoretical foundation for the relation between sample size (n) and lot size (N). As the lot size increases, the sample size increases, but at a lesser rate such that $n/N \to 0$ as $N \to \infty$.

The user of the table may choose between various 'levels' of this relationship, called inspection levels. Next, one has to choose from practical considerations an AQL (given in percent defectives or defects per hundred) and find from the table the acceptance number. For given AQL, the acceptance number is determined so that the producer's risk is reasonably small and is decreasing with lot size.

The acceptance number (c) has been arrived at such that it remains same at a given sample size multiplied by the AQL to ensure that the Poisson probability of acceptance (rejection) remains same in such cases. This ensures a desired producer's risk.

There are 13 sets of $n.AQL$ values and 11 acceptance numbers chosen in such a way that the producer's risk (excepting for c =0) varies from a maximum of 9% to a minimum of 1.44%. With increase in the n. AQL values, c increases, so that producer's risk decreases upto a level of 2% and thereafter it is kept at less than 2%. To attend to utmost simplicity of the standard in numerical and administrative respects, a set of five preferred numbers has been used for each decade as values of AQL and a proportional set has been used for values of the sample size. These preferred numbers constitute a geometric series with ratio equal to the fifth root of 10 i.e. 1.585 [See Hald (1981) for a detailed discussion.]

23.5.2. *Using published plans in a multiattribute situation*

In case of a multiattribute situation the Standard prescribes that a separate plan is to be chosen for each class of attributes. For example, the plan for critical defects will have generally a lower AQL than the plan for major defects and the plan for major defects will have an AQL lower than the plan for minor defects. Note that occurrence of defectives w.r.t critical, major and minor are mutually exclusive.

To examine the consequences of constructing a sampling plan by this method for our multiattribute situation we first consider the effective producer's risks. Secondly, we consider it reasonable to expect that the OC function should be more sensitive to the changes in the defect level of more important attributes, particularly, in a situation, where unsatisfactory defect level occurs due to more serious type of defects. Suppose that the defect/ defective on i th attribute is more serious than that on the j th attribute then, as the total defect level changes from low to high, the absolute value of the slope of the OC w.r.t the i th type of defect / defective as a function of the total defect / defective level should have always higher value than the absolute slope of the OC w.r.t the j th type of defect / defective, assuming that the relative contribution of a characteristic to the total defect / defective remains constant.

23.5.3. *The consequences using MIL-STD-105D table in a multiattribute situation*

We examine the cases of two manufactures: one who needs to verify the quality of plastic components/washer procured from vendors for packaging cosmetics and,

Table 23.2. Sampling plans using MIL-STD-105D (normal inspection, inspection level II)

Product	Lot Size	Sample Size	Defect category					
			Critical		Major		Minor	
			AQL	c	AQL	c	AQL	c
Plastic container	15,000	315	0.15%	1	1.0%	7	4%	21
Garment	50,000	500	0.1%	1	1%	10	2.5%	21

another, a garment manufacturer who wants to make sure the quality of his product before shipment. Both of them categorize their defect types as critical, major and minor and choose the AQL's accordingly. For a given lot size they find the sample size using the MIL-STD-105-D and the value of three acceptance numbers.

A lot is accepted only when the acceptance criterion is satisfied for each one of the three types of defects. The sampling plans employed using MIL-STD-105D inspection level II for normal inspection for a typical lot size are worked out as given in the Table 23.2.

In these examples we obtain the producer's risk at AQL as more than 10.5% for the first set of plans (plastic container) and around 22% for the second set (garment) of plans. It would appear that we are asking for the moon from the producer.

We have thereafter computed the slope of the OC in the manner as explained, at different process averages keeping the defect contribution of each type of attribute as proportional to its AQL value. Using this as a measure of sensitivity, the OC should appear to be more sensitive to the changes of major defect than to the changes of minor defect, when the overall quality level becomes unsatisfactory. But unfortunately the features observed depict a picture far from this ideal. (See Figure 23.1)

For $r = 3$, we have further constructed the sampling plans taking all 286 possible ordered triplets of $n.AQL$ values and possible n and c values as tabulated in the MIL-STD-105D table. We find that the producer's risk varies from 3.6% to 34.2%. There are 103 plans with producer's risk more than 16 % and there are only 34 plans with producer's risk less than or equal to 6%.

Further, there are 183 plans which do not always satisfy the condition of higher absolute slope of the OC for more serious type of attribute defect associated with lower value of AQL, as compared to the slope of OC for the attribute associated with higher value of AQL, in the range of p ($p = p_1 + p_2 + p_3$, $p_1 : p_2 : p_3 = AQL_1 : AQL_2 : AQL_3$), from $p = 0$ to the limiting quality, defined as the value of

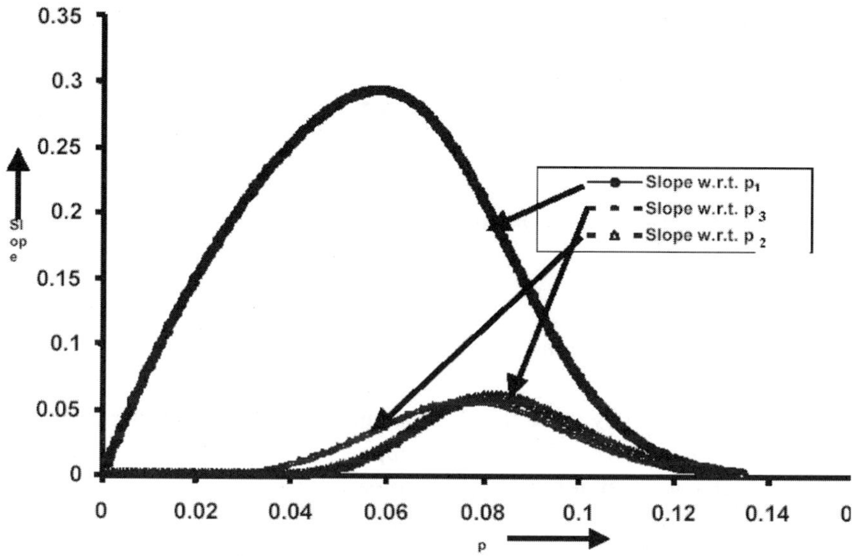

Fig. 23.1. Absolute value of Slope of the OC function of the MASSP adopted from MIL-STD-105D Inspection level II - Normal inspection used for incoming inspection of plastic containers, lot size(N) = 15000. Chosen AQL values are 0.15%, 1% and 4% for critical, major and minor defectives. Sample size = 315, $c_1 = 1$, $c_2 = 7$, $c_3 = 21$. The curve is plotted against p, for $p_1 : p_2 : p_3 = 0.15 : 1 : 4.0$

process average for which the probability of acceptance is around 0.10.

23.6. There is No Good C Kind Plan

We have further considered the class of plans where we take a sample of size n, accept the lot if and only if $x_i \leq c_i; \forall i$. We call these plans as multiattribute single sampling plans (MASSP) of C kind. Note that all MASSP's constructed from MIL-STD-105D as suggested above are C type plans.
The probability of acceptance for this plan at $(p_1, p_2, ..., p_r)$ under Poisson conditions:

$$\sum_{x_1=0}^{c_1} \sum_{x_2=0}^{c_2} \cdots \sum_{x_r=0}^{c_r} \prod_{i=1}^{r} g(x_i, m_i) = \prod_{i=1}^{r} G(c_i, m_i).$$

where

$$G(c_i, m_i) = \sum_{x_i=0}^{c_i} g(x_i, m_i); \quad m_i = np_i. \tag{23.9}$$

We obtain, writing PC'_j as the partial differential coefficient of the OC function w.r.t m_j,

$$-PC'_j = g(x_j, m_j) \prod_{i=1:i\neq j}^{r} G(c_i, m_i) \tag{23.10}$$

Defining
$m = m_1 + m_2 + ... + m_r$; $\rho_i = m_i/m$, we write,

$$Slope_j(m) = g(x_j, m\rho_j) \prod_{i=1:i\neq j}^{r} G(c_i, m\rho_i) \tag{23.11}$$

It has been shown by the author [See Majumdar (2005)] that for a given set of ρ_i, the function $H(m) = Slope_i(m) - Slope_{i+1}(m)$ $i = 1, 2, ..., r-1$ undergoes utmost one change of sign from positive to negative for $m \geq 0; i = 1, 2, ..., r-1$. Further, for $\rho_{i+1}/\rho_i > c_{i+1}/c_i$, there is exactly one real positive root for $H(m) = 0$. If we now set $\rho_{i+1}/\rho_i = AQL_{i+1}/AQL_i$ and try to ensure a reasonable producer's risk, the inequality, $\rho_{i+1}/\rho_i > AQL_{i+1}/AQL_i$ is found to be a reasonable assumption to be usually satisfied. It, therefore, follows that in such a case, the plan will fail to satisfy the condition $Slope_i(m) \geq Slope_{i+1}(m)$ for all $m > 0$. We conclude that in general there is no good C kind plan in the sense defined as above.

23.7. The A Kind Plans

We define an A kind plan as follows: we take a sample of size n, observe the number of defectives or defects in the sample for the i th attribute as x_i for all $i = 1, 2, ..., r$ and apply the following acceptance criterion: accept if $x_1 \leq a_1$,, $x_1 + x_2 \leq a_2$, ..., $x_1 + ... + x_r \leq a_r$; and reject otherwise.

Note that $a_i \leq a_j$ for $j \geq i.j = 1, 2, ..., r$.
The OC function of such a plan under Poisson conditions is given by,

$$PA(a_1, a_2, .., a_r; m_1, m_2, ..., m_r) = \sum_{x_1=0}^{a_1} \cdots \sum_{x_r=0}^{a_r - x_{(r-1)}} \prod_{i=1}^{r} g(x_i, m_i) \tag{23.12}$$

To compare the relative changes in the OC function for changes in p_i, we use the absolute value of the partial differential coefficient of the Poisson OC function as given in equation (23.12) with respect to m_i; $m_i = np_i$. This is used as a measure of discriminating power of the OC function w.r.t. j th attribute, treating it as function of $m = m_1 + m_2 + ...m_r$ and keeping m_i/m fixed for $i = 1, 2, ..., r$.

Theorem: For $i \le j$ the discriminating power of the plan (as defined) w.r.t. i th attribute is greater than or equal to that for the j th attribute for every $m = m_1 + m_2 ... + m_r$.

Proof: Denoting the partial differential coefficient of the Poisson OC function as given in equation (23.12) with respect to $m_i; m_i = np_i$ by PA_i', we get

$$-PA_i' = PA(a_1, a_2, .., a_i, ..., a_r; m_1, m_2, ..., m_r)$$
$$-PA(a_1, a_2, ..., a_i - 1, a_{i+1} - 1, ..a_r - 1; m_1, m_2, ..., m_r) \text{for} a_i > 0, \quad (23.13)$$

and

$$-PA_i' = PA(a_1, a_2, .., a_i, , a_r; m_1, m_2, ..., m_r) \quad (23.14)$$

for $a_i = 0$.

Thus, the OC function is a decreasing function of m_i or p_i. Moreover, the OC function is increasing in each a_i. Note that $a_j = 0$ implies $a_i = 0$ for all $i \le j$. It therefore follows from (23.13) and (23.14) that

$$-PA_i' \ge -PA_{i+1}' \quad \forall m, \quad m = m_1 + m_2 ... + m_r \quad (23.15)$$

This proves the theorem.

The above property of the plan A therefore allows us to order the attributes in order of relative discriminating power. If the attributes are ordered in the ascending order of AQL value, then it is possible to construct a sampling scheme ensuring an acceptable producer's risk and also satisfy the condition of higher absolute slope for the lower AQL attribute $Slope_i \ge Slope_{i+1}$, for all $m, i = 1, 2, ..., r - 1$. For comparison m_i/m is kept fixed, $i = 1, 2, ..., r$.

23.7.1. *Construction of A kind plans*

Using the set of *n.AQL* values chosen from MIL-STD-105D we establish a MASSP scheme consisting of A kind MASSP's. These plans will have same sample size as used by the MIL-STD-105D standard for a given lot size, but will have different acceptance criteria. We have done this exercise for $r = 3$, using the following procedure.

There are 13 *n.AQL* values in the MIL-STD-105D. We get 286 triplets such that $n.AQL_1 < n.AQL_2 < n.AQL_3$. We further order these 286 plans in the usual lexicographic fashion. We start with the first combination of (0.1256, 0.1991, 0.3156) and choose $a_1 = 1, a_2 = 1$, and $a_3 = 2$. This gives producer's risk of around 5.5%. All other sets of acceptance numbers are worked out such that a plan positioned later will have a lesser producer's risk ensuring that producer's risk decreases up to a level of 5%, and thereafter it is kept at less than 5%. It is heartening to note that the largest value of ratio of the limiting quality level (as defined) to the total AQL has ranged from 1.5 to 7.4, so that the OC's appear to be quite steep. For the MIL-STD-105D single sampling plans the comparable ratio varies from 1.7 to 18.3. [Hald (1981)]

We present an extract from the table constructed by the author for illustration. [See Majumdar (2005).] We may use the table as follows: Suppose the lot size is 15000 and we use Inspection Level II. From the MIL-STD-105D we get the sample size as 315. Suppose there are three Attributes viz. Attribute 1, Attribute 2 and Attribute 3. We choose their respective AQL's (say), as 0.15%, 1.0% and 4%. This gives $n.AQL_1 = 0.4725, n.AQL_2 = 3.15, n.AQL_3 = 12.6$. We may choose the plan number 145 and get the values of $a_1 = 3, a_2 = 9, a_3 = 23$.

We would, therefore, take a sample of size of 315 from a lot of size 15000, inspect these 315 pieces for all the three attributes and observe the number of defects as x_1, x_2, and x_3 for the first, second and the third attribute respectively. If $x_1 \leq 3, x_1 + x_2 \leq 9, x_1 + x_2 + x_3 \leq 23$, we shall accept the lot, else we shall reject the lot.

References

1. Ailor, R. B., Schmidt, J. W. and Bennett, G. K. (1975). The design of economic acceptance sampling Plans for a mixture of attributes and variables. *AIIE Transactions*, **7**.
2. Case, K. E., Schmidt, J. W. and Bennett, G. K. (1972). Cost-based acceptance sampling. *Industrial Engineering*, **4**, (11), 26-31.
3. Case, K. E., Schmidt, J.W. and Bennett, G. K. (1975). A discrete economic multi-attribute acceptance sampling. *AIIE Trans.*, 363-369.
4. Dodge, H. F. and Romig, H. G. (1959). "Sampling Inspection Tables", 2nd edn, *Wiley, New York*, (1st edn, 1944).
5. Hald, A. (1981). "Statistical theory of Sampling Inspection by Attributes", *ACADEMIC PRESS INC., New York.*
6. Hansen, B. L. (1957). Acceptance Plan for Automotive Vehicles., *Industr. Qual. Contr.*, **14**, 18-24.
7. Juran, J. M. and Gryna, F. M.(1996). "Quality Planning and Analysis". *Third Edition. Tata McGRAW-HILL Edition.*
8. Gadre, M. P. and Rattihalli, R. N. (2004). Exact Multi-Attribute Acceptance Single Sampling Plan.*Communications in Statistics-Theory and Methods* **33(1)**, 165-180.
9. Majumdar, A. (1990). Some acceptance criteria for single sampling multiattribute plans, *Sankhyā B,* Pt. 2., **52**, 219-230.

10. Majumdar, A. (1997). A generalization of single sampling multiattribute plans, *Sankhyā B*, Pt. 2, **59**, 256-259.
11. Majumdar, A. (2005) "Multiattribute acceptance sampling plans". *Ph.D thesis, Indian statistical Institute, Kolkata.*
12. MIL-STD-105D (1963). "Sampling Procedures and Tables for Inspection by Attributes", *U.S. Government Printing Office, Washington D.C.*
13. Moskowitz, H., Plante, R. and Tang K. (1986). Multistage multiattribute acceptance sampling in serial production systems. *IIE transaction*, **18**, 130-137.
14. Moskowitz, H., Plante, R., Tang, K. and Ravindran A. (1984). Multiattribute Bayesian Acceptance sampling plans for screening and scrapping rejected Lots. *IIE Transactions*, **16**, no. 23.
15. Schmidt, J. W. and Bennett, G. K. (1972). Economic Multiattribute Acceptance Sampling, **AIIE Transactions, 4**(3).
16. Tang, K., Plante, R. and Moskowitz, H.(1986) Multiattribute Bayesian acceptance sampling plans under nondestructive Inspection. *Management Science*, **32**(6) ,739-750.

Table 23.3. Three attribute A kind single sampling plans for normal inspection [Extract from Majumdar (2005)]

SL. No.	$n.AQL_1$	$n.AQL_2$	$n.AQL_3$	a_1	a_2	a_3	Overall Prod.risk	LQ / Total AQL
1	0.1256	0.1991	0.3155	1	1	2	0.055	7.9
2	0.1256	0.1991	0.5	1	2	2	0.054	6.4
3	0.1256	0.3155	0.5	1	2	3	0.026	6.9
4	0.1991	0.3155	0.5	1	2	3	0.039	6.4
5	0.1256	0.1991	0.7924	1	2	3	0.034	5.9
6	0.1256	0.3155	0.7924	1	2	3	0.045	5.4
7	0.1991	0.3155	0.7924	2	2	3	0.05	5.1
8	0.1256	0.5	0.7924	1	2	4	0.038	5.4
9	0.1991	0.5	0.7924	2	2	4	0.043	5.1
10	0.1256	0.1991	1.256	1	2	4	0.03	5
11	0.3155	0.5	0.7924	2	3	4	0.03	4.9
12	0.1256	0.3155	1.256	1	2	4	0.04	4.7
13	0.1991	0.3155	1.256	2	2	4	0.044	4.5
14	0.1256	0.5	1.256	1	3	4	0.049	4.2
15	0.1991	0.5	1.256	2	4	4	0.049	4.1
16	0.3155	0.5	1.256	2	3	5	0.027	4.4
17	0.1256	0.7924	1.256	1	3	5	0.037	4.2
18	0.1991	0.7924	1.256	2	3	5	0.038	4.1
19	0.1256	0.1991	1.991	1	2	5	0.039	4
20	0.3155	0.7924	1.256	2	3	5	0.05	3.8
21	0.1256	0.3155	1.991	1	2	5	0.049	3.8
22	0.1991	0.3155	1.991	2	3	5	0.044	3.7
23	0.5	0.7924	1.256	3	4	5	0.048	3.6
24	0.1256	0.5	1.991	1	2	6	0.043	3.9
25	0.1991	0.5	1.991	2	2	6	0.048	3.8
26	0.3155	0.5	1.991	2	3	6	0.033	3.7
27	0.1256	0.7924	1.991	1	3	6	0.043	3.6
28	0.1991	0.7924	1.991	2	3	6	0.044	3.5
29	0.3155	0.7924	1.991	2	4	6	0.043	3.4
30	0.5	0.7924	1.991	4	6	6	0.05	3.2

Table 23.4. (Table 23.3 Contd.) Three attribute A kind single sampling plans for normal inspection [Extract from Majumdar (2005)]

SL. No.	$n.AQL_1$	$n.AQL_2$	$n.AQL_3$	a_1	a_2	a_3	Overall Prod.risk	LQ / Total AQL
31	0.1256	1.256	1.991	1	4	7	0.036	3.4
32	0.1991	1.256	1.991	1	4	7	0.049	3.3
33	0.1256	0.1991	3.155	1	2	7	0.035	3.4
34	0.3155	1.256	1.991	2	4	7	0.045	3.2
35	0.1256	0.3155	3.155	1	2	7	0.043	3.2
36	0.1991	0.3155	3.155	2	2	7	0.046	3.2
37	0.5	1.256	1.991	3	5	7	0.041	3.1
38	0.1256	0.5	3.155	1	3	7	0.047	3.1
39	0.1991	0.5	3.155	2	3	7	0.047	3
40	0.3155	0.5	3.155	4	4	7	0.05	3
41	0.7924	1.256	1.991	3	5	8	0.038	3.2
42	0.1256	0.7924	3.155	1	3	8	0.04	3.2
43	0.1991	0.7924	3.155	2	3	8	0.04	3.1
44	0.3155	0.7924	3.155	3	3	8	0.05	3
45	0.5	0.7924	3.155	3	4	8	0.044	2.9
46	0.1256	1.256	3.155	2	4	8	0.05	2.8
47	0.1991	1.256	3.155	2	5	8	0.048	2.8
48	0.3155	1.256	3.155	2	4	9	0.041	3
49	0.5	1.256	3.155	2	5	9	0.044	2.9
50	0.7924	1.256	3.155	4	5	9	0.05	2.7
51	0.1256	1.991	3.155	2	6	9	0.045	2.7
52	0.1256	0.1991	5	2	2	9	0.049	2.7
53	0.1991	1.991	3.155	2	6	9	0.049	2.7
54	0.1256	0.3155	5	1	2	10	0.037	2.8
55	0.3155	1.991	3.155	2	5	10	0.048	2.8
56	0.1991	0.3155	5	1	2	10	0.05	2.8
57	0.1256	0.5	5	1	3	10	0.038	2.7
58	0.5	1.991	3.155	2	6	10	0.048	2.7
59	0.1991	0.5	5	1	3	10	0.05	2.7
60	0.3155	0.5	5	2	3	10	0.045	2.6

Table 23.5. (Table 23.3 Contd.) Three attribute A kind single sampling plans for normal inspection [Extract from Majumdar (2005)]

SL. No.	$n.AQL_1$	$n.AQL_2$	$n.AQL_3$	a_1	a_2	a_3	Overall Prod.risk	LQ/ Total AQL
61	0.1256	0.7924	5	2	3	10	0.05	2.6
62	0.7924	1.991	3.155	3	7	10	0.048	2.6
63	0.1991	0.7924	5	2	4	10	0.045	2.6
64	0.3155	0.7924	5	3	5	10	0.048	2.5
65	0.5	0.7924	5	2	4	11	0.045	2.6
66	0.1256	1.256	5	1	4	11	0.046	2.6
67	1.256	1.991	3.155	4	7	11	0.045	2.6
68	0.1991	1.256	5	2	4	11	0.045	2.6
69	0.3155	1.256	5	2	5	11	0.042	2.5
70	0.5	1.256	5	3	5	11	0.049	2.4
71	0.7924	1.256	5	3	5	12	0.046	2.5
72	0.1256	1.991	5	2	5	12	0.045	2.5
73	0.1991	1.991	5	3	5	12	0.05	2.4
74	0.3155	1.991	5	2	6	12	0.044	2.4
75	0.5	1.991	5	5	6	12	0.05	2.4
76	0.7924	1.991	5	3	6	13	0.05	2.4
77	1.256	1.991	5	5	8	13	0.045	2.3
78	0.1256	0.1991	7.924	2	2	13	0.046	2.3
79	0.1256	3.155	5	2	8	13	0.046	2.3
80	0.1991	3.155	5	2	8	13	0.05	2.3
81	0.1256	0.3155	7.924	2	3	13	0.047	2.3
82	0.1991	0.3155	7.924	2	4	13	0.05	2.2
83	0.3155	3.155	5	2	7	14	0.047	2.3
84	0.1256	0.5	7.924	1	3	14	0.038	2.3
85	0.1991	0.5	7.924	1	3	14	0.049	2.3
86	0.5	3.155	5	2	8	14	0.05	2.3
87	0.3155	0.5	7.924	2	3	14	0.043	2.3
88	0.1256	0.7924	7.924	2	3	14	0.048	2.3
89	0.1991	0.7924	7.924	2	4	14	0.042	2.3
90	0.7924	3.155	5	3	9	14	0.049	2.2

Table 23.6. (Table 23.3 Contd.) Three attribute A kind single sampling plans for normal inspection [Extract from Majumdar (2005)]

SL. No.	$n.AQL_1$	$n.AQL_2$	$n.AQL_3$	a_1	a_2	a_3	Overall Prod.risk	LQ / Total AQL
91	0.3155	0.7924	7.924	2	4	14	0.049	2.2
92	0.5	0.7924	7.924	4	6	14	0.049	2.2
93	0.1256	1.256	7.924	1	4	15	0.045	2.3
94	0.1991	1.256	7.924	2	4	15	0.044	2.3
95	1.256	3.155	5	4	9	15	0.045	2.2
96	0.3155	1.256	7.924	2	5	15	0.04	2.2
97	0.5	1.256	7.924	3	5	15	0.046	2.2
98	0.7924	1.256	7.924	4	7	15	0.049	2.1
99	0.1256	1.991	7.924	2	5	16	0.045	2.2
100	0.1991	1.991	7.924	2	5	16	0.049	2.2
101	1.991	3.155	5	5	10	16	0.049	2.2
102	0.3155	1.991	7.924	2	6	16	0.042	2.2
103	0.5	1.991	7.924	3	6	16	0.048	2.1
104	0.7924	1.991	7.924	4	8	16	0.048	2.1
105	1.256	1.991	7.924	5	7	17	0.049	2.1
106	0.1256	3.155	7.924	1	8	17	0.047	2.1
107	0.1991	3.155	7.924	2	8	17	0.044	2.1
108	0.3155	3.155	7.924	3	8	17	0.048	2.1
109	0.5	3.155	7.924	3	10	17	0.05	2
110	0.7924	3.155	7.924	4	8	18	0.048	2.1
111	1.256	3.155	7.924	6	10	18	0.049	2
112	0.1256	0.1991	12.56	1	2	19	0.049	2
113	0.1256	0.3155	12.56	1	3	19	0.05	2
114	0.1256	5	7.924	2	11	19	0.047	2
115	1.991	3.155	7.924	6	11	19	0.05	2
116	0.1991	0.3155	12.56	2	3	19	0.047	2
117	0.1991	5	7.924	2	11	19	0.05	2
118	0.1256	0.5	12.56	2	4	19	0.048	2
119	0.3155	5	7.924	3	13	19	0.05	2
120	0.1991	0.5	12.56	1	3	20	0.049	2

Table 23.7. (Table 23.3 Contd.) Three attribute A kind single sampling plans for normal inspection [Extract from Majumdar (2005)]

SL. No.	$n.AQL_1$	$n.AQL_2$	$n.AQL_3$	a_1	a_2	a_3	Overall Prod.risk	LQ / Total AQL
120	0.1991	0.5	12.56	1	3	20	0.049	2
121	0.3155	0.5	12.56	2	3	20	0.043	2
122	0.5	5	7.924	4	10	20	0.05	2
123	0.1256	0.7924	12.56	2	3	20	0.047	2
124	0.1991	0.7924	12.56	2	4	20	0.04	2
125	0.3155	0.7924	12.56	2	4	20	0.046	2
126	0.7924	5	7.924	4	11	20	0.049	2
127	0.5	0.7924	12.56	3	5	20	0.046	2
128	0.1256	1.256	12.56	2	5	20	0.049	1.9
129	0.1991	1.256	12.56	2	6	20	0.05	1.9
130	0.3155	1.256	12.56	2	5	21	0.039	2
131	1.256	5	7.924	6	11	21	0.049	2
132	0.5	1.256	12.56	3	5	21	0.043	2
133	0.7924	1.256	12.56	4	6	21	0.047	1.9
134	0.1256	1.991	12.56	2	6	21	0.048	1.9
135	0.1991	1.991	12.56	2	7	21	0.048	1.9
136	0.3155	1.991	12.56	3	8	21	0.05	1.9
137	1.991	5	7.924	6	12	22	0.05	1.9
138	0.5	1.991	12.56	3	6	22	0.045	1.9
139	0.7924	1.991	12.56	4	7	22	0.046	1.9
140	1.256	1.991	12.56	5	7	23	0.047	1.9
141	0.1256	3.155	12.56	2	7	23	0.048	1.9
142	0.1991	3.155	12.56	2	8	23	0.041	1.9
143	0.3155	3.155	12.56	2	8	23	0.047	1.9
144	3.155	5	7.924	9	14	23	0.049	1.9
145	0.5	3.155	12.56	3	9	23	0.045	1.9
146	0.7924	3.155	12.56	5	11	23	0.049	1.8
147	1.256	3.155	12.56	7	9	24	0.05	1.9
148	0.1256	5	12.56	2	10	25	0.048	1.8
149	1.991	3.155	12.56	7	10	25	0.049	1.8
150	0.1991	5	12.56	2	11	25	0.044	1.8

Chapter 24

Multivariate Quality Management in Terms of a Desirability Measure and a Related Result on Purchase Decision: A Distributional Study

Dilip Roy

Department of Business Administration
Burdwan University
E-mail: dr.diliproy@gmail.com

For a multi-processing system, processing effects may get captured through a vector measure of the quality characteristics of the product. However, when the end user of the product receives the same, use-quality of the product is mostly perceived in terms of a combined level of satisfaction arising out of the individual levels of satisfaction and their interrelationships. In this sense, a user examines the overall utility or the desirability of the product.

The proposed work examines different aspects of the desirability measure and obtains the distribution of a combined measure of desirability, both under the state of control and the out of control state. The concept of zonal polynomials has been used in this process to derive and present the probability functions of the overall desirability value.

Further, suitability of the combined measure has been examined to install a control mechanism for ensuring quality in the production system. Use of the desirability value has also been explored to describe the process of buying from a product category through a binary modelling approach.

Contents

24.1. Introduction

In a real life problem we often measure the quality characteristics of a product by a multidimensional vector variable. But, at the end a buyer looks at the product in totality. For example, consider the simple case of manufacturing a TV set. It may be needed to measure its quality characteristics in terms of the screen size, life of the picture tube, clearness of the sound system, potential number of channels etc. But we take the purchase decision based on an overall merit of the offer. Thus, there arises the importance of examining how the multiple quality characteristics of a product, conceived during the product designing and realized during the product manufacturing, give rise to unidimensional desirability value. It has been pointed out by Hauser et al (1983) that the utility that the purchase of a product will provide is usually compared with the same of a benchmark product or of existing stock. Corresponding models are binary probit model and binary logit model (see Lilien et al, 1999). This concept can be and need to be dealt with in a multivariate setup with utility viewed as a market measure of quality. The corresponding concept of multivariate quality management is the focus of our attention in this paper.

Earlier, Hotelling (1947) proposed a multivariate process control technique based on his celebrated T^2-statistic and studied the subject of quality control from location point of view. In support of his approach, Hotelling (1947) gave an illustration on air testing of sample bombsights. Latter, Kariya (1981) studied the robustness property of Hotelling's T^2 under a possible deviation from multivariate normality of the characteristic vector. Unfortunately, one can hardly take corrective action on relevant processing(s) in case the T^2 control chart shows lack of control.

Using the concept of principal components, Jackson and Morris (1956) developed multivariate control charts. In this sense, their approach was a characteristic vector approach that reduced the dimension of the problem. In support of their approach, Jackson and Morris (1956) gave an illustration in respect of a photographic processing. Here again, one can hardly take corrective action on relevant processing(s) in case the control chart shows lack of control.

Roy (1982) proposed a new approach in which the execution phase of controlling procedure can get adequate attention so that one can take corrective measure on relevant processing(s) when the control device raises alarm for lack of control. In case the processing system lands up in an out of control state that approach can identify the processing(s) on which corrective actions are to be initiated. Following the same approach, Roy and Basu (1989) examined the problem of controlling dispersion parameters and suggested an optimum procedure. To demonstrate the applicability of their approach, they considered a problem of mining. The problem of controlling the location parameters was studied by Roy and Basu (1991) in a more or less similar fashion.

The sequential procedure suggested in Roy (1982) and followed in Roy and Basu (1989, 1991) is of regression type obtained through invariance principle. For this regression approach one may refer to subsequent but similar works of Hawkins (1991) who have used regression-adjusted variables. Later, Hayter and Tsui (1994) addressed the identification and quantification issues in a multivariate framework. Mason et al (1995, 1997) examined the decomposition of T^2 -statistic in multivariate control chart for the pinpointed interpretation of the out of control state. Interpretation of out of control signals was also highlighted in Mason et al (1997) for multivariate T^2 control chart. Latter, the sensitivity of T^2 -statistic has been improved upon by Mason and Young (1999) for better signaling of an assignable cause of variation. Recently, Roy (2003) has proposed optimal invariant test for simultaneous testing/ control of location and dispersion in a multivariate normal set up. But in all these approaches more emphasis has been given on arrive at the corrective measures than on the overall desirability of a multivariate process or a product.

However, the seed of composite measure of satisfaction was indirectly present in some of the above-mentioned works. In 1947, when Hotelling proposed his unidimensional measure from a multidimensional set up there was an inbuilt concept of overall satisfaction. But the seedling remained unexplored or uncared. The approach of Hotelling has remained to be more suitable for maintaining univariate process control chart for a higher dimensional problem. Also, in Jackson and Morris (1956)'s principal component approach, the first principal component may work as a proxy for composite unidimensional mapping.

The first pronounced measure for overall satisfaction came up in the literature nearly a decade latter. This alternative unidimensional mapping, introduced by Harrington (1965), takes into consideration the desirability scale and specification limits. According to desirability value approach, the magnitude of each quality characteristic can be transformed on a dimensionless scale and combined to arrive at an overall desirability value. Mukherjee (1971) developed large sample chart based on desirability value and examined the suitability of a Pearsonean curve for representing the desirability value function. Exact distribution of the same has not been tried out so far.

To bridge this gap, the present work examines the distributional aspects of the desirability measure and derives the same for the overall desirability value, both under the null and non-null set-ups, i.e., under the state of control and the out of control state. Further, use of the overall desirability value has been explored for controlling the overall quality in any production system. An additional role of the desirability value has also been indicated through its application in binary modelling that depicts the purchase process from a choice set/ product category.

24.2. Desirability Determination

Ideally, the curve of desirability should have value one if the observed quality measure falls within the specification limits. It should take the value zero if the quality measure lies outside the specification limits. However, because of the recording imprecision such an ideal curve may lose its special appeal. It may be more realistic if the ideal curve is replaced by a curve which is a flat topped continuous function taking values between zero and one. Following Harrington(1965)'s concept we may define the desirability curve, $d^*(X)$ for any quality character, X, in the following way:

$$d^*(X) = exp[-|(2X - X_{max} - X_{min})/(X_{max} - X_{min})|m]. \qquad (24.1)$$

where m is a nonnegative constant, preferably an integer. It is easy to note that $0 > |(2X - X_{max} - X_{min})/(X_{max} - X_{min})|m > 1$ so that $d*(X)$ takes values between $e^{-1}(> 0)$ and 1, attaining the peak value 1 at the point $X = (X_{max} + X_{min})/2$. If in this definition we replace X_{max} by the Upper Specification Limit (USL) and X_{min} by Lower Specification Limit (LSL) then the desirability value may be used for the task of product control. Alternatively, if we replace X_{max} by the Upper Control Limit (UCL) and X_{min} by Lower Control Limit (LCL) then the desirability value may be used for the purpose of process control. In particular, let us consider $k\sigma$-limit control limits so that UCL$= \mu + k\sigma$ and LCL$= \mu - k\sigma$, where μ and σ are respectively the mean and standard deviation of X. Then the desirability curve simplifies to the following form:

$$d^*(X) = exp[-|(X - \mu)|/k\sigma^m]. \qquad (24.2)$$

Mukherjee (1971) proposed the following alternative form based on Pearsonean Type II curve

$$d^{**}(X) = (1 - \pi^2 X^2/4)^{1/2}, \text{for} - 2/\pi < X < 2/\pi \qquad (24.3)$$

$$0 \qquad , otherwise.$$

This is a symmetric curve, symmetric with respect to the point $X = 0$, and taking values between 0 and 1. Its kurtosis measure equals 2.0 indicating the flat-topped nature of this curve.

24.2.1.

In a multivariate set up let $X = (X_1, X_2, ..., X_p)^{/}$ be the vector variable of order $px1$ representing the multivariate measure of quality characteristics. Let us assume that X follows multivariate normal distribution with mean vector μ and dispersion matrix $\sum = ((\sigma_{ij}))$. Also, denote by σ_i the standard deviation of X_i, and by ρ_{ij} the product moment correlation coefficient between X_i and X_j. Under the state of control or the aimed at state, let $\mu = \mu^0$. Following the definition of Harrington as mentioned above at (24.2.2), let us express the i-th desirability value, d_i^*, corresponding to the i-th quality characteristic X_i, by

$$d_i^* = exp[-\{|(X_i - \mu_i^0)|/k\sigma_i\}^m]. \tag{24.4}$$

The overall desirability, D^*, can be viewed as the geometric mean of the individual desirabilities so that if the desirability value is very small for any one characteristic the overall desirability becomes significantly affected. Symbolically,

$$D^* = (d_1^* ... d_i^* ... d_p^*)^{1/p}. \tag{24.5}$$

Writing $d_i = -k^m log d_i^*$, and $D = -pk^m log D^*$, we get the transformed measures of individual and overall desirabilities with the following relationship

$$D = \sum_{i=1}^{p} d_i. \tag{24.6}$$

To install a control mechanism based on desirability-transform D and the conventional 3-sigma control limits we need to know the mean and standard deviation of D under the state of control. If under the state of control X follows $N_p(\mu_0, \sum)$, then each $(di)^{2/m}$ will be distributed as a χ^2 variate with degrees of freedom 1. Hence, one can write

$$E(d_i) = 2^{m/2}\Gamma((m+1)/2)\Gamma^{-1}(1/2),$$

$$Var(d_i) = 2^m\Gamma^{-2}(1/2)[\Gamma((2m+1)/2)\Gamma(1/2) - \Gamma^2((m+1)/2)].$$

To evaluate $E(d_i d_j)$, let us write $Y_i = (X_i - \mu_i^0)/\sigma_i$. Then

$$E(d_i) = 2^{m/2}\Gamma((m+1)/2)\Gamma^{-1}(1/2),$$
$$Var(d_i) = 2^m\Gamma^{-2}(1/2)[\Gamma((2m+1)/2)\Gamma(1/2) - \Gamma^2((m+1)/2)].$$

To evaluate $E(d_i d_j)$, let us write $Y_i = (X_i - \mu_i^0)/\sigma_i$. Then

$$E(d_i d_j) = \frac{1}{2\pi\sqrt{1-\rho_{ij}^2}}\int_{-\infty}^{\infty}\int_{-\infty}^{\infty}|y_i|^m|y_i|^m exp[-\frac{1}{2}\begin{pmatrix}y_i\\y_j\end{pmatrix}'\begin{pmatrix}1 & \rho_{ij}\\ \rho_{ij} & 1\end{pmatrix}\begin{pmatrix}y_i\\y_j\end{pmatrix}]dy_i dy_j$$
$$= \frac{1}{2\pi\sqrt{1-\rho_{ij}^2}}\sum_{k=0}^{\infty}\frac{1}{k!}(\frac{\rho_{ij}}{1-\rho_{ij}^2})2\int_0^{\infty}\int_0^{\infty}+\int_0^{\infty}\int_0^{\infty}y_i^{m+k}y_j^{m+k}$$
$$exp[-\frac{2}{2(1-\rho_{ij}^2)}]exp[-\frac{2}{2(1-\rho_{ij}^2)}]dy_i dy_j$$
$$= 2^m(1-\rho_{ij}^2)^{m-1/2}\sum_{k=0}^{\infty}\frac{(2\rho_{ij})^{2k}}{2k!}\Gamma^2(\frac{m+1+k}{2})\Gamma^2(\frac{1}{2}).$$

Then from the relationship (24.6) we get

$$E(D) = p2^{m/2}\Gamma((m+1)/2)\Gamma^{-1}(1/2), \tag{24.7}$$

and

$$Var(D) = p2^m\Gamma^{-1}(1/2)\Gamma((2m+1)/2) + 2^m\Gamma^{-2}(1/2)\sum\sum_{i\neq j}$$

$$(1-\rho_{ij}^2)^{m-1/2}\sum_{k=0}^{\infty}\frac{(2_{ij})^{2k}}{2k!}\Gamma^2(\frac{m+1+k}{2}) - p^2\Gamma^2((m+1)/2) \tag{24.8}$$

Given n subgroup observations $\{X_\alpha, \alpha = 1,2,,n\}$, with corresponding desirability-transforms as $\{D_\alpha, \alpha = 1,2,,n\}$, one can construct desirability chart based on

$$\overline{D} = \frac{1}{n}\sum_{\alpha=1}^{n} D_\alpha \tag{24.9}$$

The 3-sigma limits will be $E(D) \pm 3Var(D)/n^{1/2}$, which can be obtained from (24.8) and (24.9). The control rule will be to raise alarm if an observed $\overline{D}$ value falls outside the control limits or a rear pattern occurs within the control limits. It may be remarked that Mukherjee (1971) obtained 3-sigma control chart based on a simplified assumption of independence of the quality characteristics. Under independence, he approximated $E(D)$ by p and Var(D) by $pm^2/2$. These expressions are exact for m equal to 2 but vary widely as m increases. Our exact expressions for the same are

$$E(D) = p2^{m/2}\Gamma((m+1)/2)\Gamma^{-1}(1/2),$$

$$Var(D) = p2^m\Gamma^{-2}(1/2)\Gamma((2m+1)/2)\Gamma(1/2) - \Gamma^2((m+1)/2),$$

which may be closely approximated by Stirling's formula as

$$E(D) \approx pe\{\sqrt{2e}[(2m-1)/e]^m - [(m-1)/e]^m\},$$

24.2.2.

Let us examine the robustness of the above-mentioned 3-sigma limit control chart by working out the mean and variance of D under certain relaxed conditions. To do so, we consider the special case of $m = 2$ and do away with the distributional assumption of multivariate normality. In its place, let us introduce the idea of multivariate mesokurtic property as studied in Roy (1988). A multivariate distribution possesses multivariate mesokurtic property if

$$\sigma_{ijkl} = \sigma_{ij}\sigma_{kl} + \sigma_{ik}\sigma_{jl} + \sigma_{il}\sigma_{jk} \;\; forall(i,j,k,l) \in (1,2,...,p) \tag{24.10}$$

where

$$\sigma_{ijkl} = E(X_i - \mu_i^0)(X_j - \mu_j^0)(X_k - \mu_k^0)(X_l - \mu_k^0)$$

exists for all possible combinations of i, j, k, and l. This definition reduces to univariate definition of mesokurtic property for univariate case, i.e., $p = 1$. We can now determine the mean and variance of D under the assumptions of symmetry of X around mean (around μ_0 under the state of control) and its multivariate mesokurtic property. It is easy to note that for $m = 2$

$$E(d_i) = 1,$$
$$E(d_i)^2 = E(X_i - \mu_i^0)^4/(\sigma_i)^4 = \sigma_{iiii}/(\sigma_{ii})^2 = 3$$

under multivariate mesokurtic property,and

$$E(d_id_j) = \sigma_{iijj}/(\sigma_i)^2(\sigma_j)^2 = (\sigma_{ii}\sigma_{jj} + \sigma_{ij}\sigma_{ij})/(\sigma_{ii}\sigma_{jj}) = 1 + 2\rho_{ij}^2,$$

under multivariate mesokurtic property.
Thus,

$$E(D) = p, \tag{24.11}$$

and

$$Var(D) = 2p + 2\sum_{i \neq j}\sum \rho_{ij}^2, \tag{24.12}$$

and hence one can construct the same 3-sigma limit control chart under symmetry and multivariate mesokurtic property but without the assumption of multivariate normality. It is possible to work out, in a similar way, the mean and variance of D under an out of control state represented by $E(X) = \mu$. Writing

$$\Delta_i = (\mu_i - \mu_i^0)/\sigma_i, and \Delta = \sum_{i=1}^{p} \Delta_i,$$

we can show that

$$E(D) = p + \Delta, \tag{24.13}$$

and

$$Var(D) = 2(p+\Delta) + 2\sum_{i \neq j}\sum (\rho_{ij}^2 + 2\rho_{ij}\Delta_i\Delta_j). \tag{24.14}$$

24.3. Exact Distribution and Probability Limit Control Chart

24.3.1.

To construct the probability limit control chart based on $\overline{D}$, one has to know the distribution of $\overline{D}$ under the state of control. While the exact distribution is difficult to get, the asymptotic distribution is easy to work out. From a direct application of the central limit theorem on $\overline{D}$, which is the sample mean of subgroup observations $\{D_\alpha, a = 1, 2, ..., n\}$, we can claim that for large n

$$[\overline{D} - E(D)][Var(D)/n]^{-1/2} \xrightarrow{L} N(0,1). \tag{24.15}$$

where $E(D)$ and $Var(D)$ are as in (24.7) and (24.8) respectively. Then, from (24.15) probability limit control chart can be constructed with plotted points as the standardized $\overline{D}$ values, i.e., $[\overline{D} - E(D)][Var(D)/n]^{-1/2}$ and upper and lower control limits as $\tau_{\alpha/2}$ and $-\tau_{\alpha/2}$, with a level of significance.
For $m = 2$, under the assumptions of symmetry of X around mean (around μ^0 under the state of control) and multivariate mesokurtic property of X, one may be in a position to do away with the distributional assumption of multivariate normality.

With the plotted points as $[\overline{D} - E(D)][Var(D)/n]^{-1/2}$, which can be simplified, under (24.11) and (24.12), as

$$T = (\overline{D} - p)[(2p + 2\sum\sum_{i \neq j} \rho_{ij}^2)/n]^{-1/2}, \tag{24.16}$$

One can still ensure asymptotic normality under the state of control. Hence the upper and lower control limits can be fixed at $\tau_\alpha/2$ and $-\tau_{\alpha/2}$ respectively, with α as the level of significance. It can also be observed that under an out of control state T will follow asymptotic normal distribution except for suitable changes in the location and scale parameters to be dictated by (24.13) and (24.14) respectively. Then the OC surface can be worked out easily in terms of an alternative . Assuming independence among the p quality characteristics, Mukherjee (1971) derived the probability limit control chart under normality for each component of X. According to him, the asymptotic distribution of D as p tends to infinity can be expressed as

$$[D/p - 1][m^2/2p]^{-1/2} \xrightarrow{L} N(0,1).$$

But from a practical point of view this assumption of independence is not very appealing. Also, p, being the number of quality characteristic under study, can hardly be increased arbitrarily. From the point of view of cost of inspection too, it is also not desirable to increase the number of quality characteristic, as it would involve high cost due to usage of a large number of measuring instruments.

24.3.2.

In case neither p value nor n value can be made large, the need for exact distribution of $\overline{D}$ assumes importance. We propose to address this problem for a specific choice of $m = 2$ under the assumption of multivariate normality for the distribution of X. However, before we proceed further let us introduce the following notations and results to be used hereafter. Let S be a symmetric matrix of order pxp. Then $(trS)^b$ can be uniquely expanded in terms of zonal polynomials,$C_\beta(S)$, in the following manner

$$(trS)^b = \sum_\beta C_\beta(S)$$

where b is a positive integer and β is an integer partition of b and summation over β covers all possible such partitionings. Stated below are some important results (see James, 1961, 1964) involving zonal polynomials:

(a)$exp[(trS)] = \sum_b \sum_\beta C_\beta(S)/b!$
(b)$[det(I-S)]^{-a} = \sum_b \sum_\beta (a)_\beta C_\beta(S)/b!$
(c) $\int_{S>0} exp\{trSM\}[det(S)]^{r-(p+1)/2} C_\beta(ST) = [det(M)]^{-r} C_\beta(M^{-1}S)\Gamma_p(r,\beta)$,
where $\Gamma_p(r,\beta)$, is a multivariate incomplete gamma function.
We are now in a position to derive the exact distribution of $\overline{D}$, the sample mean of the desirability-transforms by evaluating its characteristic function and applying the concept of uniqueness of the same or inversion theorem on it.
Result 3.1. Under the given set up and the state of control, $T = n\overline{D}$ is distributed as a linear combination of p independent and identically distributed χ^2 variates with degrees of freedom n each and weights as the eigen values of the correlation matrix $P = ((\rho_{ij}))$.
Proof. Writing $R = ((r_{ij}))$, where

$$r_{ij} = \sum_{\alpha=1}^{n} (X_{i\alpha} - \mu_i^0)(X_{j\alpha} - \mu_i^0)\sigma_i^{-1}\sigma_j^{-1} \tag{24.17}$$

for $i, j = 1, 2, .., p$, we note that, under the multivariate normality for X, R follows the Wishart distribution with dimension p, degrees of freedom n and parameter matrix P. Symbolically, R follows (n, P). Now, writing $\overline{D}$ in terms of the elements of R we have

$$\overline{D} = \frac{1}{n}\sum_{\alpha=1}^{n} = \frac{1}{n}\sum_{i=1}^{p} r_{ii} = \frac{1}{n} trR,$$

and hence from the characteristic function of the Wishart distribution we get the characteristic function of $T = n\overline{D}$ as

$$\varphi(\theta) = [det(I - 2i\theta P)]^{-n/2} \tag{24.18}$$

Since P is a symmetric positive definite matrix there exists an orthogonal matrix H such that $HPH' = diag(\rho^1, \rho^2, ..., \rho^p)$, where ρ^j 's are the eigen values of P. Then, pre-multiplying $(I - 2i\theta P)$ by H and post multiplying the same by H' and taking the determinant we can write

$$\begin{aligned}\varphi(\theta) &= [det(I - 2i\theta diag(\rho^1, \rho^2, ..., \rho^p))]^{-n/2} \\ &= \prod_{i=j}^{p} (1 - 2i\theta\rho^j)^{-n/2}.\end{aligned} \tag{24.19}$$

From the uniqueness property of the characteristic function we can claim that T is distributed as a linear combination of p independent and identically distributed χ^2 variates with degrees of freedom n each and weights as ρ^j's. The distribution of

T can be approximated by a $c\chi_v^2$ variate where

$$C = \sum_{j=1}^{p}(\rho^j)^2 / \sum_{j=1}^{p}\rho^j(>0), v = n\{\sum_{j=1}^{p}\}^2 / \sum_{j=1}^{p}(\rho^j)^2(>0).$$

Result 3.2. Under the given set up and the state of control, the probability density function of $T = n\overline{D}$ will be given by

$$f(t) = [det(P)]^{-n/2}\sum_{b=0}^{\infty}\sum_{\beta}\frac{(n/2)_\beta}{b!}C_\beta(P*)exp[-t/2](t/2)^{b+np/2-1}/\{2\Gamma(np/2+b)\} \tag{24.20}$$

where the matrix $P* = I - P^{-1}$.

Proof. From (24.18) we have

$$\begin{aligned}
\varphi(\theta) &= [det(I - 2i\theta P)]^{-n/2} \\
&= (1-2i\theta)^{-np/2}[det(P)]^{-n/2}[I-(1-2i\theta)(I-P-1)]^{-n/2} \\
&= (1-2i\theta)^{-np/2}[det(P)]^{-n/2}\sum_{b=0}^{\infty}\sum_{\beta}\frac{(n/2)_\beta}{b!}C_\beta(P*)(1-2i\theta)^{-b} \\
&= [det(P)]^{-n/2}\sum_{b=0}^{\infty}\sum_{\beta}\frac{(n/2)_\beta}{b!}C_\beta(P*)(1-2i\theta)^{-(np/2+b)} \qquad (24.21)
\end{aligned}$$

Applying inversion theorem on (24.21) and noting that $(1-2i\theta)^{-(np/2+b)}$ is the characteristic function of a χ^2 variate with degrees of freedom $(np+2b)$, we have finally

$$f(t) = [det(P)] - n/2\sum_{b=0}^{\infty}\sum_{\beta}\frac{(n/2)_\beta}{b!}C_\beta(P*)exp[-t/2](t/2)^{b+np/2-1}/\{2\Gamma(np/2+b)\}$$

Hence follows the result.

It may be noted that with w_b as the weight function, where

$$w_b = [det(P)]^{-n/2}\sum_{\beta}\frac{(n/2)_\beta}{b!}C_\beta(P*), b = 0, 1...$$

the sum of the weights come out as

$$\begin{aligned}
\sum_{b=0}^{\infty}w_b &= \sum_{b=0}^{\infty}[det(P)]^{-n/2}\sum_{\beta}\frac{(n/2)_\beta}{b!}C_\beta(P*) \\
&= [det(P)]^{-n/2}[det(I-P*)]^{n/2} = [det(P)]^{-n/2}[det(P)]^{n/2} = 1.
\end{aligned}$$

Thus, (24.21) can be viewed as a weighted sum of χ^2 densities and is similar to mixture distribution.

The above-mentioned approach can also be adopted to derive the distribution of $T = n\overline{D}$ under an out of control state. Observing that under an out of control state, $E(X) = \mu, R$ follows noncentral Wishart distribution of the form $\omega_p(n, P, \Omega)$, where the noncentrality parameter matrix is given by

$$\Omega = nP^{-1}\Delta\Delta', \text{where} \Delta = (\Delta_1, ..., \Delta_i, ..., \Delta_p). \tag{24.22}$$

Then, the characteristic function of T under an out of control state works out as

$$\varphi(\theta) = [det(I - 2i\theta P)]^{-n/2} exp[-tr\Omega/2 + tr\Omega(I - 2i\theta P)^{-1/2}]. \tag{24.23}$$

Expanding $tr\Omega/2$ with a choice of orthogonal matrix, H, as in (24.19) that diagonalizes P we have

$$tr\Omega/2 = (n/2)trHP^{-1}H'H\Delta\Delta'H' = (n/2)(H\Delta)/(HPH')^{-1}H\Delta$$

and noting that $(HPH')^{-1} = diag(1/\rho^1, 1/\rho^2, .., 1/\rho^p)$ we have

$$tr\Omega/2 = (n/2)\sum_{j=1}^{p} \tau_j^2/\rho^j, \tag{24.24}$$

where $H\Delta = \tau = (\tau_1, \tau_2, ..., \tau_p)$, say. Similarly, expanding $tr\Omega(I - 2i\theta P)^{-1/2}$ we have

$$\begin{aligned}
tr\Omega(I - 2i\theta P)^{-1/2} &= (n/2)trH\Omega H'H(I - 2i\theta P)^{-1}H' \\
&= (n/2)tr\text{diag}(1/\rho^1, 1/\rho^2, ..., 1/\rho^p)\tau\tau' diag(1/(1 - 2i\theta\rho^1), \\
&\quad 1/(1 - 2i\theta\rho^2), ..., 1/(-2i\theta\rho^p) \\
&= (n/2)\tau'\text{diag}(1/\rho^1(1 - 2i\theta\rho^1), \\
&\quad 1/\rho^2(1 - 2i\theta\rho^2), ..., 1/\rho^p(-2i\theta\rho^p)\tau \\
&= (n/2)\sum_{j=1}^{p} \tau_j^2/(1 - 2i\theta\rho^j).
\end{aligned} \tag{24.25}$$

Thus, from (24.23) using (24.24) and (24.25) we get

$$\varphi(\theta) = \prod_{j=1}^{p}(1 - 2i\theta\rho^j)^{-n/2} exp[-(n/2)\sum_{j=1}^{p} \tau_j^2/\rho^j 1 - (1 - 2i\theta\rho^j)^{-1}.]. \tag{24.26}$$

But (24.26) implies that T is of the form

$$T = \sum_{j=1}^{p} \rho^j Y_j,$$

where Y_j's are independently distributed noncentral χ^2 variates with degrees of freedom n and noncentrality parameter $\tau_j^2/\rho^j, j = 1, 2, ..., p$. This generalizes the Result 3.1 for the out of control state and is summarized below.

Result 3.3. Under the given set up and the out of control state, $E(X) = \mu, T = n\overline{D}$ is distributed as a linear combination of p independently distributed noncentral χ^2 variates with weight function ρ^j, degrees of freedom n and noncentrality parameter $\tau_j^2/\rho^j, j = 1,2,...,p$.

Similarly Result 3.2 can be generalized as stated below.

Result 3.4. Under the given set up and the out of control state, $E(X) = \mu$, the probability density function of $T = n\overline{D}$ will be given by

$$f(t) = [det(P)]^{-n/2} exp[-tr\Omega/2]$$
$$\sum_{b=0}^{\infty}\sum_{\beta} \frac{(n/2)_\beta}{b!} C_\beta(P*) \sum_{r=0}^{\infty} \frac{1}{r!}$$
$$\{\sum_{s_1=0}^{\infty} \cdots \sum_{s_1=0}^{\infty} \prod_{i=l}^{r} a_{s_i} exp[-t/2](t/2)^{b+r+\Sigma_{i=1}^{r} s_i + np/2-1} /$$
$$\{2\Gamma(np/2+b+r+\sum_{i=1}^{r} s_i)\}\}$$

where $a_m = tr(P*)^m \Omega P^{-1}$.

Proof. From (24.23) we have

$$\varphi(\theta) = [det(I-2i\theta P)]^{-n/2} exp[-tr\Omega/2 + tr\Omega(I-2i\theta P)^{-1/2}].$$
$$= exp[-tr\Omega/2][det(P)]^{-n/2} \sum_{b=0}^{\infty}\sum_{\beta} \frac{(n/2)_\beta}{b!} C_\beta(P*)(1-2i\theta)^{-b-np/2}$$
$$exp[(1-2i\theta)^{-1} \sum_{m=0}^{\infty} (1-2i\theta)^{-m} tr(P*)^m \Omega P^{-1}/2$$
$$= exp[-tr\Omega/2][det(P)]^{-n/2} \sum_{b=0}^{\infty}\sum_{\beta} \frac{(n/2)_\beta}{b!} C_\beta(P*)(1-2i\theta)^{-b-np/2}$$
$$exp[(1-2i\theta)^{-1} \sum_{m=0}^{\infty} (1-2i\theta)^{-m} tr(P*)^m \Omega P^{-1}/2$$
$$= exp[-tr\Omega/2][det(P)]^{-n/2} \sum_{b=0}^{\infty}\sum_{\beta} \frac{(n/2)_\beta}{b!} C_\beta(P*)(1-2i\theta)^{-b-np/2}$$
$$\sum_{r=0}^{\infty} \frac{1}{r!} \{\sum_{m=0}^{\infty} (1-2i\theta)^{-m-1} a_m\}^r$$
$$= [det(P)]^{-n/2} exp[-tr\Omega/2] \sum_{b=0}^{\infty}\sum_{\beta} \frac{(n/2)_\beta}{b!} C_\beta(P*) \sum_{r=0}^{\infty} \frac{1}{r!}$$
$$\{\sum_{s_1=0}^{\infty} \cdots \sum_{s_1=0}^{\infty} \prod_{i=1}^{r} a_{s_i} (1-2i\theta)^{-b-np/2-r-} \sum_{i=1}^{r} S_i\} \qquad (24.27)$$

Applying inversion theorem on (24.27) and noting that last term is the characteristic function of a χ^2 variate with degrees of freedom $(np+2b+2r+2\sum_{i=1}^{r} S_i)$, we have finally

$$f(t)=[det(P)]^{-n/2}exp[-tr\Omega/2]\sum_{b=0}^{\infty}\sum_{\beta}\frac{(n/2)_\beta}{b!}C_\beta(P*)\sum_{r=0}^{\infty}\frac{1}{r!}\{\sum_{s_i=0}^{\infty}\cdots$$

$$\sum_{s_i=0}^{\infty}\prod_{i=1}^{r}a_{s_i}exp[-t/2](t/2)^{b+r+\sum_{i=1}^{r}s_i+np/2-1}/\{2\Gamma(np/2+b+r+\sum_{i=1}^{r}S_i)\}\}$$

This completes the proof of the result.

24.3.3.

The exact distribution of the statistic T under the state of control can be used for the construction of probability limit control chart. Its distribution in the out of control state can be used for examining the operating characteristic surface. Determination of probability limits means determination of U and L values that satisfy, under the state of control, the conditions

$$\int_U^{\infty} f_T(t)dt=\alpha/2=\int_0^L f_T(t)dt \tag{24.28}$$

where according to Result 3.2, the probability density of T can be described as a weighted sum of χ^2 variates with degrees of freedom $np+2b, b=0,1,\ldots$ and weights as

$$w_b=[det(P)]^{-n/2}\sum_{\beta}\frac{(n/2)_\beta}{b!}C_\beta(P*)$$

Let us in general consider the problem of determining t_0 value against a given probability measure satisfying the condition $Pr[T\leq t_0]=\delta$, or,

$$\sum_{b=0}^{\infty}w_b\int_0^{t_0}f_{\chi^2_{np+eb}}(t)dt=\delta,$$

$$\text{or}\sum_{b=0}^{\infty}w_bF_{\chi^2_{np+eb}}(t_0)=\delta$$

Then t_0 can be considered as a solution of the equation:

$$t=F_{np}^{-1}[(\delta-\sum_{b=1}^{\infty}w_bF_{\chi^2_{np+eb}}(t_0))/w_b] \tag{24.29}$$

This equation can be solved numerically following the iteration method:
It t_n is the trial solution at the n-th stage then t_{n+1} is given by

$$t_{n+1} = [t_n + F_{np}^{-1}[(\delta - \sum_{b=1}^{\infty} w_b F_{\chi^2_{np+eb}}(t_n))/w_b]]/2$$

24.4. Application in Binary Choice Model

24.4.1.

Hauser, et al. (1983) considered binary choice model for forecasting sales of a new durable, by taking into consideration the utility of purchase and utility of no purchase. Lilien et al (1999) have presented a detailed survey on different choice models including binary logit and binary probit models. Following Ben-Akiva and Lerman (1985), who considered the concept of utility of each brand in a consideration set for making a purchase decision, we may use a generalized concept of desirability under a multi-character set up. For simplicity, we may start with a two-brand consideration set and the same can be generalized for the k-brand consideration set.

Let there be two competing brands A and B offered by two competitors in a given market of binary choice. By binary choice we mean either brand A or brand B will be purchased. It rules out the other two situations that neither A nor B will be purchased (i.e., the case of lack of need arousal) and both A and B will be purchased (i.e., the case of lack of assessment or over arousal). With utility vectors of assessment as X_A and X_B for the brands A and B respectively, we can write the overall desirability of A as D_A and of B as D_B, with both the brands targeting the ideal brand character value μ^0. Brand quality varies with respect to correlation matrices P_A and P_B under the assumption of multivariate normality. Assuming that a decision-maker searches for initial information about the brands from the existing users, let there be n and m random observations on X_A and X_B respectively with mean desirability values as

$$\overline{D}_A = \frac{1}{n}\sum_{\alpha=1}^{n} D_{A\alpha} \text{and} \overline{D}_B = \frac{1}{m}\sum_{\alpha=1}^{m} D_{B\alpha}$$

following (24.9). Then, the chance of purchase of brand A, which is given by $Pr[D_A > D_B]$, may be approximated by $Pr[\overline{D}_A > \overline{D}_B]$. The latter probability can be evaluated for a general choice of m and n and examined for $n = m = 1$ to arrive at the former probability value. For this, we make use of Result 3.2 along the following line:

$$
\begin{aligned}
Pr[\overline{D}_A > \overline{D}_B] &= \int_0^\infty [\int_y^\infty f_{\overline{D}_A}(x)dx f_{\overline{D}_B}(y)dy] \\
&= [det(PA)]^{-n/2}[det(PB)]^{-m/2} \sum_{a=0}^{\infty}\sum_{\alpha} \frac{(n/2)_\alpha}{a!} C_\alpha(P*_A) \\
&\sum_{b=0}^{\infty}\sum_{\beta} \frac{(m/2)_\beta}{b!} C_\beta(P*_B) nm \int\int_{x>y} f_{\chi_{np+2\alpha}}(nx) f_{\chi_{2np+\beta}}(my) dxdy \\
&= [det(PA)]^{-n/2}[det(PB)]^{-m/2} \sum_{a=0}^{\infty}\sum_{\alpha} \frac{(n/2)_\alpha}{a!} C_\alpha(P*_A) \\
&\sum_{b=0}^{\infty}\sum_{\beta} \frac{(m/2)_\beta}{b!} C_\beta(P*_B) B_{\frac{m}{n+m}} \\
&(n/2+a, m/2+b)/B(n/2+a, m/2+b) \qquad (24.30)
\end{aligned}
$$

where $B_{\frac{m}{n+m}}(n/2+a, m/2+b)$ is the incomplete beta function corresponding to complete beta function $B(n/2+a, m/2+b)$. If this probability is greater than 0.5 we may conclude that brand A is preferred more often than brand B. also the respective market shares can be obtained in this process, share of brand A being $Pr[\overline{D}_A > \overline{D}_B]$ and the share of brand B being $Pr[\overline{D}_A < \overline{D}_B]$.
In case $n = m = 1$, we get

$$
Pr[\overline{D}_A > \overline{D}_B] == [det(P_A)]^{-1/2}[det(P_B)]^{-1/2} \sum_{a=0}^{\infty}\sum_{\alpha} \frac{(1/2)_\alpha}{a!} C_\alpha(P*_A)
$$

$$
\sum_{b=0}^{\infty}\sum_{\beta} \frac{(1/2)_\beta}{b!} C_\beta(P*_B) B_{\frac{1}{2}}(1/2+a, 1/2+b)/B(1/2+a, 1/2+b) \qquad (24.31)
$$

In case of the k-brand consideration set, $A_1, A_2, ..., A_k$, one may sequentially examine the brands by undertaking pair-wise comparison between the brand under consideration, A_i, and the most preferred one of the earlier brands $\{A_1, A_2, ..., A_{i-1}\}, i = 2, 3, ..., p$.. In this process we can arrange the brands according to the order of preference.

References

1. Ben-Akiva, M. and Lerman, S. R. (1985). *Discrete Choice Analysis: Theory and Application to Travel Demand*, Cambridge, Massachusetts Institute of Technology Press.
2. Harrington, E.C. Jr. (1965). The desirability function. *IQC*, 21, 494-498.

3. Hauser, J. R., Roberts, J. H., and Urban, G. L. (1983): Forecasting sales of a new durable, *In F. S. Zufryden, ed., Advances and Practices of Marketing Science Management*, 115-126, The Institute of Management Science.
4. Hawkins, D. M. (1991). Multivariate quality control based on regression adjusted variables, *Technometrics*, **33**, 61-75.
5. Hayter, H. G. and Tsui (1994). Identification and quantification in multivariate quality control problems, *J. Quality Technology*, **26**, 198-208.
6. Hotelling, H. (1947). Multivariate quality control; Illustrated by the air testing of sample bombsights, Techniques of Statistical Analysis, *McGraw Hill Book Company Inc.*, 111-184.
7. Jackson, J.E. and Morris, R.H. (1957). An application of multivariate quality control to photographic processing, *J. Amer. Statistical Assoc.*, **52**, 186-199.
8. James, A. T. (1961). Zonal polynomials of the real positive definite symmetric matrices, *Ann. Math. Statistics*, **74**, 456-469.
9. James, A. T. (1964). Distribution of matrix variates and latent roots derived from normal samples, *Ann. Math. Statistics*, **76**, 475-501.
10. Kariya, T. (1981). A robustness test for Hotelling's T2 test, *Ann. Statistics*, **9**, 211-214.
11. Lilien, G. L., Kotler, P., and Moorthy, K. S. (1999). *Marketing Models, Prentice Hall of India, India.*
12. Mason, R. L. and Young, J. C. (1999). Improving the sensitivity of T2-statistic for multivariate process, *J. Quality Technology*, **31**, 155-165.
13. Mason, R. L., Tracy, N. D. and Young, J. C. (1995). Decomposition of T2 for multivariate control chart interpretation, *J. Quality Technology*, **27**, 99-108.
14. Mason, R. L., Tracy, N. D. and Young, J. C. (1997). A practical approach for interpreting multivariate T2 control chart signals, *J. Quality Technology*, **29**, 396-406.
15. Mukherjee, S. P. (1971). Control of multiple quality characteristics, *ISQC Bull.*, **13**, 11-16.
16. Roy, D. (1982). *Contribution to Quality Control: The general case of Multivariate Quality Characteristics, Doctoral Dissertation, IIM Calcutta, India.*
17. Roy, D. (1988). Multivariate extension of mesokurtic property and a test for the same: *Aligarh Journal of Statistics*, vol. 8, 14-25.
18. Roy, D. (2003). An optimal multivariate inference procedure for simultaneous testing of some location and dispersion parameters, *Statistical Methods*, vol.5, 57-69.
19. Roy, D. and Basu, S.K. (1989). Control of dispersion parameters in a multivariate situation, *Rep. Stat. Appl. Res.*, JUSE, **36**, 21-31.
20. Roy, D. and Basu, S.K. (1991). Control of location parameters in a multivariate situation, *Rep. Stat. Appl. Res., JUSE*, **38**, 10-18.

Chapter 25

Time Series of Categorical Data Using Auto-Mutual Information with Application of Fitting an AR(2) Model

Atanu Biswas[1] and Apratim Guha[2]

[1]*Applied Statistics Unit, Indian Statistical Institute, 203 B. T. Road, Kolkata - 700 108, India*
E-mail:atanu@isical.ac.in
[2] *Department of Statistics and Applied Probability*
National University of Singapore, Singapore
E-mail:staag@nus.edu.sg

Time series data of categorical nature is a much less explored research area in statistical literature. In this paper, we first present a detailed review of the existing works in this area. Then we consider the framework based on the Pegram's(1980) operator that was originally proposed only to construct discrete AR(p) processes. We use the concept of mutual information, and introduce auto-mutual information to define the time series process of categorical data. Some inferential aspects are discussed. The procedure is then illustrated by some real data. In that process we discuss the fit of the AR(1) and AR(2) models. We also establish some theoretical results for the AR(1) model.

Contents

25.1. Introduction

Mutual information is used as a measure of association between variables. It has certain advantages over the traditional association measure, correlation coefficient, which is primarily a measure of linear association. Mutual information works well when the dependence structure is non-linear. Moreover, for certain types of data structure, where the correlation does not make any sense, mutual information may be a suitable measure of association. For example, if X and Y be two categorical variables where X takes the categories $1,2,\cdots,k_1$ and Y takes the categories $1,2,\cdots,k_2$, the usual product-moment correlation coefficient does not make any sense. Even if both X and Y are ordinal variables, the value of the correlation coefficient depends on the numbering of the categories. If, instead of $1,2,\cdots,k_1$, the values associated with the corresponding categories be $C_1,C_2,\cdots,C_{k_1}$, where not all $C_i = a+bi$ for some a,b, then the value of the correlation coefficient will change dramatically. Moreover, the concept of correlation coefficient is not applicable if at least one of X and Y is nominal categorical. Similar problem will occur if one of X and Y is discrete and the other is continuous. But, the concept of mutual information will work in all these situations, and will be invariant of any ordinal numbering of the categories.

In the present paper, we first consider time series framework of discrete data. We discuss some mutual information based modeling and analysis. We discuss some estimation procedure. Then we discuss mutivariate time series modeling. Specifically we discuss bivariate hybrid (mixed discrete and continuous) time series, and the possible use of mutual information in that situation.

The rest of the paper is organized as follows. In Section 25.2, we provide a detailed literature review of different approaches of modeling time series of categorical data. We also provide some literature review of mutual information in Section 25.3. Mutual information based time series modeling is discussed in Section 25.4. We provide detailed discussion of AR(1) and AR(2) approaches in this context. In Section 25.5, we discuss some estimation procedure and some theoretical results are derived. Section 25.6 provides the data analysis and some simulation results in that context. Section 25.7 concludes.

25.2. Time Series of Discrete Data

Box and Jenkins' ARMA models (see Box and Jenkins) have been playing a central role in modeling of time series data. Box and Jenkins' models are widely used to analyze time series of continuous data in various real situations. Here the Gaussian assumption is very crucial, and maximum likelihood estimation for parameters of the models is derived and the subsequent analyses are usually done under the assumption that data are generated from a Gaussian ARMA process. However, when the data type of a time series is known to be non-Gaussian, the Box and Jenkins' ARMA models fail. For discrete data structure, the Gaussian assumption naturally fails. There are some existing alternative ways for modeling non-Gaussian time series data. For example, a stationary AR(1) time series of counts can be modeled by the thinning operator (see Steutel and van Harn). Joe extended the definition of the thinning operator to a family of infinitely divisible distributions, and presented a unified approach to the construction of stationary ARMA processes with marginal distributions in this family, including the Box and Jenkins' Gaussian ARMA model as a special case. Stationary processes with nonnormal marginals including Poisson, gamma, negative binomial, and inverse Gaussian are special cases of the model of Joe. Also see Jørgensen and Song in this context. However, the distributions which are not infinitely divisible are excluded from this class. For example, the binomial or categorical processes are not infinitely divisible, and they do not fall in the formulation of Joe.

For discrete stationary processes, Jacobs and Lewis proposed a mixing scheme for obtaining a stationary discrete series with a given marginal probability mass function and an autocorrelation structure. Later, Jocobs and Lewis applied this method to define a class of stationary discrete ARMA processes that has non-negative correlations and a possibly countably infinite state space. The representation of Jacobs and Lewis' discrete ARMA processes is different from that of Box and Jenkins', in which the time indices are a sequence of independently and identically distributed random variables. The randomness for the time index does not seem to be appropriate for a time series observed at deterministic time points, and hence the model interpretation becomes problematic. Kanter proposed a binary stationary process through a kind of stochastic operator called "addition mod 2". Due to the operation of *modulo*, this operator can only be manipulated arithmetically, and hence is not applicable for categorical stochastic processes. Furthermore, it is difficult to extend the Kanter's "addition mod 2" operator to establish a unified framework of discrete valued stationary processes including integer-valued stationary processes. Another noticeable shortcoming of Kanter's approach is that it seems very hard to interpret the resulting autocorrelation function, which equals to zero when the marginal mean of Y_t equals to $\frac{1}{2}$. Moreover, the sign of such a serial correlation, negative or positive, depends on whether the mean of Y_t is in the interval $(\frac{1}{2}, \frac{1}{2\gamma})$ or not, where γ is the probability that a random indicator variable equals to 1 or zero. Such dependence inevitably creates some major restriction on the application of the Kanter's method to analyze real world data.

In contrast, Pegram defined a discrete AR(p) process that resembles the representation of the Box and Jenkins' ARMA processes and allows some of the correlations to be negative. Moreover, the resulting autocorrelation function is independent of the marginal mean parameters. Note that Pegram's AR models are more flexible in terms of the range of correlation and are easy to interpret. Unfortunately, Pegram's construction is only restricted to the discrete AR processes, it was not extended for general ARMA process until very recently by Song, Biswas and Freeland.

In this context it may be worthwhile to discuss the developments of nonstationary discrete valued processes. The Pegram's operator can be suitably redefined to construct a unified framework of nonstationary discrete processes. In a recent work, Qaqish proposed a transition model for binary nonstationary processes that takes the form analogous to that of the best linear unbiased predictor (BLUP), in which time-varying covariates are allowed to enter the marginal means through, for example, a logistic model.The coefficients in Qaqish's transition model are determined by a pre-fixed autocorrelation matrix, such as that of the AR(1). Unfortunately, such a model only works for processes with binary margins. This restriction can be lifted by relaxing the Pegram's operator in such a way that the resulting nonstationary processes are applicable for any discrete margins, and furthermore Qaqish's binary model becomes a special case.

Pegram's model: Pegram defined a discrete AR(p) process that resembles the representation of Box and Jenkins' ARMA processes and allows some of the correlations to be negative. The Pegram's operator is denoted by '$*$'. In fact, '$*$ is basically a mixing operator. Precisely, for two independent discrete random variables U and V, and for a given coefficient $\phi \in (0,1)$, the Pegram's operator defines a new random variable Z as a mixture of U and V with the mixing coefficients ϕ and $1-\phi$ respectively. This is defined as

$$Z = (U,\phi) * (V, 1-\phi),$$

where the marginal probability function of Z is given by

$$P(Z=j) = \phi P(U=j) + (1-\phi)P(V=j),\ j=0,1,\ldots.$$

The mixing operator $*$ can be easily extended to handle more than two discrete variables in the same way. For example, for three independent discrete random variables U, V and W, and for given coefficients $\phi_1, \phi_2 \in (0,1)$ such that $\phi_1+\phi_2 < 1$, the Pegram's operator defines the new random variable Z as a mixture of U, V and W with the mixing coefficients ϕ_1, ϕ_2 and $1-\phi_1-\phi_2$ respectively. Formally, we can write

$$Z = (U,\phi_1) * (V,\phi_2) * (W, 1-\phi_1-\phi_2),$$

where the marginal probability function of Z is given by

$$P(Z=j) = \phi_1 P(U=j) + \phi_2 P(V=j) + (1-\phi_1-\phi_2)P(W=j),\ j=0,1,\ldots.$$

Unfortunately, Pegram's construction is only restricted to the discrete AR processes.

Raftery describes another approach to model categorical time series data. Their approach is to define a special model for higher order Markov chains with the following transition probabilities:

$$P[X_t = j_0 | X_{t-1} = j_1, \cdots, X_{t-l} = j_l] = \sum_{i=1}^{l} \lambda_i q_{j_0 j_l}$$

where $\sum_{i=1}^{l} \lambda_i = 1$ and $Q = \{q_{jk}\}$ is a non-negative square matrix with column sums 1, and $0 \leq \sum_{i=1}^{l} \lambda_i q_{jk_i} \leq 1$. However, the Pegram's model is simpler and provides a ready extension of the Box-Jenkins type models to discrete processes. In the present paper, we analyze the English county cricket championship data in a suitable discrete time series framework. In the county championship data, we feel that the last year's champion starts the new season with some sort of champion's edge, and hence the Pegram's type of modeling might be very useful in such a scenario.

25.3. Mutual Information

When the concept of "correlation" between variables is appropriate, Song, Biswas and Freeland extended the Pegram's operator for stationary ARMA processes. But, it is important to note that the concept of *correlation* is not quite appropriate for many categorical set up. If the categorical data are of the nominal type, the correlation is not defined. Even if the data are ordinal, the value of the correlation changes dramatically with the numbering of the categories. If $C_1, \cdots, C_k$ be the k categories of the categorical process $\{Y_t,\ t = 1, 2, \cdots, T\}$, then the correlation between Y_t and $Y_{t'}$ are different for two sets of (not linearly related) numberings of the categories. Biswas and Guha introduces a modification of the usual concept of auto-correlation, which fails in this context to illustrate the nature of association, using *mutual information* (MI) which does not depend on the numbering of the categories.

For two random variables X and Y with joint density function $f_{XY}(x,y)$ with respect to some dominating measure, the mutual information, first introduced by Shannon, is defined as

$$H_{XY} = \int\int \log\left(\frac{f_{XY}(x,y)}{f_X(x) f_Y(y)}\right) f_{XY}(x,y)\, dx\, dy; \tag{25.1}$$

where $f_X(x)$ and $f_Y(y)$ are the respective marginals of X and Y. By Jensen's inequality, $H_{XY} \geq 0$, with equality holding if and only if X and Y are statistically independent, so that for a time series $\{X(t)\}$, $X(t)$ and $X(t+u)$ are independent if and only if $H_{X(t),X(t+u)}$ is zero. For a stationary time series, an estimate of

$H(u) = H_{X(t),X(t+u)}$ can therefore be used as a check for dependence at lag u by testing whether the estimate is significantly different from zero. Using that idea, Biswas and Guha devices a method to fit an AR model to a categorical time series.

In spite of its popularity among the applied scientists, the theoretical aspects of the MI estimators have not received much attention. Among the major contributors, Antos and Kontoyiannis prove the consistency of a nonparametric plug-in MI estimator of 25.1 based on a histogram-type estimate in the discrete setup. Moddemeijer provides expressions for the asymptotic bias and variance for a histogram estimator for (25.1) in a time-delay setup.

For the parametric case, Brillinger shows that the estimator of 25.1 is $\sqrt{N}$ consistent and asymptotically normal under regularity conditions on the MLE of the parameter.

Joe shows that for bounded continuous X and Y, the plug-in estimator of 25.1 is consistent and finds an expression for the mean square error.

Fernandes obtains an asymptotic normal distribution for a family of estimates of the generalized entropic measures

$$\rho_q = \frac{1}{1-q}\int\int\left(1-\left[\frac{p_X(x)p_Y(y)}{p_{XY}(x,y)}\right]^{1-q}\right)p_{XY}(x,y)\,dx\,dy \tag{25.2}$$

where $0 < q \leq 1$. The estimates are obtained by substituting kernel based density estimators into 25.2.

25.4. AR Processes

Suppose that categorical time series Y_t has k categories; its marginal probability mass function is time-independent and given by

$$P(Y_t = C_j) = p_j, \quad j = 0, 1, \ldots, \tag{25.3}$$

and it is denoted by $Y_t \sim Cat((C_j, p_j), j = 1, \cdots, K)$. Suppose ε_t's are also independently and identically distributed as the same $Cat(C_j, p_j), j = 1, \cdots, K)$.

Definition 25.4.1 Let $\{Y_t\}$ be a discrete stochastic process such that

$$Y_t = (I[Y_{t-1}], \phi_1) * (I[Y_{t-2}], \phi_2) * \cdots * (I[Y_{t-p}], \phi_p) * (\varepsilon_t, 1-\phi_1-\cdots-\phi_p), \tag{25.4}$$

which is a mixture of $(p+1)$ discrete distributions, where $I[Y_{t-1}], \cdots, I[Y_{t-p}]$ are p point masses, $I[E]$ being the indicator which is 1 or 0 according as E occurs or not, and $\varepsilon_t \sim Cat(C_j, p_j), j = 1, \cdots, K$, with respective mixing weights being $\phi_1, \phi_2, \cdots, \phi_p$ and $1-\phi_1-\cdots-\phi_p$, $\phi_j \in (0,1), j = 1, \ldots, p$ and $\sum_{j=1}^{p}\phi_j \in (0,1)$.

This implies that for every $t \in 0, \pm 1, \pm 2, \ldots$, the conditional probability function takes the form

$$P(Y_t = C_j | Y_{t-1}, Y_{t-2}, \ldots) = P(Y_t = C_j | Y_{t-1}, Y_{t-2}, \ldots, Y_{t-p})$$

$$= (1 - \phi_1 - \cdots - \phi_p) p_j + \phi_1 I[Y_{t-1} = C_j] + \cdots + \phi_p I[Y_{t-p} = C_j], \tag{25.5}$$

where ϕ_j, $j = 1, \ldots, p$, are chosen such that the polynomial equation $1 - \phi_1 z - \cdots - \phi_p z^p = 0$ has roots lying outside of the unit disc.
By the definition of operator $*$, if $\phi_k = 0$ then the corresponding term $\phi_k I[Y_{t-k} = j] = 0$ and hence vanishes in 25.5. The above AR(p) process is stationary.
Biswas and Guha studied the general theoretical properties of such a process in details, with particular attention to the fit of a data set which is of the AR(1)-type. But, the fitting of a more difficult model are very non-trivial extension which needs a lot of additional efforts. For an AR(1) process,

$$p_{j|i}(t-1, t) = P(Y_t = C_j | Y_{t-1} = C_i) = (1 - \phi) p_j + \phi I[i = j].$$

Consequently, the MI at lag 1 reduces to

$$\begin{aligned}
\mu(1) = \mu_{t-1,t} &= \sum_{i=1}^{K} \sum_{j=1}^{K} \log \left(\frac{(1-\phi) p_j + \phi I[i=j]\} p_i}{p_i p_j} \right) \{(1-\phi) p_j + \phi I[i=j]\} p_i \\
&= (1-\phi) \log(1-\phi) \sum_{i=1}^{K} p_i (1 - p_i) \\
&+ \sum_{i=1}^{K} \log\left(((1-\phi) p_i + \phi) p_i\right) \{(1-\phi) p_i + \phi\} p_i.
\end{aligned} \tag{25.6}$$

Write $\mu(1) = f(\phi)$.
It is observed in Biswas and Guha that estimating the parameters and fitting the model are very non-trivial tasks, even if for an AR(1) model. For higher order models the situation becomes worse. In this present paper we concentrate on some properties and fitting of more difficult models. In particular we pay attention to the AR(2) model. We also analyze a real data set which is observed to fit the AR(2) model.
For an AR(2) process, writing

$$p_{j|i}(h) = P(Y_t = j | Y_{t-h} = i),$$

$$p_{j|i}(1) = (1-\phi_1-\phi_2)p_j + \phi_1 I[i=j] + \phi_2 \frac{p_j}{p_i} p_{i|j}(1),$$
$$p_{i|j}(1) = (1-\phi_1-\phi_2)p_i + \phi_1 I[i=j] + \phi_2 \frac{p_i}{p_j} p_{j|i}(1),$$
$$p_{j|j}(1) = \frac{1-\phi_1-\phi_2}{1-\phi_2} p_j + \frac{\phi_1}{1-\phi_2},$$
$$p_{j|i}(2) = (1-\phi_1-\phi_2)p_j + \phi_1 p_{j|i}(1) + \phi_2 I[i=j],$$
$$p_{j|i}(h) = (1-\phi_1-\phi_2)p_j + \phi_1 p_{j|i}(h-1) + \phi_2 p_{j|i}(h-2),\ h>2.$$

Consequently, the auto-MIs can be obtained. For $h > 2$,

$$\mu(h) = \sum_{i=1}^{K}\sum_{j=1}^{K} \log\left(\frac{\{(1-\phi_1-\phi_2)p_j + \phi_1 p_{j|i}(h-1) + \phi_2 p_{j|i}(h-2)\}p_i}{p_i p_j}\right)$$
$$\times\{(1-\phi_1-\phi_2)p_j + \phi_1 p_{j|i}(h-1) + \phi_2 p_{j|i}(h-2)\}p_i.$$

As a general ARMA process can easily be converted to an AR process, we restrict our discussion to AR models only. For details of the modeling and the theoretical results, we refer to Biswas and Guha.

25.5. Parameter Estimation

The present paper is largely motivated towards the modeling of time series of categorical data through auto-mutual information, and studying the nature of auto-mutual information of different stationary processes. The natural question of model selection and parameter estimation is important in this context. As an MA or ARMA process can be easily converted to an AR process, in this section we restrict our discussion within AR process only.

For the model selection, one can use the standard model selection criteria like AIC or BIC. However, in the context of time series analysis by Box and Jenkins' ARMA models, both AIC and BIC perform poorly due to severe bias caused by autocorrelation. Hurvich and Tsai proposed a bias-corrected AIC that has proven to be a much better procedure for order selection in AR(p) models. Such a bias-correction can be attempted in this present context. The details are under investigation, and is beyond the scope of the present paper.

As an alternative, we can compute the partial auto-mutual information (PAMI) of different orders.

The PAMI, as defined below, is developed following the same idea as that of the partial auto correlation function (PACF).

The PAMI is a special case of what is more commonly known as the conditional mutual information (CMI), see Cover and Thomas, pg 22.

The CMI between two random variables X and Y given Z is defined as

$$\mu(X,Y|Z) = E\left(\log\frac{P(X,Y|Z)}{P(X|Z)P(Y|Z)}\right),$$

which can be easily generalized. See Yeung and Zhang and Yeung for some properties of the CMI. The PAMI $\alpha(h)$ at lag $h > 0$ for a stationary time series $\{Y_t\}$ is defined as the CMI between Y_t and Y_{t+h} given $Y_{t+1},\cdots,Y_{t+h-1}$:

$$\alpha(h) = E\left(\log\frac{P(Y_t,Y_{t+h}|Y_{t+1},Y_{t+2},\cdots,Y_{t+h-1})}{P(Y_t|Y_{t+1},Y_{t+2},\cdots,Y_{t+h-1})P(Y_{t+h}|Y_{t+1},Y_{t+2},\cdots,Y_{t+h-1}))}\right)$$
$$= \sum_{i,j;j_1,\cdots,j_{h-1}} P[Y_t = i, Y_{t+h} = j, Y_{t+1} = j_1,\cdots,Y_{t+h-1} = j_{h-1}]$$
$$\times (\log(P[Y_t = i, Y_{t+h} = j|Y_{t+1} = j_1,\cdots,Y_{t+h-1} = j_{h-1}])$$
$$-\log(P[Y_t = i|Y_{t+1} = j_1,\cdots,Y_{t+h-1} = j_{h-1}])$$
$$-\log P[Y_{t+h} = j|Y_{t+1} = j_1,\cdots,Y_{t+h-1} = j_{h-1}]).$$

Notice that the PAMI for lags $h > p$ equal to zero for an AR(p) process as defined earlier, and hence its estimates can be used to check the order of an AR process. For example, for an AR(1) process, the argument of the log term in the expression for $\alpha(h)$ for $h > 1$ can be written as

$$\frac{P[Y_t = i, Y_{t+h} = j, Y_{t+1} = j_1,\cdots,Y_{t+h-1} = j_{h-1}]P[Y_{t+1} = j_1,\cdots,Y_{t+h-1} = j_{h-1}]}{P[Y_t = i, Y_{t+1} = j_1,\cdots,Y_{t+h-1} = j_{h-1}]P[Y_{t+h} = j, Y_{t+1} = j_1,\cdots,Y_{t+h-1} = j_{h-1}]}$$
$$= \frac{P[Y_{t+h} = j|Y_{t+h-1} = j_{h-1}]P[Y_t = i, Y_{t+1} = j_1,\cdots,Y_{t+h-1} = j_{h-1}]}{P[Y_t = i, Y_{t+1} = j_1,\cdots,Y_{t+h-1} = j_{h-1}]P[Y_{t+h} = j|Y_{t+h-1} = j_{h-1}]} = 1$$

and hence $\alpha(h) = 0$ for all $h > 1$.

The PAMI equals with the AMI for $h = 1$. For $h > 1$, the PAMI may be estimated by

$$\widehat{\alpha}(h) = \sum_{t=1}^{T-h} \sum_{i,j;j_1,\cdots,j_{h-1}} \hat{P}[Y_t = i, Y_{t+h} = j, Y_{t+1} = j_1,\cdots,Y_{t+h-1} = j_{h-1}]$$
$$\times \log\left(\frac{\hat{P}[Y_t = i, Y_{t+h} = j|Y_{t+1},\cdots,Y_{t+h-1}]}{\hat{P}[Y_t = i|Y_{t+1},\cdots,Y_{t+h-1}]\hat{P}[Y_{t+h} = j|Y_{t+1},\cdots,Y_{t+h-1}]}\right)$$
$$= \sum_{i,j}\sum_{j_1,j_2,\cdots,j_{h-1}} \frac{n_{i,j_1,\cdots,j_{h-1},j}}{T-h}\log\left(\frac{n_{i,j_1,\cdots,j_{h-1},j}n_{j_1,\cdots,j_{h-1}}}{n_{j_1,\cdots,j_{h-1},j}n_{i,j_1,\cdots,j_{h-1}}}\right)$$

where

$$n_{i_1,i_2,\cdots,i_k} = \sum_{t=1}^{T-k} I[Y_t = i_1, Y_{t+1} = i_2,\cdots,Y_{t+k-1} = i_k]$$

for any k-tuple $(i_1,i_2,\cdots,i_k)$ for any k.

Once the order of the time series is settled, the next obvious question is the estimation of model parameters. This is often a highly non-trivial task, involving high computation costs. To demonstrate some of the difficulties in estimating the model parameters, let us first discuss the simpler case of AR(1). MLE can be obtained in the usual manner, but this is computationally an extensive job, if the number of parameters are large. Instead, one can use a partial MLE, where we estimate the marginal probabilities via the sample frequencies, providing consistent estimates, and then treat them as known values in likelihood. In this present situation, we can think of a third alternative. First, we estimate the marginal probabilities via the sample frequencies, and then estimate the AMI of order 1, which is an increasing function of ϕ. Thus we immediately get the estimate of ϕ using these estimates. The estimates are

$$\widehat{p}_j = \frac{1}{T}\sum_{t=1}^{T} I[Y_t = C_j],$$

$$\widehat{p}_{ij} = \frac{1}{T-1}\sum_{t=2}^{T} I[Y_t = C_j, Y_{t-1} = C_i].$$

Under $\phi = 0$, the Y_ts are independent, and hence $E(\widehat{p}_j) = p_j$ and $Var(\widehat{p}_j) = p_j(1-p_j)/T$. But, even under $\phi = 0$, the successive $I[Y_t = C_j, Y_{t-1} = C_i]$s have one Y in common. After some routine steps, we have $E(\widehat{p}_{ij}) = p_{ij}$ and

$$Var(\widehat{p}_{ij}) = \frac{p_{ij}(1-p_{ij}) - 2(1-\phi)^{-1}p_{ij}^2 + 2(p_{iii} + \phi(1-\phi)^{-1})I(i=j)}{T-1} + O_p(T^{-2}),$$

where $p_{iii} = p_{i|i}^2 p_{i\cdot}$. See Biswas and Guha (2007) for details. We also have the following theorem.

Theorem 1: The estimate of the AMI of order 1 is asymptotically a linear combination of independent and identically distributed χ^2s of 1 degree of freedom each. The weights are given by the eigenvalues of some matrix $\Sigma^{(1)}$, which is a function of $\widehat{\phi}$ and ϕ_0.

The proof of this theorem is available in Biswas and Guha(2007).

Theorem 1 can be used to test the AR(1) model under the Pegram set up. For the other set up, the result changes.

To find the estimates of the parameters of an AR(2) process we proceed as follows. We propose to find the estimates of ϕ_1 and ϕ_2 from the sample estimates of the first two PAMI's. Here

$$\alpha(1) = \mu(1) = \sum_{i=1}^{k}\sum_{j=1}^{k} \log\left(\frac{p_{ij}(1)}{p_i p_j}\right) p_{ij}(1)$$

$$= \sum_{i=1}^{k}\sum_{j=1}^{k} \log\left(\frac{p_{j|i}(1)}{p_j}\right) p_{j|i}(1) p_i$$

$$= \sum_{i,j} \log\left(\frac{(1-\phi_1-\phi_2)p_j + \phi_1 I(i=j)}{(1-\phi_2)p_j}\right) \times \frac{(1-\phi_1-\phi_2)p_i p_j + \phi_1 I(i=j) p_i}{1-\phi_2}. \quad (25.7)$$

We obtain 25.7 by solving the following pair of equations:

$$p_{j|i} = (1-\phi_1-\phi_2)p_j + \phi_1 I(i=j) + \phi_2 \frac{p_j}{p_i} p_{i|j}(1),$$

$$p_{i|j} = (1-\phi_1-\phi_2)p_i + \phi_1 I(i=j) + \phi_2 \frac{p_i}{p_j} p_{j|i}(1),$$

which gives

$$p_{j|i} = \left(\frac{1-\phi_1-\phi_2}{1-\phi_2}\right) p_j + \left(\frac{\phi_1}{1-\phi_2}\right) I(i=j).$$

Hence

$$\alpha(1) = \sum_{i \neq j} \log\left(1 - \frac{\phi_1}{1-\phi_2}\right) \cdot \left(1 - \frac{\phi_1}{1-\phi_2}\right) p_i p_j$$

$$\sum_{i} \log\left(1 - \frac{\phi_1}{1-\phi_2}\left(1 - \frac{1}{p_i}\right)\right) \cdot$$

$$\left(\left(1 - \frac{\phi_1}{1-\phi_2}\right) p_i^2 + \frac{\phi_1}{1-\phi_2} p_i\right), \quad (25.8)$$

which is increasing in $\frac{\phi_1}{1-\phi_2}$.

Now, denote

$$p_{ikj} = P(Y_t = C_i, Y_{t+1} = C_k, Y_{t+2} = C_j),$$

$$p_{j|ik} = P(Y_{t+2} = C_j | Y_{t+1} = C_k, Y_t = C_i).$$

We write $p_{ii}(2) = \sum_k p_{iki}$. Then some routine steps provide

$$\alpha(2) = \sum_{i,j,k} \log\left(\frac{p_{ikj} p_k}{p_{ik} p_{kj}}\right) p_{ikj}$$

$$= \log(1-\phi_2) + \sum_i \log\left(1 + \frac{\phi_2}{(1-\phi_1-\phi_2)p_i + \phi_1}\right) p_{iii}$$

$$+ \sum_i \log\left(1 + \frac{\phi_2}{(1-\phi_1-\phi_2)p_i}\right) (p_{ii}(2) - p_{iii}). \quad (25.9)$$

From 25.8 and 25.9, we obtain the estimates of ϕ_1 and ϕ_2 in an iterative way. This method can be extended for more complicated models, in principle.
We saw that in the case of AR(1) models, the AMI of order 1 is an increasing function of the parameter ϕ, which simplifies the estimation procedure to a great extent. However, such nice relationships do not exist in the higher order cases, and hence the estimation involves more complicated calculations. It makes the task of obtaining an asymptotic distribution considerably harder, if not impossible. Hence, in this paper we use data-driven methods to estimate standard errors of the parameter estimates, as is demonstrated in the next section.

25.6. Data Analysis and Simulations

We consider the champion teams of the English County Cricket Championship during 1889-2005. Total sample size is 116 of which 13 are missing (League was suspended during 1915-1919 due to World War One, and during 1939-1946 due to World War Two). Twice there were joint winners: 1950 and 1977. We consider six classes: Surrey, Yorkshire, Middlesex, Lancashire, Warwickshire and others, which are denoted $C_1, \cdots, C_6$ respectively. The five teams mentioned above having won the championship most frequently, this is a natural classification.
We estimate p_i's by the sample proportions of six categories. They are 0.1748, 0.0971, 0.2913, 0.0583, 0.0680 and 0.3107. Assuming an AR(1) model, we obtain $\widehat{\phi} = 0.4386$. To check whether this is a good fit or not, we obtain the estimates of PAMI as: $\widehat{\alpha}_1 = 0.3033$, $\widehat{\alpha}_2 = 0.3775$, $\widehat{\alpha}_3 = 0.2959$, $\widehat{\alpha}_4 = 0.2191$. As we have seen in our extensive simulation studies, for such categorical AR processes the $\widehat{\alpha}$'s are highly biased. But, the amount of bias might give us an idea about the order of the AR process.
For bias correction, we need to know the correct order of the AR process, but that is precisely what we are trying to establish here. To circumvent this circularity, we employ a trial-and-error method.
First, we simulate 100 observations from an AR(1) model assuming using the above $\widehat{p}_i$'s and $\widehat{\phi}$. AR(1) with $\phi =$ above $\widehat{\phi}$. For such a process the expected first four $\widehat{\alpha}$ (sd in parantheses) are: $\widehat{\alpha}_1 = 0.5319$ (0.0918), $\widehat{\alpha}_2 = 0.2189$ (0.0704), $\widehat{\alpha}_3 = 0.3140$ (0.0749), and $\widehat{\alpha}_4 = 0.2376$ (0.0557), which are highly different from the obtained esimates. From this we have enough reason to doubt the AR(1) model. We next try to check for the AR(2) model.
From the data, using the AR(2) model, we obtain $\widehat{\phi}_1 = 0.1084$ and $\widehat{\phi}_2 = 0.7495$. The estimates and standard error of PAMI from the simulated data (100 simulations) using these estimates $\widehat{\phi}_1$ and $\widehat{\phi}_2$, and assuming an AR(2) model is $\widehat{\alpha}_1 = 0.3647$ (0.0801), $\widehat{\alpha}_2 = 0.2953$ (0.0883), $\widehat{\alpha}_3 = 0.5531$ (0.0910), and $\widehat{\alpha}_4 =$

0.3161 (0.0578). Estimate of $\phi_1/(1-\phi_2)$ from real data is 0.4327. The estimates obtained here suggest no significant difference from the fitted values, with the exception of $\widehat{\alpha}_3$, which is significantly lower for our data set than the simulation. This suggests that AR(2) may be a reasonable model for the county champions data.
Figure 1 gives 3-D plots of PAMI of orders 1 and 2 respectively for different ϕ_1 and ϕ_2. This might be helpful for identifying AR(2) processes in general.

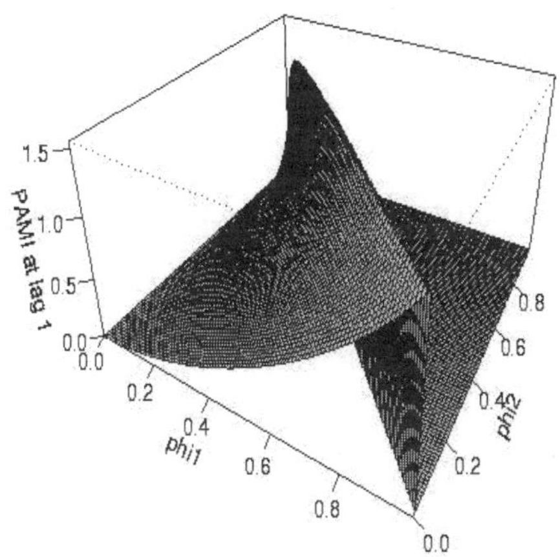

Fig. 25.1. Plot of PAMI of order 1 for AR(2) processes for different values of ϕ_1 and ϕ_2.

25.7. Concluding Remarks

The motivation of the present paper is the most important. There has not been much work in the area of time series for categorical data with more than two categories. The few that are available mostly deal with the traditional concept of autocorrelaton. But, as we see, the numbering of the categories even in ordinal categorical data is quite ad hoc, which largely influences the correlation. Thus we provide the time series based on category-free measure of association, namely auto-mutual information. This is also applicable for nominal categorical time se-

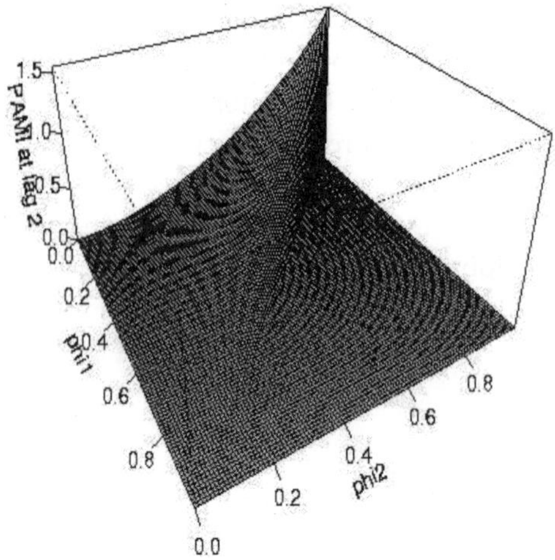

Fig. 25.2. Plot of PAMI of order 2 for AR(2) processes for different values of ϕ_1 and ϕ_2.

ries, where the concept of correlation does not exist.

We discussed model selection and parameter estimation briefly in this paper, although we admit that a lot more work is needed in this context. In particular, the problem of bias in the PAMI estimates needs to be settled conclusively so that a direct decision could be given just by looking at the PAMI estimates, the method employed in this paper perhaps being a bit ad-hoc, and computationally expensive. The computational cost is also the main reason we restrict ourselves to a small sample size for the simulation routine. However, we hope to conduct a more detailed study in the future. But that is beyond he scope of this present paper, as the paper will explode with all these. We intend to provide all the details in a future communication.

25.8. Acknowledgment

Apratim Guha's research was partially supported by NUS grants R-155-050-058-101 and R-155-050-058-133.

References

1. Agresti, A. (2000) *Categorical Data Analysis*, 2nd Ed. (Wiley, New York).
2. Antos, A. and Kontoyiannis, Y.(2001) *Random Structures & Algorithms.* **19**,163 .
3. Biswas, A. and Guha, A. (2007)Time series of categorical data using auto-mutual information. Indian Statistical Institute Technical Report Number ASD/2007/01 .
4. Box, G.E.P. and Jenkins, G.M.(1976) *Time Series Analysis: Forecasting and Control,* revised edition. (Holden-Day, San Francisco).
5. Brillinger, D.R. (2004) *Brazilian Journal of Probability and Statistics* 18, 163.
6. Cover, T. and Thomas, J.(2001) *Elements of Information Theory.* (John Wiley & Sons, New York).
7. Fernandes, M. (2001)*Unpublished manuscript.*
8. Hurvich, C.M. and Tsai, C.L. (1989) **76**, 297 .
9. Jacobs, P.A. and Lewis, P.A.W. (1978) *J. R. Statist. Soc. B* **40**, 94 .
10. Jacobs, P.A. and Lewis, P.A.W. (1983) *J. Time Series Analysis* **4**, 19 .
11. Joe, H.(1989) *Ann. Inst. Statist. Math.* **16**, 683 .
12. Joe, H. (1996) *J. Appl. Probab.* **33**, 664 .
13. Jørgensen, B. and Song, P.X.-K.(1998) *J. Appl. Probab.* **35**, 78 .
14. Kanter, M. (1975) *J. Appl. Probab.* **12**, 371 .
15. Mardia, K.V., Kent, J.T., and Bibby, J.M.(1980) *Multivariate Analysis.* (Academic Press, New York).
16. Moddemeijer, R. (1989) *Signal Processing.* **16**, 233 .
17. Pegram, G.G.S. (1980) *J. Appl. Probab.* **17**, 350 .
18. Qaqish, B.F. (2003) *Biometrika* **90**, 455).
19. Raftery, A.E. (1985). *J. R. Statist. Soc. B* 47, 528-539.
20. Rao, C.R. (2002). *Linear Statistical Inference and Its Applications*, second edition. (Wiley, New York, 2002).
21. Shannon, C. E. (1948) *Bell System Tech. J.* **27**, 379 .
22. Shannon, C. E. (1948) *Bell System Tech. J.* **27**, 623 .
23. Song, P.X.-K., Biswas, A. and Freeland, R.K.(2007) Discrete-valued ARMA processes. *Commuicated for publication.* .
24. Steutel, F.W. and van Harn, K.(1979) *Ann. Prob.* **7**, 893 .
25. Yeung, R.W. (1991) *IEEE Transactions on Information Theory* **37**, 466 .
26. Zhang, Z. and Yeung, R.W. (1998) *IEEE Trans. Inform. Theory* **44**, 1440 .

Chapter 26

Estimation of Integrated Covolatility for Asynchronous Assets in the Presence of Microstructure Noise

Rituparna Sen and Qiuyan Xu

University of California at Davis, Davis, CA, USA 95616

E-mail:rsen@wald.ucdavis.edu

The covariance for multiple price processes is of great interest in many financial applications. The naive estimator is the *realized covariance*(RC) which is the analogue of realized variance for a single process. When data on two securities are observed non-synchronously, then RC is highly biased towards zero as we increase the sampling frequency. Hayashi and Yoshida propose an alternative estimator, *cumulative covariance*(CC) that takes care of this problem using tick-by-tick data. However they solve the problem in the absence of microstructure noise. We observe that CC has very high variance when noise is present. We introduce a new estimator, random lead-lag estimator(RLLE), which coincides with the CC estimator at very high frequency and with the RC at low frequency. This estimator is asymptotically unbiased like the CC. But unlike the CC, it can also be evaluated at moderate frequencies to keep the variance under control in the presence of microstructure noise. We studied the performance of RLLE both with and without noise for non-synchronous data and obtained the optimal sampling frequency with good bias-variance trade-off. Our result is confirmed by simulation.

Contents

26.1. Introduction

The use of high frequency return data has led to dramatic improvements in financial research, especially in the area of volatility estimation. This is motivated by the recent availability of an almost continuous tick-by-tick record of quotes and transaction prices for a variety of financial assets. There is a rich literature on volatility estimation which is triggered by the work of Andersen and Bollerslev (1998). A vast amount of research followed, such as Andersen et al.(2001a and b), Barndorff-Nielsen and Shephard (2001).

Under microstructure noise, the process is 'contaminated' due to discretization or bid-ask spreads, and recently, extensive work has been done under this situation. Some references are Corsi et al(2001), At-Sahalia et al(2005), Zhang, et al(2005), Bandi and Russell (2006), Barndorff-Nielsen et al(2006), Hansen and Lunde (2006).

The covariance for multiple price processes is of great interest in most financial applications, such as portfolio selection and risk management. There are quite a few estimators constructed to estimate this quantity under various situations and assumptions. The naive estimator is the *realized covariance*(RC) as in Jacod and Shiryaev (1987) or Karatzas and Shreve (1991), which is the analogue of realized variance for a single process. This estimator is well studied for synchronous data without microstructure noise, see Jacod and Protter (1998), Barndorff-Nielsen and Shephard (2002), and Mykland and Zhang (2000). RC is a consistent estimator for integrated covolatility, with mixed normal distributed estimation error.

Under non-synchronous trading, this estimator becomes highly biased due to the Epps effect (Epps, 1979). Other estimators have been proposed to correct the bias under the situation of no microstructure noise. Scholes and Williams (1977) add one lead and lag to the realized covariance, and Dimson (1979) and Cohen et al. (1983) generalized the number of leads and lags to k. Hayashi and Yoshida (2004, 2005) propose their *cumulative covariance* (CC) estimator for non-synchronous data, which is unbiased without microstructure noise.

When microstructure noise plays a role in the market, the CC estimator becomes biased. Most recently, Griffin and Oomen (2006) study the CC estimators under i.i.d. noise scheme. Voev and Lunde (2006) have bias-corrected CC estimator and examine the sub-sampling version of that estimator. We found that also the variance of the CC estimator is very big and it is computationally demanding. This motivated us to propose a new estimator, which we call the random lead-lag

(RLLE) estimator. This can be viewed as a finite frequency version of the CC estimator. Another similar quantity is the Cohen(1983) estimator where the leads and lags are fixed. In contrast, the RLL estimator allows them to vary randomly depending on trading times.

The rest of the paper is organized as follows. In section 26.2 we present the underlying model and background material on non-synchronicity, microstructure noise and the two existing estimators: Realized Covariance and Cumulative Covariance. In section 26.3 we present the Random lead lag estimator and derive some of its properties. In section 26.4 we describe a simulation study to compare the various estimators in the absence and presence of microstructure noise. Section 26.5 presents the conclusions and some directions for ongoing and future research.

26.2. Modeling Financial Returns And Realized Covariance

26.2.1. *Basic formulation*

In 1973, Fischer Black and Myron Scholes proposed the celebrated Black-Scholes model for log stock price. This model became the root of a lot of models and technics applied by today's financial analysts. The model looks like this:

$$dP(t,\omega) = \mu dt + \sigma dW(t,\omega), \quad t \geq 0 \tag{26.1}$$

Where $P(t,\omega)$ denotes the price process of a security on the log scale at time t and under scenario ω, W is a standard Wiener process, $\sigma > 0$ is called the volatility , and μ is the drift.

In most financial applications, not only the volatility, but also the variance-covariance matrix of multiple securities is of interest, such as application in portfolio selection or risk management, etc. Generally, the volatilities are time-varying. Hence, an extension to multiple asset prices and time-varying volatility is necessary:

$$dP^{(i)}(t,\omega) = \mu^{(i)}dt + \sigma^{(i)}(t)dW^{(i)}(t,\omega), \quad i = 1,\ldots,N \quad t \geq 0 \tag{26.2}$$

where the quadratic variation between the two Wiener processes is $d < W^{(i)}, W^{(j)} >_t = \rho^{(i,j)}(t)$. From now on, we shall refer to $\rho^{(i,j)}(t)$ as the correlation. N denotes the number of securities and the superscript (i) denotes the i-th security. $\{P^{(i)}(t,\omega)\}$ is the price process for security i in logarithmic scale. $\sigma_t^{(i)} > 0, i = 1,\ldots,N$ are the volatilities , and $\mu_t^{(i)}, i = 1,\ldots,N$ are the drift terms. All of them, generally, will be (continuous) stochastic processes.

The quantity of interest is the integrated covariance

$$< P^{(i)}, P^{(j)} >_T = \int_0^T \sigma^{(i)}(t)\sigma^{(j)}(t)\rho^{(i,j)}(t)dt \tag{26.3}$$

For high frequency data, consider a fixed period of time of length $T > 0$. For concreteness we typically consider $T = 23,400$ seconds in a six and a half hour (NYSE) trading day. Traditionally daily returns are computed as

$$r_i = P(iT) - P((i-1)T), i = 1,2,...,n \tag{26.4}$$

where i indexes the day. However for high frequency data, we have additionally intra-day observations during each time period T. These observations usually are not observed at equal intervals of time. Most existing estimation methods require to convert them artificially to regularly spaced data. Let us denote the common interval by h (seconds). $h \in \{3600, 1800, 600, 300, 60, 30, 10, 1\}$ are used through our study. Let m be the number of intra-day intervals. Then $h = T/m$. The j-th intra-period return for the i-th period is defined as:

$$r_{j,i} = P((i-1)T + jh) - P((i-1)T + (j-1)h), \quad j = 1, \ldots, m, \quad i = 1, \ldots, n. \tag{26.5}$$

To convert the irregularly spaced observations to equal spaced intervals, two common schemes of data interpolation are used: 1. Previous tick interpolation, which uses the most recent values for each equal spaced sampling point. 2. Linear interpolation, which uses observations bracketing the desired time.

26.2.2. *Realized variance/covariance estimator*

Following the setup in section 26.2.1, suppose we have two assets with prices: $(P^{(i)}(t_k), P^{(j)}(t_k))_{k=0,1,\ldots,m}$ and observed at times $\pi(m) = \{0 \le t_0 \le t_1 \le \cdots \le t_m = T\}$. These correspond to times at which trades take place. Then the Realized covariance can be formulated as following:

$$V_{i,j}(\pi(m)) := \sum_{k=1}^{m} (P^{(i)}(t_k) - P^{(i)}(t_{k-1}))(P^{(j)}(t_k) - P^{(j)}(t_{k-1})).$$

While $i = j$, the realized covariance becomes the realized variance in the univariate case. While the observations are synchronous and there is no microstructure noise present, the estimator is consistent for the integrated covariance as defined in equation 26.2.1 (Jacod and Shiryaev (1987), Karatzas and Shreve (1991)), with a mixed Gaussian distribution (Jacod and Protter (1998), Barndorff-Nielsen and Shephard (2002), and Mykland and Zhang (2001)). That is:

$$V_{i,j}(\pi(m)) \xrightarrow{P} < P^{(i)}, P^{(j)} >_T \text{ as } \pi(m) := \max | t_k - t_{k-1} | \to 0 \tag{26.6}$$

Barndorff-Nielsen and Shephard (2004) derive the asymptotic distribution for the realized covariance under certain general assumptions: The price is assumed to be a continuous stochastic volatility semimartingale ($SVSM^C$). They found that the

drift term does not affect the result. They also derive the asymptotic distribution of realized correlation and realized beta. These two statistics are transformations of the realized covariance matrix defined as follows:
Realized correlation:

$$\frac{\sum_{k=1}^{m}(P^{(i)}(t_k) - P^{(i)}(t_{k-1}))(P^{(j)}(t_k) - P^{(j)}(t_{k-1}))}{\sqrt{\sum_{k=1}^{m}(P^{(i)}(t_k) - P^{(i)}(t_{k-1}))^2}\sqrt{\sum_{k=1}^{m}(P^{(j)}(t_k) - P^{(j)}(t_{k-1}))^2}}$$

Realized beta:

$$\frac{\sum_{k=1}^{m}(P^{(i)}(t_k) - P^{(i)}(t_{k-1}))(P^{(j)}(t_k) - P^{(j)}(t_{k-1}))}{\sum_{k=1}^{m}(P^{(i)}(t_k) - P^{(i)}(t_{k-1}))^2}$$

Under certain conditions, both statistics have asymptotic normal distribution. Due to the broad uses of integrated covariance, correlation, beta in Derivative Pricing, Management, Portfolio Optimization and other applications, the estimation of these quantities is of great interest. The asymptotic distribution theory provides a very useful theoretical background for estimation methods.

26.2.3. *Non-synchronicity*

In real data, the observations happen at random times corresponding to when trades take place in the market. These times do not coincide for multiple securities. For eg Asset A may records trades at 9:47 am, 9:52am, 9:56am etc whereas asset B records trades at 9:48am, 9:57am etc. This phenomenon is termed non-synchronicity.
For multivariate problems, non-synchronous trading becomes a real concern. When data on two securities are observed non-synchronously, Fisher effect (Fisher 1966) occurs irrespective of the underlying correlation structure. That is, returns sampled at equal spaced time will correlate with previous and successive returns on other assets. Also the covariation measure becomes smaller as we increase the sampling frequency. This was observed empirically by Epps(1979). Hayashi and Yoshida (2005) show that even under very mild assumptions realized covariance is biased and inconsistent if the data is non-synchronous. In fact, under certain assumptions, it converges to zero.

26.2.4. *Microstructure noise and jumps*

For the purpose of volatility estimation, Equation 26.2.2 is far from true in practice. One possible explanation is microstructure noise, such as discreteness of prices, bid - ask spread, etc. The presence of noise does not pose any particular problem if data is observed less frequently, say daily. Modern technology, however, permits almost continuous observation of data. When transaction prices or bid-ask quotes are contaminated with microstructure noise, the efficient price is not observed. The observed prices then lack martingale property. From now on, we suppose the observed log transaction or bid-ask quote price is P^*, and the underlying efficient log-price is P. For two assets, we will set their observed prices as $P_t^{(1)*}$ and $P_t^{(2)*}$, their efficient prices are $P^{(1)}$ and $P^{(2)}$ respectively. We write $P^{(i)*}$ as

$$P_t^{(i)*} = P_t^{(i)} + u_t^{(i)} \qquad i = 1,2. \tag{26.7}$$

Where $u_t^{(i)}$ is the noise component due to imperfection of the trading process, and it is a summary of a variety of microstructure noise, which can be roughly grouped into the following categories: frictions inherent in the trading process, informational effects, and measurement or data recording errors, see At-Sahalia, Mykland and Zhang (2005). (Note: they have more explicit expressions for these three kinds of microstructure noise.) .

We assume $u_t^{(i)}$ is independent of $P_t^{(i)}$. The efficient log-price $P_t^{(i)}$ follows the SDE 26.2.1. For high frequency data, the length of time intervals h is really small, usually measured in seconds. The drift term is irrelevant under this circumstance, hence we will set $\mu_t^{(i)} = 0, i = 1,2$, throughout the study.

Recently, impressive work has been done studying the properties and performance of the realized variance estimate of integrated volatility under the presence of microstructure noise. It is a fact that the estimator depends only on the noise up-to first order approximation. The following are some references in the area: Zhang, Mykland and Ait-Sahalia (2005); Ait-Sahalia, Mykland and Zhang (2005); Barndorff-Nielsen and Shephard (2003, 2004a,b); Hansen and Lunde (2006). Some suggested solutions to the problems caused by the microstructure noise are: to apply variance reduction techniques like sub-sampling, or use alternative estimators like bipower variance.

Another suggested reason for failure of Equation 26.2.2 in real data is presence of jumps in stock price process. See Fan and Wang (2005). They advocate use of wavelets to separate jumps from the continuous process.

26.2.5. *Hayashi-Yoshida Estimator*

Recently, Hayashi and Yoshida (2004,2005) proposed an alternative estimator that takes care of non-synchronicity problem using tick-by-tick data. We called it Cumulative Covariance (CC) estimator as in Hayashi and Yoshida (2005).

Similar to the notation in the section 26.2.2, suppose that we have two assets with prices: $((P^{(1)*}(t_k), P^{(2)*}(t_k))_{k=0,1,\dots,m})$, where $(P^{(i)*}(t_k), i = 1,2$ follow equation 26.2.4, and their observations are observed at times $\pi^{(1)}(m_1) = \{0 \leqslant t_0^{(1)} < t_1^{(1)} < \dots < t_{m_1}^{(1)} \leqslant T\}$, and, $\pi^{(2)}(m_2) = \{0 \leqslant t_0^{(2)} < t_1^{(2)} < \dots < t_{m_2}^{(2)} \leqslant T\}$. These usually correspond to the occurrence of transactions or quote-revisions. We assume these arrival times follow Poisson processes independent of the price process with constant arrival intensity over time but can vary across assets (see e.g. Hayashi and Yoshida, 2005).The Hayashi-Yoshida CC estimator accumulates the cross product of all overlapping (fully and partially) returns. It is specified as follows:

$$\mathrm{CC} = \sum_{i=1}^{m_1} \sum_{j \in A_i} R_i^{(1)} R_j^{(2)} \tag{26.8}$$

Where $R_i^{(l)} = P_i^{(l)*} - P_{i-1}^{(l)*}, l = 1,2$ and $A_i = \{j | (t_{i-1}^{(1)}, t_i^{(1)}) \cap (t_{j-1}^{(2)}, t_j^{(2)}) \neq \emptyset\}$.

They propose two versions of cumulative correlation estimators(we call them CCOR1 and CCOR2):

- First, if the volatilities of the two processes $\sigma^{(1)}$ and $\sigma(2)$ are known, then the estimator is:

$$\mathrm{CCOR1} = \frac{1}{T} \sum_{i=1}^{m_1} \sum_{j \in A_i} \frac{R_i^{(1)} R_j^{(2)}}{\sigma^{(1)} \sigma^{(2)}} \tag{26.9}$$

  where $R_i^{(j)}, j = 1,2$ and A_i are the same the those in the expression of Cumulative Covariance estimator.

- Second, no matter if the volatilities $\sigma^{(1)}$ and $\sigma(2)$ are known or not, the estimator is:

$$\mathrm{CCOR2} = \frac{\sum_{i=1}^{m_1} \sum_{j \in A_i} R_i^{(1)} R_j^{(2)}}{(\sum_{i=1}^{m_1} R_i^{(1)})^{1/2} (\sum_{i=1}^{m_2} R_i^{(2)})^{1/2}} \tag{26.10}$$

  where $R_i^{(j)}, j = 1,2$ and A_i are the same as the those in the expression of Cumulative Covariance estimator.

In the absence of microstructure noise, these estimators of covariance and correlation are unbiased, consistent and asymptotically normal under fairly general assumptions. Hayashi and Yoshida prove consistency when the price process is a continuous semi-martingale and the arrival processes are a sequence of stopping times.

However when microstructure noise is present, the performance of the estimators change. While the noise process is i.i.d. and independent of the efficient price process, and the two processes have independent Poisson arrival times, the CC estimator is still unbiased but the noise makes it inconsistent. It is consistent when the arrival intensities $\lambda^{(1)}$ and $\lambda^{(2)}$ go to infinity, but then this is reduced to the Realized Covariance case(Griffin and Oomen (2006)). Under other noise specifications, the estimators are not only inconsistent, but they can also be biased(Voev, Lunde (2006)). Also, computation of the estimators takes very long. Another downside of the estimators is that they require the exact timing of transactions. Thus one may still be forced to rely on realized covariance kind of estimator if they are not available (Griffin and Oomen (2006)).

26.3. Random Lead-Lag Estimator

After meriting on the upside and downside of the two major different estimators – Realized Covariance Estimator and Cumulative Covariance estimator, we found the following: under i.i.d. noise, the Realized Covariance estimator gets bigger bias while the frequency gets higher, and this is not surprising due to the Epps effect. The Cumulative Covariance estimator is unbiased but it has big variance, and it is computationally demanding. This motivates us to propose our new estimator, the random lead-lag (RLL) estimator as:

$$\mathrm{RLL} = \sum_{i=1}^{m} \sum_{j=k_1}^{k_2} (P_i^{(1)*} - P_{i-1}^{(1)*})(P_j^{(2)*} - P_{j-1}^{(2)*}) \tag{26.11}$$

where $k_1 = \max\{k \le i-1 : | P_k^{(1)*} - P_{k-1}^{(1)*} |> 0\} + 1$ and $k_2 = \min\{k \ge i : | P_k^{(2)*} - P_{k-1}^{(2)*} |> 0\}$, $P_i^{(l)*}$ denotes the observed log-price of stock l at the end of the i-th interval using previous tick interpolation.

This estimator can be interpreted as follows: wait till there is a trade for either stock. If there is a trade for both stocks, take the product of log returns. If there is a trade for stock 1 and not 2, impute the last non-zero log return for stock 2 and take product. And vice versa.

We summarize our main results in the following theorems:

Theorem 1 The random lead-lag estimator coincides with the previous-tick realized covariance estimator if there is at least 1 trade in each interval for both assets (High trading rate).

Proof When there is at least 1 trade in each interval for process 1, then for each fixed i, for any $k \le i-1$,

$$| P_k^{(1)*} - P_{k-1}^{(1)*} |> 0$$

hence,

$$k_1 = \max\{k \le i-1 : \mid P_k^{(1)*} - P_{k-1}^{(1)*} \mid > 0\} + 1 = i - 1 + 1 = i.$$

likewise, if there is at least 1 trade in each interval for process 2, then for fixed i, for any $k \ge i$,

$$\mid P_k^{(2)*} - P_{k-1}^{(2)*} \mid > 0$$

hence,

$$k_2 = \min\{k \ge i : \mid P_k^{(2)*} - P_{k-1}^{(2)*} \mid > 0\} = i.$$

Combine these two sides, we have:

$$\begin{aligned}
\text{RLL} &= \sum_{i=1}^{m} \sum_{j=k_1}^{k_2} (P_i^{(1)*} - P_{i-1}^{(1)*})(P_j^{(2)*} - P_{j-1}^{(2)*}) \\
&= \sum_{i=1}^{m} (P_i^{(1)*} - P_{i-1}^{(1)*})(P_i^{(2)*} - P_{i-1}^{(2)*}) \\
&= \text{PTRV}
\end{aligned}$$

Theorem 2 The random Lead-lag estimator coincides with the Hayashi-Yoshida estimator if there is at most one observation in each interval for either process. (Computing interval/bandwidth small).

The probability of two observations happen at the same time for different processes is zero. Hence, if the frequency is sufficiently high, intervals will be small enough so that in each interval, there is at most one observation for both assets, i.e., the two assets are not observed in the same interval, and for each one, it is not observed more than once in an interval.

Proof Suppose we have m intervals and the time points are $\pi(m) = \{0 = t_0 < t_1 < \cdots < t_m = T\}$, while the observation times for asset 1 is $\pi^{(1)}(m_1) = \{0 \leqslant t_0^{(1)} < t_1^{(1)} < \cdots < t_{m_1}^{(1)} \leqslant T\}$, and the observation time for asset 2 is $\pi^{(2)}(m_2) = \{0 \leqslant t_0^{(2)} < t_1^{(2)} < \cdots < t_{m_2}^{(2)} \leqslant T\}$. Under Assumption that there is at most one observation in each interval, $m_1 < m$ and $m_2 < m$. With probability one, we have

$$\begin{aligned}
\text{RLL} &= \sum_{i=1}^{m} \sum_{j=k_1}^{k_2} (P_i^{(1)*} - P_{i-1}^{(1)*})(P_j^{(2)*} - P_{j-1}^{(2)*}) \\
&= \sum_{i=1}^{m_1} \sum_{j=b_1}^{b_2} (P_{t_i^{(1)}}^{(1)*} - P_{t_{i-1}^{(1)}}^{(1)*})(P_j^{(2)*} - P_{j-1}^{(2)*})
\end{aligned}$$

Where $b_1 = \max\{k : t_k \leq t_{i-1}^{(1)} and \mid P_{t_k}^{(1)*} - P_{t_{k-1}}^{(1)*} \mid > 0\} + 1$ and $b_2 = \min\{k : t_k \geq t_i^{(1)} and \mid P_{t_k}^{(2)*} - P_{t_{k-1}}^{(2)*} \mid > 0\}$ Since there is at most one observation for both assets, for each i, we have

$$\sum_{j=b_1}^{b_2} (P_j^{(2)*} - P_{j-1}^{(2)*}) = \sum_{j\in A_i} (P_j^{(2)*} - P_{j-1}^{(2)*})$$

Where $A_i = \{j | (t_{i-1}^{(1)}, t_i^{(1)}) \cap (t_{j-1}^{(2)}, t_j^{(2)}) \neq \}$ Hence

$$\begin{aligned}
\text{RLL} &= \sum_{i=1}^{m_1} (P_{t_i^{(1)}}^{(1)*} - P_{t_{i-1}^{(1)}}^{(1)*}) \sum_{j=b_1}^{b_2} ((P_j^{(2)*} - P_{j-1}^{(2)*}) \\
&= \sum_{i=1}^{m_1} (P_{t_i^{(1)}}^{(1)*} - P_{t_{i-1}^{(1)}}^{(1)*}) \sum_{j\in A_i} ((P_j^{(2)*} - P_{j-1}^{(2)*}) \\
&= \sum_{i=1}^{m_1} \sum_{j\in A_i} (P_{t_i^{(1)}}^{(1)*} - P_{t_{i-1}^{(1)}}^{(1)*})((P_j^{(2)*} - P_{j-1}^{(2)*}) \\
&= \text{CC}
\end{aligned}$$

26.4. Simulation Study With No Microstructure Noise

26.4.1. *Basic Setup*

First, we carry out simulation with no microstructure noise. We shall compare the random lead-lag(RLL) estimator with the previous tick realized covariance(PTRC) estimator. In the simulation, we set the volatilities of the two processes as; $\sigma^{(1)} = 1$ and $\sigma^{(2)} = 2$. Arrivals for process 1 are Poisson Process with 1 arrival every minute on average and arrivals for process 2 are Poisson Process with 1 arrival every 5 minutes on average, which means the arrival intensities are $\lambda^{(1)} = 60$ and $\lambda^{(2)} = 300$ in seconds. $T = 23,400$ seconds in a 6.5 hour (NYSE) trading day. The simulation sample size is 10000, which can be considered as the number of trading days. Estimates are calculated with $h \in \{3600, 1800, 600, 300, 60, 30, 10, 1\}$ seconds that is the sampling frequencies considered are 60 min, 30 min, 10 min , 5min, 1 min, 30 sec, 10sec, 1 sec. The correlation coefficient of the two process runs from 0.1 to 0.9 with step equals to 0.1. ($\rho = 0.1 : 0.1 : 0.9$).

Figure 26.1, figure 26.2 and figure 26.3 plot bias, standard deviation and RMSE of the RLL(Star), PTRC(Circle) estimators for various values of frequency and ρ. The 9 panels in each figure represent the 9 values of ρ. The horizontal axis in each panel represents the index of the sampling frequency h in the set $h \in \{3600, 1800, 600, 300, 60, 30, 10, 1\}$.

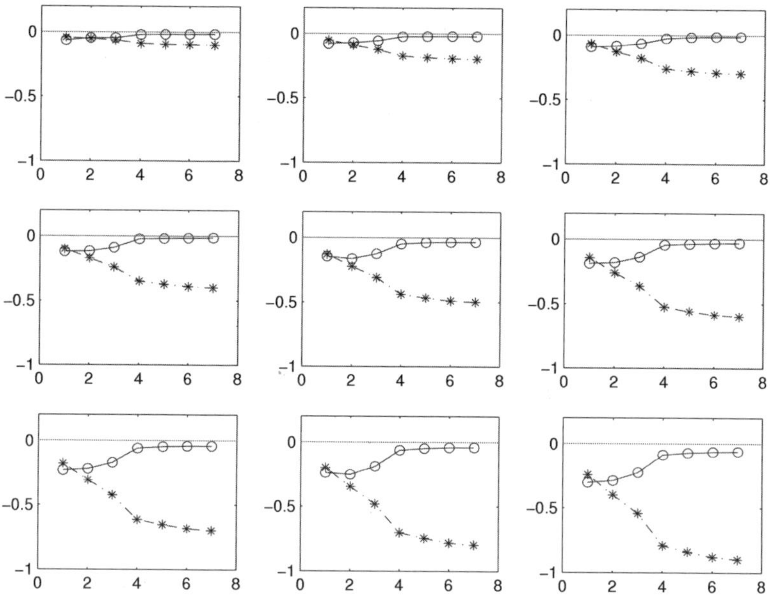

Fig. 26.1. Bias: Star RLL, Circle PTRC

26.4.2. *Choosing optimal frequency of RLL*

From the Figures 26.1, 26.2 and 26.3, we notice that the bias and variance of the RLL estimator both decrease with increasing frequency. Since the RLL estimator with the highest frequency coincides with Hayashi-Yoshida's CC estimator, the CC estimator is the best among all RLL estimators. However, higher the frequency, higher is the computational time. After frequency 1 minute, the change in both bias and variance of the RLL estimator is negligible. So we can choose 1 minute to be the optimal frequency.

26.4.3. *Bias, Variance and RMSE when microstructure noise is absent*

For all ρ's, if frequency is higher than 30 minutes, then the bias of RLL estimator is lower than that of PTRC. The RLL estimator is asymptotically unbiased as frequency increases while PTRC goes to zero as frequency increases, hence it is highly biased. For all values of ρ, the minimum (taken over frequency) bias of RLL estimator is smaller than the minimum bias of PTRC. The difference among different values of bias for the RLL is very small after frequency hits 1 minute.
The variance of PTRC is smaller than variance of RLL estimator for the same

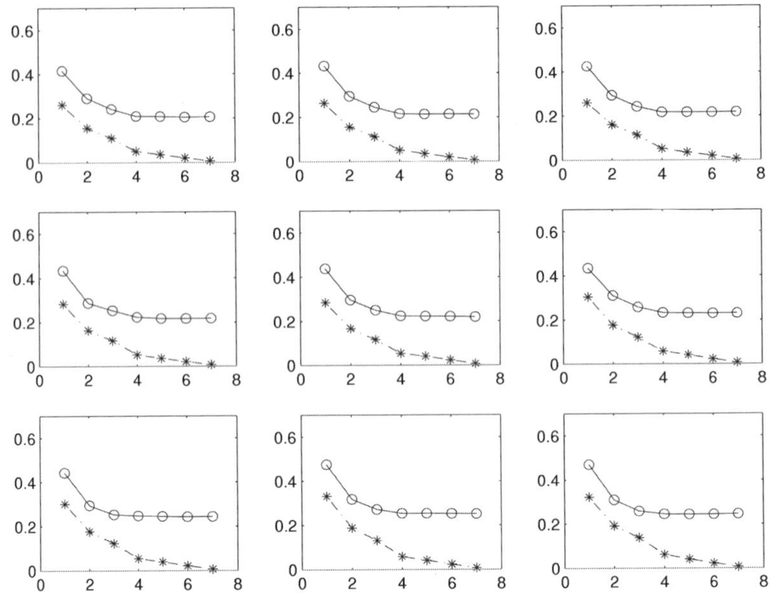

Fig. 26.2. Standard Deviation: Star RLL, Circle PTRC

frequency. If we compare the variance of RLL estimator at frequency 1 min (because that is the optimal choice clearly) with variance of PTRC for frequency 30 min (because that has comparable bias to the optimal RLL estimate and also minimum bias among the PTRC estimates), then the variance of the RLL estimate (1 min) is smaller than that of PTRC (30 min) for all ρ's.

For small ρ (≤ 0.3), the minimum rmse of PTRC is smaller than minimum rmse of RLL estimator. This is because PTRC is biased towards 0 and as freq increases goes to 0 making the variance very small. So when the parameter is itself close to zero, this is advantageous. For higher ρ, rmse of PTRC is minimized at freq 30 min and that of RLL estimator is minimized as freq decreases, but as per comment above it is enough to consider frequency 1 min. Rmse of RLL estimator at frequency 1 min is far smaller tan rmse of PTRC at 30 min for all values of ρ greater than 0.3.

26.4.4. *Bias, Variance and RMSE when microstructure noise is present*

We carry out simulation under i.i.d. Gaussian microstructure noise with standard deviation $\xi_1 = \xi_2 = 120$. All the other conditions are the same as those in the case while no noise present. Figure 26.4, figure 26.5 and figure 26.6 plot bias, stan-

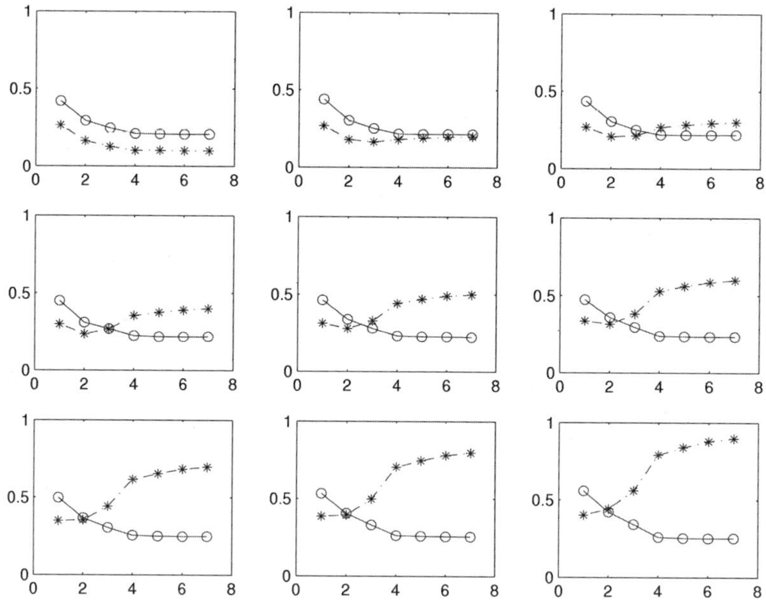

Fig. 26.3. Root mean squared error(RMSE): Star RLL, Circle PTRC

dard deviation and RMSE of the RLL(Star), PTRC(Circle) estimators for various values of frequency and ρ while noise is present.

The performance of bias while microstructure noise is present is pretty similar to those without noise. In addition to the observations made about bias in the no-noise setting, we observe that roughly, the bias of RLL decreases with increasing frequency while the bias of PTRC increases.

The performance of standard deviations of the estimators are very different from those under the situation of no noise present. The standard deviation of RLL increase with increasing frequency, and almost stays constant after the frequency hits 1 minute. The standard deviation of PTRC has a bell shape, it increases with increasing frequency first, and hits the maximum value around 5 minutes, and then starts to decrease. It reaches the overall smallest value while the frequency is the highest(1 second). While the frequency is lower than 5 minutes, the standard deviation of RLL is lower than that of PTRC; and while the frequency is higher than 5 minutes, it is the other way around.

The performance of root mean square error (RMSE) is very similar to that of the standard deviation while noise is present. This is because the magnitude of bias is relatively much smaller than the magnitude of the standard deviation. Since

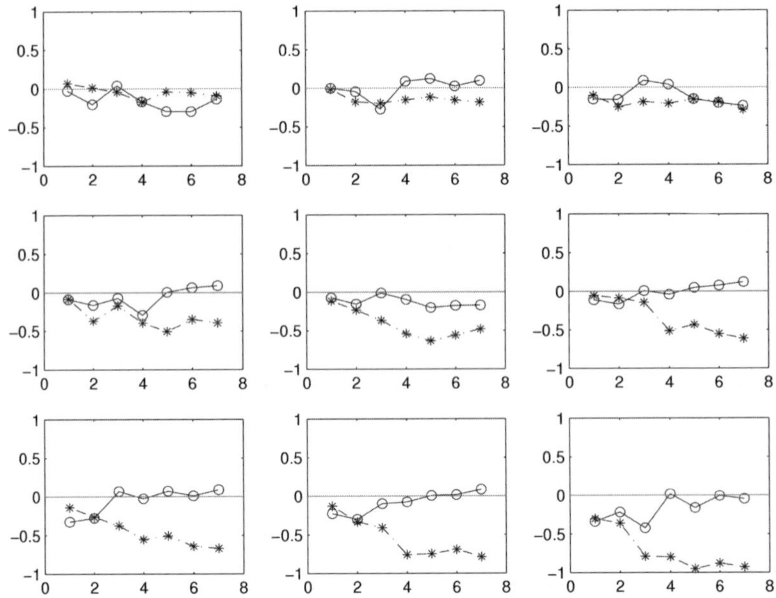

Fig. 26.4. Bias in presence of noise: Star RLL, Circle PTRC

$RMSE = \sqrt{\text{bias}^2 + \text{SD}^2}$, the value of RMSE should be very close to Standard deviation when bias is really small.

According to the discussion above, we will still choose 1 minute as our optimal frequency while microstructure noise is present. This gives us a good bias-variance trade-off.

In summary, we found from the simulations that when no microstructure noise is present, the bias and variance of the RLL estimator both decrease while the frequency increases, which implies that the Hayashi-Yoshida estimator is still the best in the sense of small bias and variance. But it is computationally demanding while the frequency gets higher. As the frequency hits one minutes, the change in bias and variance of RLL estimator is negligible, hence we choose one minute as the optimal frequency. When microstructure is present, the variance of the RLL estimator actually increases with increasing frequency which opens up the scope for choosing an optimal frequency using bias variance trade-off.

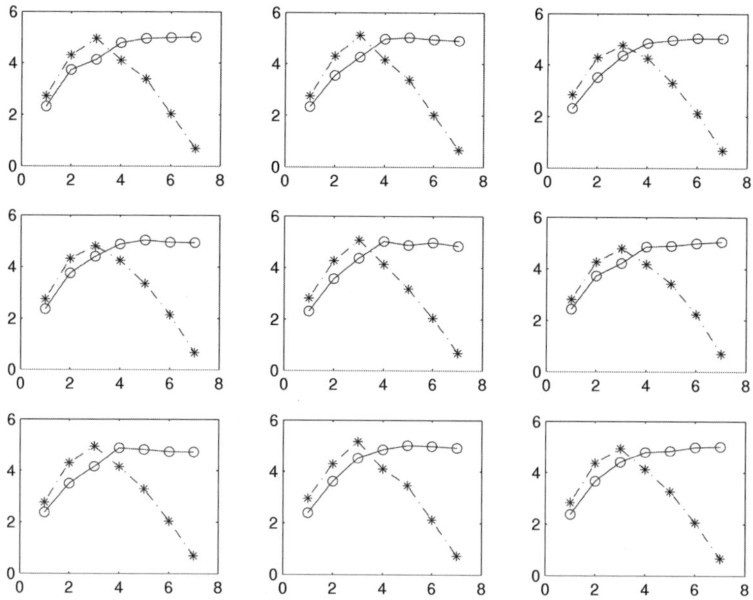

Fig. 26.5. Standard Deviation in presence of noise: Star RLL, Circle PTRC

26.5. Conclusions And Future Work

Although RC consistent for integrated covariance in the ideal situation, it is highly biased for non-synchronous data at high frequencies which is required for the asymptotics to be valid. On the other hand, CC is unbiased in the absence of noise, but has very high variance when noise is present. The contribution of this paper is in proposing a discrete-time version of CC which converges to CC at high frequency and is hence asymptotically unbiased even for nonsynchronous data. On the other hand our estimator can be calculated at lower frequencies, so that we can choose a lower frequency to keep the variance in control in the presence of microstructure noise. Our estimator is also a generalization of the class of fixed lead-lag estimators. It thus serves as a bridge between interval-based estimators like RC and tick-based estimators like CC.

The advantages of the RLL estimator over tick-based estimators like CC is that the former does not require exact timing of events and is computationally much faster. The variance can be reduced a lot by choosing a proper lower frequency at the cost of increasing the bias. However, unlike the other realized variance-type interval based estimators, the RLL estimator is asymptotically unbiased at very

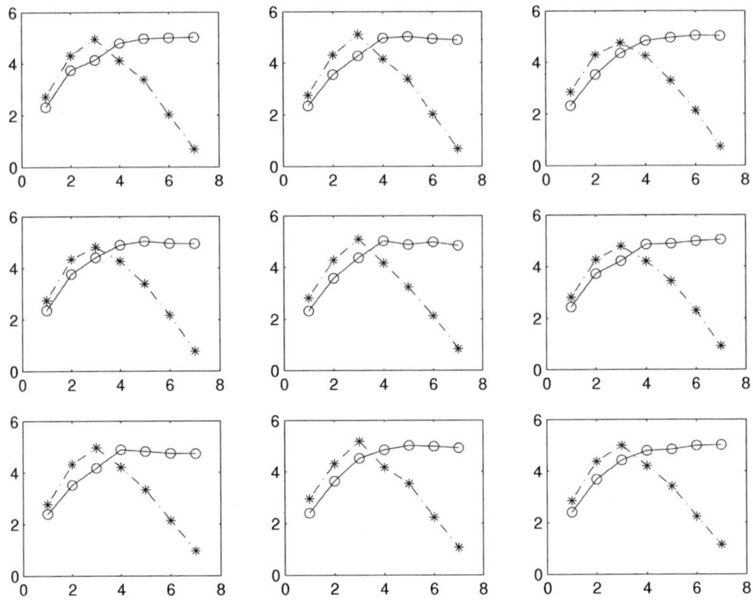

Fig. 26.6. RMSE in presence of noise: Star RLL, Circle PTRC

high frequency. So we can suitably choose a frequency that optimizes our bias-variance preferences. This is a unique feature of our estimator which is absent in any other estimator in the literature.

Here are some directions for ongoing and future research:

- We plan to apply variance reduction techniques like sub-sampling. Then we can compare the subsampled version of RLL with the subsampled version of PTRC and obtain the optimal frequency to gain good bias-variance trade-off.
- We can derive parametric bootstrap estimates for the bias and variance of the RLL estimator under specific models. If these estimates are good in simulations, then we can use this to choose the optimal frequency for real data applications.
- We are in the process of deriving and interpreting theoretical expressions for bias and variance of RLL.
- We can also use alternative estimators based on bi-power variation and study the effect of noise and asynchronicity.

References

1. Ait-Sahalia, Y., Mykland, P.A., and Zhang, L. (2005): How often to sample a continuous-time process in the presence of market microstructure noise, *Review of Financial Studies* 18 351-416.
2. Andersen, T. G., and Bollerslev, T.(1998): Answering the Skeptics: Yes, Standard Volatility Models Do Provide Accurate Forecasts, *International Economic Review*, 39 (4), 885905.
3. Andersen, T. G., Bollerslev, T., Diebold, F. X. and Ebens, H. (2001a): The distribution of realized stock return volatility, *Journal of Financial Economics* 61(1), 4376.
4. Andersen, T. G., Bollerslev, T., Diebold, F. X. and Labys, P. (2001b): The distribution of exchange rate volatility, *Journal of the American Statistical Association* 96(453), 42 55.
5. Bandi, F. M., and J. R. Russell, (2005): Realized Covariation, Realized Beta, and Microstructure Noise, manuscript University of Chicago.
6. Bandi, F. M., and J. R. Russell (2006): Separating Microstructure Noise from Volatility, *Journal of Financial Economics*, 79, 655692.
7. Barndorff-Nielsen O. E., and N. Shephard. (2002a): Estimating Quadratic Variation Using Realised Variance, *Journal of Applied Econometrics* 17, 457477.
8. Barndorff-Nielsen, O. E. and Shephard, N. (2002b): Econometric analysis of realised volatility and its use in estimating stochastic volatility models, *Journal of the Royal Statistical Society B* 64, 253280.
9. Barndorff-Nielsen, O. E., and N. Shephard. (2003): Realized Power Variation and Stochastic Volatility, *Bernoulli* 9, 243–265 [correction available at www. levyprocess.org].
10. Barndorff-Nielsen,O. E., Shephard,N. (2004a): Econometric Analysis of Realized Covariation: High Frequency Based Covariance, Regression, and Correlation in Financial Economics, *Econometrica* 72 (3), 885925.
11. Barndorff-Nielsen,O.E., Shephard,N. (2004b): Power and bipower variation with stochastic volatility and jumps (with discussion), Journal of Financial Econometrics, (2004), 2, 1-48.(
12. Barndorff-Nielsen, O. E., P. R. Hansen, A. Lunde, and N. Shephard, (2006): Designing Realised Kernels to Measure the Ex-Post Variation of Equity Prices in the Presence of Noise, manuscript University of Oxford, Nuffield College.
13. Cohen, K. J., G. A. Hawawini, S. F. Maier, R. A. Schwartz, and D. K. Whitcomb (1983): Friction in the Trading Process and the Estimation of Systematic Risk, *Journal of Financial Economics*, 12, 263278.
14. Corsi, F., G. Zumbach, U. A. Muller, and M. M. Dacorogna, (2001): Consistent High-Precision Volatility from High-Frequency Data, *Economic Notes*, 30 (2), 183204.
15. Dimson, E., (1979): Risk Measurement When Shares are Subject to Infrequent Trading, *Journal of Financial Economics*, 7, 197226.
16. Epps, T. W., (1979): Comovements in Stock Prices in the Very Short Run, *Journal of the American Statistical Association*, 74 (366), 291298.
17. Fan, J. and Wang, Y. (2005): Multiscale jump and volatility analysis for high-frequency financial data, Preprint.
18. Fisher, L., (1966): Some New Stock-Market Indexes, *Journal of Business*, 39 (1-2),

191225.
19. Griffin, J. E., and R. C. Oomen (2005): Covariance Measurement in the Presence of Non-synchronous Trading and Market Microstructure Noise, working paper.
20. Hansen, P. R. and Lunde, A. (2006): Realized variance and market microstructure noise, *Journal of Business and Economic Statistics* .
21. Hayashi, T. and Yoshida, N.(2004): "On Covariance Estimation for High-frequency Financial Data", Financial Engineering and Applications. edited by M.H. Hamza), no. 437-801, ACTA Press (2004). ISBN: 0-88986-417-9.
22. Hayashi T. and N. Yoshida. (2005): On Covariance Estimation of Non-Synchronously Observed Diffusion Processes. *Bernoulli* 11, 359379.
23. Jacod, J., and Protter, P. (1998): Asymptotic Error Distributions for the Euler Method for Stochastic Differential Equations, *The Annals of Probability*, 26, 267307.
24. Karatzas, I., and Shreve, S. E. (1991): Brownian Motion and Stochastic Calculus, New York: *Springer-Verlag*.
25. Karatzas, I., and Shreve, S. E. (1991): Brownian Motion and Stochastic Calculus, New York: *Springer-Verlag*.
26. Martens, M. (2002): Measuring and Forecasting S&P 500 Index-Futures Volatility Using High-Frequency Data, *Journal of Futures Markets* 22, 497518.
27. Mykland, P.A. and Zhang, L. (2001):Comment: A Selective Overview of Nonparametric Methods in Financial Econometrics, *Statistical Science* 20, no. 4 (2005), 347350.
28. Scholes, M., and J.Williams (1977): Estimating Betas from Nonsynchronous Data, *Journal of Financial Economics* 5, 309327.
29. Voev, V. and Lunde,A. (2006): Integrated Covariance Estimation using High-frequency Data in the Presence of Noise, *Journal of Financial Econometrics*, 2006, Vol. 5, No. 1, 000000.
30. Zhang L. (2006): Estimating Covariation: Epps Effect and Microstructure Noise. Working Paper.
31. Zhang, L., Mykland, P.A., and Ait-Sahalia, Y. (2005): A tale of two time scales: Determining integrated volatility with noisy high-frequency data, *Journal of American Statistical Association* 100 1394-141.

Chapter 27

Improving the Hansen-Hurwitz Estimator in PPSWR Sampling

Arun Kumar Adhikary

Indian Statistical Institute, Kolkata

E-mail: arun@isical.ac.in

An attempt has been made to improve upon the Hansen-Hurwitz (1943) estimator based on PPSWR sampling scheme through Rao-Blackwellisation. In order to derive the sampling variance of the improved estimator obtained by Rao-Blackwellisation, it is of interest to derive the probability distribution of the number of distinct units in the sample drawn according to PPSWR sampling scheme, as the improved estimator is based solely on the distinct units in the sample. It has been possible to write down the exact distribution of the number of distinct units in the sample drawn by PPSWR sampling scheme in a closed form. Assuming a super-population model, the model-expected design-variance of the improved estimator is worked out and it is compared with that of the Hansen-Hurwitz estimator under the same super-population model. The percentage gain in efficiency in using the improved estimator over the Hansen-Hurwitz estimator is worked out for certain selected values of the model parameters.

Contents

27.1. Introduction

It is well-known that in case of SRSWR sampling scheme, the sample mean based on distinct units which is obtained through Rao-Blackwellisation has smaller variance than the mean based on all the sampled units drawn by SRSWR sampling scheme, which has been considered by Basu (1958), Raj and Khamis (1958) and Pathak (1962). But, similar results on Rao-Blackwellised version of Hansen-Hurwitz estimator based on PPSWR sampling are not available in the literature. We only know the form of the estimate when $(n-1)$ units are distinct in a sample drawn according to PPSWR sampling scheme in n draws. Murthy (1967) has claimed that

$$t^* = \frac{1}{3}\left[\frac{y_{i_1}}{p_{i_1}} + \frac{y_{i_2}}{p_{i_2}} + \frac{y_{i_1}+y_{i_2}}{p_{i_1}+p_{i_2}}\right]$$

which is obtained from the Hansen-Hurwitz (1943) estimator through Rao-Blackwellisation, when a sample is selected in 3 draws according to PPSWR sampling scheme and 2 units are distinct, is an unbiased estimator of the population total

$$Y = \sum_{i=1}^{N} y_i$$

where y_i is the value of the study variable y assumed on the ith unit and N is the population size. In fact, t^* is not unbiased, even t^* is not an estimator. It is only an estimate in the above mentioned case. To settle this issue, we first consider the Rao-Blackwellised version of the Hansen-Hurwitz (1943) estimator as it should be under the present set-up and try to compare its performance over the Hansen-Hurwitz (1943) estimator assuming a super-population model.
We also generalize the form of the improved estimator based on distinct units when a sample is selected in n draws according to PPSWR sampling scheme. In order to derive the sampling variance of the improved estimator, we derive the distribution of the number of distinct units in the sample.

27.2. Formulation of The Problem

Let $y_i, x_i (i = 1, 2, \cdots, N)$ be the values of positively correlated variates y and x defined on a finite population of size N, where y is the study variable and x_i is the size measure and

$$p_i = \frac{x_i}{X}.$$

Here $X = \sum_{i=1}^{N} x_i$, p_i is the normed size measure for the ith unit of the population, $i =, 1, 2, \cdots, N$.
If $s = (i_1, i_2, \cdots, i_n)$ be a sample of size n drawn by PPSWR sampling scheme,

then Hansen-Hurwitz (1943) estimator of the population total $Y = \sum_{i=1}^{N} y_i$ is given by

$$\hat{Y}_{HHE} = \frac{1}{n}\sum_{j=1}^{n}\frac{y_{i_j}}{p_{i_j}}.$$

The variance of $\hat{Y}_{HHE}$ is given by

$$Var(\hat{Y}_{HHE}) = \frac{1}{n}\sum_{i=1}^{N} p_i(\frac{y_i}{p_i} - Y)^2.$$

Let ν denote the number of distrinct units in the sample and $p(s)$ denote the probability of selecting the sample s by PPSWR sampling scheme. Clearly ν can take any value between 1 and n. We may note that

$$p(s) = p_{i_1} p_{i_2} \cdots p_{i_n}.$$

Then Rao-Blackwellisation of Hansen-Hurwiz (1943) estimator yields the estimator

$$t^*(s) = \frac{1}{n}\sum_{i=1}^{\nu} y_{j_i}\frac{b_{j_i}}{P(s)} \tag{27.1}$$

where (i) $j_1, j_2 \cdots, j_\nu$ are distinct units in the sample s
(ii) $P(s) = \sum_{s' \sim s} p(s')$ where $s' \sim s$ stands for all possible samples s' which are equivalent to s in the sense of sharing the same set of distinct units in s
(iii) $b_{j_i} = \frac{\partial P(s)}{\partial p_{j_i}}, i = 1, 2, \cdots, \nu$.

$t^*(s)$ can also be written as

$$t^*(s) = \frac{1}{n}[\sum_{i=1}^{\nu}\frac{y_{j_i}}{p_{j_i}} + \sum_{i=1}^{\nu} y_{j_i}\frac{a_{j_i}}{P_*(s)}]$$

where (i) $P_*(s) = \frac{P(s)}{p_{j_1}p_{j_2}\cdots p_{j_\nu}}$ and (ii) $a_{j_i} = \frac{\partial P_*(s)}{\partial p_{j_i}}$.
Finding the variance of $t^*(s)$ is a difficult task. Even showing unbiasedness of $t^*(s)$ directly is difficult although it is known that Rao-Blackwellised version of any unbiased estimator is also unbiased.
To start with, we consider the simplest case i.e. the case of a sample s drawn according to PPSWR sampling scheme in 3 draws.
From 27.1, it follows that

$$\begin{aligned} t^*(s) &= \frac{1}{3}[\frac{y_i}{p_i} + \frac{y_j}{p_j} + \frac{y_k}{p_k}] \text{ when } \nu = 3 \\ &= \frac{1}{3}[\frac{y_i}{p_i} + \frac{y_j}{p_j} + \frac{y_i + y_j}{p_i + p_j}] \text{ when } \nu = 2 \end{aligned}$$

$$= \frac{y_i}{p_i} \text{ when } \nu = 1$$

and

$$\hat{Y}_{HHE} = \frac{1}{3}[\frac{y_i}{p_i} + \frac{y_j}{p_j} + \frac{y_k}{p_k}] \text{ when } \nu = 3$$
$$= \frac{1}{3}[\frac{2y_i}{p_i} + \frac{y_j}{p_j}] \text{ if } i \text{ occurs twice and } \nu = 2$$
$$= \frac{1}{3}[\frac{y_i}{p_i} + \frac{2y_j}{p_j}] \text{ if } j \text{ occurs twice and } \nu = 2$$
$$= \frac{y_i}{p_i} \text{ when } \nu = 1.$$

We may note that

$$p(s) = 3!p_i p_j p_k \text{ when } \nu = 3$$
$$= \frac{3!}{2!} p_i^2 p_j \text{ when } i \text{ occurs twice and } \nu = 2$$
$$= \frac{3!}{2!} p_i p_j^2 \text{ when } j \text{ occurs twice and } \nu = 2$$
$$= p_i^3 \text{ when } \nu = 1.$$

Thus,

$$\begin{aligned} E[t^*(s)] = & \sum\sum\sum_{i<j<k=1}^{N} \frac{1}{3}[\frac{y_i}{p_i} + \frac{y_j}{p_j} + \frac{y_k}{p_k}]6p_i p_j p_k \\ & + \sum\sum_{i<j=1}^{N} \frac{1}{3}[\frac{y_i}{p_i} + \frac{y_j}{p_j} + \frac{y_i + y_j}{p_i + p_j}]3p_i p_j (p_i + p_j) \\ & + \sum_{i=1}^{N} \frac{y_i}{p_i} . p_i^3 \\ = & \frac{1}{3} \sum\sum\sum_{i \neq j \neq k=1}^{N} [\frac{y_i}{p_i} + \frac{y_j}{p_j} + \frac{y_k}{p_k}] p_i p_j p_k \\ & + \frac{1}{2} \sum\sum_{i \neq j=1}^{N} [\frac{y_i}{p_i} + \frac{y_j}{p_j} + \frac{y_i + y_j}{p_i + p_j}] p_i p_j (p_i + p_j) \\ & + \sum_{i=1}^{N} y_i p_i^2 \end{aligned} \tag{27.2}$$

We may note that the first term on the right hand side of 27.2 equals

$$\sum\sum\sum_{i\neq j\neq k=1}^{N} y_i p_i p_k = \sum_{i=1}^{N} y_i[1+p_i^2-2p_i-\sum_{\substack{j=1\\ \neq i}}^{N} p_j^2]$$

and the second term on the right hand side of 27.2 equals

$$\begin{aligned}\sum\sum_{i\neq j=1}^{N}[y_i p_j(p_i+p_j)+y_i p_i p_j] &= \sum\sum_{i\neq j=1}^{N}[2y_i p_i p_j + y_i p_j^2] \\ &= 2\sum_{i=1}^{N} y_i p_i(1-p_i)+\sum_{i=1}^{N} y_i \sum_{\substack{j=1\\ \neq i}}^{N} p_j^2.\end{aligned}$$

Thus we get

$$E[t^*(s)] = \sum_{i=1}^{N} y_i = Y.$$

Hence $t^*(s)$ is unbiased for Y.

Now we derive the variance of $t^*(s)$ and obtain the gain in efficiency due to Rao-Blackwellisation.

$$\begin{aligned}\text{Var}\,[t^*(s)] = &\sum\sum_{i<j<k=1}^{N}\frac{1}{9}[\frac{y_i}{p_i}+\frac{y_j}{p_j}+\frac{y_k}{p_k}]^2 6p_i p_j p_k \\ &+\sum\sum_{i<j=1}^{N}\frac{1}{9}[\frac{y_i}{p_i}+\frac{y_j}{p_j}+\frac{y_i+y_j}{p_i+p_j}]^2 3p_i p_j(p_i+p_j) \\ &+\sum_{i=1}^{N}(\frac{y_i}{p_i})^2 p_i^3 - Y^2 \end{aligned} \tag{27.3}$$

$$\begin{aligned}\text{Var}\,(\hat{Y}_{HHE}) = &\sum\sum_{i<j<k=1}^{N}\frac{1}{9}[\frac{y_i}{p_i}+\frac{y_j}{p_j}+\frac{y_k}{p_k}]^2 6p_i p_j p_k \\ &+\sum\sum_{i<j=1}^{N}[\frac{1}{9}(\frac{2y_i}{p_i}+\frac{y_j}{p_j}^2)3p_i^2 p_j+\frac{1}{9}(\frac{y_i}{p_i}+\frac{2y_j}{p_j})^2 3p_i p_j^2] \\ &+\sum_{i=1}^{N}(\frac{y_i}{p_i})^2 p_i^3 - Y^2 \end{aligned} \tag{27.4}$$

Fortunately the first term and the third term on the right hand sides of 27.3 and 27.4 are equal. Now the second term on the right hand side of 27.4 equals

$$\begin{aligned}A &= \frac{1}{3}\sum\sum_{i\neq j=1}^{N}(\frac{2y_i}{p_i}+\frac{y_j}{p_j})^2 p_i^2 p_j \\ &= \frac{1}{3}\sum\sum_{i\neq j=1}^{N}[(\frac{2y_i}{p_i})^2+(\frac{y_j}{p_j})^2+\frac{4y_i y_j}{p_i p_j}]p_i^2 p_j\end{aligned}$$

and the second term on the right hand side of 27.3 equals

$$
\begin{aligned}
B &= \frac{1}{3}\sum_{i\neq j=1}^{N}\sum\left[\frac{y_i}{p_i}+\frac{y_j}{p_j}+\frac{y_i+y_j}{p_i+p_j}\right]^2 p_i^2 p_j \\
&= \frac{1}{3}\sum_{i\neq j=1}^{N}\sum\left[\left(\frac{y_i}{p_i}\right)^2+\left(\frac{y_j}{p_j}\right)^2+\left(\frac{y_i+y_j}{p_i+p_j}\right)^2+\frac{2p_i y_j}{p_i p_j}\right. \\
&\left.+2\frac{y_i(y_i+y_j)}{p_i(p_i+p_j)}+2\frac{y_j(y_i+y_j)}{p_j(p_i+p_j)}\right]p_i^2 p_j
\end{aligned}
$$

The task of simplifying the right hand side of 27.3 is not an easy one and hence getting the sign of

$$\text{Var}\,(\hat{Y}_{HHE}) - \text{ Var}\,[t^*(s)] = A - B$$

is difficult.
So we now consider a super-population model M:

$$y_i = \beta x_i + \in_i, i = 1, 2, \cdots, N$$

where β is an unknown real constant and $\in_i$'s are such that

$$
\begin{aligned}
&E_m(\in_i | x_i) = 0\ \forall i \\
&V_m(\in_i | x_i) = \sigma^2 x_i^g, 0 \leq g \leq 2 \\
&\text{and } E_m(\in_i \in_j | x_i, x_j) = 0\ \forall i \neq j
\end{aligned}
$$

where E_m and V_m denote respectively the expectation and the variance with respect to the super-population model M.
Then

$$
\begin{aligned}
E_m(y_i) &= \beta x_i\ \forall i \\
E_m(y_i y_j) &= \beta^2 x_i x_j \forall i \neq j \\
E_m(y_i^2) &= \beta^2 x_i^2 + \sigma^2 x_i^g \forall i
\end{aligned}
$$

We may note that

$$E_m[\frac{y_i^2}{p_i^2}p_i^2p_j] = \frac{1}{X}[\sigma^2x_i^gx_j+\beta^2x_i^2x_j]$$

$$E_m[\frac{y_j^2}{p_j^2}p_i^2p_j] = \frac{1}{X}[\sigma^2x_j^{g-1}x_i^2+\beta^2x_i^2x_j]$$

$$E_m[\frac{y_iy_j}{p_ip_j}p_i^2p_j] = \frac{1}{X}\beta^2x_ix_j$$

$$E_m[\frac{(y_i+y_j)^2}{(p_i+p_j)^2}p_i^2p_j] = \frac{1}{X}[\sigma^2\frac{x_i^{g+2}x_j+x_i^2x_j^{g+1}}{(x_i+x_j)^2}+\beta^2x_i^2x_j]$$

$$E_m[\frac{y_i(y_i+y_j)}{p_i(p_i+p_j)}p_i^2p_j] = \frac{1}{X}[\sigma^2\frac{x_i^{g+1}x_j}{(x_i+x_j)}+\beta^2x_i^2x_j].$$

Hence, the difference between the model-expected variances is

$$\begin{aligned}D_g &= E_m\mathrm{Var}(\hat{Y}_{HHE})-E_m\mathrm{Var}[t^*(s)]\\ &= E_m(A-B)\\ &= \frac{\sigma^2}{3X}\sum\sum_{i\neq j=1}^{N}[3x_i^gx_j-2\frac{x_i^{g+1}x_j+x_i^2x_j^g}{(x_i+x_j)}\\ &\quad -\frac{x_i^{g+2}x_j+x_i^2x_j^{g+1}}{(x_i+x_j)^2}]\\ &= \frac{\sigma^2}{3X}\sum\sum_{i\neq j=1}^{N}[\frac{3x_i^gx_j^3+4x_i^{g+1}x_j^2-2x_i^3x_j^g-3x_i^2x_j^{g+1}}{(x_i+x_j)^2}]\end{aligned}$$

Even for general $g, 0\leq g\leq 2$, it is difficult to get the sign of D_g. But for $g=2$, it is possible to get the sign of D_g directly.

$$\text{For } g=2, D_2 = \frac{\sigma^2}{3X}\sum\sum_{i\neq j=1}^{N}\frac{2x_i^3x_j^2}{(x_i+x_j)^2} > 0 \text{ since } x_i>0\ \forall i.$$

For general $g, 0\leq g\leq 2$, for obtaining the gain in efficiency due to Rao-Blackwellisation, we investigate numerically.
We note that

$$D_g = E_m\mathrm{Var}(\hat{Y}_{HHE})-E_m\mathrm{Var}[t^*(s)] \Rightarrow E_m\mathrm{Var}[t^*(s)] = E_m\mathrm{Var}(\hat{Y}_{HHE})-D_g.$$

Now, $\mathrm{Var}(\hat{Y}_{HHE}) = \frac{1}{3}\sum_{i=1}^{N}p_i(\frac{y_i}{p_i}-Y)^2 = \frac{1}{3}[\sum_{i=1}^{N}\frac{y_i^2}{p_i}-Y^2].$

Hence,

$$
\begin{aligned}
E_m\mathrm{Var}(\hat{Y}_{HHE}) &= \frac{1}{3}[X\sum_{i=1}^{N}(\beta^2 x_i + \sigma^2 x_i^{g-1}) \\
&\quad -\{\sum_{i=1}^{N}(\sigma^2 x_i^g + \beta^2 x_i^2) + \sum\sum_{i\neq j=1}^{N} \beta^2 x_i x_j\}] \\
&= \frac{\sigma^2}{3}[X\sum_{i=1}^{N} x_i^{g-1} - \sum_{i=1}^{N} x_i^g].
\end{aligned}
$$

Hence the gain in efficiency due to Rao-Blackwellisation under the super-population model M is given by

$$
\begin{aligned}
\Delta &= \frac{E_m\mathrm{Var}(\hat{Y}_{HHE}) - E_m\mathrm{Var}[t^*(s)]}{E_m\mathrm{Var}[t^*(s)]} \times 100\% \\
&= \frac{D_g}{E_m\mathrm{Var}[t^*(s)]} \times 100\% \\
&= \frac{D_g}{E_m\mathrm{Var}(\hat{Y}_{HHE}) - D_g} \times 100\%.
\end{aligned}
$$

27.3. Results of Numerical Investigation

Population-1: Size Measures: 23,18,27,32,14

Population-2: Size Measures: 0.1,0.2,0.3,0.4,0.5,0.6,0.7

Population-3: Size Measures: 239,122,390,230,340,172,333,421,149,362,220,397

Population-4: Size Measures: 1,2,3,4,5,6,7,8,9,10,11,12,13,14,15

From the table below we see that for any $g, 0 \leq g \leq 2$, the gain in efficiency due to Rao-Blackwellisation is always positive and is likely to be more when sampling fraction is large. Hence Rao-Blackwellisation ensures improvement over Hansen-Hurwitz (1943) estimator based on PPSWR sampling scheme.

27.4. Probability Distribution of The Total Number of Distinct Units in a Sample Drawn According to PPSWR Sampling Scheme

In order to derive an exact experession for the sampling variance of the improved estimator obtained from the Hansen-Hurwitz (1943) estimator through Rao-Blackwellisation, we intend to derive the probability distribution of the total number of distinct units (ν) in a sample drawn according to PPSWR sampling scheme in n draws. To find the probability distribution of ν, let us consider the following occupancy problem.

Suppose n objects are distributed over N pigeon holes in such a way that the probability that the ith object occupies the jth pigeon hole is $p_j \forall i = 1, 2, \cdots, n$ where

Table 27.1. The percentage gain in efficiency due to Rao-Blackwellisation

g	Δ=Gain in efficiency(%)			
	Population-1	Population-2	Population-3	Population-4
0.0	10.43	5.57	3.91	2.18
0.1	10.47	5.75	3.95	2.30
0.2	10.52	5.93	3.99	2.43
0.3	10.56	6.11	4.03	2.56
0.4	10.61	6.29	4.07	2.68
0.5	10.65	6.47	4.12	2.81
0.6	10.69	6.66	4.15	2.93
0.7	10.74	6.84	4.18	3.05
0.8	10.78	7.01	4.22	3.16
0.9	10.83	7.18	4.26	3.27
1.0	10.87	7.35	4.30	3.37
1.1	10.91	7.51	4.33	3.47
1.2	10.96	7.66	4.37	3.57
1.3	11.00	7.81	4.40	3.64
1.4	11.04	7.95	4.44	3.72
1.5	11.08	8.08	4.47	3.79
1.6	11.12	8.20	4.50	3.86
1.7	11.17	8.32	4.53	3.92
1.8	11.21	8.43	4.56	3.98
1.9	11.24	8.53	4.59	4.03
2.0	11.28	8.62	4.62	4.08

$0 < p_j < 1 \forall j = 1,2,\cdots,N, \sum_{j=1}^{N} p_j = 1$. Let ζ be the total number of pigeon holes which are occupied. Intuitively, it is clear that the probability distribution of ν and ζ are the same i.e. they are identically distributed.
Now our aim is to find the probability distribution of ζ.
We note that, $\text{Prob}[\zeta = m]$
=Prob. [exactly m cells are occupied when n objects are distributed over N pigeon holes according to the stated distribution].
Let A_i denote the event that the ith cell is occupied, $i = 1,2,\cdots,N$. Then, $P(A_i) = 1-(1-p_i)^n$.
Let B_m denote the event that exactly m cells are occupied when n objects are distributed over N pigeon holes according to the stated distribution.
Now we have to find Prob.$[\zeta = m] = P(B_m) \forall 1 \leq m \leq n$. At this point of time let us state a result.
Let $A, A_2, \cdots, A_N$ be N events. Then, the probability of realization of exactly m

out of these N events is given by

$$P(B_m) = S_m - \binom{m+1}{m} S_{m+1} + \binom{m+2}{m} S_{m+2} - \cdots + (-1)^{N-m} \binom{N}{m} S_N$$
$$= \sum_{t=m}^{N} (-1)^{t-m} \binom{t}{m} S_t$$

where $S_t = \sum\sum\cdots\sum_{i_1<i_2<\cdots<i_t=1}^{N} P(A_{i_1} \cap A_{i_2} \cap \cdots \cap A_{i_t})$.
Now, in our problem

$$P(\bigcup_{i=1}^{t} A_i^c) = \sum_{i=1}^{t} P(A_i^c) - \sum\sum_{i<j=1}^{t} P(A_i^c \cap A_j^c)$$
$$+ \sum\sum\sum_{i<j<k=1}^{t} P(A_i^c \cap A_j^c \cap A_k^c) \cdots$$
$$+(-1)^{t-1} P(\bigcap_{i=1}^{t} A_i^c)$$
$$= \sum_{i=1}^{t} (1-p_i)^n - \sum\sum_{i<j=1}^{t} (1-p_i-p_j)^n$$
$$+ \sum\sum\sum_{i<j<k=1}^{t} (1-p_i-p_j-p_k)^n$$
$$- \cdots + (-1)^{t-1}(1-p_1-p_2-\cdots-p_t)^n.$$

Thus,

$$P(\bigcap_{i=1}^{t} A_i) = P(\bigcup_{i=1}^{t} A_i^c)^c$$
$$= 1 - P(\bigcup_{i=1}^{t} A_i^c)$$
$$= 1 - \sum_{i=1}^{t} (1-p_i)^n + \sum_{i<j=1}^{t} (1-p_i-p_j)^n - \cdots$$
$$\cdots + (-1)^t (1-p_i-p_2-\cdots-p_t)^n.$$

Therefore

$$S_t = \sum\sum\cdots\sum_{i_1<i_2<\cdots<i_t=1}^{N} [1 - \sum_{j=1}^{t} (1-p_{i_j})^n +$$
$$\sum\sum_{j<k=1}^{t} (1-p_{i_j}-p_{i_k})^n -$$
$$\cdots + (-1)^t (1-p_{i_1}-p_{i_2}-\cdots-p_{i_t})^n]$$

and hence we get

$$P(B_m) = \sum_{t=m}^{N}(-1)^{t-m}\binom{t}{m} \sum\sum_{i_1<i_2<\cdots<i_t=1}^{N}\cdots\sum [1 - \sum_{j=1}^{t}(1-p_{i_j})^n$$
$$+ \sum\sum_{j<k=1}^{t}(1-p_{i_j}-p_{i_k})^n - \cdots + (-1)^t$$
$$(1-p_{i_1}-p_{i_2}-\cdots-p_{i_t}^n) \quad (27.5)$$

Now, the equation 27.5 can be written as

$$P(B_m) = \sum_{t=m}^{N}(-1)^{t-m}\binom{t}{m}\binom{N}{t}$$
$$-[\sum_{t=m}^{N}(-1)^{t-m}\binom{t}{m}\binom{N-1}{t-1}]\sum_{j=1}^{N}(1-p_j)^n$$
$$+[\sum_{t=m}^{N}(-1)^{t-m}\binom{t}{m}\binom{N-2}{t-2}]\sum\sum_{j<k=1}^{N}(1-p_j-p_k)^n$$
$$-\cdots$$
$$= \sum_{t=m}^{N}(-1)^{t-m}\binom{t}{m}\binom{N}{t} - [\sum_{t=m}^{N}(-1)^{t-m}\binom{t}{m}\binom{N}{t}\frac{t^{(1)}}{N^{(1)}}]\sum_{j=1}^{N}(1-p_j)^n$$
$$+[\sum_{t=m}^{N}(-1)^{t-m}\binom{t}{m}\binom{N}{t}\frac{t^{(2)}}{N^{(2)}}]\sum\sum_{j<k=1}^{N}(1-p_j-p_k)^n$$
$$-\cdots$$

where $t^{(r)} = t(t-1)(t-2)\cdots(t-r+1)\ \forall r$
and $N^{(r)} = N(N-1)(N-2)\cdots(N-r+1)\ \forall r.$

Therefore,

$$P(B_m) = \sum_{t=m}^{N}(-1)^{t-m}\binom{t}{m}\binom{N}{t}$$
$$+\sum_{j=1}^{N}(-1)^j[\sum_{t=m}^{N}(-1)^{t-m}\binom{t}{m}\binom{N}{t}\frac{t^{(j)}}{N^{(j)}}].$$
$$\sum\sum_{i_1<i_2<\cdots<i_{j=1}}^{N}\cdots\sum (1-p_{i_1}-p_{i_2}-\cdots-p_{i_j})^n. \quad (27.6)$$

Now,

$$\begin{aligned}\sum_{t=m}^{N}(-1)^{t-m}\binom{t}{m}\binom{N}{t} &= \sum_{t=m}^{N}(-1)^{t-m}\frac{t!}{m!(t-m)!}\cdot\frac{N!}{t!(N-t)!}\\ &= \binom{N}{m}\sum_{t=m}^{N}(-1)^{t-m}\binom{N-m}{t-m}\\ &= \binom{N}{m}\sum_{p=0}^{N-m}(-1)^{p}\binom{N-m}{p}\\ &= \binom{N}{m}(1-1)^{N-m}=0.\end{aligned}$$

Therefore, we get

$$\begin{aligned}P(B_m) = \sum_{j=1}^{N}(-1)^{j}[\sum_{t=m}^{N}(-1)^{t-m}\binom{t}{m}\binom{N}{t}\frac{t^{(j)}}{N^{(j)}}].\\ \sum\cdots\sum_{i_1<i_2<\cdots<i_j=1}^{N}(1-p_{i_j}-p_{i_2}-\cdots-p_{i_j})^{n}. \qquad (27.7)\end{aligned}$$

We may note that

$$\begin{aligned}&\sum_{t=m}^{N}(-1)^{t-m}\binom{t}{m}\binom{N}{t}\frac{t^{(j)}}{N^{(j)}}\\ &= \binom{N}{m}\frac{1}{N^{(j)}}\sum_{t=m}^{N}(-1)^{t-m}\binom{N-m}{t-m}t^{(j)}\\ &= \binom{N}{m}\frac{1}{N^{(j)}}[\sum_{p=0}^{N-m}(-1)^{p}\binom{N-m}{p}(m+p)^{(j)}]. \qquad (27.8)\end{aligned}$$

Let us assume that

$$\begin{aligned}(m+p)^{(r)} &= (m+p)(m+p-1)\cdots(m+p-r+1)\\ &= a_0^r(m)p^{(0)}+a_1^r(m)p^{(1)}+\cdots+a_r^r(m)p^{(r)}\end{aligned}$$

where $p^{(0)}=1$ and $a_i^r(m)$'s are some suitable constants $\forall i=1,2,\cdots,r$.
Similarly, let us assume that

$$\begin{aligned}(m+p)^{(r+1)} &= (m+p)(m+p-1)\cdots(m+p-r+1)(m+p-r)\\ &= a_0^{r+1}(m)p^{(0)}a_1^{r+1}(m)p^{(1)}+\cdots+a_r^{r+1}(m)p^{(r)}+a_{r+1}^{r+1}(m)p^{(r+1)}.\end{aligned}$$

Now we find out a relation between the coefficients $a_i^r(m)$'s and $a_i^{r+1}(m)$'s. To get such a relation, let us consider

$$\begin{aligned}(m+p)^{(r+1)} &= (m+p)^{(r)} * (m+p-r)\\ &= [a_0^r(m)p^{(0)} + a_1^r(m)p^{(1)} + \cdots + a_r^r(m)p^{(r)}](m+p-r)\\ &= [(m-r)a_0^r(m)p^{(0)} + a_0^r(m)p^{(1)}]\\ &\quad +[(m-r+1)a_1^r(m)p^{(1)} + a_1^r(m)p^{(2)}]\\ &\quad +[(m-r+2)a_2^r(m)p^{(2)} + a_2^r(m)p^{(3)}]\\ &\quad +\cdots+[(m-r+r)a_r^r(m)p^{(r)} + a_r^r(m)p^{(r+1)}]\\ &= (m-r)a_0^r(m)p^{(0)} + [a_0^r(m) + (m-r+1)a_1^r(m)]p^{(1)}\\ &\quad +[a_1^r(m) + (m-r+2)a_2^r(m)]p^{(2)} + \cdots\\ &\quad +[a_{r-1}^r(m) + (m-r+r)a_r^r(m)]p^{(r)} + a_r^r(m)p^{(r+1)}.\end{aligned}$$

Hence we get

$$\begin{aligned}a_0^{r+1}(m) &= (m-r)a_0^r(m)\\ a_1^{r+1}(m) &= a_0^r(m) + (m-r+1)a_1^r(m)\\ a_2^{r+1}(m) &= a_1^r(m) + (m-r+2)a_2^r(m)\\ &\vdots\\ a_r^{r+1}(m) &= a_{r-1}^r(m) + (m-r+r)a_r^r(m)\\ a_{r+1}^{r+1}(m) &= a_r^r(m)\end{aligned}$$

We claim that

$$a_i^j(m) = \binom{j}{i} m^{(j-i)} \ \forall 0 \le i \le j \tag{27.9}$$

We will prove the above claim by induction on j.

For $j = 1$

$$(m+p)^{(j)} = m+p = mp^{(0)} + 1.p^{(1)}.$$

Therefore,

$$a_0^1(m) = m = \binom{1}{0} m^{(1)};\ a_1^1(m) = 1 = \binom{1}{1} m^{(0)}.$$

Hence the equation 27.9 is ture for $j = 1$.

Let us suppose that the equation 27.9 is true for $j = r$ for some $r \ge 1$. Then we will show tha the equation27.9 is true for $j = r+1$.

Since the equation 27.9 is true for $j = r$, therefore

$$a_i^r(m) = \binom{r}{i} m^{(r-i)} \; \forall 0 \leq i \leq r.$$

Now

$$\begin{aligned} a_0^{r+1}(m) &= (m-r)a_0^r(m) \\ &= (m-r)\binom{r}{0} m^{(r)} \\ &= \binom{r+1}{0} m^{(r+1)}. \end{aligned}$$

Therefore the equation 27.9 is true for $j = r+1$ and $i = 0$.
Again,

$$\begin{aligned} a_{r+1}^{r+1}(m) &= a_r^r(m) \\ &= \binom{r}{r} m^{(0)} \\ &= \binom{r+1}{r+1} m^{(r+1-r-1)}. \end{aligned}$$

Then the equation 27.9 is true for $j = r+1$ and $i = r+1$.

Now, for any $0 < i \leq r$ we get

$$\begin{aligned} a_i^{(r+1)}(m) &= a_{i-1}^r(m) + (m-r+i)a_i^r(m) \\ &= \binom{r}{i-1} m^{(r-i+1)} + (m-r+i)\binom{r}{i} m^{(r-i)} \\ &= m^{(r-i+1)}[\binom{r}{i-1} + \binom{r}{i}] \\ &= \binom{r+1}{i} m^{(r+1-i)}. \end{aligned}$$

Hence, the equation 27.9 is true for $0 < i \leq r$ and $j = r+1$. Hence the claim in the equation 27.9 is true.

Now let us recall the equation 27.8.

$$\begin{aligned}
&\sum_{t=m}^{N}(-1)^{t-m}\binom{t}{m}\binom{N}{t}\frac{t^{(j)}}{N^{j)}}\\
&=\binom{N}{m}\frac{1}{N^{(j)}}[\sum_{p=0}^{N-m}(-1)^{p}\binom{N-m}{p}(m+p)^{(j)}]\\
&=\binom{N}{m}\frac{1}{N^{(j)}}[\sum_{p=0}^{N-m}(-1)^{p}\binom{N-m}{p}\sum_{i=0}^{j}a_i^j(m)p^{(i)}]\\
&=\binom{N}{m}\frac{1}{N^{(j)}}[\sum_{i=0}^{j}a_i^j(m)\sum_{p=1}^{N-m}(-1)^{p}\binom{N-m}{p}p^{(i)}]\\
&=\binom{N}{m}\frac{1}{N^{(j)}}[\sum_{i=0}^{j}a_i^j(m)(N-m)^{(i)}\sum_{t=0}^{N-m-i}(-1)^{t+i}\binom{N-m-i}{t}]\\
&=\binom{N}{m}\frac{1}{N^{(j)}}[\sum_{i=0}^{j}(-1)^{i}a_i^j(m)(N-m)^{(i)}\sum_{t=0}^{N-m-i}(-1)^{t}\binom{N-m-i}{t}].\quad(27.10)
\end{aligned}$$

We may note that

$$\begin{aligned}
\sum_{t=0}^{N-m-i}(-1)^{t}\binom{N-m-i}{t} &= 0 \text{ if } i>N-m\\
&=(1-1)^{N-m-i}=0 \text{ if } i<N-m\\
&=1 \text{ if } i<N-m
\end{aligned}$$

Therefore

$$\begin{aligned}
&\sum_{t=m}^{N}(-1)^{t-m}\binom{t}{m}\binom{N}{t}\frac{t^{(j)}}{N^{(j)}}\\
&=\binom{N}{m}\frac{1}{N^{(j)}}[(-1)^{N-m}a_{N-m}^{j}(m)(N-m)^{(N-m)}]\\
&=\binom{N}{m}\frac{1}{N^{(j)}}(N-m)!(-1)^{N-m}\binom{j}{N-m}m^{(j-N+m)}\\
&=(-1)^{N-m}\frac{(N-j)!}{m!}\binom{j}{N-m}m^{(j-N+m)}\\
&=(-1)^{N-m}\frac{(N-j)!}{m!}\frac{m!}{(m-j+N-m)!}\binom{j}{N-m}\\
&=(-1)^{N-m}\binom{j}{N-m}\\
&=0 \text{ if } j<N-m \quad(27.11)
\end{aligned}$$

From 27.7 and 27.11 we get the following

$$P(B_m) = (-1)^{N-m} \sum_{j=N-m}^{N} (-1)^j \binom{j}{N-m} \sum\sum_{i_1<i_2<\cdots<i_{j=1}}^{N}\cdots\sum (1-p_{i_1}-p_{i_2}-\cdots p_{i_j})^2 \tag{27.12}$$

Now let us recall our original problem. Our aim is to get the probability distribution of the total number of distinct units in a sample drawn according to PPSWR sampling scheme in n draws from a population of size N i.e. to get Prob. $[\nu = m]$ for $1 \leq m \leq n$ where ν stands for the random variable representing the total number of distinct units in such PPSWR sample.

Now it is not very hard to see that

Prob. $[\nu = m] = P(B_m)$ where $P(B_m)$ is as given in equation 27.12

Hence the equation 27.12 and the above statement give us the following

$$\begin{aligned} \text{Prob. } [\nu = m] &= (-1)^{N-m} \sum_{j=N-m}^{N} (-1)^j \binom{j}{N-m} \sum\sum_{i_1<i_2<\cdots<i_{j=1}}^{N}\cdots\sum \\ &\quad (1-p_{i_1}-p_{i_2}-\cdots p_{i_j})^n \\ &\quad \text{for } m = 1,2,\cdots,n. \end{aligned} \tag{27.13}$$

The above expression in 27.13 shows the form of the p.m.f of the variable ν, the total number of distinct units in a sample drawn according to PPSWR sampling scheme in n draws.

From the expression in 27.13, it is immediate that the expression for the p.m.f. of ν is always non-negative. Hence the non-negativity property of the p.m.f is verified.
Now the possible values that ν can take are $1,2,\cdots,n$. So we have to verify that

$$\sum_{m=1}^{n} \text{Prob. } [\nu = m] = 1.$$

If we can show that the above is true, it will immediately follow that the value of the expression for the p.m.f. of ν always lies between 0 and 1 since the non-negativity is already verified.
We may note that

$$\sum_{m=1}^{n} \text{Prob. } [\nu = m] = \sum_{m=1}^{n} (-1)^{N-m} \sum_{j=N-m}^{n} (-1)^j \binom{j}{N-m}$$
$$\sum\sum\cdots\sum_{i_1<i_2<\cdots<i_{j=1}}^{N} (1 - p_{i_1} - p_{i_2} - \cdots - p_{i_j})^n$$
$$= \sum_{u=N-n}^{N-1} (-1)^u \sum_{j=u}^{N-1} (-1)^j \binom{j}{u}$$
$$\sum\sum\cdots\sum_{i_1<i_2<\cdots<i_{j=1}}^{N} (1 - p_{i_1} - p_{i_2} - \cdots - p_{i_j})^n$$
$$= \sum_{j=N-n}^{N-1} (-1)^j [\sum_{u=N-n}^{j} (-1)^u \binom{j}{u}]$$
$$\sum\sum\cdots\sum_{i_1<i_2<\cdots<i_{j=1}}^{N} (1 - p_{i_1} - p_{i_2} - \cdots - p_{i_j})^n$$
$$= \sum_{j=N-n}^{N-1} (-1)^j [\sum_{u=N-n}^{j} (-1)^u \binom{j}{u}]$$
$$\sum\sum\cdots\sum_{i_1<i_2<\cdots<i_{N-j}=1}^{N} (p_{i_1} + p_{i_2} + \cdots + p_{i_{N-j}})^n \quad (27.14)$$

(replacing $N-m$ by u and noting that $p_{i_1} + p_{i_2} + \cdots + p_{i_N} = 1$

Now let us consider the expansion

$$(p_1 + p_2 + \cdots + p_N)^n = \sum_{k=1}^{n} \sum\sum\cdots\sum_{i_1<i_2<\cdots<i_{k=1}}^{N} \sum_{n_1\geq 1} \sum_{n_2\geq 1} \cdots \sum_{n_k\geq 1}$$
$$\sum_{\sum_{i=1}^{k} n_i = n} \frac{n!}{n_1!n_2!\cdots n_k!} p_{i_1}^{n_1} p_{i_2}^{n_2} \cdots p_{i_k}^{n_k}. \quad (27.15)$$

Since $p_1 + p_2 + \cdots + p_N = 1$, we can write

$$\sum_{k=1}^{n} \sum\sum\cdots\sum_{i_1<i_2<\cdots<i_{k=1}}^{N} \sum_{n_1\geq 1} \sum_{n_2\geq 1} \cdots \sum_{n_k\geq 1} \sum_{\sum_{i=1}^{k} n_i = n} \frac{n!}{n_1!n_2!\cdots n_k!} p_{i_1}^{n_1} p_{i_2}^{n_2} \cdots p_{i_k}^{n_k} \quad (27.16)$$

From 27.16 it is clear that for any fixed $k(\leq n)$ and for fixed $(n_1, n_2 \cdots, n_k)$ such that $n_i \geq 1 \ \forall i$ with $\sum_{i=1}^{k} n_i = n$, the co-efficient of

$$p_{i_1}^{n_1} p_{i_2}^{n_2} \cdots p_{i_k}^{n_k}$$

in the left hand side of 27.16 is the same for any choice of $1 \le i_i < i_2 < \cdots < i_k \le N$ and for any permutation $(n_1^*, n_2^*, \cdots, n_k^*)$ of $(n_1, n_2, \cdots, n_k)$ and the coefficient is

$$\frac{n!}{n_1!n_2!\cdots n_k!}.$$

Now, let us fix any $k(\le n)$ and $(n_1, n_2, \cdots, n_k)$ such that $n_i \ge 1\ \forall i$ with $\sum_{i=1}^{k} n_i = n$.

Then for any fixed $1 \le i_1 < i_2 < \cdots < i_k \le N$, the co-efficient of $p_{i_1}^{n_1} p_{i_2}^{n_2} \cdots p_{i_k}^{n_k}$ in the right hand side of 27.15 would be

$$\sum_{j=N-n}^{N-k} (-1)^j [\sum_{u=N-n}^{j} (-1)^u \binom{j}{u}] \binom{N-k}{j} \frac{n!}{n_1!n_2!\cdots n_k!} \tag{27.17}$$

since $p_{i_1}, p_{i_2}, \cdots, p_{i_k}$ can appear simultaneously in the right hand side of 27.10 if $N - j \ge k$ i.e. if $j \le N - k$. Also we may note that for any fixed j, the number of ways that we can choose $p_{k_1}, p_{k_2}, \cdots, p_{k_{N-j}}$ under the restriction that $p_{i_1}, p_{i_2}, \cdots, p_{i_k}$ appear in such choices is

$$\binom{N-k}{N-k-j} = \binom{N-k}{j}.$$

Now, on the basis of equation27.15, equation 27.16 and the expression for the co-efficient in 27.17 we can say that it is enough to show

$$\sum_{j=N-n}^{N-k} (-1)^j [\sum_{u=N-n}^{j} (-1)^u \binom{j}{u}] \binom{N-k}{j} = 1 \tag{27.18}$$

for showing

$$\sum_{m=1}^{n} \text{Prob. } [\nu = m] = 1.$$

We may note that for any $a \le b$, where a and b are positive integers

$$\begin{aligned}
\sum_{u=a}^{b} (-1)^u \binom{b}{u} &= (-1)^a [\binom{b}{a} - \binom{b}{a+1} + \binom{b}{a+2} - \binom{b}{a+3} + \cdots + (-1)^{b-a} \binom{b}{b}] \\
&= (-1)^a [\binom{b}{a} - \{\binom{b-1}{a} + \binom{b-1}{a+1}\} + \{\binom{b-1}{a+1} + \binom{b-1}{a+2}\} - \\
&\quad \cdots + (-1)^{b-a} \{\binom{b-1}{b-1} + \binom{b-1}{b}\}] \\
&= (-1)^a [\binom{b}{a} - \binom{b-1}{a}] \\
&= (-1)^a \binom{b-1}{a-1}
\end{aligned}$$

[applying the result $\binom{n}{r}+\binom{n}{r-1}=\binom{n+1}{r}$ and noting that $\binom{b-1}{b}=0$].

Replacing $b=j$ and $a=N-n$ in the above expression we get

$$\sum_{u=N-n}^{j}(-1)^u\binom{j}{u}=(-1)^{N-n}\binom{j-1}{N-n-1}.$$

Therefore, we get

$$\begin{aligned}&\sum_{j=N-n}^{N-k}(-1)^j\Big[\sum_{u=N-m}^{j}(-1)^u\binom{j}{u}\Big]\binom{N-k}{j}\\ &=\sum_{j=N-n}^{N-k}(-1)^{j+N-n}\binom{j-1}{N-n-1}\binom{N-k}{j}\end{aligned}\tag{27.19}$$

Again, for any $b\geq a$, a,b being integers, we have

$$\begin{aligned}&\sum_{j=a}^{b}(-1)^{j+a}\binom{j-1}{a-1}\binom{b}{j}\\ &=\sum_{p=0}^{b-a}(-1)^p\binom{a+p-1}{a-1}\binom{b}{a+p}\\ &=\sum_{p=0}^{b-a}(-1)^p\binom{a+p-1}{p}\binom{b}{a+p}\end{aligned}\tag{27.20}$$

(substituting p for $j-a$)

Now, we note the following expansions

$$(1+t)^b=\sum_{p=0}^{b}\binom{b}{p}t^p$$

$$(1+t)^{-a}=\sum_{p=0}^{\infty}(-1)^p\binom{a+p-1}{p}t^p\tag{27.21}$$

Therefore the co-efficient of t^p in $t^{-a}(1+t)^b$ is

$$\binom{b}{a+p}$$

and the coefficien of t^{-p} in $(1+t^{-1})^{-a}$ is

$$(-1)^p\binom{a+p-1}{p}$$

(replacing t by t^{-1} is 27.21 we make the above conclusion).

Hence

$$\sum_{p=0}^{b-a}(-1)^p\binom{a+p-1}{p}\binom{b}{a+p}$$

is the coefficient of t^pt^{-p} in the expansion of

$$t^{-a}(1+t)^b(1+t^{-1})^{-a}.$$

Now, we note that

$$\begin{aligned} t^{-a}(1+t)^b(1+t^{-1})^{-a} &= t^{-a}(1+t)^b(\frac{1+t}{t})^{-a} \\ &= t^{-a}(1+t)^b\frac{t^a}{(1+t)^a} \\ &= (1+t)^{b-a}. \end{aligned}$$

Hence we get

$$\sum_{p=0}^{b-a}(-1)^p\binom{a+p-1}{p}\binom{b}{a+p}$$

is the constant term in the expansion of $(1+t)^{b-a}$ and the constant term is 1.

Therefore, we get the following identity

$$\sum_{j=a}^{b}(-1)^{j+a}\binom{j-1}{a-1}\binom{b}{j} = 1 \tag{27.22}$$

for any $b \geq a, a, b$ being integers.

Replacing b by $N-k$ and a by $N-n$ in the above identity 27.22 we get

$$\sum_{j=N-n}^{N-k}(-1)^{j+N-n}\binom{j-1}{N-n-1}\binom{N-k}{j} = 1 \tag{27.23}$$

which immediately proves 27.18 and hence we get

$$\sum_{m=1}^{n} \text{Prob. } [\nu = m] = 1$$

Thus the probability distribution given by the equation (27.4.14) is a valid probability distribution.

27.5. General Expression for The Variance of The Improved Estimator Obtained Through Rao-Blackwellisation

For general n, the Rao-Blackwellised version of the Hansen-Hurwitz (1943) estimator can be written as

$$t^*(s) = \frac{1}{n}\sum_{j=1}^{\nu} b_{i_j}\frac{y_{i_j}}{P(s)}$$

where $P(s) = \sum_{s'\sim s} p(s'), p(s) = p_{i_1}p_{i_2}\cdots p_{i_n}$
and $b_{i_j} = \frac{\partial P(s)}{\partial p_{i_j}}$.

We may note that $t^*(s)$ can alternatively be written in the form

$$t^*(s) = \sum_{i=1}^{N} d_{si}y_i$$

$$\begin{aligned} \text{where} dsi &= \tfrac{b_i}{nP(s)} \text{ if } i \,\varepsilon\, s \\ &= 0 \text{ otherwise} \end{aligned}$$

Also we note that Var $[t^*(s)] = 0$ if $y_i = Cp_i$ for some constant $C \neq 0$ because

$$\sum_{j=1}^{\nu} p_{i_j}\frac{\partial p(s)}{\partial p_{i_j}} = nP(s)$$

by Euler's Theorem as $P(s)$ is a homogeneous function of degree n.

Then, following Rao (1979), the variance of $t^*(s)$ can be written as

$$\text{Var }[t^*(s)] = -\sum\sum_{i<j=1}^{N} d_{ij}p_ip_j\left(\frac{y_i}{p_i} - \frac{y_j}{p_j}\right)^2$$

where $d_{ij} = E_p[d_{si}d_{sj}] - 1$. An unbiased estimator of Var $[t^*(s)]$ is given by

$$\hat{V}\text{ar}[t^*(s)] = -\sum\sum_{i<j=1}^{N} d_{ij}(s)p_ip_j\left(\frac{y_i}{p_i} - \frac{y_j}{p_j}\right)^2$$

where $d_{ij}(s) = 0$ if $s \not\ni i, j$ and

$$E_p[d_{ij}(s)] = d_{ij} \text{ i.e. } \sum_{s\ni i,j} d_{ij}(s)p(s) = d_{ij}.$$

Using the probability distribution of the total number of distinct units (ν) given by equation 27.14, the variance of $t^*(s)$ for $n = 3$ comes out to be

$$\begin{aligned} \text{Var }[t^*(s)] &= \frac{1}{3}\sum\sum_{i<j=1}^{N}\left(1 - \frac{p_ip_j}{p_i+p_j}\right)p_ip_j\left(\frac{y_i}{p_i} - \frac{y_j}{p_j}\right)^2 \\ &< \frac{1}{3}\sum\sum_{i<j=1}^{N} p_ip_j\left(\frac{y_i}{p_i} - \frac{y_j}{p_j}\right)^2 = \text{ Var }[\hat{Y}_{HHE}]. \end{aligned}$$

Numerical Illustration

Population 1 $N=4, n=3$

Unit	1	2	3	4
y_i	84	168	252	336
p_i	0.2	0.3	0.1	0.4

$$\text{Var}\,[\hat{Y}_{HHE}] = 113680,\ \text{Var}\,[t^*(s)] = 104616.4$$

$$\text{Percentage gain in efficiency} = \frac{\text{Var}\,[\hat{Y}_{HHE}] - \text{Var}\,[t^*(s)]}{\text{Var}\,[t^*(s)]} \times 100\% = 8.66\%$$

Population 2 $N=4, n=3$

Unit	1	2	3	4
y_i	5	10	12	17
p_i	0.2	0.3	0.1	0.4

$$\text{Var}\,[\hat{Y}_{HHE}] = 228.2767,\ \text{Var}\,[t^*(s)] = 202.88963$$

$$\text{Percentage gain in efficiency} = \frac{\text{Var}\,[\hat{Y}_{HHE}] - \text{Var}\,[t^*(s)]}{\text{Var}\,[t^*(s)]} \times 100\% = 12.51\%$$

Thus the probability distribution of the total number of distinct units (ν) in a sample drawn according to PPSWR sampling scheme in n draws enables us to derive the exact variance of the improved estimator $t^*(s)$ obtained through Rao-Blackwellisation and it is possible to calculate the percentage gain in efficiency for any n.

27.6. Acknowledgement

The author is grateful to a referee for the valuable comments on the paper.

References

1. Basu, D. (1958). On sampling with and without replacement. *Sankhyā*, 20, 287-294.
2. Hansen, M.H. and Hurwitz, W.N. (1943). On the theory of sampling from finite populations. *An. Math. Stat.* 14, 333-362.
3. Murthy, M.N. (1967). Sampling Theory and Methods. *Statistical Publishing Society, Calcutta, India.*
4. Pathak, P.K. (1962). On simple random sampling with replacement. *Sankhyā*, 24, Ser.A. 287-302.
5. Pathak, P.K. (1962). On sampling units with unequal probabilities. *Sankhyā*, 24,Ser A, 315-326.

6. Raj, D. and Khamis, H.S. (1958). Some remarks on sampling with replacement. *Ann. Math. Statist.*, 29, 550-557.
7. Rao, J.N.K (1979). On deriving mean square errors and their non-negative unbiased estimators in finite population sampling. *Jour. Ind. Statist. Assoc.*, 17, 125-136.